HIGH ENERGY
SOLAR PHYSICS

HIGH ENERGY SOLAR PHYSICS

Greenbelt, MD   August 16–18, 1995

EDITORS
Reuven Ramaty
Goddard Space Flight Center

Natalie Mandzhavidze
USRA at Goddard Space Flight Center

Xin-Min Hua
NRC at Goddard Space Flight Center

**AIP CONFERENCE
PROCEEDINGS 374**

American Institute of Physics   **Woodbury, New York**

phys
Sep/ae

L.C. Catalog Card No. 96-84513
ISBN 1-56396-542-9
DOE CONF- 9508217

Printed in the United States of America

CONTENTS

Preface ix

HISTORICAL OVERVIEW

Evolution of Our Understanding of Solar Flare Particle Acceleration:
(1942−1995)... **3**
    E. L. Chupp

ENERGETIC CHARGED PARTICLES

Energetic Particles from Solar Flares and Coronal Mass Ejections **35**
    D. V. Reames
Solar Flare Gamma-Ray Emission and Energetic Particles in Space.......... **45**
    E. W. Cliver
Coronal Mass Ejections and Solar Energetic Particle Events **61**
    S. W. Kahler
Energetic Solar Electron Spectra and Gamma-Ray Observations **78**
    W. Dröge
Measurements of the Ionic Charge States of Solar Energetic Particles
at 15−70 MeV/Nucleon Using the Geomagnetic Field **86**
    R. A. Leske, J. R. Cummings, R. A. Mewaldt, E. C. Stone,
    and T. T. von Rosenvinge
HIIS Results on the Mean Ionic Charge State of SEP Fe Above
200 MeV per Nucleon **96**
    A. J. Tylka, P. R. Boberg, J. H. Adams, Jr., L. P. Beahm, W. F. Dietrich,
    and T. Kleis
High Energy Particles in Interplanetary Space on 11 June 1991 106
    D. F. Smart and M. A. Shea
Neutron Decay Electrons after the Solar Flare of 1980 June 21............. 116
    D. Ruffolo, W. Dröge, and B. Klecker
Longitudinal Extents of Coronal/Interplanetary Shocks 124
    H. V. Cane
Unusual Intensity-Time Profiles of Ground-Level Solar Proton Events....... 131
    M. A. Shea and D. F. Smart
On the Formation of Relativistic Particle Fluxes in Extended
Coronal Structures... 140
    L. I. Miroshnichenko, J. Pérez-Peraza, E. V. Vashenyuk,
    M. D. Rodríguez-Frías, L. del Peral, and A. Gallegos-Cruz

GAMMA RAYS AND NEUTRONS

Thin Target γ-Ray Line Production During the 1991 June 1 Flare 153
  G. Trottet, C. Barat, R. Ramaty, N. Vilmer, J. P. Dezalay, A. Kuznetsov,
  N. Mandzhavidze, R. Sunyaev, R. Talon, and O. Terekhov
Gamma-Ray Line Measurements and Ambient Solar Abundances 162
  G. H. Share, R. J. Murphy, and J. G. Skibo
Solar Atmospheric Abundances from Gamma Ray Spectroscopy 172
  R. Ramaty, N. Mandzhavidze, and B. Kozlovsky
Abundance Study of the 4 June 1991 Solar Flare Using *CGRO*/OSSE
Spectral Data 184
  R. J. Murphy, G. H. Share, J. E. Grove, W. N. Johnson, R. L. Kinzer,
  J. D. Kurfess, M. S. Strickman, and G. V. Jung
Spectral Evolution of an Intense Gamma-Ray Line Flare 194
  E. Rieger
COMPTEL Solar Flare Measurements 200
  J. M. Ryan and M. M. McConnell
Observations of Gamma-Ray Spectra and X-Ray Images of Solar Flares 210
  M. Yoshimori, K. Morimoto, K. Suga, T. Matsuda, and N. Saita
Extended γ-Ray Emission in Solar Flares 219
  G. Rank, K. Bennett, H. Bloemen, H. Debrunner, J. Lockwood,
  M. McConnell, J. Ryan, V. Schönfelder, and R. Suleiman
Pion Decay and Nuclear Line Emissions from the 1991 June 11 Flare 225
  N. Mandzhavidze, R. Ramaty, D. L. Bertsch, and E. J. Schneid
Origin of the High Energy Gamma-Ray Emission in the March 26, 1991
Solar Flare 237
  V. Kurt, V. V. Akimov, and N. G. Leikov
The 1990 May 24 Solar Flare and Cosmic Ray Event 246
  L. Kocharov, G. Kovaltsov, J. Torsti, I. Usoskin, H. Zirin, A. Anttila,
  and R. Vainio
Solar Neutrons on May 24th, 1990 256
  Y. Muraki and S. Shibata

X-RAYS

Solar Flare Energy Release and Particle Acceleration as Revealed
by Yohkoh HXT 267
  T. Kosugi
Reconnection Dynamics in Cusp-Shaped Flare Loops 275
  T. G. Forbes
Yohkoh Observations of Flares with Superhot Properties 285
  H. S. Hudson and N. Nitta
Conditions for Energetic Flares 294
  N. Nitta
Hard X-Ray Timing 300
  M. J. Aschwanden

Subsecond Time Variations in Solar Flares around 100 keV:
Diagnostics of Electron Acceleration 311
    N. Vilmer, G. Trottet, H. Verhagen, C. Barat, R. Talon, J. P. Dezalay,
    R. Sunyaev, O. Terekhov, and A. Kuznetsov
Inferring the Accelerated Electron Spectrum in Solar Flares 320
    R. P. Lin and C. M. Johns-Krull
Solar Hard X-Ray Albedo due to Compton Scattering 329
    T. Bai and R. Ramaty
What is the Spatial Relationship between hard X-Ray Footpoints
and Vertical Electric Currents? 336
    J. Li, T. R. Metcalf, R. C. Canfield, J.-P. Wülser, and T. Kosugi
Solar Coronal Abundances: Some Recent X-Ray Flare Observations 343
    A. C. Sterling
Solar Flare Dynamics as Revealed by *Yohkoh* Observations 353
    G. A. Doschek
Energetics of the DC-Electric Field Model 359
    D. M. Zarro and R. A. Schwartz
Hard X-Ray Polarimetry of Solar Flares with BATSE 368
    M. McConnell, D. Forrest, W. T. Vestrand, and M. Finger

RADIO EMISSION

Submillimeter/IR Solar Bursts from High Energy Electrons 379
    P. Kaufmann
Microwave and Hard X-Ray Sources in Two X-Class Limb Flares 393
    H. Nakajima and T. R. Metcalf
Nonthermal Radio Emission from Coronal X-Ray Structures 402
    M. R. Kundu, J.-P. Raulin, and N. Nitta
Radio and X-Ray Manifestations of a Bright Point Flare 408
    N. Gopalswamy, M. R. Kundu, Y. Hanaoka, S. Enome, and J. R. Lemen
Razin Suppression in Solar Microwave Bursts 416
    L. Belkora
A Summary of Three-Year Observations with the Nobeyama
Radioheliograph 424
    S. Enome

THEORY OF PARTICLE ACCELERATION

Selective Enrichment of Energetic Ions in Impulsive Solar Flares 435
    M. Temerin and I. Roth
Acceleration and Heating by Turbulence in Flares 445
    V. Petrosian
Heavy Ion Acceleration by Cascading Alfvén Waves in Impulsive
Solar Flares 450
    J. A. Miller and D. V. Reames

**Accelerated Particle Composition in Impulsive Events: Clues
to the Conditions of Acceleration** 461
    J.-P. Meyer
**Particle Acceleration by DC Electric Fields in the Impulsive Phase
of Solar Flares.** 479
    G. D. Holman
Reconnection and Acceleration to High Energies in Flares 493
    B. V. Somov
On the Formation of the Helium-3 Spectrum in Impulsive Solar Flares 498
    Y. E. Litvinenko
Charged Particle Diffusive Transport 505
    G. Lenters and J. A. Miller

RAPPORTEUR PAPERS

**X-Ray Aspects of the High Energy Solar Physics Workshop:
Rapporteur Paper I** 519
    B. R. Dennis
**Implications of Solar Flare Charged Particle, Gamma Ray and Neutron
Observations: Rapporteur Paper II for the High Energy Solar Physics
Workshop.** 533
    R. Ramaty and N. Mandzhavidze

Author Index 545

PREFACE

This volume contains articles based on the talks presented at the High Energy Solar Physics Workshop held in August 1995 at the Goddard Space Flight Center in Greenbelt, Maryland. The workshop was organized by Reuven Ramaty (Chair), Gordon Holman, Natalie Mandzhavidze, and Donal Reames. Financial support was provided by NASA, and very important logistical support came from the Universities Space Research Association (USRA). The organizers wish to express their gratitude to Denise Dunn (USRA) and LaShawn Davis (Goddard) for their invaluable help. The organizers would also like to thank John Brown, Astronomer Royal for Scotland, for providing a very entertaining and informative banquet speech.

The purpose of the workshop was to discuss the exciting recent developments in the study of solar flares, emphasizing work carried out at high energies. The key problems were, and still are, the processes of flare energy release, particle acceleration, and accelerated particle transport and interaction. These problems can be best addressed by observing the high energy photons which are produced by the accelerated particles, as well as the particles themselves whose properties can be studied in detail by direct observations. For the high energy photons, currently the principal observables are hard X-ray images, and hard X-ray and gamma ray time profiles and energy spectra, including line spectroscopy in the gamma ray region. For the particle observations, detailed abundance and charge state studies have emerged as crucial diagnostics for testing acceleration mechanisms. The talks at the Workshop, as well as the papers in this volume, address all of these issues, as well as multiwavelength studies and theoretical investigations.

We have organized the volume into five major chapters dealing with charged particles, gamma rays, X-rays, radio emission, and particle acceleration theories. We fully realize that the solution of the overall problem of flare energy release and particle acceleration should consider all of these aspects together. Efforts in achieving this have been made by many of the contributors. The volume also contains a historical overview by Professor Chupp and two rapporteur papers. We hope that the book will constitute an important contribution to the solar physics and astrophysics literature.

Reuven Ramaty
Natalie Mandzhavidze
Xin-Min Hua

HISTORICAL OVERVIEW

Evolution of our Understanding of Solar Flare Particle Acceleration: (1942–1995)

Edward L. Chupp

Department of Physics and Space Science Center,
Institute for the Study of Earth, Oceans and Space
University of New Hampshire, Durham, New Hampshire 03824

We review the evidence for the acceleration of charged particles to relativistic energies associated with solar flares from 1942, the time of their discovery, to the present. First, we present a brief summary of early attempts to identify the mechanism which accelerates the charged particles in solar flares. Second, we describe observational progress made before 1970 which revealed additional properties of the solar flare particle acceleration process. Also, some relevant, significant pre-1970 theoretical advances are mentioned. Third, the neutral high-energy observational data, obtained since 1972, are reviewed as diagnostics for the properties of the accelerated electrons and ions. These provide constraints for acceleration theories. Finally, the capabilities of stochastic, direct electric field, and shock acceleration mechanisms are briefly discussed.

The early concept that the particles are accelerated in a two-phase process in which electrons are first accelerated to ~100 keV, followed by a longer term process which accelerates ions to relativistic energies is not consistent with observations. Rather, ions and electrons are accelerated together and the basic acceleration mechanisms, i.e., stochastic, direct electric field, and shock acceleration may all be operative in a flare.

INTRODUCTION

Transient production of relativistic particles, associated with solar flares, first became evident with the discovery, in 1942, of dramatic increases in the count rates of several ground level cosmic-ray intensity monitors.[1] However, with the exception of a report in 1942 by Lange and Forbush (95), the observations from the global cosmic-ray monitor network, relating to intensity increases associated with solar activity on 1942 February 28 and March

---

[1]This phenomenon is often referred to as a ground level event (GLE), an enhancement of the count rate of a cosmic-ray intensity monitor. Hereafter, we will use the designation GLE to refer to solar related monitor enhancements due to charged particles.

7, were not to appear in the published literature until 1946, after World War II. Also, in 1942, British radar stations were to record, for the first time, intense radio noise in meter waves (4–6) m from the direction of the Sun (72). This emission was associated with the central meridian passage of an active region which had produced a solar flare, the radiation having been first detected on 1942 February 26 two days before the first GLE was observed (72). Subsequently, on 1946 July 25 (51), another cosmic-ray monitor intensity increase was recorded following an intense flare near central solar meridian.

The first three GLE observations were made with ionization chambers and Geiger-Mueller counters which detect predominantly the charged secondary components produced in the atmosphere by a spectrum of energetic extraterrestrial particles. After the end of World War II neutron monitors were developed which respond to the secondary nucleonic component produced by the lower energy particles in the spectrum incident at the top of the atmosphere (147). Using this new technique, Adams and Braddick (3,4), on 1949 November 19, observed a GLE following a large solar flare. The intensity recorded during the maximum of the event was six times the normal value (2), a relative increase which was considerably larger than that for earlier bursts.

In 1952, in a detailed paper, Elliot (51), reviewed all published cosmic-ray monitor records through 1949 which showed rate increases associated with solar activity (50,95,19,34,61,62,49,43,41,2,3). Analysis of the relative intensity increase recorded by monitors at different geomagnetic latitudes demonstrated that these increases were due to charged particles from the Sun, with energies as high as 5 GeV. It is interesting to note that many other intensity increases were also observed between 1941 and 1943 (50,35), however, in the literature, most frequently cited are only the "famous" four GLE events, mentioned above.

In this paper we first, briefly, review theoretical attempts made by 1950 to explain the GLE phenomena. We then discuss observations made during the next four and a half decades that relate to the acceleration of the ions and electrons which *produce the neutral diagnostic radiations* associated with solar flares. These include radio, optical, ultra-violet, X ray, γ ray and high-energy neutron emissions. In some cases it will be necessary to discuss also the solar energetic particles (SEP) directly observed in space when they are closely correlated with the observation of neutral radiations. However, due to space limitations for this paper we do not discuss the connection of particles producing the neutral radiation with those producing the "large proton" events in space or the large particle events on the Earth, the so-called "solar flare myth" which is currently under debate.[2] Other papers in these Proceedings

---

[2]See EOS Trans. **76**, No. 41, October 10, 1995. An understanding of the acceleration processes which produce the SEP and the GLE events, which are attributed by some to coronal mass injection, may or may not be causally related to the charged particles which produce the neutral radiations associated with solar flares.

discuss various aspects of this problem (37,136). Finally, we discuss the basic acceleration mechanisms as currently understood in application to the solar flare particle acceleration problem.

EARLY THEORETICAL WORK (1942–1950)

Since there was close time association of cosmic-ray intensity increases with visual brightenings on the Sun it was tacitly assumed that charged particles were accelerated in the associated sunspot regions or in the neighborhood of the Sun during the optical flare. Actually, the first theoretical efforts inspired by the solar GLEs were directed toward explaining how the primary cosmic rays were accelerated, assuming they, in fact, originated at the Sun. Though a solar origin of the ubiquitous cosmic rays was proposed earlier (40) than 1942, the first suggestion that the particles responsible for the recently discovered GLEs were accelerated by electric fields associated with solar corpuscular streams, polarized as a result of their motion through an assumed permanent solar magnetic field, was presented by Alfvén (9). Following an earlier suggestion by Swann (157), Ehmert (50) proposed that acceleration by the betatron mechanism in the changing magnetic field of a growing sunspot could explain the GLE effect. In a variation Bagge and Biermann (12) suggested that the changing magnetic field could result from the relative motion of a pair of sunspots. In 1948 Menzel and Salisbury (109) proposed that the electric fields associated with low frequency (0.01–1000) Hz electromagnetic waves propagating away from the Sun could accelerate charged particles. These waves were presumed to be produced by fluctuating sunspot magnetic fields resulting from turbulence. Richtmyer and Teller (137) assumed such a solar mechanism was operative and produced the "ordinary" cosmic rays which were then confined to the solar system by an extensive magnetic field. McMillan (106) investigated the Menzel and Salisbury (109) proposed mechanism in detail and attempted to explain both the "ordinary" cosmic rays and the flare associated GLEs by the action of these waves. Kiepenheuer (89) argued that the electric fields implied in the above mechanisms could not exist in the lower corona because of the high electrical conductivity. Instead, he imagined a magnetized beam leaving the sunspot region which carried a thin shielding current caused by a potential drop of $10^{10} - 10^{11}$ volts.

Because of the limited observational data available on the GLEs at this time these speculations only deepened the mystery and therefore, it was not until continuous observations of the Sun were carried out by US and USSR satellites that the full dimensions of the solar flare acceleration phenomena could be revealed.

OBSERVATIONS OF HIGH-ENERGY SOLAR FLARE EMISSIONS
(1950–1995)

The Years 1950–1960

In the late 1940s observation of the Sun in the far ultra-violet (UV) (15,167) and soft X rays (66) was carried out with rocket borne instruments by the Naval Research Laboratory (NRL). These programs greatly stimulated study of the Sun and solar flares. Major activity in the early part of 1950 was low, then on 1956 February 26 a large GLE occurred. This event became a powerful stimulus to research on the high-energy charged particles and photons associated with solar flares. The neutron monitor increase reported by Meyer, Parker, and Simpson (110) showed that protons with energies $> 15 - 30$ GeV were produced in close time association with the flare optical and radio emissions. Parker (120) hypothesized that the protons causing this GLE were accelerated by the Fermi mechanism in the turbulent magnetic field produced by the chromospheric solar flare.

During the International Geophysical Year (IGY), (1957 July 1–1959 December 31), frequent flights of X-ray, γ-ray, and particle detecting instruments, were launched on balloon, rocket, and satellite platforms. It is from a balloon borne experiment that the first observations of energetic photons from a solar flare were made on 1958 March 20 by Peterson and Winckler (123,124). The instruments used were a small Geiger counter and an ionization chamber, each of which recorded a burst in the event rates in a time interval of < 1 minute during a class 2 flare. These detector increases coincided exactly with the maximum intensity of the optical flare (which began 5 minutes earlier) and with a 800 MHz microwave radioburst! Analysis of the individual detector rates, considering their different relative responses to photons of different energy, indicated that the burst was caused by bremsstrahlung from $\sim 10^{35}$ electrons with energies $\sim$0.5 MeV (124).

After the launch of Sputnik (1957 October 4), by the USSR and with initiation of the US satellite program, Morrison (115) published a stimulating paper predicting the fluxes expected for γ-ray emission from several celestial objects and phenomena including solar flares. Importantly, Morrison's initial estimates would become the basis for a detailed theoretical study of high-energy neutral emissions from solar flares by Lingenfelter and Ramaty (101).

It was also in the post World War II period that the systematic study of radio emission associated with solar flare "eruptions" began. By the mid-1950s radio spectrographs were observing solar radio bursts. Eventually, interferometers were added, which could give the spatial locations of the sources and a concrete picture of the cause of the radio phenomena developed. This was to lead in the 1960s to a two phase model which explained the radio bursts as due to accelerated, $\sim$100 keV electrons followed, in large flares, by Fermi acceleration of protons and electrons to very high energies, thereby explaining the GLE events and the low-energy flare associated protons detected by satellite, balloon and "riometer" detectors (179). As will be discussed below, this model was to exist until the 1980s before new observations complicated the picture.

The Years 1960-1970

In 1960 soft X-ray studies were carried out by the first NRL solar radiation, *SR I*, satellite followed by *SR III* in 1961. The results were the first to demonstrate that increases in the solar X-ray flux are a sensitive indicator of solar activity (66). Also, in this decade the first Orbiting Solar Observatory *(OSO 1)* was launched carrying detectors searching for solar X (67) and γ rays (59) with energies >100 MeV. The X-ray detector, which covered the energy range (20–100) keV, observed five solar flares with hard X-ray bursts, well correlated with microwave bursts in 3 of the events (67). The γ-ray detector yielded no evidence for radiation from the Sun with an upper limit for the quiet Sun flux of 2×10^{-3} photons cm^{-2} s^{-1} and an upper limit during a Type III flare, on 1962 March 22, of 10^{-2} photons cm^{-2} s^{-1} (59). There were many additional experiments in the 1960s studying flare associated X-ray bursts extending up to 500 keV (76), however, there were no clear observations of either γ-ray lines or neutrons (26) from solar flares. Several hard X-ray flare observations made from the Orbiting Geophysical Observatory *(OGO)* satellites, i.e., *OGO-1*, *OGO-3*, and *OGO-5* were reviewed by Kane (86). Gamma-ray observations from the Environmental Research Satellite (ERS-18), during the large solar flare on 1967 May 23, implied an upper limit to the strength of the neutron capture line (2.223 MeV) of 0.52 photons cm^{-2} s^{-1} (69).

Even though the prospects for observing the predicted γ-ray lines (115) were bleak, the fact that copious ions were often accelerated in association with solar flares gave impetus to further refine the theory of γ ray production (45). The most detailed treatment of the production of nuclear reactions in the solar atmosphere was given by Lingenfelter and Ramaty (101) who calculated the yield of neutrons, positrons, and γ-ray lines as well as the light isotopes ^{2}H, ^{3}H, and ^{3}He using normalized rigidity spectra for the accelerated ions. These calculations were applied to nuclear fragment, neutron and γ-ray observations made during the 1960 November 12 flare. At this time it was expected that ions observed in space were directly related to the population of accelerated particles which remained at the Sun to produce γ-ray lines and neutrons. Since there were no positive observations of these neutral radiations for this event, the calculations served to constrain the mean density of the interaction region and the number of accelerated particles for different assumed accelerated particle spectra.

It is important to mention here one of the earliest flare models developed in 1960 which meets the several primary requirements proposed by Sturrock (154) to explain a flare. In summary, his requirements are energy storage, energy release, acceleration of particles, mass ejection, and locating the temperature minimum region (154). A model which comes close to meeting these requirements is that of Gold and Hoyle (68). This consists of twisted flux tubes anchored in the photosphere, carrying field aligned currents. The tubes have opposite longitudinal magnetic fields, so when they are pressed together, by photosphere motions,the longitudinal fields annihilate and the toroidal fields tend to link and pinch the plasma within. This model does not address the question of particle acceleration but does satisfy, in principle, the requirements of energy storage and energy release. We mention it here because some current models, which do treat acceleration, often use this basic model.

The Years 1970–1980

In this decade solar observations continued with the successful series of *OGO* and *OSO* satellites. Experiments on *SKYLAB* as well as rocket and balloon experiments were dedicated to study the Sun. Instruments on High Energy Astronomical Observatories *HEAO-1* and *HEAO-3* dedicated to cosmic observations could also record high-energy solar emissions. The University California San Diego (UCSD) Group's X-ray spectrometers on the *OSO 7* satellite obtained, in the early 1970s, copious X-ray observations in the 2–300 keV range (125). From studying the spectra of about 200 flares it was concluded that for events with both thermal and nonthermal spectral components, nonthermal electrons do not have sufficient energy to power the thermal emission. A few other events had pure thermal spectra, several had long duration (from 10 minutes to hours) and one, the 1972 March 3 event, had initially a 1-minute long nonthermal hard X-ray burst, which was followed by a thermal phase which lasted as long as 15 minutes.

An auspicious development in this period was the appearance of the complex active region (McMath 11976) on 1972 July 11 (164) which heralded the occurrence of the 1972 August series of major flares with concurrent intense high-energy emissions and associated terrestrial effects. On 1972 August 2, 4, and 7 major flares from a single active region produced the most intense and long lasting radiations over the full electromagnetic spectrum observed up to then with associated intense long duration charged particle fluxes in space. During these events the *OSO 7* satellite instruments were often saturated by the intense ultra-violet and X-ray fluxes. Fortuitously, during a 10-minute period in the rising phase of the August 4 flare, the *OSO 7* γ-ray spectrometer recorded, for the first time, an intense γ-ray line and continuum spectrum (27) which dramatically confirmed the 1958 prediction by Morrison (115) that nuclear reactions in the solar atmosphere during flares could produce a 2.223 MeV line detectable at the Earth (31). The spectrum obtained, also gave evidence for nuclear lines at 0.511 MeV, 4.4 MeV, and 6.1 MeV, with continuum photons to 10 MeV. Unfortunately, the satellite entered eclipse, so only the rising phase of the nuclear emissions was recorded. These observations im-

8

plied that acceleration of protons above 30 MeV was closely associated with the production of the relativistic electrons which produced the bremsstrahlung continuum. This concept was not taken seriously though, probably because of the three-minute time resolution of the observations. Therefore, the two-phase model, mentioned above (179), was generally considered valid. Shortly after the *OSO 7* observations were reported, Ramaty and collaborators (128) reviewed their earlier work (101) and extended the calculations on γ-ray and neutron production in solar flares which were instrumental in planning future experiments. Also, based on the published *OSO 7* data Svestka in his 1976 monograph on solar flares (156) proposed for the first time that electrons and ions were accelerated in the "same (one-step) acceleration process, or that the second-step acceleration (giving rise to > 30 MeV protons) immediately follows the process of preacceleration."

During the decay phase of the August 7 flare from the same active region, 40 minutes after the flare's onset, significant delayed emissions at 0.511 MeV and at 2.223 MeV were recorded (27). For some time after these results, no new observations of nuclear γ-ray emissions were reported. Later in the decade, when new satellite missions were launched, the 2.223 MeV line was observed by scintillation spectrometers on *Prognoz 6* (162,25) and by *HEAO-1* (79) during flares on 1977 November 22 and 1978 July 11, respectively. During the 1979 November 9 flare a 2.223 MeV line was observed, of width limited at 2 keV by the resolution of the germanium spectrometer, on the *HEAO-3* satellite (126). Because the neutron capture line is produced in the photosphere where the temperature is about 10^4 K it is expected that the line width would be ≲100 eV (128) which is well below the 2 keV energy resolution of the best γ-ray spectrometers currently available for space observations.

Other developments in this period have been comprehensively discussed in *Solar Flares (A Skylab monograph)* (154).

The Years 1980–1990

In 1973 at a meeting at Woods Hole, Massachussetts, the Solar Maximum Mission *(SMM)* was conceived to carry out a comprehensive study of solar flares during the maximum phase of sunspots *cycle 21*. Near the same time, in Japan, the solar mission *Hinotori* was planned. Also, in 1973 plans were made for an international global study of solar activity which was to lead to the Solar Maximum Year (SMY) (44). With the launch of the *SMM* satellite on 1980 February 14 and the Japanese *Hinotori* satellite on 1981 February 21, the continuing International Sun Earth Explorer/International Comet Explorer *(ISSE 3/ICE)* and Interplanetary Monitoring Probe *(IMP)* operations, and the coordinated ground based solar observations of the SMY, an avalanche of new information on solar flares became available. Satellite observations by the hard X-ray and γ-ray spectrometers on these new missions provided dramatic new diagnostic tools for determining the properties of particle acceleration

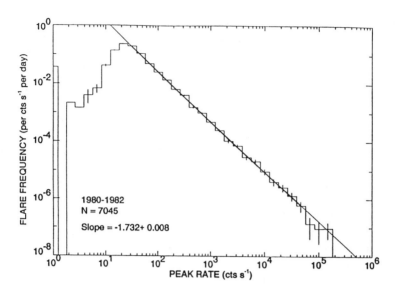

FIG. 1. The size distribution is shown for peak intensities of HXRBS hard X-ray flares during solar activity *cycle 21* (39).
Reprinted by permission of Kluwer Academic Publishers.

associated with solar flares.

The *SMM* Hard X-Ray Burst Spectrometer (HXRBS) recorded over 12,000 hard X-ray bursts associated with solar flares which generally have a nonthermal spectrum. The events in this vast data base dramatically demonstrate that acceleration of electrons is a principal characteristic of a solar flare. Another characteristic of these events, which reveals a fundamental property of electron acceleration, is the size distribution. Figure 1 (39) shows that the distribution of the number of hard X-ray flares versus their peak intensity, over ≈ more than six orders of magnitude, is a power law with slope ≈1.7. This fact may be related to the concept that the observed intensity of a flare burst may be a convolution of many fast elementary injections at various repitition rates (88), or to the more recent suggestion (103) that a given flare burst is a result of an avalanche of many small reconnection events. It has been evident from the detailed calculations of Ramaty and collaborators (101,128), and the *OSO-7* results (27) that a deeper understanding of the acceleration of protons associated with solar flares could result from extensive observations of the γ-ray lines and neutrons. Therefore, in the following we will focus on hard X-ray, γ-ray and neutron observations which contribute *new insights in understanding of the particle acceleration phenomena associated with solar flares.*

Within the first few months of flare observations in 1980 with the gamma-ray spectrometer (GRS) on *SMM* it became clear that acceleration of ~50 MeV protons and relativistic electrons was simultaneous to within the in-

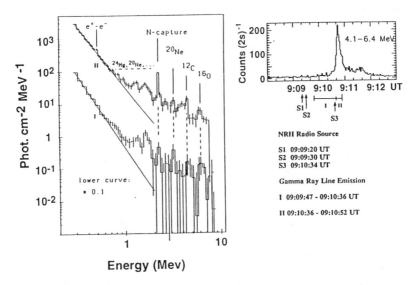

FIG. 2. The γ-ray spectra for the 1989 September 9 flare (*left*) at the two intervals shown in the time history (*right*, roman numerals). Interval I is in the pre-flash; interval II is the flash phase of the flare. The arrows labeled S1, S2 under the time history mark the time of appearance of new radio sources (18).

Reprinted by permission of Kluwer Academic Publishers.

strument time resolution of 1–16 s (64), confirming the earlier suggestion of Svestka (156). We emphasize that this observation is contrary to the two-phase acceleration model proposed by Wild and collaborators (179). During an intense limb flare on 1980 June 21 the *SMM* GRS detected, for the first time, a burst of energetic neutrons at the Earth (28), following a 1–minute long burst of γ-ray lines and electron bremsstralung which extended to over 100 MeV in photon energy. This confirmed the 1951 prediction (21) that relativistic protons, accelerated during a solar flare, could produce a flux of high-energy neutrons observable at the Earth. Then during the impressive east-limb flare on 1982 June 3, which produced intense nuclear line emission, it was shown (65) that the photon spectrum above 10 MeV included contributions from meson-decay γ rays and electron bremsstralung extending to over 100 MeV. The energetic emissions from this flare occurred in a one minute duration "impulsive" phase followed by a second distinct "extended" phase which lasted over 20 minutes, with 80% of the meson-decay γ-ray emission in the later phase. As expected, a strong flux of neutrons was observed by the GRS (30) and ground level neutron monitors in Europe recorded, for the first time, secondary neutrons produced in the atmosphere by primary solar neutrons (42). The protons resulting from the decay of solar neutrons in space were observed by *ISEE 3/ICE* and *IMP* detectors (56–58). Several flares observed by the *SMM* GRS show rich γ-ray line spectra, from which it is possible, in principle, to determine the composition of the ambient solar atmosphere where the nuclear reactions occur. The analysis of the spectrum obtained for the 1981 April 27 flare, a long duration γ-ray flare, indicates that the ambient medium composition differs from that of both the photosphere and the corona, and requires an enhanced neon abundance (116). For this flare it is suggested (116,105) that significant γ-ray line production could take place in the corona. In the case of impulsive flares most of the line emission is expected to peak in the lower chromosphere with essentially no contribution from the corona (78). The γ-ray line composition analysis of the 1981 April 27 flare also requires enhancement in helium and heavy element accelerated particle abundances over the photosphere as in the case of SEP composition from the so called "impulsive flares" (23,84). In general the *SMM* GRS observations were confirmed by the γ-ray spectrometer (GRS) on the *Hinotori* satellite (180–182).

By the end of solar activity *cycles 20* and *21* in 1986, all of the high-energy radiations predicted (115,101) had been observed revealing new insights into the solar flare particle acceleration enigma. Table 1 summarizes the principal characteristics of the emissions associated with solar flares. *These may be considered as observational constraints which any theory of particle acceleration must satisfy.* The time evolution of a flare may be separated into three phases: *onset, impulsive and extended phase.* A particular flare may be of short duration (a few seconds or minutes) or of long duration (hours), each having emission characteristics as shown in the three columns of the table. Flare characteristics have been previously discussed by several authors (156,154,32,105). Since our main interest here is *acceleration of ions and electrons* we emphasize the characteristics of the high-energy neutral emissions that they produce: hard X-rays, γ-ray lines and continuum, and neutrons. A flare with many of the properties listed in the first two columns of Table 1 occurred at 09:09 UT on 1989 September 9. The right panel in Figure 2 (33) shows the time history of the *SMM* GRS emission in the (4.1–6.4) MeV energy band. The event is initiated by a radio noise storm (S1) commencing at 09:09:20 UT. A second radio source (S2) appears at a different location at 09:09:30 UT. As the flare develops in energetic photon emissions, the hard X-ray emission increases sporadically and low-level nuclear line emission appears in the time interval (09:09:47–09:10:36) UT. The deconvolved *SMM* GRS γ-ray spectrum for this time interval, shown by the lower plot in the left panel of Figure 2, clearly indicates the presence of nuclear lines above 0.8 MeV. A third distinct coronal radio source (S3) appears at 09:10:34 UT just before the major increase in intensity of energetic photons commencing at 09:10:36 UT. The GRS spectrum for the 16.384 s interval ending at 09:10:52 UT, during the major burst, is shown by the upper plot in the left panel of Figure 2. The fact that γ-ray line, hard X-ray and neutron production (the latter revealed by the presence of the neutron capture line) is detectable before the major burst of energetic photons clearly shows that the accelerator of relativistic

TABLE 1. Observational constraints for acceleration theory time evolution.

[I.] Onset Phase	[II.] Impulsive Burst	[III.] Extended Emission
Duration is typically minutes.	Sudden enhancement of bremsstrahlung to several hundred MeV.	Impulsive events with simple decay.
Brightening occurs in UV, soft X rays, or $H\alpha$.	Simultaneous enhancement of γ-ray line emission.	Impulsive event with continued production of high-energy emissions.
Type III burst at coronal height $\sim (0.4 - 1)\ R_\odot$.	Occasional Bursts of > 10 MeV photons.	Succession of impulsive events with γ-ray line emission dominate over bremsstrahlung.
Type I noisestorms.	Emission of meson decay γ rays.	Succession of impulsive events with bremsstrahlung dominate over γ-ray line emission.
Increasing intensity of hard X-ray bremsstrahlung.	Arrival of high-energy neutrons at Earth ~ 10 MeV $< E_n > 2$ GeV.	
Low-level γ-ray line emission develops.	Production of escaping particles (SEP)?	
Bursts of > 10 MeV photons sometimes occur.		
New radio sources appear as new energetic emissions appear.	Type III/V Radio Burst.	Type II and IV radio bursts

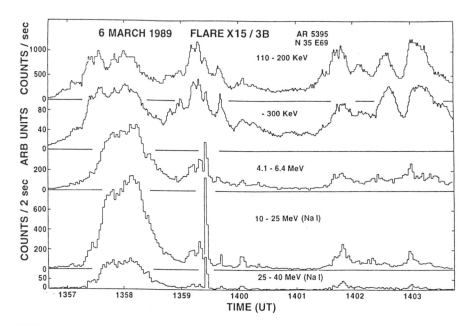

FIG. 3. Time history of the first 7 minutes of the 1989 March 6 flare is shown which contains "two electron dominated" events, one around 1358 UT, lasting about 70 s and the other at 1359:24 UT of 4 s duration (139).

electrons and 50 MeV ions is already fully developed. This means that the accelerator is operating initially at, apparently, low intensity or is confined to a smaller volume than is involved in the major burst.

While the flare just discussed is characteristic of many short (minute duration) impulsive flares there are several cases of extended emission where there is a succession of impulsive bursts (cf columns 2 and 3 in Table 1). However, the ratio of bremsstrahlung to nuclear line emission is highly variable in different individual bursts. Such an event occurred on 1989 March 6 in which nearly all known forms of high-energy emission were displayed. Figure 3 shows (139) the time history of the energetic photon emissions during this event which lasted over 1 hour but only the first six minutes are shown. This illustrates two types of bursts in which bremsstrahlung dominates. The first burst at (13:57-13:59) UT has a spectrum which extends to over 40 MeV and is clearly dominated by electron bremsstrahlung as shown by the spectrum in Figure 4 (139). Nevertheless, a careful examination of this spectrum shows the distinct presence of nuclear lines; for example at 2.2 and 4.4 MeV suggesting again the simultaneous acceleration of ions and relativistic electrons. This spectrum can be compared with the well studied γ-ray line rich flare on 1981 April 27 which provided the unique results on the abundances in the nuclear interaction region (116). A second type of bremsstrahlung burst is seen in Figure 3 at about 13:59:30 UT which lasts less than 10 s and shows no delay

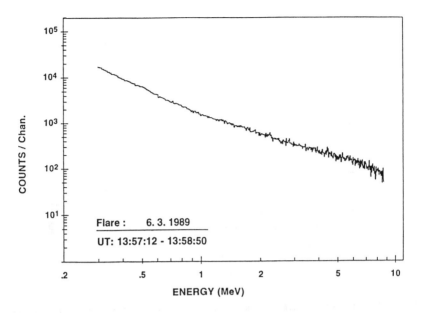

FIG. 4. The time integrated background subtracted events spectrum is shown for time period 13:57:12–13:58:50 UT. Note the flattening of the spectrum around 1 MeV and the weak indication of line structure above this energy (139).

in initiation of the burst from X ray to γ ray energies (139). Both types of bursts are strongly suggestive of the sudden appearance of a transient potential drop as large as 100 MV apparently accelerating relatively more electrons than ions. About 40 minutes later in this same event an extended burst of duration about 15 minutes occurred in which nuclear lines were dominate as illustrated in Figure 5 (139). Earlier in this event, at ∼1407 UT, 10 minutes after the event initiation, meson decay γ rays and high-energy neutrons which reached Earth were detected (46) implying acceleration of ions to >300 MeV for protons.

An important characteristic of high-energy photon emission which bears on the geometry of the flare acceleration process is the longitude distribution of hard X-ray and γ-ray flares. Observations with the *SMM* GRS have shown a statistical solar limb excess in the number of γ-ray emitting flares (>0.3 MeV) (174) which seems more clearly evident for flares with emissions >10 MeV (138). On the other hand studies of the longitude distribution of hard X-ray flares observed on *SMM* and *Venera 13* and *14*, respectively (98,99), do not show a limb enhancement. Vilmer (176) reviewed the now conflicting longitude distributions for the flares with emissions >10 MeV observed by several spacecraft.

Space does not permit us to review the relationship of the SEP events to the γ-ray line producing flares as done primarily by Cliver (36) and Reames

15

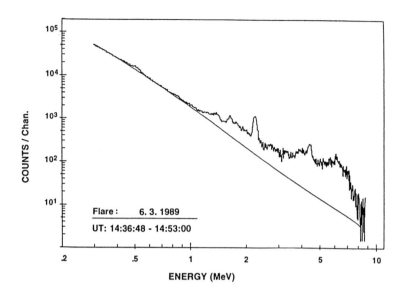

FIG. 5. The time integrated, background subtracted counts spectrum is shown for the time period 14:36:38–14:53:00 UT. The drop in the spectrum above 7 MeV is due to the absence of nuclear deexcitation lines above this energy (139).

(135). However, it is important to refer, briefly, to the large SEP events observed by *Helios 1* (173), during the large γ-ray line flares on 1980 June 21 and 1982 June 3, which produced intense high-energy photons (> 1) MeV and neutrons at the Earth as discussed above. It is of interest that the SEP intensities and energy spectra of these two events typical of the so called "large gradual" events (135), while the particle composition is like that for the so called "impulsive, ^{3}He-rich" flares (135). Because of the proximity of *Helios 1* to the Sun (∼0.5 AU) and the excellent magnetic connection to the flare site, small precursor particle increases were observed prior to both events, suggesting particle storage at the Sun from earlier flares (173). This observation may be a vital clue to understand when and how the high-energy particles are accelerated. There is strong evidence that the energy resources of an active region may not be sufficient to supply the energetic particles produced in some large flares and larger coronal structures may be involved (87).

The Years 1990–1995

On 1989 December 1 the French/Russian satellite *GRANAT* was launched with two instruments with the capabilities of observing high-energy flare neutral emissions. The PHEBUS experiment which consists of six BGO spectrometers covering the energy range ∼ (0.1–100) MeV and the anti-coincidence shield of the SIGMA coded aperture γ-ray telescope which can give crude spectra of solar γ-ray bursts between (0.25–15) MeV (171). PHEBUS recorded spectra of several flares, many with γ-ray line emission. Two flares on 1990 May 15 and 1990 June 11 had simultaneous Nançay radioheliograph observations which show that new radio sources appear in association with new (i.e., different successive) peaks in the hard X-ray and/or γ-ray emission (170) consistent with earlier observations (33). PHEBUS also provides the only satellite γ-ray data (122,163) during the large flare on 1990 May 24 at 20:48 UT, which gave a strong neutron signal detectable at ground level (127,146). It is interesting to note that there was not a clear excess of prompt (4–7) MeV line emission during the most intense emission (122), however, a strong 2.22 MeV line has been reported (166) during the later phase of this event so prompt line emission must have been swamped by intense bremsstrahlung continuum from accelerated electrons with energies above 10 MeV. This high-energy emission probably consists of two components i.e., predominately relativistic electron bremsstrahlung during the intense initial phase with low-level emission after 200 s apparently due to the arrival of high-energy neutrons at the spacecraft. The time behavior of the high-energy emissions indicate that this event is similar to the 1982 June 3 flare in its emission characteristics (166). Kocharov (92) has made a detailed analysis of all of the data for this event and concluded that a shock seen in Hα (a Moreton wave) is more likely to be responsible for the prompt acceleration of both electrons and protons with trapping in magnetic loops rather than a continual acceleration process!

The Russian spacecraft *GAMMA-1* provides the only γ-ray data available on the intense flare at 20:28 UT on 1991 March 26 (7). These observations provide γ-ray time histories above 30 MeV and show impulsive emission lasting for about 10 s which is due predominately to γ rays with energies below 100 MeV and a component delayed by 8 minutes from the flare onset which is apparent in γ rays above 100 MeV. The impulsive component is attributed (7) to bremsstrahlung from electrons accelerated by a direct electric field produced to at least 300 MeV as proposed by Alfvén and Carlqvist (10). On the other hand the delayed component, which has a much harder photon spectrum, is attributed to the decay of neutral pions. The high-energy photon properties suggest this is a so called "electron dominated"[3] event (139). PHEBUS also recorded γ-ray line emissions associated with a flare, which apparently occurred behind the east limb on 1991 June 1, from AR 6659 (13). The γ-ray spectrum from this flare (13) shows prompt lines in the 1–2 MeV and the 4–7 MeV energy ranges with no evidence for a neutron capture line at 2.2 MeV, as normally expected from limb flares. (An exeption to this rule is the limb flare on 1989 September 29, which apparently shows evidence for a line at 2.2 MeV (175).) The source of the prompt γ-ray emissions in the

---

[3]The short intense bursts of γ rays extending well above 10 MeV which characterize the "electron dominated" events sometimes occur in the time evolution of an otherwise normal γ-ray flare (172).

1991 June 1 flare is assumed to be in the lower corona at heights greater than 3000–7000 km since emission from lower altitudes was occulted. This implies that the ambient density in the ion interaction region where the γ-ray lines are produced is $< (10^{11}$–$5\times10^{11})$. In earlier work (29) the ambient density in the interaction region was calculated as $n_H > 10^{12}$ cm^{-3}. On the other hand Hulot and colleagues (81) indicate that the maximum production rate of prompt γ-ray lines can be at densities as high as $10^{14} - 10^{15}$ cm^{-3}. Nevertheless it has been shown (82) that in some flares as much as 30%, or more, of the prompt emission could originate in the corona at a mean density of 5×10^{10} cm^{-3}. This point can be verified for the 1991 June 1 flare since the hard X-ray detector (15–150) keV on Ulysses, observed the emission from the full hard X-ray source (87).

Since April 1991 additional solar flare observations were made by the four instruments on the *Compton Gamma Ray Observatory (CGRO)*. After the 1991 June 1 flare, discussed above, the next major flare from 6659 at N34E75 was an X12 beginning in soft X-rays at 03:37 UT on 1991 June 4 followed by frequent additional flaring after June 8. The Sun was designated as a *CGRO* target of opportunity (ToO) placing the Sun in the direct field-of-view of all *CGRO* instruments. Thus all *CGRO* instruments directly viewed the three additional X-class flares from AR 6659 on 1991 June 9, 11 and 15. Independently, the Sun was made an OSSE ToO in early June so it was able to make excellent observations of the June 4 flare (119). The flare was unusually intense and of long duration with the *GOES* satellite detectors saturated for more than 30 minutes. OSSE observed the flare for more than 110 minutes and lower limit estimates for the 2.223 and 4.44 MeV line fluences are respectively, 692 ±9 and 68±4 photons cm^{-2} not correcting for data gaps due to detector saturation and satellite night. Assuming that the 2.223 to 4.44 MeV line fluence ratio does not depend strongly on correcting for data gaps, Murphy and colleagues (132,119) use the measured ratio (10.2±0.6) and recent calculations (93) to obtain the spectrum shape of the interacting ions responsible for the γ-ray line emission. If the spectrum has a power-law form an exponent $\sim$2.8 is implied which is unusually hard with only the 1988 December 16 flare at 08:20 UT having a flatter exponent of 2.7 for the interacting particles if 1981 April 27 compositions are used for both the ambient medium and the accelerated particles. Presumably, at least in the 1988 flare, most of the pion and high-energy neutron production took place in the "delayed" or "extended" phase of the event (46,47).

During the intense initial phase of the 1991 June 4 flare all X-ray and γ-ray instruments in space were saturated with the exception of the thin plastic scintillator anticoincidence charged particle detectors (CPDs) which cover the BATSE large area detectors. By using EGRET and OSSE observations of nuclear lines after the peak intensity period (143,118), Ramaty and colleagues (133) have shown that the CPDs respond to MeV photons that are predominately relativistic electron bremsstrahlung. Also, by comparing the CPD time history with the 80 GHz flux density the γ ray and μwave observations are consistent with one another if the latter is gyrosynchrotron radiation, produced by trapped electrons, and the γ rays are the thick-target bremsstrahlung from the same population of electrons leaking from the footpoints of the trap. This interpretation also suggests that the trapping magnetic field be about 200 to 300 gauss and the electron spectral index be between 3 and 5.

Active region 6659 produced several additional flares which were observed by instruments on *CGRO, GRANAT,* and *GAMMA 1* and comprehensive data were, therefore, obtained from large X-class flares on June 6, 9, 11, and 15. The high-energy γ-ray and neutron emissions from these flares have been described in several papers (140,117,141,85,170,144,5,6,97,8). Because the instruments on the *CGRO*, particularly, were much larger than those on the *SMM* and *Hinotori*, extended (hours) emissions of γ rays and neutrons at low flux levels could be identified following some of these flares. The most dramatic example of extended emission was obtained by the *CGRO* EGRET, following the 1991 June 11 flare when high-energy γ rays ($>$100 MeV) from meson decay and ultrarelativistic bremsstrahlung were observed to continue more than 8 hours after the peak of the flare (85,20,144). The PHEBUS spectrometers on *GRANAT* also recorded γ-ray line and continuum emissions (169) throughout the intense part of the flare when the EGRET sparkchamber and the COMPTEL were disabled because of deadtime. The EGRET/TASC however can also function idependently of the spark chamber, so high-energy emissions could be recorded throughout the flare. After the intense phase of this flare, analysis of the COMPTEL data indicates a flux of neutrons in the energy range (40–60) MeV (134) and the EGRET/TASC also shows possible evidence for even higher energy neutrons (144). An important question posed by the long-term high-energy γ-ray emission concerns the question – are the particles accelerated in a short impulsive phase of duration a few minutes and trapped in magnetic loops, subsequently precipitating into the lower corona or – are the particles continually accelerated by other processes than that which produced the particles causing the initial burst of high-energy radiation? The first possibility was considered by Mandzhavidze and Ramaty (104) who fit the observed high-energy photon spectrum (85) with combined primary electron bremsstrahlung and meson decay γ-ray decay radiation. By comparing the 1991 June 11 flare observations with those during the 1982 June 3 (65,30) and 1991 June 15 (5,97) solar flares, these authors (104) conclude that the trapping model can explain the observations. This conclusion imposes a requirement of a low-level of plasma turbulence and coronal densities less than 5×10^{11} cm^{-3}. The second possibility has been considered by Akimov and collaborators (8) in interpreting the *GAMMA 1* and COMPTEL data for the 1991 June 15 flare. They use correlated emission time histories in the Hα, soft X-rays, meter wave and microwave radio emissions with the γ-ray line histories to determine the behavior of the electrons and protons which produce the observed emissions. Electric fields produced in a reconnecting current sheet (behind a rising CME) are suggested to explain particularly the delayed microwave emission, the 1–8 MeV and 100 MeV and 1 GeV γ rays (8). An earlier

TABLE 2. Some extreme particle properties.

Parameter	Electrons	Ions (Protons)	Reference
Number	$10^{41}(> 20\text{ keV})$		(87)
	$10^{36}(> 100\text{ keV})$		(112)
	$5 \times 10^{34}(> 300\text{ keV})$		(112)
		$3 \times 10^{35}(> 30\text{ MeV})$	(130,131)
		$\sim 10^{32}(> 300\text{ MeV})$	*
Risetime (s)	10^{-2}	> 1	*
Duration (s)	$10 \rightarrow$	$60 \rightarrow$	*
Total Energy (ergs)	10^{34} (>20 keV)		(87)
	$10^{29}(> 100\text{ keV})$		(112)
	10^{28} (> 300 keV)		(112)
		$10^{30}(> 30\text{ MeV})$	(130,131)
		$3 \times 10^{28}(> 300\text{ MeV})$	*
Power (ergs s^{-1})			
	$10^{32}(>20\text{ keV})$		(87)
		$2 \times 10^{28}(> 30\text{ MeV})$	*

*Author's estimates

analysis of essentially the same data concluded (91) that the delayed (gradual phase) emission was a result of prolonged production due to stochastic acceleration. Further interpretations of flares from AR 6659 and the question of storage or continuous acceleration of charged particles has recently been reviewed (80).

ACCELERATION MECHANISMS

Observations referred to in the previous sections provide the constraints any particle acceleration mechanism(s) must satisfy (see Table 1). In Table 2. we summarize extreme properties of the accelerated ions and electrons inferred from various observations. Several noteworthy, detailed discussions of the many flare models have been put forward to explain the complete solar flare, e.g., (156,154), but here we focus on the microscopic mechanisms which can give individual particles the properties necessary to explain the observations. Consideration of Maxwell's equations and the Lorentz force expression reminds us that, fundamentally, only a direct or induced electric field can energize a charged particle, but the necessary macroscopic field itself depends on the particle's trajectory (83) and ultimately the effect of the accelerated particles on the field must be considered.

It is usual in astrophysics to consider three separate physical mechanisms: stochastic, direct electric field and shock acceleration. Reviews of the theoretical work on these processes up to 1980 were made by Forman, Ramaty and Zweibel (63). A general discussion of solar flare accelerating mechanisms was given by Heyvaerts (73) and Spicer (151), the latter reviewing the role of electrical currents in flares from a global point of view. A study of solar flare particle acceleration during the *SMM* era was conducted at the *Solar Maximum Mission Flare Workshops* in 1983 and 1984 (178) and at other conferences e.g., *Particle Acceleration in Cosmic Plasmas* (183).

Stochastic Acceleration

Fermi (60), in order to explain the origin of the galactic cosmic ray energy spectrum, proposed acceleration by collision of ambient interstellar protons with randomly moving intragalactic magnetized clouds. A turbulent plasma threaded by magnetic fields also supports a number of wave modes such as Alfvén waves. Forman, Ramaty and Zweibel (63) have discussed several stochastic acceleration processes with some of their primary references cited below. For example, particles whose gyroradius is of the order the wavelength of the mode can undergo resonant pitch-angle scattering. Acceleration can then occur for particles in a region where the waves are propagating both parallel and antiparallel to the average magnetic field (107,148,14). Magnetosonic waves whose wavelengths are longer than the gyroradius of the particle can produce acceleration if pitch angle scattering maintains isotropy (94,108,1). Stochastic acceleration has also been suggested by Langmuir waves whose phase velocities are near the speed of the particle (108) and by lower hybrid waves (17). A detailed treatment of the stochastic acceleration of ions appropriate for the production of the SEP was made in 1979 by Ramaty (129). A basic discussion of stochastic acceleration processes can be found in Melrose (108).

Recently Miller and Roberts (113) have argued that the injection requirement for resonant pitch angle scattering from Alfvén waves is artificial, because it includes only wave frequencies much less than the proton gyrofrequency mentioned above. If higher frequency waves are included the injection energy can be negligible (113). Indeed, Smith and Miller (149) have shown that Alfvén wave turbulence alone can accelerate protons from the ambient medium to 30 MeV in a few seconds and in sufficient quantity to explain the γ-ray line observations. In this (149) analysis nonlinear Landau damping of Alfvén waves (96) can efficiently energize protons from the ambient medium to above the Alfvén velocity so they may be further accelerated by the gyroresonance interaction. A thorough treatment of the acceleration of protons by long wavelength Alfvén waves alone has been recently presented (113) with numerical simulation results. Stochastic acceleration of electrons to relativistic energies by resonance with fast mode broad band MHD waves has also been proposed (114). This reference (114) also gives detailed results for a variety of initial conditions for the waves assuming the initial temperature of the flare plasma is $\sim 3 \times 10^6$ K consistent with the heavy ion abundance ratios in the SEPs. Thus it may be possible that MHD turbulence alone can provide for both electron and ion acceleration, provided it can be developed (or preexists) with sufficient energy density in a time scale as short as 0.1 s. It has been shown that (112) the turbulent energy-density must be of the order $\sim 10^3$ ergs cm^{-3} and Mandzhavidze and Ramaty (105) show that as much as 10% of the particles, contained in a volume of 10^{27} cm^3, at an average density of 10^{10} cm^{-3}, must be accelerated to the energies required for the production of the γ-ray lines.

The similarity of particle acceleration times scales in the terrestrial aurora to those in solar flares has lead to the suggestion that other plasma wave modes, believed to be operative locally, may be applicable to the flare problem. In order to explain production of excess ^{3}He and heavy ion enhancements observed in impulsive SEP events, it has been (165) suggested that "electromagnetic hydrogen cyclotron waves," which resonate with the ^{3}He gyrofrequency could be responsible. Stochastic acceleration of ^{3}He has also been proposed (111). A wave packet consisting of a spectrum of lower hybrid (LH) waves could conceivably resonate with electrons and produce transient bursts of electrons (168).

In Elliot's precipitation model of a solar flare (52,53) it was assumed that the stochastic Fermi process accelerated protons to relativistic energies over a time scale as long as a day.

Direct Electric Field Acceleration

Several features of the hard X-ray and γ-ray emissions associated with solar flares are suggestive of acceleration by intense transient electric fields. Generally, electric fields can be perpendicular or parallel to the local magnetic field. The perpendicular case is usually associated with reconnection near neutral points and current sheets such as in Syrovatskii's model (158–161,150). Detailed discussions and critiques of this early acceleration process have been given by Svestka (156) and Sturrock 1980 (154). As these authors suggest many of the requirements for ion and electron acceleration can be met but there are critical questions concerning the escape of the accelerated particles if they are to escape into interplanetary space. In this regard Speiser (153) calculated trajectories of particles in fields associated with neutral sheets applicable to the Earth's magnetosphere. Parker (121) also discusses the development of strong electric fields in the formation of current sheets.

Electrical currents parallel to the magnetic field (field aligned currents) can give rise to strong transient electric fields if current interruption occurs as a result of an instability producing anomalous resistivity in the confined plasma. Alfvén and Calquist (10) pointed out that if the current is interrupted at some point in an inductive circuit the full magnetic energy available is dissipated in that region. A large inductive voltage ($\sim 10^9$ V) could appear over the region of current interruption, accelerating electrons and ions, at the same time, to the same energy! Colgate (38) in an extensive analysis of a single twisted loop flare model and taking into account the laboratory experience with linear and toroidal plasma confinement geometries describes the consequences of enhanced dissipation of a field aligned current. This model was applied to the 1972 August 4 solar flare to explain the X-ray and γ-ray line observations. By some unspecified process the interruption of an inferred current of 7×10^7 A in 10^3 s leads to an inductive voltage of 10^9 V and dissipation of the magnetic field energy. The dissipation of the current leads to heating the loop volume by collisional thermal conduction giving an electron temperature of $\sim 2 \times 10^7$ K and a thermal X-ray emission measure consistent with observations ($\sim 5 \times 10^{49}$ cm^{-3}. The twisted flux tube current is assumed to be carried by runaway ions with energies ($E > 4$ MeV) sufficient to produce the γ-ray line intensities observed in this flare, rather than the current being carried by electrons as is usually assumed. The motive for this flare model comes from laboratory observations where several physical conditions, magnetic field, current, and particle density, are similar to these in the flare environment.

In order to explain the microwave and hard X-ray emissions, from a flare, Holman (75) investigated the conditions necessary for acceleration of runaway electrons and heating of the plasma when a DC electric field is imposed. By considering the number of runaway electrons that can be accelerated by an inductive electric field it is found that it is possible to supply the $10^{31} - 10^{32}$, 100 keV electrons necessary to produce a typical microwave flare. The production rate of runaway electrons depends on whether electrons are resupplied for successive acceleration by plasma flowing into the current sheet from the sides or are scattered into the runaway region by Coulomb collisions. For example, if the hard X-ray emission is due to nonthermal bremsstrahlung then the accelerator must supply a beam of $> 10^{35}$ electrons s^{-1}. It has been pointed out frequently (77,38,152,70,55) that such a high flux of electrons in a single current channel would have a self magnetic field several orders of magnitude greater than observed. Therefore, the bulk of the hard X-ray emission (100 keV), if from nonthermal bremsstrahlung, must be due to at least 10^4 oppositely directed current channels (75). The alternatives are, that another process such as a stochastic mechanism (114), that does not require a high current, accelerates the electrons, or else most of the hard X-ray emission is thermal! (See Holman (75) for further discussion.) According to Holman[4], the currents responsible for electron runaway can also cause ions to runaway. In the case of electric fields, less than the Driecer field, as a result of electron drag, ions will be pulled out of the thermal distribution and drift with the field while the heavier ions move opposite to the field, being dragged by the electrons. Of course strong fields ($E > E_d$) should accelerate all ions to energies limited only by the scale length of the ordered field.

Haerendel (70) has hypothesized the existence in the corona of a large num-

---

[4] As discussed in the *Flares 22* section of *Solar Physics* 153 (18).

ber of independent magnetic twisted flux tubes, $\sim 10^5$, which carry a total field aligned current as large as 10 A/m^2 and which by spontaneous untwisting could supply the energy for a flare. Even stronger currents are postulated that can lead to a GV field aligned potential drop providing further heating and acceleration of particles to the energies implied by the observations. (See also reference (71).)

More recent theoretical work involving electric field acceleration includes the loop coalescence model of Sakai (142) and the reconnecting current sheet model of Litvinenko and Somov (102). The model of Holman has been applied by Benka and Holman (16) to explain the hard X-ray observations of the 1980 June 27 flare (100) and Zarro, Mariska and Dennis (184) have applied it to a *YOHKOH/CGRO* flare on 1982 September 6 with some success. Strong support for acceleration, of at least electrons, by direct electric fields is provided by a recent study (11) of simultaneous hard X-ray and radio bursts which give evidence for bidirectional beams of electrons. These may be produced by an acceleration mechanism involving oppositely directed electric fields or, alternately, by stochastic acceleration near an X-type reconnection region where the electrons would have simultaneous access to the corona and the chromosphere below the acceleration site (11).

Acceleration by Shock Waves

Because of the sudden release of energy by the flare process in the solar atmosphere the development of shock waves is expected. In fact Type II radio bursts have long been the proof of the existence of coronal shocks (179) and they appear to be well correlated with SEP events (155), (24). Shocks moving laterally through the chromosphere with velocities between 500 and 2500 km s^{-1} called Moreton waves (156) have been proposed (92) as the accelerator producing the high-energy neutral emissions in the strong 1990 May 24 flare discussed above. The ability of shocks to accelerate particles in interplanetary space to MeV energies is well established (145) but their role in accelerating the electrons and ions which produce the high energy neutral emissions associated with solar flares has not been clarified. The two fundamental types of shock acceleration, which are first order Fermi processes, are scatter-free and diffusive. In the first case particles in a single crossing of the shock front can only increase their energy by at most a factor of 2.5 (83) while in the second case multiple scattering of a particle from the turbulent media upstream and downstream of the shock can lead to a very large energy gain since every shock crossing leads to an energy gain. Therefore, diffusive shock acceleration is potentially a strong contender for the solar flare accelerator. Several extensive reviews emphasizing diffusive shock acceleration are available (83,48,54,63,178) but most work is generally directly applicable to ion (SEP) acceleration. An exception are the studies which treat the acceleration of electrons which are presumed to generate the Type II radio emission (74) or that in the stationary or moving Type IV bursts (177). In this regard, a recent study (90) of hard X-ray and wide band radio observations for several flares concludes that extended coronal shock waves play a minor role in the acceleration of relativistic electrons observed in the low corona. In spite of the attractiveness of diffusive shock acceleration, if it is to be relevant for the fastest γ ray flares the shocks must develop in < 0.1 s and for ions to be accelerated they must have a threshold velocity which exceeds the local Alfvèn velocity or the mean square velocity of the turbulent scatterers (63).

CONCLUSIONS

Even though our understanding of the particle acceleration process in solar flares has evolved dramatically during the past half-century, there are still many unanswered questions, thus continuing to present a yet to be solved enigma. Major advances, in the past fifty years, have come from observations which give the characteristics of the high-energy neutral emissions, i.e., hard X-rays, γ rays and neutrons. These observations have revealed new insights into the particle acceleration question and have provided constraints which must be met by any proposed acceleration mechanism(s). In this paper we have attempted to show the great variety of spectral and temporal characteristics in the neutral emissions. In the discussion of the γ ray spectra presented we have emphasized the strong variability of the bremsstrahlung component compared to the nuclear line component in a flare of extended duration. A special challenge is posed by transient bursts of bremsstrahlung (>10 MeV) which can occur any time in a flare. Further, simultaneous ion and electron acceleration appears fully developed and at variable intensity level during any phase of the flare. It seems clear that all three basic acceleration mechanisms could be simultaneously involved in some flares. The size distribution of hard X-ray flares and requirements on the number of accelerated electrons seems to be a basic characteristic of the flare process and this suggests flares may result from conditions on a larger scale, such as a corona in a self-organized critical state, rather than in the relatively small volume of any active region. (See e.g., references (22) and (87).)

ACKNOWLEDGEMENTS

The author wishes to acknowledge the support of a NATO Collaborative Research Fellowship at the Max-Planck-Institut für Extraterrestrial Physik and support from Dr. Berrien Moore III, director of the University of New Hampshire Institute for the Study of Earth, Oceans and Space. The use of the Max-Planck-Institute facilities and the hospitality of directors Drs. Gerhard Haerendel and Joachim Trümper is profoundly appreciated. Thanks go to Mary M. Chupp for her significant contribution preparing and editing the manuscript and Philip P. Dunphy for reviewing the final draft of the paper. Special thanks go to R. Ramaty and N. Mandzhavidzhe for helpful suggestions and comments on a draft of this paper. I also commend them and all members of the *Organizing Committee* for arranging and conducting this timely workshop.

REFERENCES

1. Achterberg, A., Astron, Astrophys. **97**, 259 (1981).
2. Adams, N., Phil. Mag. **41**, 503 (1950).
3. Adams, N. and Braddick, H. J. J., Phil. Mag. **41, 505** (1950).
4. Adams, N. and Braddick, H. J. J., Zeits. für Naturforschung **6a**, 592 (1951).
5. Akimov, V. V. and 32 Co-authors, 22nd ICRC **3**, 73 (1991).
6. Akimov, V. V., Akimov, V. V., Leikov, N. G., Belov, A. V., Chertok, I. M., Kurt, V. G., Magun, A., and Melnikov, V. F., Proc. 23rd Internat. Cosmic. Ray Conf. **3**, 111 (1993).
7. Akimov, V. V., Leikov, N. G., Kurt, V. G., and Chertok, I. M. in *AIP Conf. Proc. # 294,* eds. Ryan, J. M. and Vestrand, W. T. (AIP:New York)130 (1994).
8. Akimov, V. V. and 12 Co-authors, Submitted to Solar Physics, January 1995. (1995).
9. Alfvén, H., Nature **158**, 618 (1946).
10. Alfvén, H., and Carlqvist, P., Solar Phys. **1**, 220 (1967).
11. Aschwanden, M. J., Benz, A. O., Dennis, B. R., and Schwartz, R. A., ApJ **455**, 347 (1995).
12. Bagge, E., Biermann, L., Naturwissenshaften **35**, 120 (1948).
13. Barat, C., Trottet, G., Vilmer, N., Dezalay, J.- P., Talon, R., Sunyaev, R., Terekhov, O., and Kuznetsov, A., ApJ (Letters) **425**, L109 (1994).
14. Barbosa, D. D., ApJ **233**, 383 (1979).
15. Baum, W. A., Johnson, F. S., Oberly, J. J., Rockwood, C. C., Strain, C. V., and Tousey, R., Phys. Rev. **70**, 781 (1946).
16. Benka, S. and Holman, G. D., ApJ **435**, 469 (1994).
17. Benz, A. and Smith, D. F., Solar Phys. **107**, 299 (1987).
18. Benz, A. O. *et al.*, Solar Physics **153**, 33 (1994).
19. Berry, E. B., Hess, V. F., Terrestrial Magnetism And Atmospheric Electricty **47, 251** (1942).
20. Bertsch, D. L., *et al.* in preparation (1996).
21. Biermann, L., Haxel, O., and Schlüter, A., Z. Naturforsch. **6a**, 47 (1951).
22. Bromund, K. R., McTiernan, J. M., and Kane, S. R., ApJ **455**, 733 (1995).
23. Cane, H. V., McGuire, R. E., and von Rosenvinge, T. T., ApJ **301**, 448 (1986).
24. Cane, H. V., Reames, D. V. and von Rosenvinge, T. T., J. Geophys. Res., **93**, 9555 (1988).
25. Chambon, G. Hurley, K., Niel, M. Talon, R., Vedrenne, G. Likine, O. B., Kouznetsov, A. V., and Estouline, I. V., in *Gamma Ray Spectroscopy in Astrophysics*, eds. Cline, T. L. and Ramaty, R., NASA TM 79619, p. 70 (1978).
26. Chupp, E. L., Space. Sci. Rev. **12**, 486 (1971).
27. Chupp, E. L., Forrest, D. J., Higbie, P. R., Suri, A. N. Tsai, C., and Dunphy, P. P., Nature **241**, 333 (1973).
28. Chupp, E. L., Forrest, D. J., Ryan, J. M., Heslin, J., Reppin, Pinkau, K., Kanbach, G., Rieger, E., Share, G. H., Ap. J. (Letters) **263**, L95 (1982).

29. Chupp, E. L., Ann. Rev of Astron. Astrophys. **22** 359 (1984).
30. Chupp, E. L., Debrunner, H., Flückiger, E., Forrest, D.J., Golliez, F., Kanbach, G., Vestrand, W. T., Cooper, J., and Share, G., ApJ **318**, 913 (1987).
31. Chupp, E. L., in *AIP Conf. Proceedings # 170*, eds. Gehrels, N., and Share, G., p. 24 (AIP:New York) (1988).
32. Chupp, E. L., Science **250**, 229 (1990).
33. Chupp, E. L., Trottet, G., Marschhäuser, H., Pick, M., Soru-Escaut, I., Rieger, E. and Dunphy, P. P., A & A **275**, 602 (1993).
34. Clay, J., Proc. K. Ned. Akad. Wet. **52**, 906 (1949a).
35. Clay, J., Proc. K. Ned. Akad. Wet. **52**, 923 (1949b).
36. Cliver, E. W., Forrest, D. J., Cane, H. V., Reams, D. V., von Rosenvinge, T. T., McGuire, R. E., Kane, S. R., and MacDowell, R. J., ApJ **343**, 953 (1989).
37. Cliver, E. W., in these Proccedings (1996).
38. Colgate, S. A., ApJ **221**, 1068 (1978)
39. Crosby, N. B., Aschwanden, M. J., and Dennis, B. R., Solar Physics **143**, 275 (1993).
40. Dauvillier, M. A., J. de phys. et rad. **5**, 640 (1934).
41. Dauvillier, A., Compte Rendu Acad. Sci. Paris **229**, 1096 (1949).
42. Debrunner, H., Flückiger, E., Chupp, E. L, and Forrest, D. J., Proc. 18th International Cosmic Ray Conf. **4**, 75 (1983).
43. Dieminger, W. J., *et al.*, Atmos. Terr. Phys. **1**, 37 (1950).
44. de Jager, C. in *The Physics of Solar Flares*, Advances in Space Res., eds. de Jager, C. and Svestka, Z. (Pergamon:Oxford) **6**, No 6, p. 353 (1986).
45. Dolan, J. F., and Fazio, G. G., Rev. Geophys. **3**, 319 (1965).
46. Dunphy, P. P. and Chupp, E. L., Proc. of 22nd Int. Cosmic Ray Conference **3**, 65 (1991).
47. Dunphy, P. P. and Chupp, E. L. in *AIP Conf. Proc. # 294*, eds. Ryan, J. M. and Vestrand, W. T. (AIP:New York) p. 112 (1994).
48. Drury, L. O'C., Axford, W. I. and Summers, D., Monthly Notices Roy. Astron. Soc. **198**, 833 (1982).
49. Duperier, A., Proc. Phys. Soc. **57**, 468 (1945).
50. Ehmert, A., Zeitschrift für Naturforschung **3a**, 264 (1948).
51. Elliot, H. in *Progress in Cosmic Ray Physics* ed. J. G. Wilson (North-Holland:Amsterdam), p. 502 (1952).
52. Elliot, H., Planet. Space Sci. **12**, 657 (1964).
53. Elliot, H., in *Solar Flares and Space Research*, eds. de Jager, C. and Svestka, Z., (North-Holland:Amsterdam) p. 356 (1969).
54. Ellison, D. C. and Ramaty, R., ApJ **298**, 400 (1985).
55. Emslie, A. G., Henoux, Jean-C., ApJ **446**, 371 (1995).
56. Evenson, P., Kroeger, R., Meyer, P., and Müller, D., Proc. 18th ICRC **4**, 97 (1983).
57. Evenson, P., Meyer, P., and Pyle, K. R., ApJ **274**, 875 (1983).
58. Evenson, P., Kroeger, R., and Meyer, P., Proc. 19th ICRC **4**, 130 (1985).
59. Fazio, G. G., in *AAS-NASA Symposium on the Physics of Solar Flares* ed. Hess, W. N., NASA SP-50, p. 145 (1963).
60. Fermi, E., Phys. Rev. **75**, 1169 (1949).
61. Forbush, S. E., Phys. Rev. **70**, 771 (1946).
62. Forbush, S. E., Strichcomb, T. B., and Schein, M., Phys Rev. **79**, 501 (1950).
63. Forman, M. A., Ramaty, R., and E. G. Zweibel in *Physics of the Sun* ed.

Sturrock, P. (D. Reidel:Dordrecht) p 249. (1986).

64. Forrest, D. J. and Chupp, E. L., *et al.*, Nature **305**, 291 (1983).

65. Forrest, D. J., Vestrand, W. T., Chupp, E. L., Rieger, E., Cooper, J., and Share, G., Proc. 19th International Cosmic Ray Conf. **4**, 146 (1985).

66. Friedman, H., Ann. Rev. Astron. Astrophys. **1** 59 (1963).

67. Frost, K. J., in *AAS-NASA Symposium on the Physics of Solar Flares*, ed. Hess, W. N., NASA SP–50, p. 139 (1963).

68. Gold, T. and Hoyle, F., MNRAS **120,** 89 (1960).

69. Gruber, D. E., Peterson, L. E. and Vette, J. I., in *High Energy Phenomena on the Sun*, eds. Ramaty, R. and Stone, R. G., NASA SP-342, p. 143 (1973).

70. Haerendel, G., in *Proceedings of the 21st ESLAB Symposium,*, ESA SP-275 (ESA Publications Division:Noordwiijk) p. 205. (1987).

71. Haerendel, G. in *Proceedings of the Joint Varenna-Abastumani International School & Workshop on Plasma Astrophysics* ESA SP-285 **1** 37 (1988).

72. Hey, J. S., Nature **157**, 47 (1946).

73. Heyvaerts, J. in *Solar Flare Magnetohydrodynamics* ed. Priest, E. R. (Gordon and Breach:New York), p. 429. (1981).

74. Holman, G. D. and Pesses, M. E., ApJ **267**, 837 (1983).

75. Holman, G. D., ApJ **293**, 584 (1985).

76. Holt, S. S., and Cline, T. L., ApJ **154**, 1027 (1968).

77. Hoyng, P., A & A **55**, 23 (1977).

78. Hua, X. M., Ramaty, R. and Lingenfelter, R. E., ApJ **341**, 516 (1992).

79. Hudson, H. S., Bai, T., Gruber, D. E., Matteson, J. L., Nolan, P. L., and Peterson, L. E., ApJ (Letters) **236**, L91 (1980).

80. Hudson, H. and Ryan, J. M., Ann. Rev. Astron. and Astrophys. **33**, 239 (1995).

81. Hulot, Vilmer, N. and Trottet, G. A & A **213**, 383 (1989).

82. Hulot, E., Vilmer, N., Chupp, E. L., Dennis, B. R., and Kane, S. R., A & A **256**, 273 (1992).

83. Jokipii, J. R., in *AIP Conf. Proc. # 56*, eds. Arons, J., Max, C., and McKee, C. (AIP:New York) p. 1 (1979).

84. Kallenrode, M.-B., Cliver, E. W., and Wibberenz, G. ApJ **391**, 370 (1992).

85. Kanbach, G. O., Bertsch, L., Fichtel, C. E., Hartman, R. C., Hunter, S. D., *et al.* A & A Suppl. **97**, 349 (1993).

86. Kane, S., in *High Energy Phenomena on the Sun*, eds. Ramaty, R. and Stone, R. G., NASA SP-342, p. 55 (1973).

87. Kane, S. R., Hurley, K., McTiernan, J. M., Sommer, M. Boer, M., and Neil, M., Astrophysical Journal (Letters) **446**, L47 (1995).

88. Kaufmann, P., Solar Physics **102**, 97 (1985).

89. Kiepenheuer, K. O., Phys. Rev. **78**, 809 (1950).

90. Klein, K-L, Trottet, G., Benz, A. O. zand Kane, S. R., in Proc. of Joint Varenna-Abastumani Int. School and Workshop on Plasma Astrophysics. eds. Guyenne, T. D., and Hunt, J. J., **ESA- SP-285**, p. 157, (1988).

91. Kocharov, L. G., Kocharov, G. A., Kovaltsov, G. E., Chuikin, E. I., Usoskin, I. G., *et al.* Solar Physics, **150**, 267 (1994).

92. Kocharov, L. G., Lee, J. W., Zirin, H., Kovaltsov, G. A., Usoskin, I. G., Pyle, K. R., Shea, M. A., and Smart, D. F., Solar Physics **155**, 149 (1994).

93. Kozlovsky, B., Murphy, R. J., and Ramaty, R. in preparation (1994).

94. Kulsrud, R. M., and Ferrari, A., Astrophys. Space Sci. **12**, 302 (1971).

95. Lange, I., Forbush, S. E., Terrestrial Magnetism and Atmospheric Electricty

47, 331 (1942).

96. Lee, M. A., Volk, H. J., Astrophys. and Space. Sci. **24**, 31 (1973).
97. Leikov, N. G. and 16 Co-authors, A & A Suppl. **97**, 345 (1993).
98. Li, P., ApJ **421**, 381 (1993).
99. Li, P., Hurley, K., Barat, C., Niel, M., Talon, R., and Kurt, V., ApJ **426**, 758 (1994).
100. Lin, R. P., Schwartz, R.A., Pelling, R. M., and Hurley, K. C., ApJ **25**, L109 (1981).
101. Lingenfelter, R., and Ramaty, R., in *High-Energy Nuclear Reactions in Astrophysics* ed. Shen, B. S. P. (Benjamin:New York) p. 99 (1967).
102. Litvinenko, Y. E. and Somov, B. V., Solar Physics **158**, 317 (1995).
103. Lu, E. T. and Hamilton, R. J., ApJ. **380**, L89 (1991).
104. Mandzhavidze, N. and Ramaty, R., ApJ (Letters) **396**, L111 (1992).
105. Mandzhavidze, N., Ramaty, R., Nuclear Phys. (B) **33A, B**, 141 (1993).
106. McMillan, E. M., Phys. Rev. **79**, 498 (1950).
107. Melrose, D. B., Solar Physics **37**, 353 (1974).
108. Melrose, D. B., Plasma Astrophysics II (Gordon and Breach:New York), (1980).
109. Menzel, D. H., Salisbury, W. W., Nucleonics **2**, 67 (1948).
110. Meyer, P., Parker, E. N., Simpson, J. A., Phys. Rev. **104**, 768 (1956).
111. Miller, J. and Vinas, A. F., ApJ **412**, 386 (1983).
112. Miller, J. M. and Ramaty, R., in *AIP Conf. Proc. # 264* eds. Zank, G. P. and Gaisser, T. (AIP:New York) p. 223 (1992).
113. Miller, J. A. and Roberts, D. A., ApJ **452**, 912 (1995).
114. Miller, J. A., LaRosa, T. N., and Moore, R. L., ApJ **461**, 445 (1996).
115. Morrison, P. Nuovo Cimento **7**, 858 (1958).
116. Murphy, R. J., Ramaty, R., Kozlovsky, B., and Reames, D. V., ApJ **371**, 793 (1991).
117. Murphy, R. J., Share, G. H., Grove, J. E., Johnson, W. N., Kinzer, R. L., Kroeger, R. A., Kurfess, J. D., Strickman, M. S., Matz, S. M. *et al.* in *AIP Conf. Proc. # 280*, eds. Friedlander, M., Gehrels, N., and Macomb, D. J. (AIP:New York) p. 619 (1993).
118. Murphy, R. J. *et al.* in *Proc. 23rd ICRC* **3**, 99 (1993).
119. Murphy, R. J., Share, G. H., Grove, J. E., Johnson, W. W., Kinzer, R. L., *et al.* in *AIP Conf. Proc. # 294* eds. Ryan, J. M., and Vestrand, W. T. (AIP:New York) p. 15. (1994).
120. Parker, E. N., Phys. Rev. **107**, 830 (1957).
121. Parker, E. N., *Cosmical Magnetic Fields* (Clarendon Press:Oxford) (1979).
122. Peleaze, F., Mandrou, P., Niel, M., Mena, B., Vilmer, N., Trottet, G., Lebrun, F., and Paul, J., Solar Physics **140**, 121 (1992).
123. Peterson, L. and Winkler, J. R., Phys. Rev. (Letters) **1**, 205 (1958).
124. Peterson, L. E., and Winckler, J. R., J. Geophys. Res. **64**, 697 (1959).
125. Peterson, L. E., Datlowe, D. W., and McKenzie, D. L. in *High Energy Phenomena on the Sun* eds. Ramaty, R. and Stone, R. G., NASA SP-342, p. 132 (1973).
126. Prince, T., Ling, J. C., Mahoney, W. A., Riegler, G. R., and Jacobson, A. S., ApJ **255**, L81 (1982).
127. Pyle, K. R., Shea, M. A., and Smart, D. F., Proc. of 22nd ICRC **3**, 57 (1991).

128. Ramaty, R., Kozlovsky, B., and Lingenfelter, R. E., Space Sci. Rev. **18**, 341 (1975).

129. Ramaty, R., in *AIP Conference Proc # 56,* eds. Arons, J., Max, C., and McKee, C. (AIP:New York) p. 138 (1979).

130. Ramaty, R. and Murphy, R. J., Space Sci. Rev. **45**, 213 (1987).

131. Ramaty, R., Dennis, B. R., and Emslie, A. G., Solar Physics **118**, 17 (1988).

132. Ramaty, R., Mandzhavidze, N., Kozlovsky, B., and Skibo, J. B., Adv. Space Res. **13**(9), 275 (1993).

133. Ramaty, R., Schwartz, R. A., Enome, S. and Nakajima, H., ApJ **436**, 941 (1994).

134. Rank, G., Diehl, R., Lichti, G. G., Schönfelder, V., Varendorff, M., *et al.*, in *AIP Conf. Proc. # 294* eds. Ryan, J. M., and Vestrand, W. T. (AIP:New York) p. 100. (1994).

135. Reames, D. V., ApJ Suppl. **73**, 235 (1990).

136. Reames, D. V. in these Proceedings (1996).

137. Richtmyer, R. D., Teller, E., Phys. Rev. **75**, 1729 (1949).

138. Rieger, E., Reppin, C., Kanbach, G., Forrest, D. J., Chupp, E. L., *et al.*, Proc. 18th ICRC **4** 83 (1983).

139. Rieger, E., and Marschhäuser, H. in *Proc. of Max91 Workshop 3,* eds. Winglee R. M. and Kiplinger, A. L. (U. Colorado:Boulder) p. 68, (1990).

140. Ryan, J., Forrest, D., Lockwood, J., Loomis, M., McConnell, M., Morris, D., *et al.* in Proc. of First Compton Symposium, eds. Friedlander, M., Gehrels, N., Macomb, D. J., in *AIP Conf Proc # 280.* (AIP:New York) p. 631 (1993).

141. Ryan, J. M. in Proc. of 2nd Compton Symposium eds. Fichtel, C. E., Gehrels, N., Norris, J. P., *AIP Conf. Proc. # 304,* (AIP:New York) p. 12. (1994).

142. Sakai, J., ApJ Supplement **73**, 321 (1990).

143. Schneid, E. J. and 16 Co-authors, Paper presented at 1994 AAS Winter Meeting (1994).

144. Schneid, E. J., *et al.* in AIP Conference Proceedings # 294 eds. Ryan, J. M., and Vestrand, W. T. (AIP:New York) p. 94 (1994).

145. Scholer, M. in *AIP Conf. Proc #264,* eds. Zank, G. P. and Gaisser, T. K., p. 125 (1992).

146. Shea, M. A., Smart, D. F., and Pyle, K. R., Geophys. Res. Lett. **18**, 1655 (1991).

147. Simpson, J. A., Fonger, W., and Treiman, S. B., Phys. Rev. **90**, 934 (1953).

148. Skilling, Monthly Notice Royal Astron. Soc. **172**, 557 (1975). Peter Sturrock, (Colorado Associated University Press:Boulder), (1980).

149. Smith, D. F., Miller, J. A., ApJ **446**, 390 (1995).

150. Somov, B. V. and Syrovatskii, S., Solar Physics **55**, 393 (1977).

151. Spicer, D., Space Sci. Rev. **31**, 351 (1982).

152. Spicer, D. S., Adv. Space Res. **2**, 135 (1982).

153. Speiser, T. W., J. Geophys. Res. **70**, 4219 (1965).

154. Sturrock, P. A., in *Solar Flares* ed. Sturrock, P. A. (Colorado Associated U. Press:Boulder) p. 411 (1980).

155. Svestka, Z. and Fritzova-Svestkova, L., Solar Physics **36**, 417 (1974).

156. Svestka, Z., *Solar Flares,* (D. Reidel:Dordrecht) (1976).

157. Swann, W. F. G., Phys. Rev. **43**, 217 (1933).

158. Syrovatskii, S. A., J. Exp. Theor. Phys. **50** 1133 (USSR) **23**, 754 (AIP) (1966).

159. Syrovatskii, S. A., in *Solar Flares and Space Research* eds. de Jager, C., and

Svestka, Z. (North-Holland:Amsterdam) p. 346. (1969).

160. Syrovatskii, S. A., in *The Sun – Part I of Solar Terrestrial Physics/1970*, ed. C. deJager, p. 119 (1972).

161. Syrovatskii, S. A., Ann. Rev. of Astron. Astrophys. **19**, 163 (1981).

162. Talon, R. Vedrenne, G., Melioransky, A. S., Pissankerov, N. E., Shamolin, V. M., and Likin, O. B., IAU Symp. **68**, 315 (1975).

163. Talon, R., Trottet, G., Vilmer, N., Barat, C., Dezalay, J.-P., Sunyaev, R., Terekhov, O., and Kuznetsov, A., Solar Physics **147**, 137 (1993).

164. Tanaka, K. and Zirin, H., *High Energy Phenomena on the Sun*, NASA SP-342, p. 26. (1993).

165. Temerin, M. and Roth, I., ApJ (Letters) **391**, L105 (1992).

166. Terekhov, O., Sunyaev, R. A., Kuznetsov, A. V., Barat, C., Talon, R., Trottet, G. and Vilmer, N., Astron. Lett. **19, 65** (1993).

167. Tousey, R., Space Science Rev. **2**, 3 (1963).

168. Treuman, R. A., Dubouloz, N., Pottelette, R., MPE Preprint (1995).

169. Trottet, G., Vilmer, N., Barat, C., Dezalay, J. -P., Talon, R., Sunyaev, R., Kuznetsov, A., and Terekhov, O., A & A Suppl. **97**, 337 (1993).

170. Trottet, G., Vilmer, N., Barat, C., Dezalay, J.-P., Talon, R., Sunyaev, R., Terekhov, O., and Kuznetsov, A. in *Advances in Space Research.* **Vol. 13.** (Pergammon Press:NY) p. 285 (1993).

171. Trottet, G., in *AIP Conf. Proc. # 294*, eds. Ryan, J. M. and Vetsrand, W. T., p. 3 (AIP:New York) (1994).

172. Trottet, G., Chupp, E. L., Marschhäuser, Pick, M., Soru-Escaut, I., Rieger, E. and Dunphy, P. P., A & A **288**, 647 (1994).

173. Van Hollebeke, M.A.I., McDonald, F. B. and Meyer, J. P., ApJ Suppl. **73**, 285 (1990).

174. Vestrand, W. T., Forrest, D. J., Chupp, E. L., Rieger, E., and Share, G. H., ApJ **322** (1987).

175. Vestrand, W. T. and Forrest, D. J., in *AIP Conf. Proc. # 294* eds. Ryan, J. M. and Vestrand, W. T. (AIP:New York) p. 143 (1994).

176. Vilmer, N., Astrophysical Journal Supplement Series **90**, 611 (1994).

177. Vlahos, L., Gergely, E. T., and Papadopoulos, K., ApJ **258**, 812 (1982).

178. Vlahos, L. *et al.* in *Energetic Phenomena on the Sun*, NASA Conference Publication 2439, (1986).

179. Wild, J. P. and Smerd, S. F. and Weiss, A. A., Ann. Rev. of Astron. and Astrophys. **1**, p. 291 , (1963).

180. Yoshimori, M., J. Phys. Soc. Japan **54**, 487 (1985).

181. Yoshimori, M., J. Phys. Soc. Japan **54**, 1205 (1985).

182. Yoshimori, M., ApJ Suppl. **73**, 227 (1990).

183. Zank, G. P., and Geisser, T., eds. *AIP Conf. Proc. # 264* (AIP:New York) (1992).

184. Zarro, D. M., Mariska, J. T., Dennis, B. R., Ap.J. **440**, 888 (1995).

ENERGETIC CHARGED PARTICLES

Energetic Particles from Solar Flares and Coronal Mass Ejections

Donald V. Reames[1]

NASA, Goddard Space Flight Center
Greenbelt, MD 20771

We review the recent evidence that distinguishes particles accelerated in flares and in shock waves driven by coronal mass ejections (CMEs). CME-driven shocks, not flares, produce most of the large particle events at 1 AU and can accelerate protons up to 20 GeV. In contrast, flare-accelerated ions have characteristic abundances produced by resonant wave-particle interactions in the flare plasma. Only the direct particle observations have allowed us to study this new physics of ion acceleration in flares, since the energetic ion abundances have been largely invisible in photons.

INTRODUCTION

For nearly 30 years it was thought that all solar energetic particles (SEPs) were accelerated in solar flares and that they could somehow diffuse across magnetic field lines to distant longitudes by a mysterious mechanism known only as "coronal diffusion." During recent years, a new class of observations has revealed two distinct populations of SEPs, with completely different origins, based upon the abundances, ionization states and time profiles of the particles as well as the longitude distribution and the radio, optical, X-ray and γ-ray associations of the events (23, 25, 27, 28). It has become clear that the largest and most energetic particle events at Earth are associated with shock waves driven out into interplanetary space by coronal mass ejections (CMEs). Energetic particles in these events have, on average, the same element abundances and ionization states as those in the ambient plasma of the corona or solar wind; these particles have *not* come from the hot plasma of a flare or reconnection region where they would have become highly ionized. The other class of SEP events have strong associations with impulsive Hα and X-ray flares and type III radio bursts. The energetic particles from these events have distinctive 1000-fold enhancements in the $^3He/^4He$ ratio and 10-fold enhancements in heavy element abundance such as Fe/O. Elements up to Si are fully ionized and Fe has charge 20 indicating heating or other ionization of the plasma. The strong evidence of electron beams in these events has led to the suggestion that the enhancements come from resonant interactions of the particles with waves generated by electron-beam-driven in-

---

[1] Also Solar-Terrestrial Environment Lab., Nagoya, Japan.

stabilities in the flare plasma (33, 19, 20). The properties of gradual and impulsive events are summarized in Table 1.

TABLE 1. Properties of Impulsive and Gradual Events

	Impulsive	**Gradual**
Particles:	Electron-rich	Proton-rich
$^3He/^4He$	~ 1	~ 0.0005
Fe/O	~ 1	~ 0.1
H/He	~ 10	~ 100
Q_{Fe}	~ 20	~ 14
Duration	Hours	Days
Longitude Cone	<30 deg	~180 deg
Radio Type	III, V (II)	II, IV
X-rays	Impulsive	Gradual
Coronagraph	-	CME (96%)
Solar Wind	-	IP Shock
Events/year	~1000	~10

Historically, the terms *impulsive* and *gradual* referred to the time duration of the soft X-rays in the event. However, the X-ray duration gives only a poor, statistical distinction of the underlying mechanisms, while the particle abundances, for example, distinguish them cleanly. Therefore, we now use the terms impulsive and gradual to refer to the underlying acceleration mechanisms, independently of the actual X-ray duration in an event. Of course, there are events in which both impulsive and gradual phenomena occur (24, 3)

More recent observations have extended these conclusions to particles of very high energy. Even in ground-level events (GLEs) particles of ~20 GeV have a clear association with CME-driven shocks. These new observations are discussed in the next section. One of the important consequences of the new paradigm for SEP events is the demise of the concept of "coronal diffusion" of particles across magnetic field lines. The consequences of this demise are not fully appreciated by the authors of some recent papers, as discussed in the subsequent section. Finally, element abundances in gradual and impulsive SEP events are considered.

NEW RESULTS ON HIGH-ENERGY PARTICLES

Ionization states of Fe and lighter elements have now been measured in large SEP events by 4 experiments on 3 different spacecraft, spanning an energy range from 0.3 to 600 MeV/amu (14, 15, 13, 34). These measurements are summarized in Table 2. While there are some differences, it is clear that the Fe has not been subjected to electron temperatures above 2-3 MK. These ionization states are consistent, in magnitude and variance, with those found for Fe, and lighter elements, in the solar wind.

TABLE 2. Mean Ionization States of Energetic Fe in Large Gradual Events

MeV/amu	Q_{Fe}	Events	S/C	Reference
0.3 - 2	14.1±0.2	12	ISEE 3	Luhn et al. 1987 (14)
0.5 - 5	11.0±0.2	2	SAMPEX	Mason et al. 1995 (15)
15 - 70	15.2±0.7	2	SAMPEX	Leske et al. 1996 (13)
200 - 600	14.1±1.4	3	LDEF	Tylka et al. 1995 (34)

Neither has the Fe been ionized by passing through material at high velocity, so the acceleration must have taken place in a low-density region. The ionization states of high-energy Fe would be measurably altered in less than 1 sec at a density of 10^{10} atoms/cm^3 that is typical of the low corona where flare acceleration occurs. It is extremely unlikely that Fe could be accelerated to >200 MeV/amu in such a short time. Hence, the ionization states suggest acceleration of Fe from low-density ambient plasma in the high corona or in the solar wind.

Conversely, $Q_{Fe} \approx 20$ in impulsive events suggests that the acceleration must occur in a sufficiently dense plasma that collisions with electrons can occur. It is not clear whether the Fe is ionized by thermal electrons or by collisions with the electron beam in these events. We will see in a later section that ionization probably occurs during or after acceleration in impulsive flares.

In the meantime, Kahler (8, 9) studied the injection altitude of protons up to 21 GeV in large gradual SEP events and found that the proton intensities peak when the CME-driven shock is at >5 solar radii, *i.e.*, outside the corona, in the solar wind. The event of 1989 September 29 was one event where the 21 GeV protons reached maximum intensity when the shock at the leading edge of the CME was at ~6 $R_\odot$. Protons of ~1 GeV generally reached maximum when the shock was at 12 $R_\odot$ or more. The 1989 September 29 event was also one of the events included in the Tylka *et al.* (34) observation of Fe at 200-600 MeV/amu with an average charge of 14.1. This event occurred behind the west limb at W105° . These observations are quite consistent with a shock wave from the CME propagating across the high corona to accelerate high-energy protons and elements up through Fe from the ambient plasma at ~6 $R_\odot$ near the base of the field line connected to Earth (35).

THE LEGACY OF "CORONAL DIFFUSION"

"Coronal diffusion" is an artifact of the "flare myth" (6, 29). If particles are only accelerated at a flare, *i.e.*, a point source in space and time, a mechanism must be invented to transport them to longitudes of 90° or more from that source, where they are observed, as seen in Fig. 1(a). It is an irony of history that there was abundant radio evidence of type II bursts produced by shock waves and shock acceleration of protons was proposed by Wild, Smerd and Weiss (36) before the birth of "coronal diffusion" (28). Shock waves were also observed in the interplanetary plasma at that time. Shock waves easily cross magnetic field lines but charged particles do not.

Evidently there is still some fascination in fitting mathematical "diffusive" forms to time profiles that have a fast rise and slow decline, by varying several adjustable parameters, without the necessity of having to consider physical processes underlying acceleration and transport. The practice persists to this day. Those time profiles that rise slowly or remain constant (over half of the events) are simply ignored by the coronal-diffusion advocates as not being "diffusive."

The evidence against "coronal diffusion" is as follows:

1) Particles in large proton events are *not* accelerated in point-source flares. Ionization states show no flare heating and events with the same time profiles come from "disappearing filament" events where there is a CME but no flare (10). Cross-field diffusion is not necessary from an *extended* CME source.

2) Particles that *are* injected by point-source impulsive flares have a narrow distribution of longitudes of full width ≤30° (see Fig. 1). Most of this width comes from variation of the connection longitude with solar wind speed. These particles do not diffuse in longitude far from their injection point, and they stream rapidly out to 1 AU with little interplanetary scattering as well.

In fact, the particles in the gradual proton events are accelerated by the CME-driven shock that easily crosses field lines to accelerate particles as it goes. There is a 96% correlation between CMEs and proton events (11). In large events the shock has been directly observed by spacecraft near 1 AU that are separated in longitude by 160°; if the shock were symmetric about the source in that event, it would subtend a 240° longitude interval (30). Multispacecraft observations of time profiles of the particles *vs.* longitude are well organized and understood in terms of the evolution of an observer's magnetic connection to the shock.

If we grant that gradual events involve acceleration at shocks, there is still a question about the longitude distribution of impulsive events and its origin. It is difficult to confidently correct the observed longitude distribution for variation in solar wind speed, however, some estimates of the longitude spread from multispacecraft observations and persistence of a single active region suggest a distribution of width 5° to 20°. Recently Dröge (5) found electron events that were visible over a large spacecraft separation, however, with a factor of 600 decrease in intensity, this is not inconsistent with a ~15° e-folding angle for these electrons, if they are indeed from an impulsive flare.

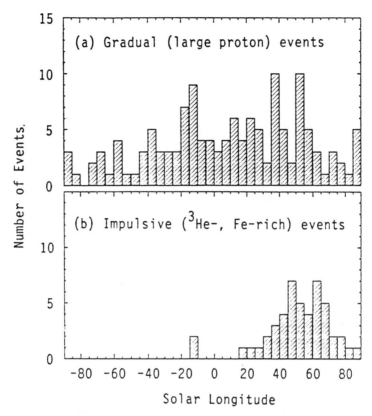

Fig. 1. Longitude distributions of gradual and impulsive events. The narrow distribution of the impulsive events shows that the broad distribution of the gradual events cannot be the result of particle diffusion in the corona.

It seems likely that the main contribution to the longitude distribution in a single impulsive event is the random walk of magnetic field lines discussed by Parker (see 22 and references therein). This process results in a distribution of the source longitudes for the field lines sampled during several hours of particle measurements near 1 AU. While the random walk of the field lines is diffusive in nature, the particles simply follow the bundle of pre-existing field lines that fan out from the flare site; they can not cross to other field lines that pass more than a few gyroradii away unless scattering is intense. Actual scattering across field lines only occurs in isolated regions of high turbulence (*e.g.*, in the immediate vicinity of flares or shocks) or after long periods of time (days).

When is an event ^{3}He-rich?

From the foregoing, we would not expect large, poorly-connected events to be ^{3}He-rich. Yet, Chen et al. (2) found 16 of 29 large events with ^{3}He/^{4}He > 0.5% at 50-110 MeV/amu. Most of these events have ^{3}He/^{4}He ~1%, a 10-fold enhancement over the nominal value, however, they show no enhancement of Fe/O and some have source longitudes like E78 and E85. Of course, the solar wind plasma itself does occasionally have values of ^{3}He/^{4}He as large as 1% (4). Are these events truly ^{3}He-rich in our context? Are the particles accelerated in an impulsive flares?

No. These are gradual events in which shock acceleration has produced enhancements of ^{3}He only at high energies. Evidence for this comes from the flatter spectra that Chen et al. (2) found for ^{3}He than for ^{4}He. The work of Mazur et al. (16) on the spectra of H, He, O and Fe in large gradual events showed a strong Q/A dependence in the spectral indices at high energies. The H/^{4}He ratio often increases by a factor of 100 between 1 and 100 MeV/amu. In this context, a 10-fold increase in ^{3}He/^{4}He between 1 and 100 MeV/amu is not surprising in large gradual events. This enhancement is of different origin than the resonant wave enhancement of ^{3}He and Fe in impulsive events.

Correlations between X-rays and protons?

Recently, Kiplinger (12) has reported a high correlation between the existence of 10 MeV protons at Earth and a characteristic pattern of X-ray spectral evolution for 18 associated flares. Again, many of these flares are far to the east. Do the protons come from the flare in these large events? What do X-rays produced by electrons in the flare have to do with protons at 1 AU?

In fact, these 18 events were among the 235 proton events that led Cane et al. (1) to propose proton acceleration at CME-driven shocks. The longitude span, time profiles and abundances show the 18 events to be typical large, gradual, proton events. It seems quite clear that the protons are accelerated at the CME-driven shock. The only possibility for a valid correlation seems to be that flares that accompany powerful CMEs occur in a unique configuration of the magnetic field, for example, that leads to a characteristic evolution of the electron spectrum. It is more likely that the correlation results from "big flare syndrome" (7), which states that, despite appearances, not all processes occurring in conjunction with big flares are causally related to the flare or to each other, even when well correlated.

It is instructive to revisit a similar type of correlation, mentioned earlier, namely, X-ray duration and proton events. The long duration of X-ray emission presumably occurs by electron acceleration and heating on "post-flare" loops that

40

are created by magnetic reconnection beneath a rising CME. The protons we see are accelerated by the shock ahead of the CME. Late in the event the reconnection may stop, but the shock and proton acceleration may continue far out in the heliosphere. The apparent correlation between X-ray properties and proton events does *not* imply that protons are accelerated on post-flare loops where the X-rays originate. In general, X-rays may reveal the presence of energetic electrons, but they tell us little about ion acceleration.

ELEMENT ABUNDANCES

Gradual Events

The event-averaged abundances of elements in gradual events, obtained from low-energy measurements, provide a direct measure of element abundances in the corona and solar wind. These abundances are almost entirely independent of the temperature and ionization state of the source plasma. It has been well known for many years (17) that the ratio of coronal and photospheric abundances of elements is a well-organized function of the first ionization potential (FIP) of the element. A recent summary of abundances is shown in Table 3 (26).

Impulsive Events

Energetic particles from impulsive flares show element abundances that differ from those in the corona in that elements with Z>8 are strongly enhanced relative to coronal abundances while He, C, N, or O. are not (31). Ne, Mg, and Si are enhanced by almost the same factor, relative to O. This pattern of enhancement is consistent with acceleration of the ions from a plasma in the temperature range of 3-5 MK (31, 18). Elements with the same charge-to-mass ratio, Q/A, have the same magnetic gyrofrequency, thus they resonate with the same part of the wave spectrum and are enhanced by the same amount. At 3-5 MK, He, C, N and O are fully ionized, with Q/A=0.5, while Ne, Mg and Si all have Q/A ~ 0.42. These ion must become fully ionized later in the event; if they were fully ionized before acceleration they would have the same Q/A as He, C and O and could not be relatively enhanced. The pattern of enhancements is discussed in terms of wave absorption in the "He valley" by Meyer (18, see also 32) and is modeled by a theory of cascading waves by Miller and Reames (20).

TABLE 3. Solar Energetic Particle Abundances

	Z	FIP	Photosphere	SEP Corona (Gradual Events)	Flares (Impulsive Events)
H	1	13.53	1.18×10^6	$(1.57 \pm 0.22) \times 10^6$	$\sim 1 \times 10^6$
He	2	24.46	1.15×10^5	57000 ± 3000	46000 ± 4000
C	6	11.22	468	465 ± 9	434 ± 30
N	7	14.48	118	124 ± 3	157 ± 18
O	8	13.55	1000	1000 ± 10	1000 ± 45
F	9	17.34	0.0351	<0.1	<2
Ne	10	21.47	161	152 ± 4	400 ± 28
Na	11	5.12	2.39	10.4 ± 1.1	34 ± 8
Mg	12	7.61	44.6	196 ± 4	408 ± 29
Al	13	5.96	3.54	15.7 ± 1.6	68 ± 12
Si	14	8.12	41.7	152 ± 4	352 ± 27
P	15	10.9	0.433	0.65 ± 0.17	4 ± 3
S	16	10.3	20.4	31.8 ± 0.7	117 ± 15
Cl	17	12.95	0.218	0.24 ± 0.1	<2
Ar	18	15.68	4.21	3.3 ± 0.2	30 ± 8
K	19	4.32	0.157	0.55 ± 0.15	2 ± 2
Ca	20	6.09	2.55	10.6 ± 0.4	88 ± 13
Ti	22	6.81	0.10	0.34 ± 0.1	<2
Cr	24	6.74	0.563	2.1 ± 0.3	12 ± 5
Fe	26	7.83	37.9	134 ± 4	1078 ± 46
Ni	28	7.61	2.05	6.4 ± 0.6	42 ± 9
Zn	30	9.36	0.0525	0.11 ± 0.04	6 ± 4

DISCUSSION AND CONCLUSIONS

Recent evidence shows that particles accelerated by CME-driven shocks, rather than by flares, produce most of the largest particle events seen at 1 AU, even the ground-level events with particles of energies as high as ~ 20 GeV (8). Protons of 1-10 MeV may be accelerated throughout a large fraction of the heliosphere in large events. During the first few days, the temporal evolution of these particles at a point in space is determined almost entirely by the changes in the locus of the observers magnetic connection to the shock (30, 27). This locus scans across the surface of the outbound shock, sampling different source intensities as a function of time. Particles of the highest energies are primarily accelerated nearer the sun, where the shock is strongest, yet chiefly outside the corona. For these particles, the radial variation of the source strength can dominate.

We can distinguish flare-accelerated particles by their unique abundances, enhancements in ^{3}He/^{4}He, Ne/O, Mg/O, Si/O and Fe/O. The resonant wave-particle processes that produce these enhancements seem to occur in all the flare-associated events we observe (now over 200 events). Of course, we can only observe ions from magnetically well-connected events. With the exception of one event where broad γ-ray lines were observed (21), all our information on the nature of the ion acceleration process comes from the particle observations in interplanetary space. Radio and X-ray observations tell us of the electron beams that may be required for the ^{3}He enhancements (33, 19), but the nature and even the existence of resonant ion physics in solar flares was revealed only by the particles at 1 AU. The γ-ray and neutron observations tell us about the time scale for ion acceleration

Abundances and charge states of the elements in SEP events have provided information on the temperature of the source plasma, the composition of the solar corona and the existence and nature of resonant ion acceleration in flares. They have also corrected serious errors in our understanding of particle origin and transport. Those errors were caused by an excessive focus on only protons and electrons and their time profiles, and by a neglect of shocks and CMEs in favor of flares.

I would especially like to thank Prof. Y. Muraki and the Solar-Terrestrial Environment Laboratory at Nagoya University for their hospitality during my recent 3-month visit. A substantial part of this paper was written during that time. I would also like to thank C. K. Ng for many pleasant and productive hours of discussion on this subject.

References

1. Cane, H. V., Reames, D. V., and von Rosenvinge, T. T., *J. Geophys. Res.* **93**, 9555 (1988).

2. Chen, J., Guzik, T. G., and Wefel, J. P., *Astrophys. J.* **442**, 875 (1995).

3. Cliver, E. W., this volume (1996).

4. Coplan, M. A., Ogilvie, K. W., Bochsler, P., and Geiss, J., *Solar Physics* **93**, 415 (1984).

5. Droge, W., this volume (1996)

6. Gosling, J. T., *J. Geophys. Res.* **98**, 18949 (1993).

7. Kahler, S. W., *J. Geophys. Res.* **87**, 3439 (1982).

8. Kahler, S. W., *Astrophys. J.* **428**, 837 (1994).

9. Kahler, S. W., this volume (1996).

10. Kahler, S. W., Cliver, E. W., Cane, H. V., McGuire, R. E., Stone, R. G., and Sheeley, N. R., Jr., *Astrophys. J.* **302**, 504 (1986).

11. Kahler, S. W., Sheeley, N. R., Jr., Howard, R. A., Koomen, M. J., Michels, D. J., McGuire, R. E., von Rosenvinge, T. T., and Reames, D. V., *J. Geophys. Res.* **89**, 9683 (1984).

12. Kiplinger, A. L., *Astrophys. J.*, **453**, 973 (1995).

13. Leske, R. A., Cummings, J. R., Mewaldt, R. A., Stone, E. C., and von Rosenvinge, T. T., *Astrophys. J.*, in press (1996)

14. Luhn, A., Klecker, B., Hovestadt, D., and Möbius, E., *Astrophys. J.* **317**, 951 (1987).

15. Mason, G. M., Mazur, J. E., Looper, M. D., and Mewaldt, R. A., *Astrophys. J.* **452**, 901 (1995)

16. Mazur, J. E., Mason, G. M., Klecker, B., and McGuire, R. E., *Astrophys. J.* **401**, 398 (1992).

17. Meyer, J. P., *Astrophys. J. Suppl.* **57**, 151 (1985).

18. Meyer, J. P., this volume (1996).

19. Miller, J.A. and Viñas, A. F., *Astrophys. J.* **412**, 386, 1993.

20. Miller, J. A. and Reames, D. V., this volume (1996).

21. Murphy, R. J., Ramaty, R., Kozlovsky, B., and Reames, D. R., *Astrophys. J.* **371**, 793 (1991).

22. Parker, E. N., *Physics Today* **40** (July) 36 (1987).

23. Reames, D. V., *Astrophys. J. Suppl.* **73**, 235 (1990).

24. Reames, D. V., *Astrophys. J. (Letters)* **358**, L63 (1990).

25. Reames, D. V., *Adv. Space Res.* **13** (No. 9), 331 (1993).

26. Reames, D. V., *Adv. Space Res.* **15** (No. 7), 41 (1994).

27. Reames, D. V., *Third SOHO Workshop: Solar Dynamic Phenomena and Solar Wind Consequences*, Ed. A. Poland, Estes Park, CO , ESA, p. 107 (1994).

28. Reames, D. V., *Revs. Geophys (Suppl.)* **33**, 585 (1995).

29. Reames, D. V., *Eos*, **75**, 405 (1995).

30. Reames, D. V., Barbier, L. M., and Ng, C. K., *Astrophys. J.*, in press (1996).

31. Reames, D. V., Meyer, J. P., and von Rosenvinge, T. T., *Astrophys. J. Suppl.* **90**, 649 (1994).

32. Steinacker, J., Meyer, J. P., Steinacker, A., and Reames, D. V., *Astrophys. J.*, in press (1996).

33. Temerin, M. and Roth, I., *Astrophys. J. (Letters)* **391**, L105 (1992).

34. Tylka, A. J., Boberg, P. R., Adams, J. H., Jr., Beahm, L. P., Dietrich, W. F., Kleis, T., and *Astrophys. J. (Letters)* **444**, L109 (1995).

35. Vestrand, W. T., and Forrest, D. J., *Astrophys. J. (Letters)* **409**, L69 (1993).

36. Wild, J. P., Smerd, S. F., and Weiss, A. A., *Ann. Rev. Astron. Ap.* **1**, 291 (1963).

Solar Flare Gamma-Ray Emission
and
Energetic Particles in Space

E.W. Cliver

Geophysics Directorate, Phillips Laboratory,
Hanscom AFB, MA 01731-3010 USA

We expand Reames' tabular summary of the two-class picture of solar energetic particle (SEP) events to include characteristics of the particles that interact at the Sun to produce gamma-ray emission. This addition underscores the contributions of gamma-ray observations to our current understanding. The broad picture that is emerging is remarkable for its simplicity: while SEP events come in two basic types depending on the duration of the associated flare, the interacting particles in impulsive and gradual flares appear to be indistinguishable and resemble the SEPs observed in space following impulsive flares. The expanded classification system includes hybrid events, i.e., flares in which the gradual/impulsive distinction is blurred and for which the SEP events contain a mixture of flare-accelerated and CME/shock-accelerated particles. We argue that SEP events associated with long duration flares can be expected to have a temporally and spatially confined "core" of flare-accelerated particles surrounded by a "halo" of CME/shock particles. Thus SEP composition should be checked in comparative studies of gamma-ray emission and particles in space to ensure that the SEPs are flare-accelerated. We discuss how recently-discovered types of gamma-ray flares (electron-dominated events, spatially and temporally extended gamma-ray events) may fit into the expanded classification scheme. We suggest that the acceleration process in the pion-rich phase of large flares (e.g., 1982 June 3) is similar to that occurring earlier in the flare, the main differences being the greater height of the acceleration region and the presence of previously accelerated seed particles.

1. INTRODUCTION

The current "two-class" picture of solar energetic particle (SEP) events is shown in Table 1 (1). This framework had its origins in the work of Wild et al. (2) based on metric radio bursts and was given fresh impetus by the work of Cane et al. (3) that focused on flare soft X-ray (SXR) emission and SEPs. In the last decade the two-class paradigm has been significantly advanced by Reames (1,4,5) and

Table 1. Properties of Impulsive and Gradual Events (1)

	Impulsive	Gradual
Particles:	Electron-rich	Proton-rich
$^3He/^4He$	~1	~0.0005
Fe/O	~1	~0.1
H/He	~10	~100
Q_{Fe}	~20	~14
Duration	Hours	Days
Longitude Cone	<30°	~180°
Radio Type	III,V(II)	II,IV
X-Rays	Impulsive	Gradual
Coronagraph	---	CME
Solar Wind	---	IP Shock
Events/Year	~1000	~10

colleagues through composition studies. Impulsive SEP events are thought to originate in flares and the gradual SEP events are linked to coronal/interplanetary shock waves driven by mass ejections.

The classification system in Table 1 represents a breakthrough for our understanding of SEP acceleration. During the last several years, a similar unifying picture has begun to take shape for the accelerated particles that interact in the solar atmosphere to produce gamma-ray emission. Simply put, while SEP events come in two basic "flavors", interacting particles appear to come in one. In this paper, we expand the current SEP classification system to include the properties of interacting particles. This broader framework highlights the contributions of gamma-ray observations to our knowledge of particle acceleration at the Sun and in space. The neglect of such observations in Table 1 arises because many gamma-ray flares do not fit neatly into either of the two classes. They are hybrids, having aspects of both impulsive and gradual events. We present the expanded classification system in Section 2. In Section 3 we discuss new types of gamma-ray flares that pose a challenge even for the expanded system and, in Section 4, we briefly discuss our results and comment on directions for further progress.

2. EXPANDED CLASSIFICATION SYSTEM FOR SEP EVENTS

The proposed scheme is given in Table 2. The key new features of the system are: (i) introduction of hybrid events (6), referred to as "mixed-impulsive" and "mixed-gradual"; (ii) allowance for the temporal evolution of composition (and charge state) of the mixed-gradual events; and (iii) a listing of the characteristics of interacting particles for the various classes of events. Other additions to the table include considerations of: (iv) coronal mass ejection (CME) width, (v) electron spectra, and (vi) the ratio of interacting (solar) to interplanetary (IP) protons. We

Table 2. Expanded SEP Classification System

	Pure Impulsive	Mixed-Impulsive	Mixed-Gradual	Pure Gradual
Sun				
Radio Type	III(km III?)	II,III,**V**	II,III,**IV**	IV(I?)
SXR Duration	<1 hr(weak)	< 1 hr	> 1 hr	> 1 hr (weak)
CME Width	---	~20°	>45°	>45°
Solar Wind				
IP Shock	---	---	yes	yes
SEPs				
0.5 MeV e^-/ 10 MeV pr	---	>100	<100	<100
H/He	~10	>10?	~100	~100
$^3He/^4He$	~1	<0.1	~ 0.0005	~0.0005
Fe/O	~1	<1?	~1 --> ~0.1	~0.1
Q_{Fe}	~20	~20	(~20 ---> ~14)	~14
Electron Spectra				
High Energy	---	Flatten	No Break	No Break
Low Energy	No Roll-over	Roll-over	No Roll-over	No Roll-over
Solar/IP pr (~10 MeV)	---	~1-100	~0.1-10	<0.1
Longitude Cone	<30°	~100°	~200°	~ 200°
Interacting Beam				
Fe/O	(~1)	(~1)	~1	?
H/He	(<10)	(<10)	<10	?
0.5 MeV e^-/ 10 MeV pr	---	>100	>100	?
Bremsstrahlung Index (~0.5 MeV)	---	~3	~3	?

will consider each of these modifications to Table 1 in turn.

The impulsive SEP events of Cane et al. (3) generally had detectable ~10 MeV protons and associated metric type II radio emission as well as high energy electrons (their basis for selection). About half of the Cane et al. impulsive events that were observed by the Gamma-Ray Spectrometer on the Solar Maximum Mission (SMM) had associated gamma-ray-line (GRL) emission (7,8). In addition, comparison of the Cane et al. impulsive events with lists of CMEs from Solwind and SMM reveals a relatively high degree of CME association (45-82%; 5-9/11 (5 certain associations, 4 possible)) for limb events with coronagraph coverage (9,10). (Impulsive GRL flares within 30° of the limb are also well associated (60-80%) with CMEs (7).) In contrast, ^{3}He-rich events, the archetypes of impulsive SEP events, often lack detectable ~10 MeV protons and characteristically are not associated with type II bursts, detectable GRL emission, or CMEs (4,11). Moreover, the ^{3}He-rich events have electron "cones of emission" < 30°, vs. ~100° for the more energetic impulsive events (6,12; cf., 5). For these reasons we have drawn a distinction between "pure impulsive" and mixed-impulsive flare/SEP events in Table 2. Comparison of the results of various composition studies (11,13,14,15) indicates a dilution of ^{3}He/^{4}He, and possible dilutions of Fe/O and H/He, for the mixed-impulsive events.

There is also reason to propose a mixed-gradual type of flare/SEP event. For certain well-connected gradual events, the Fe/O ratio evolves in time, beginning with a value ~1, characteristic of impulsive events, and ending with a value ~0.1, signifying a gradual event (16,17). Interestingly, one such event was the gradual flare/SEP event of 1981 July 20. Cliver et al. (18) reported that no gamma-ray-lines or >300 keV continuum were observed for the associated flare; yet, the Fe/O ratio shows clear evidence for flare particles early in the SEP event. Similarly, the 1980 October 15 SEP event, that was associated with a long duration SXR flare with a weak impulsive phase (S_p (9 GHz) ~50 sfu) (18), has been classified as an Fe-rich event (14). If flare-accelerated particles are observed following these relatively weak events, we are led to suggest that such particles are present in all gradual events and can be detected *providing the observer is well-connected to the flare site early in the event* (cf., 16). A schematic to illustrate what we believe is occurring in these events is shown in Figure 1. Note that because of the vagaries of the interplanetary magnetic field, nominal good connection, i.e., a satellite footpoint from 40-70° for an Earth-orbiter, may not necessarily mean good connection in fact. Conversely, sometimes nominally poorly connected events may be actually well-connected. For example, the 1980 October 15 flare located at E55 was apparently connected to Earth by magnetic fieldlines distended by an earlier CME from the active region (19). We would go so far as to argue that SEP composition itself is the best means of confidently identifying events with good connection. Although it remains to be shown, we would expect that the Fe charge states of well observed gradual events would undergo a corresponding evolution from high (~20) to low (~14) values.

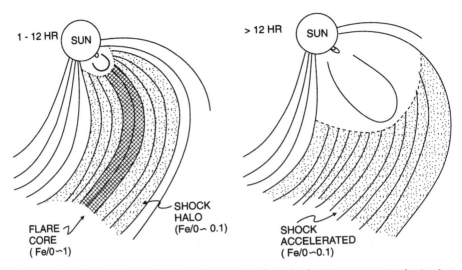

Figure 1. Core-halo model for a large mixed-gradual SEP event. Early in the event (left), a well-connected observer will see Fe-rich material. After about 12 hours (right), shock-accelerated Fe-poor material will dominate. The dashed line represents the CME-driven coronal/interplanetary shock.

The simulation study of Murphy et al. (20) yielded key insight on the composition of the interacting particles in solar flares. Those authors found that the interacting beam in the 1981 April 27 GRL flare had composition that most closely resembled that of ^{3}He-rich and high-Z enhanced (i.e., impulsive) SEP events. As they pointed out, this marked the first time that the composition of an accelerated beam of particles interacting in an astrophysical source had been compared with the composition of particles escaping from such a source. Of equal import, the fact that 1981 April 27 was a gradual SXR flare implies that the acceleration process for interacting particles in gradual flares and impulsive flares might be identical. Support for this proposition was provided by Ramaty et al. (21) who showed that electron to proton ratios of interacting particles were the same in impulsive and gradual flares and were most similar to ratios observed for impulsive SEP events. They also found no differences between the spectra of interacting electrons in impulsive and gradual flares, although for both types of events the spectra of interacting electrons were softer than those of the electrons observed in space. Additional evidence for an identical acceleration process for interacting particles in gradual and impulsive flares was recently presented by Trottet et al. (22) from an analysis of the time evolution of the ratio (R) of 1.1-1.8 MeV to 4.1-7.6 GRL emission in the 1991 June 1 flare. Trottet et al. determined that the increase in R during this long duration flare was consistent with an impulsive flare composition for the accelerated particles. In a similar vein to the above studies, Cliver et al.

(23) obtained a correlation between 4-8 MeV GRL fluences and > 50 keV hard X-ray fluences (cf., 24) and interpreted their result in terms of a common acceleration mechanism for the bulk of the ~100 MeV electrons and ~10 MeV protons that interact at the Sun in large flares of both long and short durations.

Support for a common acceleration process for interacting particles in impulsive and gradual flares can also be inferred from consideration of the 1980 October 15 event discussed above. This gradual flare lacked the strong metric type IIIGG/V(3) emission (IIIS(W) was reported by Culgoora (25)) characteristic of impulsive phase acceleration, yet the SEP event was Fe-rich (14). This suggests that the "impulsive" SEPs in this event originated during what would traditionally be called the gradual ("post-flare" loop (cf., 26)) phase of the flare, consistent with the idea that the acceleration process at the flare site is identical during impulsive and gradual flares (or, alternatively, during the impulsive and gradual phases of a hybrid flare). Thus flare (interacting) particles seem to come in a single flavor and flare duration appears to bear primarily on acceleration external to the flare, i.e., SEP acceleration at a CME-driven shock. Sheeley et al. (27) have shown that the probability of CME association increases with flare duration although, as we have noted, energetic impulsive flares are often accompanied by CMEs. A final piece of evidence consistent with a common acceleration mechanism for interacting particles in both short- and long-duration flares is the observation by Masuda et al. (28) of an impulsive flare that apparently arose in a "current sheet-reconnection" magnetic field geometry similar to that of the Kopp-Pneuman model (29) which has long been argued to apply for long duration flares. If the flare geometry is similar and the composition of the interacting particles is similar in impulsive and gradual flares, it is difficult to argue for different acceleration mechanisms (for interacting particles) in the two types (or phases) of flares.

The correlation obtained by Kahler et al. (30) between SXR flare duration and CME angular span is included in Table 2 as is a more definitive discriminator between SEP types, the spectra of high energy electrons observed in space. The finding of Moses et al. (31) that the flare SXR duration neatly divided energetic electron spectra into two classes of events was a key piece of evidence leading to the acceptance of the two-class picture following the Cane et al. (3) study. Specifically, Moses et al. found that rigidity spectra of electron events associated with impulsive flares exhibited spectral flattening above 1 MV while for SEP events associated with gradual flares, the electron spectrum could be fitted by a single power law over the range from 0.1 to 100 MV. This separator between impulsive and gradual flares was not included in Table 1 because the ^{3}He-rich events are often only accompanied by low energy electron events. (Notable examples of ^{3}He-rich events with associated MeV electrons have been reported by Kane et al. (32) and Van Hollebeke et al. (33).) Similarly, the ratio of interacting to interplanetary protons (18,34) is omitted from Table 1 because pure impulsive events typically are unaccompanied by GRL emission (4) (required to determine the number of interacting protons) and frequently lack associated ~10 MeV protons in space (11).

A reported difference between low-energy electron spectra of pure and mixed impulsive events is indicated in Table 2. Lin (35) found that certain of the Cane et al. impulsive events (mixed-impulsive events in Table 2) had intensity "roll-overs" at low energies indicating acceleration low in the corona at heights ~10^4 km while for a sample of ^{3}He-rich events (pure impulsive events in Table 2), Kahler et al. (36) found no evidence for such spectral roll-overs at low energies indicating acceleration high (> 10^5 km) in the corona. Lin (37) and Cliver and Kahler (38) have also argued that impulsive acceleration can take place high in the corona, although Reames et al. (15) find it unlikely that ^{3}He-rich events are strongly associated with high coronal flares because of the low densities at such heights. Lin (35) noted the absence of a low energy roll-over for a gradual event on the Cane et al. (3) list and interpreted it in terms of shock acceleration high in the corona.

In Table 2, the inclusion of the data on interacting particles, revealed through the mixed-types which encompass the gamma-ray-line flares, suggests that the flare-resident acceleration process for interacting particles applies to both impulsive and gradual flares and contributes to both classes of SEPs. This process largely determines the composition of the pure- and mixed-impulsive SEP events, and plays a prominent role in the early phase of (well-connected) gradual SEP events. The dilution of SEP composition ratios between pure and mixed-impulsive events is thought to result from contamination by shock-accelerated particles because of the type II bursts often seen in the mixed-impulsive events, although a depletion process may account for the lower ^{3}He/^{4}He ratios (15). The pure gradual events are those which presumably lack a flare component. The best observed example of this type of event occurred on 1981 December 5 (39) in conjunction with a disappearing solar filament (DSF). Since the 1980 October 15 event is not that far removed on the spectrum of eruptive flares (40) from the 5 December event, however, we are reluctant to rule out a flare particle component even for these supposedly pure events. We note that Yohkoh observations of coronal arcades following DSF (or more subtle (41)) events are similar to those observed for more energetic gradual events. Also, recent observations of the extended phases of flares discussed in Section 3 below reveals more energetic processes during the "post-flare loop" stage of flares than was previously thought possible.

The additions we have made in Table 2 do not fully reflect the contributions of gamma-ray observations to our understanding of SEP acceleration. For example, the linking of gamma-ray flares with electron-rich events in space (42) was a important forerunner of the Cane et al. study. The identification of large SEP events in association with flares without detectable gamma-ray-line emission (18) provided underpinning for the transition from a flare-centered to a CME-centered view of large SEP events (43,44). Forrest and Chupp's (45) observation that electrons and protons could be accelerated simultaneously (within seconds) to high energies helped gain acceptance for (and motivate research on) electric field and stochastic models of acceleration (46). The substantial body of work on solar neutrons is reviewed in Chupp's (47) article in this volume. The reader is also referred to the recent review by Vestrand and Miller (48). Gamma-ray observations continue to provide insight to acceleration of particles on the Sun. Ramaty et al. (49) have recently reported evidence for a steep ion spectrum in gamma-ray flares extending down to ~1 MeV per nucleon. This result implies that accelerated ions may contain a significant fraction of flare energy, a view long advocated by Simnett (50).

3. NEW TYPES OF GAMMA-RAY FLARES

The observation of temporally extended gamma-ray-line sources by Gamma Ray Observatory (GRO) in June 1991 and a spatially extended GRL event by SMM on 1989 September 29 rank among the more remarkable, and illuminating, discoveries in this field in recent years. Before discussing these extended events, however, we consider a class of impulsive events, the so-called electron-dominated events, reported by Rieger and Marschhäuser (51).

3.1. Electron-Dominated Gamma-Ray Events

These events are characterized by a smooth bremsstrahlung spectrum extending to ~50 MeV that shows no, or only weak, prompt emission lines. Where do these events fit in Table 2? Is a different acceleration process involved? Table 3 contains data for the four electron-dominated events observed by SMM that were

Table 3. Electron-Dominated Gamma-Ray Events from Impulsive SXR Flares

Date	SXR Max Time	Hα Longitude	HXRBS (c s^{-1})	CME	10 MeV SEP	e/p	Metric Radio
04 Jun 80	0655	E59	35,200	No	No	-	Quiet
29 Jun 80	1043	W90	9,350	N/A	0.007	~1	II,III,V
15 Jun 82	0032	W89	25,300	No	BKG	-	Quiet
14 Jun 89	1354	W78	32,070	?	0.01	~3	II(V?),III

associated with impulsive SXR flares. The e:p ratios (at ~10 MeV (3)) of SEPs associated with two of the impulsive flares are consistent with pure or mixed impulsive events. Ramaty et al. (21) have shown that the e:p ratios (lower limits) of the interacting particles were not markedly higher than those of other impulsive, or gradual, GRL flares. Thus these events seem to be essentially impulsive flares with a slight twist. A possible clue to the nature of these events is given by the absence of low frequency radio emission in two of the four listed cases. Simnett and Benz (52) reported that no metric or decimetric radio emission was observed for the 1980 June 4 event by the radio spectrograph at Zürich. The absence of type III emission indicates the absence of open field lines in the corona at the flare site. We speculate that the electron dominance in these events stems from their origin in closed magnetic loops. Simnett and Benz (52) found that ~15% of gamma-ray continuum (>300 keV) events could be "radio quiet" at meter wavelengths. Similarly, Cliver et al. (53) found that 18% of large microwave bursts had spectra that were "cut off" at low frequencies. These results imply that an even larger fraction of all flares may not involve open field lines. This is an important point, because it serves to remind that all impulsive flares may not be encompassed by the helmet-streamer type of geometry reported by Masuda et al. (28) from Yohkoh observations of the 13 January 1992 impulsive flare.

3.2. Temporally Extended Gamma-Ray Line Events

Recent reviews that covered this class of event have been given by Mandzhavidze (54) and Hudson and Ryan (55). The defining aspect is the observation of pion-rich emission extending long after the impulsive phase of the flare. While shorter duration examples of such events had been observed by SMM (56,57), the greater sensitivity of instruments on GRO made it possible in some cases to observe the high energy emission for several hours. Akimov et al. (58,59) pointed out that the coincidence of the delayed gamma ray emission with microwave peaks and argued

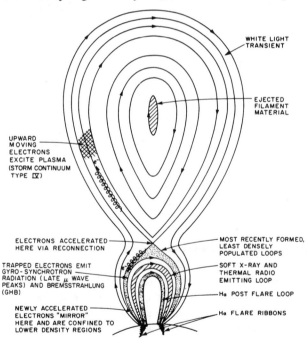

Figure 2. Proposed reconnection geometry to account for the delayed acceleration of pion-rich material in large solar flares. Field-lines distended by a CME reconnect to release energy and fuel the acceleration process.

that the short life-times of electrons responsible for the microwave emission could only mean that a delayed/prolonged acceleration had occurred, in disagreement with other models that had interpreted these events in terms of impulsive phase acceleration and long-term trapping (e.g., 60). These long duration gamma-ray events are plausibly explained in terms of prolonged acceleration via reconnection in post flare loop systems (59) as depicted in Figure 2 (61) for less energetic non-thermal emissions. Note (Figure 3) that this is a radical departure from the usual picture of particle acceleration in flares in which emissions got progressively softer (more thermal) (e.g., 62). Recall that the Kopp-Pneuman (29) model only called for Joule heating and no particle acceleration in conjunction with the reconnection process.

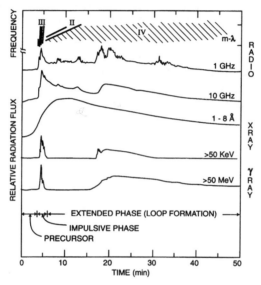

Figure 3. Time sequence of electromagnetic emissions from a large solar flare. In contrast with previous pictures, high energy particles can be accelerated late in the event.

Because any SEPs from the delayed acceleration process shown in Figure 3 come directly from the flare site, we stress the importance of good connection for comparative studies of extended GRL emission and interplanetary particles. If an event is not well-connected, one runs the risk of comparing GRL apples with SEP oranges. Even for events that appear to be well-connected, there is no guarantee of actual good connection. This can be seen in Table 4 where the characteristics of two extended phase events that were nominally well-connected to an observer are given. For the 1991 June 15 event the SEP composition appears to be that of a typical shock-dominated event. Consequently comparisons of such parameters as the spectra of interacting and interplanetary particles must be viewed with caution.

Table 4. Nominally Well-Connected Flares With Time-Extended Pion Production

Date	SXR Peak/ Max Time	SXR 10% Dur.(hr)	Hα Lon	SEP Peak Flux 10 MeV	SEP Characteristics	Reference
03 Jun 82	X8/1147	0.3	E71*	~200	Electron-Rich Fe-rich	33
					SEP & GRL Spectra Match	33,63
					$N_{int}/N_{IP} \sim 2$	18
15 Jun 91	X12/0821	1.5	W70	1400	Electron-poor Fe-poor	64
					SEP Spectrum Harder than GRL Spectrum	65

* Well-observed by Helios.

The 1982 June 3 event holds more promise. In this case, the particles have characteristics distinct from a shock component and are similar to those observed for an impulsive phase component, albeit with an interesting difference. Specifically, this event was unusual in that impulsive phase SEPs were observed at energies of ~50 MeV/nucleon (33). Several authors (e.g., 33,63,66) have suggested that these high energy SEPs resulted from shock acceleration of an impulsive phase seed population. Alternatively, we suggest that the Fe and ^{3}He enhancements observed to high energies in this event result from prolonged energy release and turbulence generation via reconnection in "post flare" loops as shown in the schematic of Figure 2. Our reasoning is as follows. The gamma-ray analyses of the long duration flares of 1981 April 27 and 1991 June 1 suggest that the acceleration process for interacting particles in gradual flares is identical to that occurring in impulsive flares. The conceptual model of Figure 2, proposed initially (67) to account for "secondary peaks" in large microwave bursts and subsequently invoked as the magnetic field geometry for gradual hard X-ray bursts (61) and the extended pion-rich sources under discussion (59), is the standard Kopp-Pneuman reconnection geometry generally accepted (e.g., 40) to describe the decay (post-flare or gradual) phase of large solar flares. Reconnection is an intermittent process and it should not be surprising if particularly powerful reconnections occasionally result in the delayed energetic emission peaks depicted in Figure 3. The basic geometry remains unchanged, however, even as the reconnection point moves upward in time. Moreover, trapped seed particles accelerated earlier in the flare can be accelerated to higher energies, accounting for the energetic "impulsive phase" particles seen in space during the 1982 June 3 event. This seems a more likely prospect than the current picture in which a shock overtakes and reaccelerates MeV flare particles, particularly in a "scatter-free" environment (33). The spectral development of gamma-ray emission in the 1982 June 3 event has been modelled by Ryan and Lee (68) in terms of turbulence in a loop although, as an assumption in their simulation, they explicitly assumed that the SEPs observed in space for this event did not escape from the loop but were accelerated elsewhere.

3.3. Spatially-Extended GRL Emission During the 1989 September 29 Flare

This flare is projected to have occurred ~8° behind the west limb of the Sun yet a strong 2.2 MeV gamma-ray-line, which is normally limb-darkened, was observed (69). Thus accelerated particles were interacting well onto (~30°) the visible disk. Vestrand and Forrest suggested that the interacting particles may have been accelerated at a coronal shock. A schematic illustrating this effect is shown in Figure 4 (70). This picture is attractive because the "transport" of precipitating particles is accomplished in exactly the same manner that has been proposed for SEPs, i.e., by widespread acceleration at a shock front. Mandzhavidze (54) has criticized the proposed scenario of Figure 4, however, because the gamma-ray spectrum was similar in some aspects to the spectra of electron-dominated events and shocks are not thought to be efficient accelerators of electrons (witness the low e:p ratios of gradual SEP events). Mandzhavidze suggested that the shock picture might apply if the spectrum were pion-dominated and, in fact, Tom Vestrand presented evidence in support of this conjecture at the workshop. Also during the workshop, Brian Dennis presented evidence for X-ray spectral hardening during the time of the GRL emission, indicating observation of

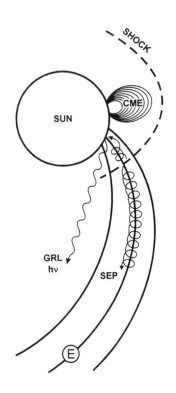

Figure 4. Shock scenario to account for widespread GRL emission on 1989 September 29.

post-flare-loops above the limb (61). For a source 8° behind the west limb the occultation height would be ~7000 km. Such a source might contribute to prompt gamma-ray-line emission but this would only tend to lower the ratio of the 2.2:4-8 MeV emission. The ~10^{11} cm^{-3} densities of an elevated source (71) would be too low to support either 2.2 MeV neutron capture emission or the pion radiation (with any efficiency). Thus the shock picture remains viable. At any rate, the source must be diffuse, because a GRL fluence ~10 γ cm^{-2} as was observed in this case is invariably accompanied by an Hα flare and no frontside emission was reported except for an eastern hemisphere subflare. In terms of Table 2, this event is a reminder that just as flare-accelerated particles may interact at the Sun or escape to the interplanetary medium, shock accelerated particles may also appear in both venues. Precipitation from a shock should be investigated as an alternative (or contributing) mechanism to acceleration via reconnection in post-flare-loop systems (Figure 2) for the long-lived (several hours) gamma-ray emission discussed in Section 3.2.

4. CONCLUSION

In this paper we have argued, on the basis of the available SEP composition and gamma-ray studies, that the process that accelerates the particles that interact at the Sun to produce gamma-ray emission is independent of flare duration. This notion is supported by the evidence for a similar reconnection-type magnetic geometry in impulsive (28) and gradual flares (61) that likely also encompasses the time-extended pion-rich phase of gamma-ray flares (59).

The actual flare-resident process appears to be complex, involving several somewhat independent stages. For example, the composite model of Temerin and Roth (72; cf., 73) and Miller and colleagues (74,75; see also 76) presupposes an unstable electron beam which generates electromagnetic ion cyclotron waves that accelerate ^{3}He ions up to a few MeV, while high-Z ions and electrons undergo stochastic acceleration by cascading Alfvén wave turbulence. Alfvén wave cascades may also account for proton acceleration in such a multi-ion plasma although this remains to be demonstrated. We suggest that the presence of (at least) two separate acceleration stages manifests itself in the distinction we have drawn between the pure and mixed-impulsive SEP classes of Table 2. In the pure events, which are less energetic, the Temerin and Roth mechanism dominates and the principal outputs are ^{3}He ions and lower energy electrons. In the mixed-impulsive events, the more general acceleration presumably occurring via cascading Alfvén waves can result in a fully developed gamma-ray spectrum as well as SEP events with MeV electrons, high-energy protons, and reduced ^{3}He/^{4}He ratios. These energetic impulsive flares are more likely to have associated CMEs and type II bursts, as indicated in Table 2.

A solar cycle ago, a principal challenge facing this field was the understanding of SEP events with unusual abundances: electron-rich, ^{3}He-rich, and high-Z enhanced. What followed during the next decade was a remarkable synthesis of these "peculiar" or "anomalous" events. As Don Reames has often reminded us, these "peculiar" events are, in fact, the most common type of SEP event and their understanding proved to be a Rosetta stone for deciphering the field as a whole.

In this brief review, the emphasis has been on the complementary role played by gamma-ray observations in developing the current picture. Equally as remarkable as the picture of two classes of SEP events is the emerging view of a single class of interacting particle event. The situation is not static, however. Similar to circumstances a decade ago, a new class of peculiar events - in this case revealed through gamma-ray observations rather than SEP composition - will likely hold the key to deeper understanding.

ACKNOWLEDGEMENTS

I thank Steve Kahler, Bob Lin, Natalie Mandzhavidze, Jim Miller, Don Reames, and Tom Vestrand for comments on the manuscript and Reuven Ramaty and Natalie Mandzhavidze for organizing a very stimulating and productive workshop.

REFERENCES

1. D.V. Reames, *Rev. of Geophys., Supplement*, 585 (1995).
2. J.P. Wild, S.F. Smerd, and A.A. Weiss, *Ann. Rev. Astron. Astrophys.*, **1**, 291 (1963).
3. H.V. Cane, R.E. McGuire, and T.T. von Rosenvinge, *Ap. J.*, **301**, 448 (1986).
4. D.V. Reames, *Ap. J. (Suppl. Ser.)*, **73**, 235 (1990).
5. D.V. Reames, these proceedings.
6. M.-B. Kallenrode, E.W. Cliver, and G. Wibberenz, *Ap. J.*, **391**, 370 (1992).
7. E.W. Cliver, H.V. Cane, D.J. Forrest, M.J. Koomen, R.A. Howard, and C.S. Wright, *Ap. J.*, **379**, 741 (1991).
8. B.R. Dennis et al., *NASA Tech. Mem. 4332*, NASA, Greenbelt, MD (1991).
9. R.A. Howard, private communication (1995).
10. J.T. Burkepile and O.C. St. Cyr, *NCAR Tech. Note 369+STR*, HAO, Boulder, CO (1993).
11. S.W. Kahler, D.V. Reames, N.R. Sheeley, Jr., R.A. Howard, M.J. Koomen, and D.J. Michels, *Ap. J.*, **290**, 742 (1985).
12. D.V. Reames, M.-B. Kallenrode, and R.G. Stone, *Ap. J.*, **380**, 287 (1991).
13. G.M. Mason, D.V. Reames, B. Klecker, D. Hovestadt, and T.T. von Rosenvinge, *Ap. J.*, **303**, 849 (1986).
14. D.V. Reames, H.V. Cane, and T.T. von Rosenvinge, *Ap. J.*, **357**, 259 (1990).
15. D.V. Reames, J.P. Meyer, and T.T. von Rosenvinge, *Ap. J. (Suppl. Ser.)*, **90**, 649 (1994).
16. G.M. Mason, G. Gloeckler, and D. Hovestadt, *Ap. J.*, **267**, 844 (1983).
17. D.V. Reames, *Ap. J. (Lett.)*, **358**, L63 (1990).
18. E.W. Cliver, D.J. Forrest, H.V. Cane, D.V. Reames, R.E. McGuire, T.T. von Rosenvinge, S.R.Kane, and R.J. MacDowall, *Ap. J.*, **343**, 953 (1989).
19. I.G. Richardson, H.V. Cane, and T.T. von Rosenvinge, *J. Geophys. Res.*, **96**, 7853 (1991).

20. R.J. Murphy, R. Ramaty, B. Kozlovsky, and D.V. Reames, *Ap. J.*, 371, 793 (1991).
21. R. Ramaty, N. Mandzhavidze, B. Kozlovsky, and J.G. Skibo, *Adv. Space Res.*, 13(No. 9), 275 (1993).
22. G. Trottet et al., these proceedings.
23. E. W. Cliver, N.B. Crosby, and B.R. Dennis, Ap. J., 426, 767 (1994).
24. D.J. Forrest, in *Positron-Electron pairs in Astrophysics*, eds., M.L. Burns, A.K. Harding, and R. Ramaty, AIP, New York, NY, p.3 (1983).
25. *Solar Geophysical Data*, No. 436, Pt I, p. 144 (1980).
26. E.W. Cliver, *Solar Phys.*, 157, 285 (1995).
27. N.R. Sheeley, Jr., R.A. Howard, M.J. Koomen, and D.J. Michels, *Ap. J.*, 272, 349 (1983).
28. S. Masuda, T. Kosugi, H. Hara, S. Tsuneta, and Y. Ogawara, *Nature*, 371, 495 (1994).
29. R.A. Kopp and G.W. Pneuman, *Solar Phys.*, 50, 85 (1976).
30. S.W. Kahler, N.R. Sheeley, Jr., and M. Liggett, *Ap. J.*, 344, 1026 (1989).
31. D. Moses, W. Dröge, P. Meyer, and P. Evenson, Ap. J., 346, 523 (1989).
32. S.R. Kane, P.A. Evenson, and P. Meyer, *Ap. J. (Lett.)*, 299, L107 (1985).
33. M.A.I. Van Hollebeke, F.B. McDonald, and J.P. Meyer, Ap. J. (Suppl. Ser.), 73, 285 (1990).
34. N. Mandzhavidze and R. Ramaty, *Nuc. Phys. B, Proc. Suppl.*, 33, 14 (1993).
35. R.P. Lin, *Proc. 21st Int. Cosmic Ray Conf.*, 5, 88 (1990).
36. S.W. Kahler, R.P. Lin, D.V. Reames, R.G. Stone, and M. Liggett, *Solar Phys.*, 107, 385 (1987).
37. R.P. Lin, *Solar Phys.*, 100, 537 (1985).
38. E. Cliver and S. Kahler, *Ap. J. (Lett.)*, 366, L91 (1991).
39. S.W. Kahler, E.W. Cliver, H.V. Cane, R.E. McGuire, R.E. Stone, and N.R. Sheeley, Jr., *Ap. J.*, 302, 304 (1986).
40. Z. Švestka and E.W. Cliver, in *Eruptive Solar Flares*, eds., Z. Švestka, B.V. Jackson, and M.E. Machado, Springer-Verlag, Berlin, p. 1 (1992).
41. A.H. McAllister, M. Dryer, P. McIntosh, and H. Singer, *J. Geophys. Res.*, in press (1996).
42. P. Evenson, P. Meyer, S. Yanagita, and D. Forrest, *Ap. J.*, 283, 43 (1984).
43. S.W. Kahler, *Ann. Rev. Astron. Astrophys.*, 30, 113 (1992).
44. J.T. Gosling, *J. Geophys. Res.*, 98, 18937 (1993).
45. D.J. Forrest and E.L. Chupp, *Nature*, 305, 291 (1983).
46. M.A. Forman, R. Ramaty, and E.G. Zweibel, in *Physics of the Sun*, eds., P.A. Sturrock, T.E. Holzer, D.M. Mihalas, and R.K. Ulrich, Dordrecht, 2, p. 249 (1986).
47. E.L. Chupp, these proceedings.
48. W.T. Vestrand, and J.A. Miller, in *The Many Faces of the Sun*, in press (1996).
49. R. Ramaty, N. Mandzhavidze, B. Kozlovsky, and R.J. Murphy, *Ap. J. (Lett.)*,

455, L193 (1995).

50. G.M. Simnett, *Space Science Rev.,* **73**, 387 (1995).
51. E. Rieger and H. Marschhäuser, in *Proc. of 3rd MAX 91 Workshop,* eds., R.M. Winglee and A.L. Kiplinger, University of Colorado, Boulder, CO, p. 68 (1990).
52. G.M. Simnett and A. Benz, *Astron. Astrophys.,* **165**, 227 (1986).
53. E.W. Cliver, L.F. McNamara, and L.C. Gentile, *J. Geophys. Res.,* **90**, 6251 (1985).
54. N. Mandzhavidze, in *Proc. 22nd Int. Cosmic Ray Conf.: Invited, Rapporteur, and Highlight Papers,* eds., D.A. Leahy, R.B. Hicks, and D.Venkatesan, World Scientific Pub. Co. Pte. Ltd., Singapore, p. 157 (1994).
55. H. Hudson and J. Ryan, *Space Science Rev.,* **33**, 239 (1995).
56. D.J. Forrest, W.T. Vestrand, E.L. Chupp, E. Rieger, J. Cooper, and G. Share, *Proc. 19th Int. Cosmic Ray Conf.,* **4**, 146 (1985).
57. P.P. Dunphy and E.L. Chupp, in *High Energy Solar Phenomena, AIP Conf. Proc. 294,* eds., J.M. Ryan and W.T. Vestrand, AIP, New York, NY, p. 112 (1994).
58. V.A. Akimov, N.G. Leikov, A.V. Belov, I.M. Chertok, V.G. Kurt, A.Magun, and V.F. Melnikov, in *High Energy Solar Phenomena, AIP Conf. Proc. 294,* eds., J.M. Ryan and W.T. Vestrand, AIP, New York, NY, p. 106 (1994).
59. V.A. Akimov et al., *Solar Phys.,* in press (1996).
60. N. Mandzhavidze, and R. Ramaty, *Ap. J. (Lett.),* **396**, L111 (1992).
61. E.W. Cliver, B.R. Dennis, A.L. Kiplinger, S.R. Kane, D.F. Neidig, N.R. Sheeley, Jr., and M.J. Koomen, *Ap. J.,* **305**, 920 (1986).
62. S.R. Kane, in *Coronal Disturbances, IAU Symp. No. 57,* ed., G. Newkirk, Jr., D. Reidel, Dordrecht, p. 105 (1974).
63. R. Ramaty, R.J. Murphy, and C.D. Dermer, *Ap. J. (Lett.),* **316**, L41 (1987).
64. D.V. Reames, private communication (1995).
65. L.G. Kocharov et al., *Solar Phys.,* **150**, 267 (1994).
66. S.W. Kahler, *Proc. 23rd Int. Cosmic Ray Conf.,* **3**, 1 (1993).
67. E.W. Cliver, *Solar Phys.,* **84**, 347 (1983).
68. J.M. Ryan and M.A. Lee, *Ap. J.,* **368**, 316 (1991).
69. W.T. Vestrand and D.J. Forrest, *Ap. J. (Lett.),* **409**, L69 (1993).
70. E.W. Cliver, S.W. Kahler, and W.T. Vestrand, *Proc. 23rd Int. Cosmic Ray Conf.,* **3**, 91 (1993).
71. R. Moore et al., in *Solar Flares,* ed., P. Sturrock, Colorado Assoc. Univ. Press, Boulder, CO, p. 341 (1980).
72. M. Temerin and I. Roth, *Ap. J. (Lett.),* **391**, L105 (1992).
73. J.A. Miller and A.F. Vinas, *Ap. J.,* **412**, 386 (1993).
74. J.A. Miller and D.V. Reames, these proceedings.
75. J.A. Miller, T.N. LaRosa, and R.L. Moore, *Ap. J.,* in press (1996).
76. J.A. Miller and D.A. Roberts, *Ap. J.,* **452**, 912 (1995).

Coronal Mass Ejections and Solar Energetic Particle Events

S. W. Kahler

GPSG, Phillips Laboratory
Hanscom AFB, MA 01731-3010

We review the observations relating solar energetic particle (SEP) events to coronal mass ejections (CMEs). Nearly every gradual SEP event is associated with a fast (v > 400 km/s) CME, which is presumed to drive a coronal shock that accelerates the SEPs. Evidence supporting the contention that all SEP ions observed in large, gradual events are shock accelerated is reviewed. Evidence for shock acceleration of electrons is found to be more ambiguous.

The following current questions in SEP/CME relationships are discussed: 1. SEP production by electric fields in post-flare loops; 2. the relationship of type II burst shocks and CME-driven shocks; 3. flare impulsive phase contributions to SEP events; and 4. the evidence for shock-accelerated (SA) events; and 5. progressively hardening X-ray spectra and SEP events.

SOLAR FLARES AS SOURCES OF SEPS

The first solar energetic particle events (SEPs) were reported by Forbush (17) as increases in cosmic ray intensities closely following solar flares. In a subsequent work Forbush et al (18) suggested that charged particles could be accelerated to energies of 10 GeV in the variable magnetic fields of flaring sunspots. The SEPs could then escape the sun and reach the earth. Later observations of lower energy (10 MeV) SEP events using riometer observations during the IGY in the 1950s also emphasized the flare associations (65).

A number of investigators subsequently assumed that the energetic protons of SEP events were accelerated in the same flare process that produced energetic electrons. They then used flare centimetric radio data to try to infer the properties of the associated SEP events observed at 1 AU. A detailed account of this early work is given by Kahler (24). Another problem was to understand how the SEPs propagated away from the flare site through the coronal magnetic fields after their production in the flare region. The lack of any correlation between SEP relative elemental abundances and the angular separation from the associated flare site (57) was a serious challenge to models of coronal propagation and suggested acceleration of SEPs by large-scale coronal shocks. The good association of SEP events with metric type II bursts found earlier (73) supported this idea.

INITIAL STUDIES WITH SKYLAB/IMP-7 OBSERVATIONS

The large set of coronal mass ejections observed with the Skylab corona-graph allowed us to study their relationships with both solar flares and SEP events. Although many CMEs were associated with flares, most were not, and a better relationship of CMEs was found with erupting prominences (62). These Skylab results have since been confirmed with SMM coronagraph observations (72).

Kahler et al (28) compared 4-23 MeV SEP events observed on IMP 7 during the Skylab era with the Skylab CMEs. Fourteen of the 16 SEP events with good coronagraph observations could be associated with CMEs. In 11 of 15 cases the associated soft X-ray event was also a long-decay event (LDE). We now understand that LDE events and Hα post flare loop prominence systems (LPS) are the coronal signatures of CMEs as proposed by Kopp and Pneuman (47). It was therefore not surprising that Bruzek (2) had earlier found a good correlation between polar cap absorption events at the Earth, caused by E > 10 MeV protons, and post flare LPS.

Reinhard and Wibberenz (66) had deduced the existence of a "fast az-imuthal propagation region" in which SEPs are rapidly distributed over a broad ($\sim$60°) region of solar longitude. Kahler et al (28) proposed that the SEPs were accelerated in a broad shock, as first suggested by Lin and Hud-son (53). The width of the shock could then be understood in terms of the widths of the CMEs driving the shocks. Nine of the 10 CME speeds exceeded 400 km/s, the approximate value of the coronal Alfven speed, and some ev-idence for a correlation between CME speed and SEP peak flux was found, supporting the shock interpretation.

SOLWIND/IMP-8 COMPARISONS

The Solwind coronagraph on the P78-1 satellite provided a large data set of CME observations over the period 1979-1985. A survey of IMP-8 and ISEE-3 SEP events by Kahler et al (29) showed that 26 of 27 SEP events could be associated with CMEs, confirming the earlier Skylab results. A good correlation (r = 0.56) was found between the peak E > 4 MeV SEP fluxes and the CME speeds (Figure 1). In addition, the probability of detecting a SEP event associated with a fast CME increased with the CME speed.

Do all fast CMEs produce SEP events? Kahler et al (32) found that 29 of 31 Solwind CMEs with speeds of v $\geq$ 800 km/s and considered to be on the frontside of the west limb were associated with SEP events at the Earth. In a similar study with SMM CMEs Kahler (26) found that at least 10 of 11 such CMEs (but with v $\geq$ 750 km/s) were associated with E > 10 MeV SEP events. Hundhausen et al (22) have presented the speed distribution of the outer loops of CMEs, the structures which should be most relevant for driving shocks (Figure 2). Only about 10% of those speeds exceed 800 km/s.

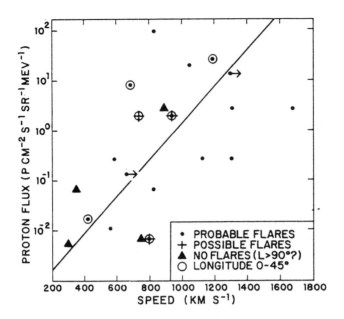

FIG. 1. Plot of the IMP-8 peak 4-22 MeV proton fluxes against the speeds of the associated Solwind CMEs (29). Solid line is the least squares best fit.

However, speeds of CMEs associated with SEP events can extend down to 400 km/s, which is less than the average value of the outer loop speeds.

Kahler et al (29) found that the SEP peak fluxes correlated with CME widths for the 1979-1982 data, but later (32) found no correlation for the widths in the 1983-85 data. The median width of the 11 CMEs of the Kahler (26) study of SMM events was 65°, well above the 44° median of all SMM CMEs. In their survey of CME speeds Hundhausen et al (22) found higher speeds for CMEs with widths above 60°, and Kahler et al (34) also found a correlation between Solwind CME speeds and widths. Thus, CME widths may be an important parameter for SEP fluxes, but perhaps only as a result of the correlation between CME widths and speeds. Some CMEs associated with SEP events lay completely out of the ecliptic plane (29), suggesting either that the shocks extended well beyond the limits of the CMEs or that mid and high latitude coronal field lines extend down to the ecliptic plane. Hundhausen et al (22) found that CME speeds are almost independent of solar latitude, indicating that high-latitude CMEs are not a "weak" form of coronal activity.

SEP events with high ^{3}He/^{4}He ratios are now known to be distinctly different from the large-flux events with long time scales (63). A comparison of the times of the ^{3}He-rich events with CMEs showed (30) that there is only a random-chance association of such SEP events with either CMEs or with metric type II bursts, which might be the manifestation of CME-driven coronal

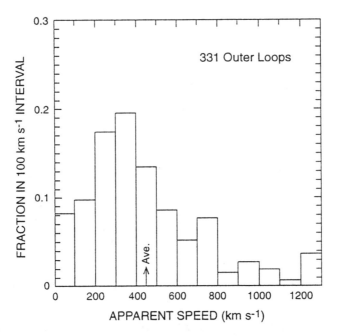

FIG. 2. Speed distribution of SMM CMEs from Hundhausen et al (22).

shocks. These events are now modelled in terms of plasma wave acceleration of the SEPs following wave generation by beams of nonrelativistic electrons (60). CMEs play no role in these models.

SEP EVENTS WITHOUT FLARES

Most gradual SEP events are associated with solar flares, as discussed above. However, if the CME-driven coronal shocks are the means for accelerating SEPs, we should find some cases in which the SEP event is associated with a fast CME, but not with a flare. Kahler (25) examined the risetimes of 15 < E < 44 MeV proton events observed on the GOES spacecraft during the SMM coronagraph observations. Of the 12 SEP events associated with well connected (20°W to 70°W) flares, only three had rise times exceeding 8 hrs. In each of those cases the SEP onset was associated with both a flare and a CME. In at least two of the three cases a second fast CME not associated with flaring activity also occurred and was followed by a substantial SEP flux increase (Figure 3). In those cases the long rise times of the SEP events were due to additional SEP acceleration by shocks driven by the second CMEs.

A simple SEP event associated with a Solwind CME but not with a flare was observed on 5 December 1981 (31). The CME source region consisted of a quiescent filament far from any plage region. The erupting filament

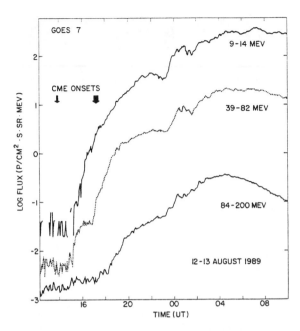

FIG. 3. GOES-7 proton fluxes on August 12-13, 1989 showing the long (17 hr) rise time to maximum (25). Arrows show the times of the two observed CMEs.

could be seen as a bright core in the CME. The gradual disappearance of the filament led to an Hα double ribbon (Figure 4), as expected in the Kopp and Pneuman (47) scenario for a post flare loop system. The proton energies detected at ISEE-3 exceeded 50 MeV in that event. If one considers low energy (< 5 MeV) SEP events, then at least 14 such events associated with solar disappearing filaments instead of large flares have been observed with ISEE-3 particle detectors (67).

RELATION OF CME HEIGHTS AND SEP INJECTION PROFILES

While the previous studies have made clear the association of gradual SEP events with fast CMEs, they did not reveal the spatial and temporal relation-ships of CMEs with the injection profiles of the SEP events. Kahler et al (35) assumed that SEPs are accelerated on open field lines by CME-driven shocks and asked how the SEP injection profiles varied as a function of CME height. Since the CME height as a function of time was known, the SEP flux profile as a function of time could be replotted as a function of CME height. An important step was to assume that the SEPs propagated with no scattering so that the "solar release times" could be deduced from the flux profiles at 1 AU by using the simple time shift of dt = 1.3 AU/v - 8.3 min where v is the

65

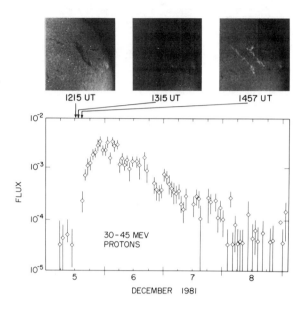

FIG. 4. The Hα erupting filament and SEP event of 5 December 1981 (31). The 0.2 to 2 MeV electron flux profile was similar to that of the 30 to 45 MeV protons.

speed of the SEP and 1.3 AU the assumed path length from the Sun to the Earth. The analysis of 10 SEP events associated with Solwind CMEs showed that at 50 and 175 MeV proton injection profiles are increasing and sometimes reaching maximum when the associated CMEs are at heights of 2 to 10 R⊙.

The use of solar release times to determine the SEP injection profiles was justified by Kahler et al (35) on the grounds that several of their events were highly anisotropic while all showed the same qualitative behavior with regard to the CME heights. It clearly was desirable to do the same kind of analysis for SEPs with higher energies and correspondingly longer mean free paths. Kahler (27) did such a comparison using SMM CMEs associated with three ground-level events (GLEs) with associated energies of 470 MeV to 21 GeV. Again, the resulting injection profiles were increasing or reaching peak flux when the leading edges of the CMEs were 5 to 15 R⊙ from the sun, as shown in Figure 5. Since this is the "solar" component of SEP fluxes, and the injection occurs in interplanetary space above the closed field regions of the corona, Kahler argued that this result supports the model of a single shock to accelerate both the solar and interplanetary components of SEP events (40).

We have seen from the event shown in Figure 4 that MeV electrons were associated with a fast CME that had no detectable microwave, metric, or hard X-ray signature, suggesting that all the electrons of that event were accelerated by a CME-driven shock. However, we also know that electrons accelerated to tens or hundreds of keV are produced in impulsive flares and escape the sun leaving metric type III burst signatures. Thus, the primary source of an electron event which is associated with both a fast CME and a flare impulsive phase is not easily resolved.

A simple spectral difference between the two kinds of events was suggested by the result of Moses et al (61) that electron events with a single power law in rigidity from 75 keV to 100 MeV were associated with long-duration (> 1 hr) soft X-ray events and those with steeper spectra at low rigidities were generally associated with short-duration soft X-ray events. Since long-duration X-ray events are much more likely than short-duration events to be associated with CMEs, and hence with coronal shocks, a coronal shock origin for the single power-law events was offered (11).

This picture of two basic classes of electron events, one of which is produced in a CME-driven shock, was criticized by Kahler et al (38) for several reasons. First, the median 200 keV flux associated with the shortest duration X-ray flares is only a factor of 3 less than that associated with the longest duration flares, although the latter should include additional shock-accelerated contributions. Second, while very few of the 19 electron events associated with Solwind CMEs should also be associated with short-duration flares and double power-law spectra, nine were so associated. Finally, when the escape efficiencies, defined as the ratios of the peak electron fluxes to the corresponding X-ray fluences (10), are calculated for 30 of the events, the single power-law events have lower, not higher, escape efficiencies. A higher escape efficiency is expected for particles produced in coronal shocks because the lower ambient density of the shock region results in less thin or thick-target X-ray emission. Using this argument to look for evidence of shock acceleration in $E >$ 75 keV electrons, Kahler et al (38) found that the escape ratios of $E > 75$ keV electron events associated with CMEs were comparable to those without CMEs. They concluded that for nonrelativistic electrons the population of shock-accelerated electrons is at most comparable to the flare-accelerated population.

A different approach to look for evidence of shock-accelerated electrons is to assume that flare-accelerated electrons are injected nearly instantaneously and that the extended duration of any shock acceleration will be reflected in an increase in the rise time of the interplanetary electron event. Thus, the time profile, rather than the peak flux or energy spectrum, is considered the diagnostic parameter of shock acceleration. Kahler et al (39) compared rise times of Helios $E > 0.3$ MeV electron events with Solwind CME observations. Although the statistics were limited to 24 events, a progressive increase in

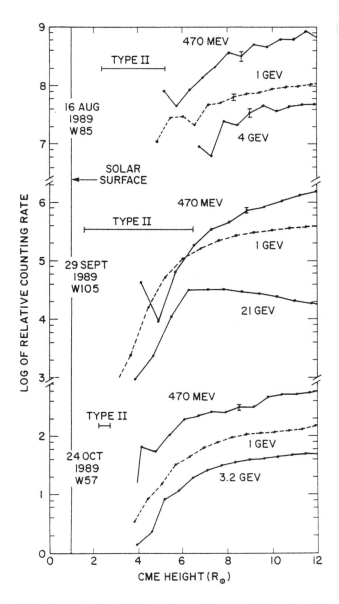

FIG. 5. The proton injection flux profiles of three GLEs as a function of the SMM CME height. Relative counting rates after background subtraction are shown. The 470 MeV fluxes are taken from the GOES HEPAD detector; other profiles are from ground-based neutron monitors. From Kahler (27).

median event rise times was seen for events with i) no CMEs, ii) slow (v <
800 km/s), and iii) fast (v > 800 km/s) CMEs. A similar result has been found
for 15 E > 0.35 MeV electron events measured on the Phobos 2 spacecraft
(71). These results, combined with those based on escape efficiencies (10,38)
suggest that the primary effect of a fast shock is to increase the rise time
and perhaps the total number of electrons in the event, but not its peak flux
relative to the flare-associated impulsive electrons.

SHOCK ACCELERATION OF ALL SEPS

The argument of Kahler (27) that all the SEPs of a gradual event are accelerated in a single coronal/interplanetary shock implies that the shock must
be capable of accelerating ions up to the GeV energies observed in GLEs.
The recent result (76) that the mean ionic charge state of Fe at 200 to 800
MeV/nucleon in GLE events is ∼14, and hence characteristic of the ambient corona, is strong supporting evidence for this view. Further recent measurements of ionic charge states in 0.5 to 5 MeV/nucleon (58) and 15 to 70
MeV/nucleon (51) ranges support the view that the SEP source is ambient
coronal or solar wind material and can not be heated ($T > 10^7$ K) plasmas of
flares or reconnection regions. The 21 August 1979 GLE was associated with
a fast CME, but the associated flare had a very weak impulsive phase (7),
again suggesting that the flare was not the source of the GeV SEPs. That
event was one of several large SEP events with weak impulsive phases in the
associated flares (8).

Recent shock modelling has suggested that with appropriate conditions
SEPs can be accelerated to GeV energies in very short time scales. Ellison
and Ramaty (14) assumed a planar shock to model first-order Fermi shock
acceleration of electrons, protons, and alpha particles. Using a limited range
of shock compression ratios (r = 1.6-3), they found good agreement between
their model and observations of a number of SEP events, in several cases extending to energies of 10 GeV. With adequate particle scattering from shock
turbulence, they found that acceleration to 100 MeV could take as little as
1 sec. Lee and Ryan (50) presented a global time-dependent model for the
coronal and interplanetary shock acceleration and propagation of SEPs. By
varying the diffusion coefficients, they were able to reproduce the SEP time
profiles observed at 1 AU. Zhang et al (81,82) suggested that a time-dependent
diffusive shock wave model can explain SEP energies extending to the TeV
range. Again, with reasonable diffusion coefficients, they find acceleration to
10 GeV in a time as short as ∼ 4 sec. Thus we see that there appears to be no
theoretical objection to rapid acceleration of SEPs to relativistic energies by
CME-driven shocks in the short time scales required by the GLE observations.

CURRENT QUESTIONS ABOUT SEP ORIGINS

The comparisons of gradual interplanetary SEP events with CMEs has led us to a simple picture in which all the SEPs are produced in a single CME-driven shock which propagates through the corona and interplanetary space. There are, however, several observations which suggest either modifications or serious challenges to this view. These are reviewed in the following five sections.

SEP Production in Post-Flare Loops

To find any alternative sources of SEPs in the corona, we should seek situations in which substantial energy release occurs on or near open field lines. Impulsive flares are an obvious candidate, and these frequently lead to the ^{3}He-rich SEP events discussed by Reames (63). However, eruptive flares, characterized by associated CMEs and post-flare loops, continue to release energy for tens of minutes to hours after their onsets. The view today is that the post-flare loops form during the process of magnetic reconnection of oppositely directed field lines (e.g., (16,15)). The reconnection leads to the reformation of coronal streamers blown open during the preceding CME (37).

The reconnection in post-flare loops sometimes results in gradual hard X-ray bursts characterized by X-ray emission with a gradually hardening spectrum and a large ratio of microwave to hard X-ray emission (9,48). The interpretation is that electrons are accelerated to relativistic energies at the tops of the reconnecting loops (Figure 6) with heights of $> 3 \times 10^4$ km (9). This general picture would explain the continuous acceleration and injection of nonthermal electrons at different altitudes implied by the similarities of gradual hard X-ray and microwave profiles (74) and by the association of hectometric emission with such bursts (44).

Clear evidence for the presence of $E > 300$ MeV ions in flare gradual phases was found in the γ-ray observations of several large flares in June 1991. A flare on 15 June was found to have an extended phase of pion-decay emission lasting at least 2 hours after the flare onset (1,49). Another flare on 11 June lasted at least 8 hours after flare onset, again with a spectrum consistent with pion decay (41). Similar extended-phase pion-decay events of shorter durations had been seen with the lower-sensitivity detector on SMM (13). While Mandzhavidze and Ramaty (55) favored long-term trapping of particles accelerated during the flare impulsive phase, Kocharov et al (46) and Akimov et al (1) interpreted the events in terms of continuous acceleration of electrons and protons. The delayed emission occurs well after the occurrence of type II bursts (46), and Akimov et al (1) discuss particle acceleration in terms of post-flare loop systems. Martens (56) found that electrons and protons can be accelerated to relativistic energies by the direct electric fields in current sheets overlying post-flare loops. A more realistic model (54) considering

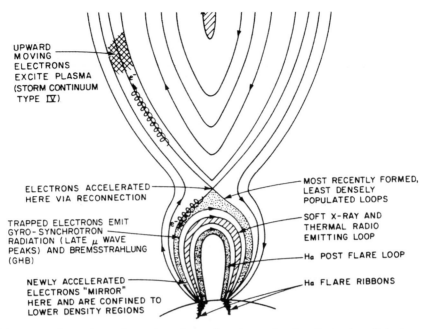

FIG. 6. A schematic configuration of post-flare loops showing the region of electron acceleration in reconnecting coronal fields (9).

the perpendicular magnetic field in the current sheet and the perpendicular electric field outside the current sheet confirms Martens' basic conclusion. The question is whether these very energetic particles observed in the gradual phases can escape to interplanetary space.

Klein et al (45) make the case that noise storms observed for several hours both in the flaring region and several tens of heliographic degrees away following the 19 October 1989 GLE mark the regions of SEP injection into interplanetary space. Since CMEs are typically $\sim 50°$ wide (21), we should expect that the reconnection shown in Figure 6 will occur not only in an active region but also along a magnetic neutral line extending several tens of degrees. A number of such events have now been observed in the Yohkoh soft X-ray images (75,59). Klein et al (45) make the unproven, but plausible, assumption that the electron signatures in these events are also representative of energetic ions. Thus one might expect that a population of energetic ions and electrons would be produced and injected into interplanetary space from reconnecting fields overlying the neutral line. The energetic noise storms, such as those on 19 October or the gradual hard X-ray bursts discussed by Cliver et al (9) generally follow CMEs within one to several hours, so the appearance of any SEPs from the noise storm will be masked by those SEPs generated in the CME-driven shock. Even in the case of the June 1991 extended γ-ray events the time scales of the GLEs appear to be too long to distinguish a second

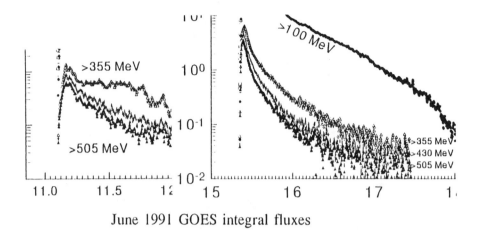

June 1991 GOES integral fluxes

FIG. 7. GOES high-energy integral flux profiles for the GLEs of 11 and 15 June 1991 showing the decay profiles during the extended pion-decay events observed with CGRO and the GAMMA-1 telescope. Adapted from Smart et al (69).

population of SEPs as shown in Figure 7 (69).

Most CMEs are too slow to drive coronal shocks, but all CMEs should be followed by reconnecting fields as shown in Figure 6, so we should encounter at least a few cases in which all the SEPs are injected from the reconnection region and no shock-accelerated SEPs confuse the observations. Such a SEP population would not correlate well with the presence of shocks, but this would seem to contradict the basic observed organization of all SEP events in terms of shocks (5). Although no such ion events have been identified, electron events associated with metric noise storms have been observed. However, both the energies (E < 10 keV) and the fluxes of those interplanetary electron events are significantly lower than for electrons observed from flares or interplanetary shocks (52,12). These particles should be confined to the vicinity of the neutral sheet extending above the reforming streamer (20). The positions of those electron streams relative to the heliospheric current sheets have not been studied. It appears that if SEP ions are injected from gradually reconnecting fields following CMEs, the fluxes and energies may be too low to measure easily.

CMEs and Coronal Shocks

The good associations of SEP events with both metric type II radio bursts (73,23) and CMEs suggested that the SEPs are accelerated in coronal shocks driven by CMEs. In that case the type II bursts should be well associated with fast (v > 400 km/s) CMEs. The Skylab results (62) supported that idea, but Wagner and MacQueen (79) suggested that coronal shock waves are produced by flares and propagate through any associated CME. The current view is that type II shocks originate only in flares (3,19), and recent work (78) identifies the beginning of the rapid rise of the first burst of the associated microwave burst as the origin of the shock. Thus the shock driven by the CME, which accelerates SEPs and propagates into interplanetary space, may often be initiated at a frequency too low to be detected on earth (19). This further suggests that MHD models based on the assumption that the interplanetary shock is an extension of the type II burst shock (68,70) may use incorrect boundary conditions. The observational problem here is to distinguish clearly the flare shock from the CME-driven shock. This can only be done by satellite observations in the 2 to 20 MHz band (19).

Flare Impulsive Phase Contributions to Gradual SEP Events

When a gradual SEP event is also associated with a well connected flare, a hybrid event consisting of Fe-rich ions early in the event followed later by Fe-poor ions may be seen (64). In some events the Fe-rich ions reach energies of tens of MeV per nucleon, suggesting that impulsive phase ions are shock-accelerated to the higher observed energies (77,64). The SEP events of 3 June 1982 and 21 June 1980 observed on Helios-1 were unusual in that they were dominated by Fe-rich abundances at $E > 50$ MeV/nucleon throughout each event. They were also associated with impulsive soft X-ray flares, which are generally associated with narrow CMEs (34). Kahler (26) has suggested that the associated narrow CMEs drive correspondingly narrow shocks which may preferentially accelerate the Fe-rich ions near the flare site rather than the ambient coronal material away from the flare site.

If Fe-rich ions from impulsive flares can dominate the SEP fluxes up to 50 MeV/nucleon, can this process take place at GeV energies? Kahler et al (36) discussed this possibility for six GLEs associated with impulsive soft X-ray flares. The presence of type IV radio bursts following 5 of the 6 associated flares suggests that CMEs were associated with those events. However, two recent results raise some doubt about this idea of additional shock acceleration of impulsive SEPs. The first is that Dunphy and Chupp (13) find that the 3 June 1982 and 21 June 1980 flare events were followed by extended phases of high energy γ-ray emission, thus allowing alternative candidate sources for the observed SEP events, as discussed above. The second is that Chen et al (6) find that $^3He/^4He$ ratios of events observed in the 50 to 110 MeV/nucleon range are high ($> .005$) and independent of solar source longitude for 16 events observed on the CRRES mission. In several cases the ratios exceed 0.01 for events associated with eastern hemisphere flares. This result suggests that ^3He enhancement may not arise exclusively in impulsive flares at high energies, contrary to the situation for the low energy (few Mev/nucleon) range.

Shock-Accelerated Events

If electrons are accelerated in coronal shocks, the beams of energetic electrons might appear as fast-drift kilometric bursts originating in the herringbone structure of metric type II bursts. A class of intense, long-duration bursts at 1980 kHz were found by Cane et al (4) to be well associated with metric type II bursts, and became known as shock-associated (SA) bursts. Defining an SA burst as 1980 kHz emission temporally associated with a metric type II burst, but not associated with a metric type III burst, Kahler et al (33) found that nearly half of all type II bursts have associated SA bursts. Using Culgoora Observatory records, they found that the SA bursts corresponded to the most intense features of the metric type II bursts. They also found a generally poor correspondence between the flux profiles of SA bursts and of simultaneous microwave bursts, suggesting that SA emission is not due to energetic electrons escaping from the microwave emission region.

Klein (43) has argued that since all of a group of 12 SA events generally show metric storm continuum or flare continuum during each SA event, electron acceleration and injection occurs over a range of altitudes, including open field lines at the hectometric level, and is not due to shock acceleration. Subsequently, Klein and Trottet (44) found that 5 gradual hard X-ray/radio bursts were also accompanied by hectometric emission similar to the SA events, but not necessarily accompanied by type II bursts. In their view, all the hectometric emission other than type III bursts is due to electron escape from the middle corona and is not the result of large-scale shock acceleration. In our view they have made a good case for electron injection from the coronal gradual events observed in the metric and hard X-ray range. However, their analysis of a small number of very energetic events does not preclude the shock acceleration scenario. It is also possible that electrons accelerated in shocks are not always manifested as herringbone emission. If we accept that protons are accelerated by shocks, then the striking spatial and temporal similarities between 1 MeV electrons and 50 MeV protons found by Wibberenz and Cane (80) in their multispacecraft study suggests shock acceleration of the electrons, as well.

Progressively Hardening X-Ray Events and SEPS

If SEPs are produced only in CME-driven shocks, we do not expect to see any relationship between the SEP event and the hard X-ray burst of the associated flare. Recently, however, Kiplinger (42) has found such a relationship with the $30 < E < 500$ keV X-ray bursts detected by the HXRBS detector on the SMM spacecraft. The characteristic signature for a good association with SEP events is a progressive hardening of the X-ray spectrum either during an individual flux peak or during the burst decay. The analysis used large HXRBS events and asked whether they were associated with observed SEP events. No inverse study of associations based on observed SEP events was done. An algorithm was developed for the X-ray bursts with which associated SEP events could be forecast with about 95% accuracy. Besides the excellent event correlation, a correlation was found between the FWHM of the longest progressively hardening burst peak and the cube of the peak proton flux of the SEP event.

Progressive spectral hardening of X-ray bursts was one of the characteristics of the gradual hard X-ray bursts studied by Cliver et al (9) and Kosugi et al (48). For those events Cliver et al argued for acceleration and trapping of radiating electrons in post flare loop systems following CMEs. The progressively hardening X-ray bursts of the Kiplinger study (42) were also interpreted as high coronal source regions of particle acceleration, but in addition, as the source of the interplanetary SEPs. If CME-driven shocks are the main source of SEPs, we might expect that SEP events are well associated with post flare loop systems following the CMEs, but the results of the Kiplinger study suggest a much closer relationship between the hard X-ray characteristics of those systems and the SEP events, suggesting that the post flare loops, not the shock, are the primary SEP source.

REFERENCES

1. Akimov, V.V., A.V. Belov, I.M. Chertok, V.G. Kurt, N.G. Leikov, A. Magun, V.F. Melnikov, 23rd ICRC (Calgary) **3**, 111, 1993.
2. Bruzek, A., Ap.J. **140**, 746, 1964.
3. Cane, H.V., and D.V. Reames, Ap.J. **325**, 895, 1988.
4. Cane, H.V., R.G. Stone, J. Fainberg, J.L. Steinberg, and S. Hoang, Geophys. Res. Letters 8, 1285, 1981.
5. Cane, H.V., D.V. Reames, and T.T. von Rosenvinge, J. Geophys. Res. **93**, 9555, 1988.
6. Chen, J., T.G. Guzik, and J.P. Wefel, Ap.J. **442**, 875, 1995.
7. Cliver, E.W., S.W. Kahler, H.V. Cane, M.J. Koomen, D.J. Michels, R.A. Howard, and N.R. Sheeley, Jr., Solar Phys. **89**, 181, 1983.
8. Cliver, E.W., S.W. Kahler, and P.S. McIntosh, Ap.J. **264**, 699, 1983.
9. Cliver, E.W., B.R. Dennis, A.L. Kiplinger, S.R. Kane, D.F. Neidig, N.R. Sheeley, Jr., and M.J. Koomen, Ap.J. **305**, 920, 1986.
10. Daibog, E.I., V. Kurt, Yu.I. Logachev, and V.G. Stolpovskii, Cosmic Research **27**, 97, 1989.
11. Droge, W., P. Meyer, P. Evenson, and D. Moses, Sol. Phys. **121**, 95, 1989.
12. Dulk, G.A., Sol. Phys. **130**, 139, 1990.
13. Dunphy, P.P., and E.L. Chupp, AIP Conf. Proc. **294** (eds J.M. Ryan and W.T. Vestrand), 112, 1994.
14. Ellison, D.C., and R. Ramaty, Ap.J. **298**, 400, 1985.
15. Forbes, T.G., Geophys. Astrophys. Fluid Dynamics **62**, 15, 1991.
16. Forbes, T.G., J.M. Malherbe, and E.R. Priest, Sol. Phys. **120**, 285, 1989.
17. Forbush, S.E., Phys. Rev. **70**, 771, 1946.
18. Forbush, S.E., P.S. Gill, and M.S. Vallarta, Rev. Mod. Phys. **21**, 44, 1949.

19. Gopalswamy, N., and M.R. Kundu, Coronal Magnetic Energy Releases (eds A. Benz and A. Kruger), 223, Springer-Verlag, 1995.

20. Hirayama, T., Proceedings of Kofu Symposium (eds. S. Enome and T. Hirayama), NRO Report Nbr **360**, 231, 1994.

21. Hundhausen, A.J., J. Geophys. Res. **98**, 13177, 1993.

22. Hundhausen, A.J., J.T. Burkepile, and O.C. St. Cyr, J. Geophys. Res. **99**, 6543, 1994.

23. Kahler, S.W., Ap.J. **261**, 710, 1982a.

24. Kahler, S.W., J. Geophys. Res. **87**, 3439, 1982b.

25. Kahler, S.W., J. Geophys. Res. **98**, 5607, 1993.

26. Kahler, S.W., 23rd ICRC (Calgary) **3**, 1, 1993.

27. Kahler, S., Ap.J. **428**, 837, 1994.

28. Kahler, S.W., E. Hildner, and M.A.I. van Hollebeke, Solar Phys. **57**, 429, 1978.

29. Kahler, S.W., N.R. Sheeley, Jr., R.A. Howard, M.J. Koomen, D.J. Michels, R.E. McGuire, T.T. von Rosenvinge, and D.V. Reames, J. Geophys. Res. **89**, 9683, 1984.

30. Kahler, S., D.V. Reames, N.R. Sheeley, Jr., R.A. Howard, M.J. Koomen, and D.J. Michels, Ap.J. **290**, 742, 1985.

31. Kahler, S.W., E.W. Cliver, H.V. Cane, R.E. McGuire, R.G. Stone, and N.R. Sheeley, Jr., Ap.J. **302**, 504, 1986.

32. Kahler, S.W., E.W. Cliver, H.V. Cane, R.E. McGuire, D.V. Reames, N.R. Sheeley, Jr., and R.A. Howard, 20th ICRC (Moscow) **3**, 121, 1987.

33. Kahler, S.W., E.W. Cliver, and H.V. Cane, Solar Phys. **120**, 393, 1989.

34. Kahler, S.W., N.R. Sheeley, Jr., and M. Liggett, Ap.J. **344**, 1026, 1989.

35. Kahler, S.W., D.V. Reames, and N.R. Sheeley, Jr., 21st ICRC (Adelaide) **5**, 183, 1990.

36. Kahler, S.W., M.A. Shea, D.F. Smart, and E.W. Cliver, 22nd ICRC (Dublin) **3**, 21, 1991.

37. Kahler, S.W., and A.J. Hundhausen, J. Geophys. Res. **97**, 1619, 1992.

38. Kahler, S.W., E.I. Daibog, V.G. Kurt, and V.G. Stolpovskii, Ap.J. **422**, 394, 1994.

39. Kahler, S.W., V.G. Stolpovskii, and E.I. Daibog, Solar Coronal Structures, IAU Coll. **144**, 479, 1994.

40. Kallenrode, M.-B., Adv. Space Res. **13**, 341, 1993.

41. Kanbach, G., D.L. Bertsch, et al, Astron. Astrophys. Supple. Ser. **97**, 349, 1993.

42. Kiplinger, A.L., Ap.J. **453**, 973, 1995.

43. Klein, K.-L., Solar Wind Seven (eds. E. Marsch and R. Schwenn), 635, Pergamon Press, Oxford, 1992.

44. Klein, K.-L., and G. Trottet, AIP Conf. Proc. **294** (eds J.M. Ryan and W.T. Vestrand), 187, 1994.

45. Klein, K.-L., G. Trottet, H. Aurass, A. Magun, and Y. Michou, Adv. Space Res. **17**, (4/5)247, 1995.

46. Kocharov, L.G., G.A. Kovaltsov, G.E. Kocharov, E.I. Chuikin, I.G. Usoskin, M.A. Shea, D.F. Smart, V.F. Melnikov, T.S. Podstrigach, T.P. Armstrong, and H. Zirin, Solar Phys. **150**, 267, 1994.

47. Kopp, R.A., and G.W. Pneuman, Solar Phys. **50**, 85, 1976.

48. Kosugi, T., B.R. Dennis, and K. Kai, Ap.J. **324**, 1118, 1988.

49. Kovaltsov, G.A., I.G. Usoskin, L.G. Kocharov, H. Kananen, and P.J. Tanskanen, Solar Phys. **158**, 395, 1995.

50. Lee, M.A., and J.M. Ryan, Ap.J. **303**, 829, 1986.

51. Leske, R.A., J.R. Cummings, R.A. Mewaldt, E.C. Stone, and T.T. von Rosenvinge, Ap.J. **452**, L149, 1995.

52. Lin, R.P., Sol. Phys. **100**, 537, 1985.

53. Lin, R.P., and H.S. Hudson, Solar Phys. **50**, 153, 1976.

54. Litvinenko, Yu.E., and B.V. Somov, Solar Phys. **158**, 317, 1995.

55. Mandzhavidze, N., and R. Ramaty, Ap.J. **396**, L111, 1992.

56. Martens, P.C.H., Ap.J. **330**, L131, 1988.

57. Mason, G.M., G. Gloeckler, and D. Hovestadt, Ap.J. **280**, 902, 1984.

58. Mason, G.M., J.E. Mazur, M.D. Looper, and R.A. Mewaldt, Ap.J. **452**, 901, 1995.

59. McAllister, A.H., M. Dryer, P. McIntosh, H. Singer, and L. Weiss, Proc. Third SOHO Workshop, ESA SP-**373**, 315, 1994.

60. Miller, J.A., and A.F. Vinas, Ap.J. **412**, 386, 1993.

61. Moses, D., W. Droge, P. Meyer, and P. Evenson, Ap.J. **346**, 523, 1989.

62. Munro, R.H., J.T. Gosling, E. Hildner, R.M. MacQueen, A.I. Poland, and C.L. Ross, Solar Phys. **61**, 201, 1979.

63. Reames, D.V., Proc. of the Third SOHO Workshop, ESA SP- **373**, 107, 1994.

64. Reames, D.V., H.V. Cane, and T.T. von Rosenvinge, Ap.J. **357**, 259, 1990.

65. Reid, G.C., and H. Leinbach, J. Geophys. Res. **64**, 1801, 1959.

66. Reinhard, R., and G. Wibberenz, Solar Phys. **36**, 473, 1974.

67. Sanahuja, B., A.M. Heras, V. Domingo, and J.A. Joselyn, Solar Phys. **134**, 379, 1991.

68. Smart, D.F., and M.A. Shea, J. Geophys. Res. **90**, 183, 1985.

69. Smart, D.F., M.A. Shea, and L.C. Gentile, AIP Conf. Proc. **294** (eds J.M. Ryan and W.T. Vestrand), 222, 1994.

70. Smith, Z., and M. Dryer, Solar Phys. **129**, 387, 1990.

71. Stolpovskii, V.G., G. Erdos, S.W. Kahler, E.I. Daibog, and Yu. I. Logachev, 24th ICRC (Rome) **4**, 301, 1995.

72. St. Cyr, O.C., and D.F. Webb, Solar Phys. **136**, 379, 1991.

73. Svestka, Z., and L. Fritzova-Svestkova, Solar Phys. **36**, 417, 1974.

74. Trottet, G., Sol. Phys. **104**, 145, 1986.

75. Tsuneta, S., T. Takahashi, L.W. Acton, M.E. Bruner, K.L. Harvey, Y. Ogawara, Publ. Astron. Soc. Japan **44**, L211, 1992.

76. Tylka, A.J., P.R. Boberg, J.H. Adams, Jr., L.P. Beahm, W.F. Dietrich, and T. Kleis, Ap.J. **444**, L109, 1995.

77. Van Hollebeke, M.A.I., F.B. McDonald, and J.P. Meyer, Ap.J. Supplement **73**, 285, 1990.

78. Vrsnak, B., V. Ruzdjak, P. Zlobec, and H. Aurass, Solar Phys. **158**, 331, 1995.

79. Wagner, W.J., and R.M. MacQueen, Astron. Astrophys. **120**, 136, 1983.

80. Wibberenz, G., and H.V. Cane, 23rd ICRC (Calgary) **3**, 274, 1993.

81. Zhang, L., J. Mu, B. Dai, and W. Zhou, 23rd ICRC (Calgary) **3**, 33, 1993a.

82. Zhang, L., B. Ai, Y. Feng, J. Mu, and W. Zhou, 23rd ICRC (Calgary) **3**, 37, 1993b.

Energetic Solar Electron Spectra and Gamma-Ray Observations

Wolfgang Dröge

Institut für Kernphysik, Universität Kiel, D-24118 Kiel, Germany

We analyze solar energetic electron events measured with particle detectors on board of the *ISEE-3 (ICE)* and *Helios 1* and *2* spacecraft. Energy spectra in the range 0.1 to tens of MeV are generated applying the results of a careful re-examination of the electron response function of the instruments. The spectral shapes of events observed simultaneously, among them five on all three s/c, are in very good agreement inspite of the sometimes considerable difference in azimuthal and radial distances of the s/c with respect to the flare. These findings suggest that transport processes at the Sun and in the interplanetary medium depend only weakly on the electron energy and that the observed spectra are representative of the accelerated electron spectra at the Sun. A comparison of the electron spectra with SMM gamma-ray spectra gives evidence for the existence of different acceleration and emission mechanism in flares with long (LDEs) and short duration (SDEs) soft X-ray emission.

INTRODUCTION

Particle acceleration in solar flares can be studied by direct measurements of the particles which escape into interplanetary space or through the observation of neutral radiations, such as radio, X-ray, and gamma-ray emission as well as energetic neutrons which are produced by the accelerated particles at or near their acceleration site. However, the interpretation of the data is often difficult because of the chain of assumptions required regarding the convolution of properties of energetic particles with properties of the photospheric and coronal plasma and the interaction between them. The simultaneous observation of flare-accelerated particles with two or more spacecraft at different positions with respect to the flare and of the associated neutral radiation can therefore provide important diagnostics of the acceleration, transport and interaction processes of energetic particles in the solar atmosphere. In the current paper we investigate the characteristics of interplanetary electron spectra in the energy range 0.1 to 50 MeV and compare for selected events electron spectra with properties of gamma-ray emission from the parent flares.

I. DATA ANALYSIS

Electron observations were made with two instruments on board the *ISEE-3/ICE* spacecraft: the ULEWAT spectrometer (MPE/University of Maryland) which measured the electron flux in the nominal energy range $0.075-1.3$ MeV, and the University of Chicago MEH spectrometer which measured the electron flux in the energy range $5-100$ MeV (for a full description of the instruments see (1)), and with the University of Kiel cosmic ray instrument onboard *Helios* (2) in the nominal energy range $0.8-4$ MeV. During the time interval of the observations presented in this study (1978 – 1982) *ISEE-3* was positioned at the Earth-Sun Lagrangian point well outside of the Earth's geomagnetic field whereas *Helios 1* and *2* operated in the inner solar system between 0.3 and 1 AU.

The gamma-ray observations used in this work were made with the gamma-ray spectrometer (GRS) of the Solar Maximum Mission *(SMM)* satellite and have been adopted from the study of Vestrand et al. (3). For the period under consideration a total of 15 events were found for which gamma-ray spectra could be constructed and energetic electrons were observed on *ISEE-3*. Five of these events were simulteaneously observed on *Helios 1*.

II. ELECTRON SPECTRA

A careful analysis of the electron response of the *Helios* particle detector (4) showed that the determination of reliable electron spectra was problematic because no calibration measurements with electrons were made and large-angle scattering and straggeling effects at energies below ~ 5 MeV lead to different geometrical factors and energy ranges of the electron channels compared to the nominal values derived from purely geometrical arguments and energy-range tables, respectively. In addition, the electron channels were severely contaminated by protons during times of high proton fluxes. Applying the results of a Monte Carlo Simulation performed with the CERN Library program GEANT 3 (5) Bialk et al. (4) were able to determine the true electron response functions of the instrument and to correct for the proton contamination. A similar analysis was carried out to obtain the electron response function of the ULEWAT detector.

Electron spectra were constructed taking the maximum differential flux $J(E)$ in each energy interval (c.f., (6)). As a result of applying the corrected response functions ULEWAT and *Helios* spectra were shifted to higher energies by a factor of ~ 2 and sometimes exhibited considerably different spectral shapes compared to spectra obtained with the nominal response functions published by Moses et al. (1) and elsewhere. The left panel of Figure 1 shows the electron spectrum of the 3 April 1979 flare, observed simultaneously on *ISEE-3* and *Helios 1* and *2*, as function of both energy (open symbols) and momentum (filled symbols). In the latter representation power laws in momentum of the phase space density $N(p) \propto J(E)$ which are predicted by current acceleration models (c.f., (7)) appear as straight lines and can be easily identified by optical inspection. To emphasize details and display possible

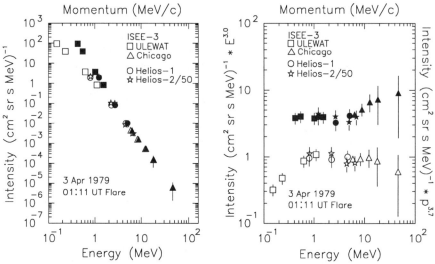

FIG. 1. Electron spectra of the 3 April 1979 solar event observed simultaneously on *ISEE-3* and *Helios* 1 and 2 plotted both as a function of energy (open symbols) and momentum (filled symbols). For details see text.

deviations between the observations from different s/c, in the right panel of the figure the spectra are shown multiplied by power laws in energy and momentum, respectively. We note that the *Helios* spectrum when multiplied with an arbitrary factor to account for the different radial and azimuthal position of the 3 s/c (c.f., (8)) is in good agreement with what one might extrapolate from the spectra obtained with the instruments on *ISEE-3*. It is found that in general the shapes of the *Helios* spectra, when adjusted in absolute intensity, for 5 events observed on all 3 s/c and a number of further events observed on *Helios* 1 and *ISEE-3* fit well into the ULEWAT and University of Chicago parts of the spectra.

Despite the changes in the low energy *ISEE-3* spectra the basic results of the survey of Moses et al. (1), which will be briefly summarized in the following, remain valid. Electron spectra can be divided into two distinct classes. In the first class power law fits in momentum (or rigidity) applied separately to the parts of the spectra below and above ∼ 3 MeV exhibit similar spectral indices, suggesting that the spectra can be modeled by single power laws over the entire range (see Fig.2 left panel). Spectra of the second class are steeper at low momenta and flatter at high momenta and hence cannot be fitted by a single power law in momentum (c.f., Fig. 2 right panel). Spectra of the first class are associated with gradual flares (LDEs) while spectra of the second class are associated with impulsive flares (SDEs), in agreement with the well known classification of flare properties with respect to the duration of their soft X-ray emission (c.f., (9)).

However, the improvements in the reconstruction of the low energy part of the spectra, and the inclusion of the *Helios* data points which fill the gap in the 1 - 8 MeV range between the two *ISEE-3* instruments have revealed

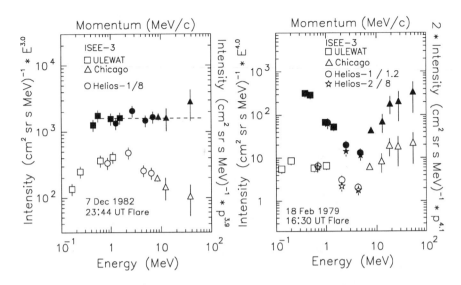

FIG. 2. Electron spectra of the 7 December 1982 (LDE, left panel) and 18 February 1979 (SDE, right panel) flares. Symbols have the same meaning as in Fig. 1.

features of the spectra not evident from the Moses et al. (1) survey. There are a number of properties all spectra analyzed so far seem to have in common: below 0.2 MeV the spectra show a flattening confirming earlier findings of Lin et al. (6), between 0.2 and 5 MeV the spectra are well characterized by power laws in momentum, above 5 MeV power laws give poorer fits. The latter effect is more pronounced for SDEs. Many LDE spectra can be modeled by single power laws in momentum above 0.2 MeV, some show a slight hardening a high energies. In the $J(E)$ vs energy representation the spectra are usually curved in one way or another and a classification based on power law fits in energy to the low and high energy parts of the spectra is not as striking. LDE spectra always exhibit a steepening in the high energy range. SDE spectra can be both flatter and steeper at high energies where the break in the spectrum is at around 3 MeV. They can in general not be modeled by a single power law. Note how the *Helios* data for the event of 21 June 1980, shown in Figure 3, follow the spectral break at ∼ 5 MeV extrapolated from the *ISEE-3* data points in the flux vs. momentum representation.

III. GAMMA-RAY SPECTRA

Solar gamma-ray continuum emission between 0.3 and 1 MeV is very likely produced by electrons with energies above ∼ 1 MeV via electron bremsstrahlung. However, the question whether the electrons which interact with the solar atmosphere and the ones which escape to interplanetary space are accelerated at the same time and originate from the same source is still unresolved. The characteristics of the gamma-ray source is usually described in terms of thin-target or thick-target emission models. In the thin-target model, electrons escape from the gamma-ray emitting region in a time short compared to their collisional lifetime. Assuming that the escape does not depend on energy the spectra of injected and escaping electrons are the same. In the thick-target model the gamma-ray emission is produced by electrons which lose all of their energy through collisions in the interaction region. In this case the spectrum of the gamma-ray emitting electrons is determined by the balance between the injection of newly accelerated electrons and the loss of electrons through collisions. Thin and thick-target processes result in different gamma-ray spectra for a given injected electron spectrum. In general, if the electron spectrum is a power law, the photon spectrum is also found to be a power law. If the number density of electrons in the thin-target source is given by $N(E) = K_1 E^{-\nu}$ electrons cm^{-3} MeV^{-1} or if the spectrum of electrons injected into a thick-target source is assumed to be $F(E) = K_2 E^{-\delta}$ electrons cm^{-2} s^{-1} MeV^{-1} then the resulting spectrum of gamma rays with energy ε is a power law of the form $J(\varepsilon) = A\varepsilon^{-\gamma}$ photons cm^{-2} s^{-1} MeV^{-1} ($K_1, K_2, A = $ const.) where at relativistic energies $\gamma \approx \nu$ in the thin-target and $\gamma \approx \delta - 1$ in the thick-target model (cf., (10)). A more realistic relation between the gamma-ray spectral index in the $0.3 - 1$ MeV range and the electron spectral index ($E < 100$ MeV), in an isotropic thick-target model with a neutral ambient gas was presented by Ramaty et al. (11) and will be shown in Figure 5.

IV. DISCUSSION

Electron spectra from solar flares observed simultaneously on *ISEE-3* and *Helios* are in very good agreement inspite the fact the s/c were located at different radial distances from the Sun and the footpoints of their connecting field lines sometimes differed by more than 100 degrees. These results indicate that processes of escape from the flare site, lateral transport in or close to the solar corona and interplanetary propagation seem to have little influence on the shapes of the spectra. Spectra of flare electrons observed in the inner Heliosphere, even on a single s/c which is not well connected to the flare, should therefore be fairly representative of the spectra of electrons released from the flare.

The observation that interplanetary electron spectra from LDEs usually are continuous power laws in rigidity suggests that the electrons are accelerated in one single process, probably by stochastic acceleration in MHD turbulence in extended coronal loops or by coronal shock waves. Based on an acceleration model including Coulomb losses Steinacker et al. (12) showed that a single acceleration process taking place in compact flare loops with a high plasma density can produce spectra which become steeper towards lower rigidities, similar to the ones observed in SDEs. However, in this model the low rigidity part rather has a shape that is curved upwards towards lower energies than being a steeper power law. On the other hand, as flare models often assume that particle acceleration occurs in different phases or steps (c.f., (13)) electron spectra in SDEs may as well be a superposition of two components from

82

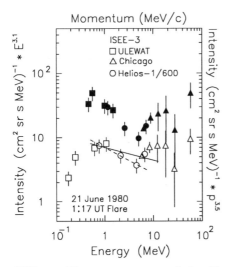

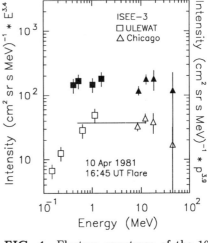

FIG. 3. Electron spectrum of the 21 June 1980 (SDE) event. The solid line represents a power law fit in energy to the *ISEE-3* data points in the 1 - 10 MeV range, the dashed line is a power law fit to the three lowest *Helios* points.

FIG. 4. Electron spectrum of the 10 April 1981 (LDE) event. The solid line represents a power law fit in energy to the *ISEE-3* data points in the 1 - 10 MeV range.

different sources or acceleration mechanisms. The rather sharp transition at ~ 5 MeV from one power law to another which becomes apparent from the combination of *ISEE-3* and *Helios* data in the momentum representation of Figures 2 (right panel) and 3 might in fact support the latter alternative.

In a previous comparison (11) of *SMM* 0.3 to 1 MeV continuum observations (3) and electron spectra it was assumed that electrons with the same spectra than those in interplanetary space impinge on a thick-target, where the electron spectra were reconstructed from the momentum power law fit spectral exponents published by Moses et al. (1). It was found that bremsstrahlung spectra calculated from the full (0.1 to 50 MeV) interplanetary electron spectra were systematically harder than those observed and concluded that electrons observed in interplanetary space and the ones producing the gamma rays cannot come from the same population without the former being affected by re-acceleration or energy-dependent escape. From the re-analyzed and combined *ISEE-3* ULEWAT and *Helios* spectra presented in this study it becomes evident that electron spectra below ~ 5 MeV reconstructed from the original (1) momentum power law exponents can bear large errors. However, these considerations do not alter the fact that the observed and calculated gamma-ray spectral indices are poorly correlated, if the assumptions of thick-target emission and a full contribution of the electron spectrum above 10 MeV are kept.

Here we will discuss the possibility that in SDEs only the steeper component which dominates the spectra in the 0.2 − 5 MeV range (but may ex-

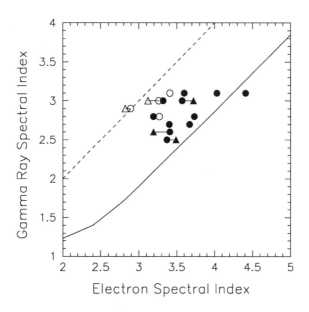

FIG. 5. Photon vs. electron spectral index for SDEs (filled symbols) and LDEs (open symbols). Dots denote *ISEE-3*, triangles *Helios* observations. Data points of events observed simultaneously on both s/c are connected. Also shown are the predictions of thin-target (dashed line) and thick-target (solid line, from (11)) emission models.

tend to higher energies) is responsible for gamma-ray emission in the $0.3 - 1$ MeV range. Based on the similarity of electron spectra observed on *ISEE-3* and *Helios* we conclude that the spectra are not affected much by transport processes in the solar corona, injection into and propagation in the interplanetary medium and hence assume that the spectra of electrons which are injected into the gamma-ray source are similar to those of the escaping electrons. According to the above hypothesis an attempt was made to reconstruct the steep SDE component in the $1 - 10$ MeV range: on *ISEE-3* from the two highest ULEWAT and the lowest Chicago data point, if a reliable value at ~ 8 MeV was available, otherwise the spectrum was extrapolated from the lowest three ULEWAT data points, and on *Helios* from the lowest three data points. Power law fits to energy spectra in the above range (c.f., Figs. 3 and 4) of this component were obtained, for LDEs fits to the full spectra were made. the corresponding spectral indices ($\nu \approx \delta$ in this energy range) are shown in Figure 5 versus the gamma-ray spectral index γ separately for SDEs (filled circles) and LDEs (open circles). All events lie approximately in a region confined by the predictions of thin-target and thick-target models. Data points for impulsive and long duration flares are clearly separated, where the LDEs tend to be closer to the prediction of the thin-target model, and SDEs to the prediction of the thick-target model. Within the observational uncertainty, electron spectral indices obtained for events which were observed simultaneously on *ISEE-3* and *Helios* are in good agreement.

V. CONCLUSIONS

Although other possibilities cannot be ruled out, our findings are consistent with the following scenario: in LDEs electron acceleration above 200 keV occurs in or is dominated by a single stage mechanism which produces a power law in momentum. Such a spectral shape is predicted by certain classes of stochastic and shock acceleration models currently under discussion (14) (7). In SDEs there are two electron components, one with a steep spectrum which is also consistent with a power law in momentum dominating below ~ 5 MeV and one with a flatter spectrum dominating above that energy. The electron populations of the single component in LDEs and of the steep component in SDEs are sources of electrons escaping into interplanetary space and at the same time streaming into the gamma-ray emitting region without undergoing substantial spectral changes. The high energy component in SDEs seems to be not directly related to the emission of gamma rays and may consist of electrons accelerated at a different location at the flare site or by a different mechanism. The emission mechanism is neither strictly thin-target nor thick-target. Recent Yohkoh observations (15) which have revealed the existence of a hard X-ray source above a soft X-ray loop (at about 20 000 km) where thick-target emission cannot be expected may support this interpretation.

ACKNOWLEDGEMENTS

I wish to thank D. Hovestadt and B. Klecker for providing *ISEE-3* particle data, and P. Evenson, M. Bialk and B. Heber for helpful discussions.

REFERENCES

1. D. Moses, W. Dröge, P. Meyer, and P. Evenson, Ap. J. **346**, 523 (1989).
2. H. Kunow et al., J. Geophys. **42**, 615 (1977).
3. W. T. Vestrand et al., Ap. J. **322**, 1010 (1987).
4. M. Bialk, W. Dröge, and B. Heber, Proc. 22nd ICRC (Dublin) **2**, 699 (1991).
5. R. Brun et al., GEANT3, CERN Data Handling Division, DD/EE/84-1 (1987).
6. R. P. Lin, R. A. Mewaldt, and M. A. I. van Hollebeke, Ap. J. **253**, 949 (1982).
7. M. A. Forman, R. Ramaty, and E. G. Zweibel, in: The Physics of the Sun, Vol. 2, ed. P. A. Sturrock et al., (Dordrecht: Reidel) 249 (1986).
8. M.-B. Kallenrode, J. Geophys. Res. **98**, 5573 (1993).
9. H. V. Cane, R. E. McGuire, and T. T. von Rosenvinge, Ap. J. **301**, 448 (1986).
10. R. J. Murphy, C. D. Dermer, and R. Ramaty, Ap. J. (*Suppl.*) **63**, 721 (1987).
11. R. Ramaty, R., N. Mandzhavidze, B. Kozlovsky, B., and J. Skibo, Adv. Space Res. **13**, 275 (1993).
12. J. Steinacker, W. Dröge, and R. Schlickeiser, Sol. Phys. **115**, 313 (1988).
13. T. Bai and P. A. Sturrock, ARA&A **27**, 421 (1989).
14. D. C. Ellison and R. Ramaty, Ap. J. **298**, 400 (1985).
15. S. Masuda et al., Nature **371**, 495 (1994).

Measurements of the Ionic Charge States of Solar Energetic Particles at 15–70 MeV/nucleon Using the Geomagnetic Field

R. A. Leske*, J. R. Cummings*, R. A. Mewaldt*, E. C. Stone*, and T. T. von Rosenvinge[†]

*California Institute of Technology, Pasadena, CA 91125
[†]NASA/Goddard Space Flight Center, Greenbelt, MD 20771

The mean charge states of abundant heavy ions with $\sim 15 - 70$ MeV nucleon^{-1} in the two large solar energetic particle events of 1992 October 30 and November 2 have been determined using measurements of the invariant latitude of the cosmic ray geomagnetic cutoffs as a function of time, particle energy, and element from the Mass Spectrometer Telescope on the polar–orbiting *SAMPEX* satellite. The deduced charge state values are in good agreement with the mean values measured directly in previous solar energetic particle events at much lower energies of ~ 1 MeV nucleon^{-1}, with inferred equilibrium source temperatures of typically 2×10^6 K. This result provides additional evidence that solar energetic particles in gradual–type events consist of accelerated coronal material.

INTRODUCTION

Recent work indicates that solar energetic particle (SEP) acceleration in gradual–type SEP events occurs primarily at a shock driven by a fast coronal mass ejection (CME), rather than at the flare site itself (1). Direct measurements of SEP ionic charge states at ~ 1 MeV nucleon^{-1} in gradual events (2) find values consistent with temperatures of $1 - 2 \times 10^6$ K (3), much cooler than the $\sim 10^7$ K flare plasma (4) but typical of coronal material, supporting the CME acceleration scenario. At higher energies, if particles traverse a longer pathlength during the acceleration process, their mean charge state may be increased by electron stripping if the density of the ambient material is sufficiently high. In principle, measurements of charge states over broader energy intervals may serve as a sensitive probe of the energy–dependent pathlengths associated with SEP acceleration and subsequent transport to Earth and could help to constrain models of these processes.

Previous studies at low energies (2) employed electrostatic deflection to directly measure ionic charge states, however this becomes impractical above energies of a few MeV nucleon^{-1}. The effects of deflection in the Earth's magnetic field are easily detectable at much higher energies, since partially stripped ions have a higher rigidity and access to lower magnetic latitudes than fully stripped ions of the same element and energy. Because of this rigidity filter effect, measurements from the Mass Spectrometer Telescope (MAST) of the spatial distribution of SEPs along the orbit of the Solar, Anomalous, and Magnetospheric Particle Explorer (*SAMPEX*) allowed us to determine SEP charge states, as reported in more detail in (5). Here we summarize the earlier report (5) and provide some additional details on our measurements of relative charge states for abundant elements at energies of $\sim 15-70$ MeV nucleon^{-1}, about a factor of 30 higher than is possible with direct measurements, for the large SEP events of 1992 October 30 and November 2. We compare our results to theoretical calculations and other measurements, and discuss the implications for SEP acceleration and transport.

DATA ANALYSIS

MAST is a silicon solid state detector telescope (6) designed to measure the elemental and isotopic composition of energetic particles using the conventional dE/dx vs residual energy technique. Since incident ions are quickly stripped in traversing the material of the instrument, MAST is not directly sensitive to a particle's ionic charge, Q. However, from measurements of the nuclear charge, Z, mass, M, and total kinetic energy, E, provided by MAST the mean Q may be found if the mean rigidity can also be determined. At any point in the 82° inclination orbit of *SAMPEX*, only those particles with rigidities greater than the local geomagnetic cutoff rigidity, R_C, are detected. Although rigidity can not be measured for each individual event, the mean rigidity of a collection of events can be found if the invariant latitude (7) of the cutoff, Λ_C, can be determined, and if the relation between Λ_C and R_C is known. Since other measurements at high latitudes and low energies (e.g., (8,9)) generally find cutoffs below simple model expectations (10), and geomagnetic disturbances which accompany solar flares often cause the cutoff location to vary from that under quiescent conditions in ways difficult to accurately model at low rigidities (11), we choose to empirically derive the necessary relation between Λ_C and R_C.

The SEP events discussed here are the largest observed by MAST to date, and were each associated with a major western–limb solar flare, magnetically well connected to the Earth. As reported in the *Daily Summaries of Solar Geophysical Activity*, a class X1.7/2B flare at 22°S, 61°W reached its peak at 1816 UT on 1992 October 30, followed less than 3 days later by a class X9.0 event from the same active region, then several degrees beyond the west limb at 0308 UT on November 2. Both of these gradual–type events were also associated with large CMEs.

The MAST counting rates shown in Figure 1 indicate that during these SEP events significantly elevated count rates were observed for more than a week. Even with fluxes this high, however, the determination of Λ_C for rarer heavy ions with sufficient statistics requires that the detected particles be

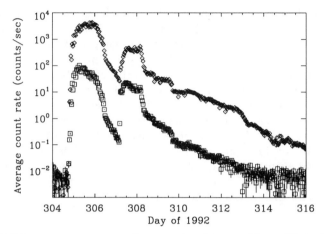

FIG. 1. MAST count rates, averaged over each polar pass ($\Lambda > 66°$) during the two SEP events, for 8-15 MeV nucleon^{-1} He (*diamonds*) and $\sim 12 - 300$ MeV nucleon^{-1} heavy ions ($Z \geq 3$, *squares*).

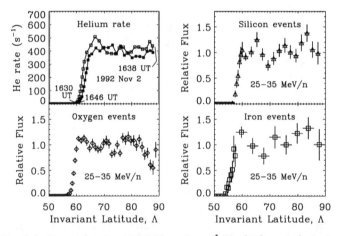

FIG. 2. *Top left:* Rate of $\sim 8 - 15$ MeV nucleon^{-1} He (Z2) vs Λ (calculated using the IGRF 1990 field model), used in measuring the time variation of the cutoff, for a single passage into (*open squares*) and out of (*filled squares*) the north polar cap. Distributions of O (*bottom left*), Si (*top right*), and Fe events (*bottom right*) summed over the second SEP event at 25–35 MeV nucleon^{-1} vs time–corrected Λ, used in determining Q. The statistical uncertainties shown are larger at the high–latitude plateau due to increased proton–induced dead time losses.

summed over the entire duration of the SEP event. Before this is done, any time variation of the cutoff must first be corrected. Using the MAST Z2 rate (which responds to He nuclei at $\sim 8-15$ MeV nucleon^{-1}), we can measure Λ_C for He for each of the four cutoff crossings per orbit during the SEP events to within $\sim 0.2°$. A typical profile of this rate as a function of invariant latitude, Λ, is shown in Figure 2 for a single polar crossing. The resulting measured time variation in the cutoff location, which is found to be strongly correlated with geomagnetic activity, is used as a template to correct the values of Λ for the heavy ion pulse height events as described in (5).

For each of the abundant elements, Λ_C is determined in small energy intervals, typically 5 MeV nucleon^{-1} for heavy ions, and 1 MeV nucleon^{-1} for He. The number of events in each time–dependence–corrected Λ bin for each energy interval is corrected for instrumental dead time, chance coincidences, and exposure time at that Λ to produce a distribution of relative flux vs Λ as illustrated for several elements in Figure 2, with a clear edge and plateau. The cutoff is taken to be the latitude where the flux falls to half its mean value above 70° and is determined from a linear fit to the cutoff edge, with an overall uncertainty based on both the uncertainties in this fit and in the flux level at the plateau.

RESULTS AND DISCUSSION

Figure 3 illustrates a small sample of these cutoff measurements, showing Λ_C for C, O, and Si as a function of energy for the second SEP event, which had better heavy ion statistics (after allowing for instrumental dead time) than the first. The cutoff latitude is seen to clearly decrease with increasing energy, qualitatively as expected. Notice that up through at least 50 MeV nucleon^{-1} the O cutoffs are consistently and significantly below the C cutoffs at the same kinetic energy per nucleon. Therefore, the O rigidity must be higher than that of C at the same velocity, and thus O (which was found to be nearly fully ionized with an average measured charge of +7.0 at ~ 1 MeV nucleon^{-1} (2)) can not be fully stripped at these energies, regardless of the charge state of C and independent of any cutoff–rigidity relation. Similar arguments show that all heavier species examined here are on average only partially stripped, as suggested by the Si data in Figure 3.

To obtain more quantitative results, we derive a cutoff–rigidity relation based on three assumptions which are justified and discussed in more detail in (5). First, we assume that SEP He is fully stripped at $8-15$ MeV nucleon^{-1}, as may be inferred from recent measurements (12). Secondly, to directly establish the cutoff–rigidity relation at the higher energies of the heavier ions without a large extrapolation in energy from He, we assume that the mean ionic charge of C is somewhere between +6 (fully stripped) and +5.7, as measured at low energies (2). Finally, guided by the Störmer model (13), we assume that R_C is linearly related to $\cos^4(\Lambda_C)$. Our data (Figure 4) support a straight line fit, but with a large offset from the origin unaccounted for by the Störmer model (13). Although it was necessary to assume values for Q in

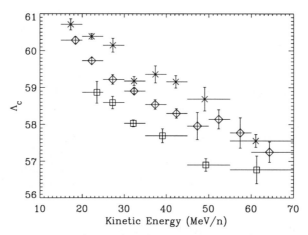

FIG. 3. Measured Λ_C for C (*crosses*), O (*diamonds*), and Si (*squares*) as a function of energy in the 1992 November 2 SEP event.

order to plot the data vs rigidity, both the existence and the magnitude of the offset are independent of the assumed Q for linear fits to any single element. The offset corresponds to free access of particles at $\Lambda \geq 65°$, which is similar to the observed boundary of the electron polar cap (14) and agrees well with other lower rigidity cutoff measurements in these same SEP events (15). The derived cutoff–rigidity relations based on these assumptions are:

$$\cos^4 \Lambda_C = 0.0250 + 8.57 \times 10^{-5} R_C \text{ for the first SEP event, if } Q(C) = +5.7$$

$$\cos^4 \Lambda_C = 0.0233 + 9.32 \times 10^{-5} R_C \text{ for the first SEP event, if } Q(C) = +6$$

$$\cos^4 \Lambda_C = 0.0298 + 7.06 \times 10^{-5} R_C \text{ for the second SEP event, if } Q(C) = +5.7$$

$$\cos^4 \Lambda_C = 0.0278 + 7.81 \times 10^{-5} R_C \text{ for the second SEP event, if } Q(C) = +6$$

where R_C is in units of MV. With the two limiting C charge states used to bound the allowed variations in the cutoff–rigidity relation, the set of measured cutoffs and energies can now be used to obtain the mean charge states for other elements.

The resulting mean values for Q are listed in Table 1, with uncertainties typically dominated by the systematic uncertainty in the C ionic charge state. These Q values are consistent with those measured at low energies (2), and are clearly less than fully stripped for all elements examined heavier than C, as illustrated in Figure 5. Note that the close agreement between our results and those of Luhn et al. (2) validates the use of the low energy Q/M values to organize SEP elemental abundances at energies up to 50 MeV nucleon^{-1} (e.g., (16)). The values are similar to those expected from equilibrium calculations of collisional ionization for a plasma at 2×10^6 K (17,18), as shown in Figures 5 and 6. The inferred source temperatures for Ne and perhaps

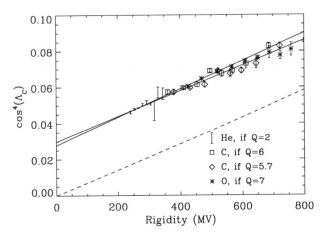

FIG. 4. Plot of $\cos^4(\Lambda_C)$ vs rigidity for He (*vertical bars*), C (*squares, diamonds*), and O (*asterisks*) for the 1992 November 2 SEP event, assuming the charge states indicated. Limiting cases for the cutoff–rigidity relation adopted here are the fits to the He and C data (*solid lines*), one for each plotted C charge state, and are compared with the Störmer model (13) western cutoff (*dashed curve*).

TABLE 1. SEP ionic charge states and corresponding source temperatures.

	Z	E (MeV/n)	SEP Event #1 (1992 Oct 30)		SEP Event #2 (1992 Nov 2)	
			$<Q>$	T (MK)[a]	$<Q>$	T (MK)[a]
He	2	$8-16$	2.00^b		2.00^b	
C	6	$15-65$	5.70–6.00^b		5.70–6.00^b	
N	7	$16-65$	6.30 ± 0.30	1.95 ± 0.30	6.49 ± 0.20	2.14 ± 0.26
O	8	$17-70$	6.93 ± 0.20	2.37 ± 0.22	6.99 ± 0.22	2.45 ± 0.25
Ne	10	$19-65$	8.68 ± 0.30	3.91 ± 0.64	8.47 ± 0.28	3.39 ± 0.66
Na	11	$19-48$	8.50 ± 0.39	1.09 ± 0.20	9.36 ± 0.37	3.67 ± 1.65
Mg	12	$20-65$	10.35 ± 0.40	4.63 ± 2.32	10.29 ± 0.35	4.38 ± 2.18
Al	13	$21-48$	11.63 ± 0.73	7.53 ± 4.96	10.66 ± 0.68	4.17 ± 2.69
Si	14	$22-65$	10.57 ± 0.39	1.76 ± 0.11	10.51 ± 0.40	1.75 ± 0.12
S	16	$24-50$	10.82 ± 0.81	2.02 ± 0.26	10.84 ± 0.44	2.04 ± 0.14
Ar	18	$25-35$			10.08 ± 0.91	1.67 ± 0.39
Ca	20	$26-50$			11.46 ± 0.49	2.05 ± 0.27
Fe	26	$28-65$	15.59 ± 0.81	3.90 ± 1.49	14.69 ± 0.86	2.59 ± 0.53
Ni	28	$31-45$			12.62 ± 1.30	2.04 ± 0.22

[a] Based on equilibrium calculations of collisional ionization (17,18)
[b] Used in normalization

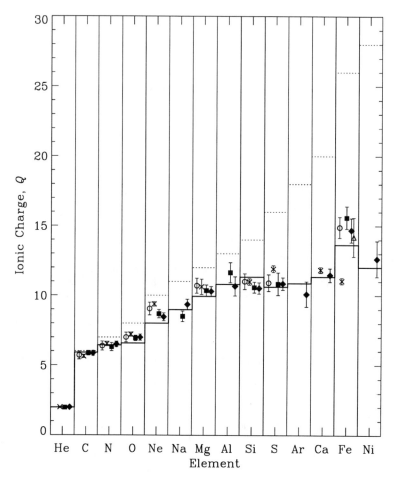

FIG. 5. Mean measured values of Q for the indicated elements, from the 1992 October 30 SEP event (*squares*) and 1992 November 2 SEP event (*diamonds*) at $\sim 15 - 70$ MeV nucleon^{-1}, from the Luhn et al. (2) 12–flare average at $\sim 0.3 - 3$ MeV nucleon^{-1} (*circles*) with 5% systematic uncertainty, from Mason et al. (15) at $\sim 0.5 - 5$ MeV nucleon^{-1} (*crosses*), and from Tylka et al. (21) at $\sim 200 - 600$ MeV nucleon^{-1} (*triangle*). The heavy solid line shows the value of Q expected for a 2×10^6 K source plasma (17,18), while the dashed line marks the fully stripped ($Q = Z$) value of Q for each element.

Mg are higher than for the other elements, confirming earlier findings (2), and suggesting that an additional mechanism such as photoionization due to flare X–rays (3) may play an important role, that equilibrium temperature calculations (particularly for 2–electron ions) may need refining, or that the assumption of equilibrium is not valid. As seen in Figure 6, some elements not measured by Luhn et al. (2), such as Na and Al, have deduced temperatures with large uncertainties since 2–electron ions exhibit a broad, relatively flat plateau where little change is expected in Q with temperature (18), while others such as Ar, Ca, and Ni are consistent with $\sim 2 \times 10^6$ K. Note also that calculations indicate that at 2×10^6 K, the charge state for Ni should actually be less than that for Fe, consistent with our measurements. This temperature indicates that the source material for large SEP events is more likely to be ambient coronal material, rather than the hotter flare plasma.

Similar measurements during these same SEP events from other instruments on *SAMPEX* at the lower energies of $0.5 - 5$ MeV nucleon^{-1} (15) and at $10 - 100$ MeV nucleon^{-1} (19) are generally in close agreement with both the MAST and Luhn et al. (2) charge state values, with the notable exception of Fe, for which Mason et al. (15) report $Q = 11.04 \pm 0.22$. Since the Fe cutoffs for MAST are at values of $\cos^4(\Lambda_C)$ about a factor of 2 beyond those of He and C used to define the cutoff–rigidity relation, the possibility exists that the slight extrapolation of the relation to Fe is inaccurate. Recent studies (20) find that R_C is 5% below the Störmer western cutoff at $\cos^4(\Lambda_C) \sim 0.25$, corresponding to rigidities well beyond those of our Fe data. Using a line through only our He data and this point reduces our Fe charge state in the first event to ~ 13.7, leaving the value in the second event unchanged as this new line is essentially identical to the cutoff–rigidity relation we derive for the second SEP event. The MAST data do not appear to support a Q as low as 11 at > 30 MeV nucleon^{-1}, and Fe measurements in other SEP events at $\sim 200 - 600$ MeV nucleon^{-1} using passive nuclear track detectors (21) yield a value of Q of 14.2 ± 1.4, similar to that obtained from MAST.

It may be possible to account for this discrepancy if there is a mixture of charge states in the source plasma, and if those with larger Q/M ratios are preferentially accelerated to higher energies than those with lower Q/M. This would appear to be consistent with the fact that these are Fe–poor events (relative to the coronal Fe/O ratio) at the MAST energies (since Q/M is ~ 0.27 for Fe and ~ 0.44 for O), yet apparently Fe–rich at lower energies (15). We find an Fe/O ratio of 0.031 in the first event and 0.071 in the second, compared with the coronal value of 0.172 (22), while the published low energy spectra (15) yield a value of ~ 0.4. While our Fe measurements are consistent with no variation in Q with energy, the uncertainties are large enough to allow a Q of 11 at 1 MeV nucleon^{-1} if linearly extrapolated, and correlative studies are underway with the other *SAMPEX* investigators to better determine any possible energy dependence and its significance.

Also in progress are studies of the path length required to produce little or no additional stripping for most elements during acceleration to these higher energies. Preliminary calculations suggest that it may be possible to place

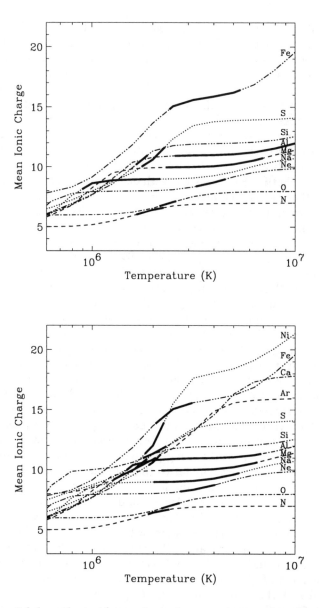

FIG. 6. Expected dependence of mean ionic charge on temperature, based on calculations of collisional ionization (17,18), for the indicated elements, with different line styles serving merely to help separate the various elements. Heavy line segments highlight those portions of the curves consistent with MAST measurements of Q in the 1992 October 30 (*top*) and November 2 (*bottom*) SEP events.

an upper limit on the amount of material SEPs encounter of as much as a factor of 50 lower than the 30 mg cm^{-2} upper limit obtained from studies of the nuclear fragmentation isotopes of H and He (23), which again points to acceleration in a low density region.

This work was supported by NASA under contract NAS5-30704 and grant NAGW-1919.

REFERENCES

1. Kahler, S. W., *Ann. Rev. Astron. and Astrophys.* **30**, 113 (1992).
2. Luhn, A., et al., *Proc. 19th Internat. Cosmic Ray Conf. (LaJolla)* **4**, 241 (1985).
3. Mullan, D. J. and Waldron, W. L., *Astrophys. J.* **308**, L21 (1986).
4. Doschek, G. A., *Phil. Trans. R. Soc. Lond. A* **336**, 451 (1991).
5. Leske, R. A., Cummings, J. R., Mewaldt, R. A., Stone, E. C., and von Rosenvinge, T. T., *Astrophys. J.* **452**, L149 (1995).
6. Cook, W. R., et al., *IEEE Trans. Geoscience Remote Sensing* **31**, 557 (1993).
7. Roederer, J. G., *Dynamics of Geomagnetically Trapped Radiation*, New York: Springer (1970).
8. Fanselow, J. L. and Stone, E. C., *J. Geophys. Res.* **77**, 3999 (1972).
9. Seo, E. S., Ormes, J. F., Streitmatter, R. E., Stochaj, S. J., Jones, W. V., Stephens, S. A., and Bowen, T., *Astrophys. J.* **378**, 763 (1991).
10. Shea, M. A. and Smart, D. F., *Proc. 18th Internat. Cosmic Ray Conf. (Bangalore)* **3**, 415 (1983).
11. Boberg, P. R., Tylka, A. J., Adams, J. H., Jr., Flückiger, E. O., and Kobel, E., *Geophys. Res. Letters* **22**, 1133 (1995).
12. Gloeckler, G., et al., *J. Geophys. Res.* **99**, 17637 (1994).
13. Smart, D. F. and Shea, M. A., *Proc. 23th Internat. Cosmic Ray Conf. (Calgary)* **3**, 781 (1993).
14. Evans, L. C. and Stone, E. C., *J. Geophys. Res.* **77**, 5580 (1972).
15. Mason, G. M., Mazur, J. E., Looper, M. D., and Mewaldt, R. A., *Astrophys. J.* **452**, 901 (1995).
16. Breneman, H. H. and Stone, E. C., *Astrophys. J.* **299**, L57 (1985).
17. Arnaud, M. and Raymond, J., *Astrophys. J.* **398**, 394 (1992).
18. Arnaud, M. and Rothenflug, R., *Astron. and Astrophys. Suppl.* **60**, 425 (1985).
19. Oetliker, M., Klecker, B., Hovestadt, D., Scholer, M., Blake, J. B., Looper, M., and Mewaldt, R. A., *Proc. 24th Internat. Cosmic Ray Conf. (Rome)* **4**, 470 (1995).
20. Selesnick, R. S., Cummings, A. C., Cummings, J. R., Mewaldt, R. A., Stone, E. C., and von Rosenvinge, T. T., *J. Geophys. Res.* **100**, 9503 (1995).
21. Tylka, A. J., Boberg, P. R., Adams, J. H., Jr., Beahm, L. P., Dietrich, W. F., and Kleis, T., *Astrophys. J.* **444**, L109 (1995).
22. Garrard, T. L. and Stone, E. C., *Proc. 23rd Internat. Cosmic Ray Conf. (Calgary)* **3**, 384 (1993).
23. Mewaldt, R. A. and Stone, E. C., *Proc. 18th Internat. Cosmic Ray Conf. (Bangalore)* **4**, 52 (1983).

HIIS Results on the Mean Ionic Charge State of SEP Fe Above 200 MeV per nucleon

Allan J. Tylka*, Paul R. Boberg**, James H. Adams, Jr.*,
Lorraine P. Beahm*, William F. Dietrich†, and Thomas Kleis††

*E. O. Hulburt Center for Space Research
Naval Research Laboratory, Washington, DC 20375-5352
**Universities Space Research Association, Washington, DC 20024
†Laboratory for Astrophysics and Space Research, Enrico Fermi Institute
University of Chicago, Chicago, IL 60637
††Institut für Reine und Angewandte Kernphysik,
Christian-Albrechts Universität zu Kiel, D-24118 Kiel, Germany

We have analyzed the geomagnetic transmission of solar energetic
Fe ions at $\sim 200 - 600$ MeV/nuc during the great solar energetic particle (SEP) events of 1989 September-October. By comparing fluences
from the Chicago charged-particle telescope on *IMP-8* in interplanetary space and from NRL's Heavy Ions in Space (HIIS) experiment
aboard *LDEF* in low-Earth orbit, we obtain a mean ionic charge $<Q>$
$= 14.2 \pm 1.4$. This result is significantly lower than $<Q>$ observed at
~ 1 MeV/nuc in impulsive events, and suggests that neither acceleration at the flare site nor flare-heated plasma significantly contributes to
the high-energy Fe ions we observe. But it agrees well with the $<Q>$
observed in gradual SEP events at lower energies, demonstrating that
acceleration by CME-driven shocks is the primary SEP production
mechanism in gradual events even at these very high energies.

INTRODUCTION

As acknowledged by many speakers at this Workshop, the classification of
solar energetic particle (SEP) events into two categories ("impulsive", in which
the particle acceleration occurs in a localized region, near the electromagnetic
flare site; and "gradual", in which fast coronal mass ejections (CMEs) drive
shocks, which accelerate particles from the ambient plasma as they propagate
through the high corona and interplanetary medium) is generally accepted,
at least in broad outline and for energies of ~ 1 MeV/nuc (1). But what is
the SEP acceleration mechanism at higher energies? Gamma-ray observations
clearly indicate that flares can accelerate protons to >1 GeV (2,3). Can CME-
driven shocks also accelerate ions to such high energies? There is much less
consensus in the community on the answers to these questions, with some

researchers apparently believing that CME-driven shocks are incapable of accelerating SEPs to more than a few, or perhaps a few tens, of MeVs (4).

Kahler (5) addressed these questions by correlating the expansion profiles of fast CMEs with solar energetic proton timelines. He showed that the acceleration of protons at 470 MeV - 21 GeV peaked when the CME reached altitudes >5 solar radii, thereby indicating acceleration by shocks in the high corona and interplanetary medium, rather than acceleration at the flare site.

Historically, a particularly powerful and cogent indicator of the two distinct acceleration mechanisms has been the ionic charge state of Fe. Averaging over 12 gradual events, Luhn et al. (6) found $<Q> = 14.1 \pm 0.2$ for Fe ions at ~ 1 MeV/nuc, corresponding to an inferred plasma temperature of ~ 2 MK, typical of coronal material (7). For impulsive events, they reported a significantly higher value, $<Q> = 20.5 \pm 1.2$, suggesting additional heating of the source plasma and/or further stripping after acceleration.

In this paper, we review our recent measurement of the mean ionic charge state of SEP Fe at $\sim 200 - 600$ MeV/nuc. We analyzed the ability of these ions to penetrate the geomagnetic field by comparing SEP Fe measurements from the U. Chicago's charged-particle telescope aboard *IMP-8* (in interplanetary space) with simultaneous observations from NRL's HIIS experiment (which was deployed in a low-inclination, low-altitude orbit, to which Fe ions of these energies can penetrate from interplanetary space only if they are partially-ionized). Our result averages over several SEP events and is dominated by those of 1989 September-October (8–10), which produced ground-level neutron-monitor enhancements (11) and some of the largest SEP fluences ever observed (12). A detailed report on our result has already been published in the refereed literature (13).

We have found $<Q> = 14.2 \pm 1.4$, consistent with previous reports in gradual events and indicating that neither acceleration at the flare site nor ions from flare-heated plasma significantly contribute to the high energy Fe we observe. Our results, combined with earlier measurements of Luhn et al. (6), recent SAMPEX $<Q>$ determinations (LICA (14) at ~ 1 MeV/nuc; HILT (15) at ~ 10 MeV/nuc; and MAST (16) at ~ 40 MeV/nuc), and the injection profile studies of Kahler (5), demonstrate that shock acceleration of the ambient, unheated, coronal or solar wind plasma is the primary SEP production mechanism at all energies in gradual events.

HIIS OBSERVATIONS

NASA's *Long Duration Exposure Facility* (*LDEF*; (17)) was deployed in a 28.5°, 476 km circular orbit on 1984 April 7 and retrieved at 332 km on 1990 January 12. *LDEF* was 3-axis stabilized, with NRL's Heavy Ions in Space (HIIS) experiment ((13) and references therein) mounted on the space-facing end. HIIS had an unobstructed view and efficient particle detection down to zenith angles of $\sim 70°$. HIIS used large stacks of plastic track detectors with a total vertical depth of ~12 g/cm². Fig. 1 shows the HIIS mission-averaged Fe flux measurements, fully corrected for energy-loss and fragmentation in the detectors, as derived from one-eighth of the HIIS detector area, with a geometry factor of ~2500 cm²-sr. Our fluxes are in good agreement with those from other *LDEF* experiments (18–20).

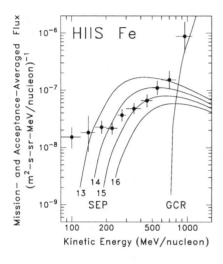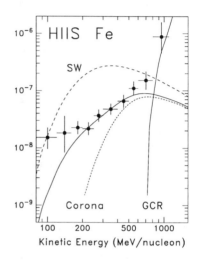

FIG. 1a. *(left)* HIIS mission- and acceptance-averaged Fe fluxes. The GCR curve is an absolute prediction of the expected flux of fully-ionized Galactic cosmic-ray Fe, propagated to the *LDEF* orbit. The SEP curves correspond to different assumed values of the mean ionic charge ($<Q>$).

FIG. 1b. *(right)* HIIS Fe fluxes and GCR curve, as in Fig. 1a. Long dashes: expected flux at HIIS if SEPs had a typical solar wind ("SW") charge state distribution, with $<Q>=11.8$ (30). Short dashes: expected flux if SEPs had a coronal charge state distribution (7) with $<Q>=14.5$. Solid curve: mixture of 10% solar wind and 90% coronal, yielding $<Q>=14.2$. See text.

An important point to understand about the HIIS detectors is that they provided no direct time-tagging of when the Fe ions were collected during *LDEF*'s six-year mission. Conclusions about the origin(s) of ions observed in HIIS must therefore be deduced from other characteristics of the data, such as fluence, energy spectrum, composition, and arrival directions.

First of all, Galactic cosmic rays (GCRs), which are fully-ionized because they have passed through an average of ~5 g/cm² of matter during their journey through the Galaxy, are geomagnetically excluded from *LDEF*'s orbit at energies below ~800 MeV/nuc. The curve labeled "GCR" in Fig. 1, which is an absolute calculation of the expected GCR Fe fluence in the *LDEF* orbit, shows this quantitatively. This calculation (described in more detail in (13)) takes into account the occasional periods of enhanced GCR access to the inner magnetosphere caused by the rare, large geomagnetic storms which occurred during the *LDEF* mission. For example, a fully-ionized GCR Fe ion at ~500 MeV/nuc requires a geomagnetic storm with $Dst \lesssim -300$ nT in order to reach the *LDEF* orbit. During the *LDEF* mission, there were only 11 hours during which $Dst < -300$ nT. The GCR fluence collected by HIIS during these 11 hours falls at least 3 orders of magnitude below the observed fluence.

Composition provides one of the most powerful indications of origin. Fig. 2 shows the observed sub-Fe to Fe ratio at various depths in the detector stack. The curves (calculated with a nuclear transport code, as described in more detail in (13)) show how a given incident source composition would be expected to evolve with increasing depth, due to fragmentation in the detector. Galactic cosmic-rays (which are incident on the detector with a sub-Fe to Fe ratio of ~0.5, because of fragmentation in the interstellar medium) and cosmic-ray albedo (in which further fragmentation has occurred in the atmosphere, before reaching *LDEF*) are clearly inconsistent with the data. But the observed composition is clearly consistent with an incident sub-Fe to Fe ratio of 0.032 ± 0.014, as observed on *IMP-8* at $\sim 100 - 400$ MeV/nuc during the 1989 September-October events, and typical of SEP events.

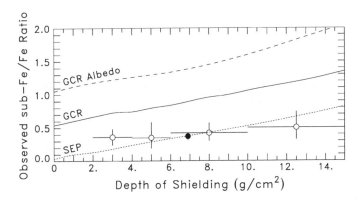

FIG. 2. Sub-Fe to Fe ratio (defined as $\Sigma(21 < Z < 25)/\Sigma(Z \geq 25)$) as observed at various depths in the detector stack. The open circles are for subsets of the data. The filled circle is for the complete dataset, with abscissa at the mean shielding depth. The curves show the expected composition from various sources, after taking into account fragmentation in the detector.

Finally, since the HIIS observations were made in the inner magnetosphere, one might wonder whether geomagnetically-trapped Fe ions could explain the HIIS fluence. However, the arrival directions of the HIIS Fe ions above ~120 MeV/nuc are inconsistent with a geomagnetically-trapped population. At these energies, the observed distribution of arrival directions is highly anisotropic and centered about west-southwest, the direction of lowest geomagnetic-cutoff in the *LDEF* orbit, just as expected for the geomagnetic transmission from a steeply-falling *interplanetary* spectrum. The arrival directions of geomagnetically-trapped ions in HIIS, on the other hand, is centered about the southeast, as predicted for *LDEF*'s path through the South Atlantic Anomaly and as confirmed by observations at lower energies (21).

Thus, partially-ionized SEPs are left as the only viable explanation of the HIIS Fe fluence at $\sim$200 - 600 MeV/nuc.

INTERPLANETARY SEP OBSERVATIONS

The University of Chicago charged-particle telescope on *IMP-8* (22) monitored SEP events throughout the *LDEF* mission. According to the Chicago data, the events of 1989 September-October accounted for >95% of the high-energy (> 100 MeV/nuc) SEP Fe fluence during the entire *LDEF* mission.

The events of 1989 September-October were one of the most extraordinary episodes of high-energy solar phenomena ever witnessed, all apparently associated with the same active region. These SEP events were also associated with Hα and X-class flares and types IV and II radio bursts. In terms of energetic (>10 MeV) protons at geosynchronous orbit, these events comprised the largest events of Solar Cycle 22 and produced by themselves more energetic protons than in all of Solar Cycle 21. Four of the outbursts caused ground level enhancements (23) in neutron monitors, and Kahler (5) measured speeds >1400 km/s for the two CMEs for which coronagraph images were available.

Fig. 3 shows GOES proton fluxes for these events in three energy channels. Six distinct abrupt increases can be indentified, the most dramatic of which occurred on 20 October due to the arrival of an interplanetary shock at Earth. (Anyone who doubts that interplanetary shocks can accelerate ions to very high energies should note that >25% of the >100 MeV protons observed at Earth in this entire one-month period came from this particular shock.) As seen in the *Dst* data in the lower panels, this shock also produced a large geomagnetic disturbance.

In the 20 October shock event, the proton time profiles are nearly identical at all energies, except for velocity dispersion in the onsets. In the other events, however, the time profiles evolve with increasing energy. At the lowest energies, the protons show the typical gradual-event profile, with fluxes growing over $\sim$6-12 hours and then persisting at enhanced levels for 1-2 days. At higher energies, however, the enhancements look much more "impulsive": in events in which the CME lifts off from west of the central meridian (29 September, 22 and 24 October) there is a sharp rise followed by decay with e-folding times of just a few hours in the highest-energy channel. This energy-dependent time-evolution is just as one would expect in these events, as the expanding shock and Archimedian spiral of the interplanetary magnetic field cause Earth's connection point to the shock to move toward the shock's eastern flank, where the shock is intrinsically weaker and less capable of accelerating

100

particles to the highest energies. In the 19 October event, in which the flare-proxy and perhaps the nose of the shock are east of the central meridian, the expansion brings the connection point closer to the nose, producing a somewhat less abrupt rise and more gradual fall off until the arrival of the shock on 20 October.

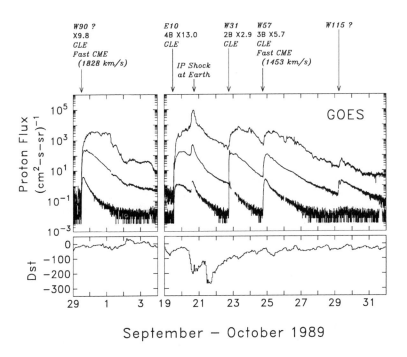

FIG. 3. Upper panels: GOES proton fluxes for the 1989 September-October SEP events in three energy channels, >5 MeV, >100 MeV, and 505-685 MeV. The labels along the top give the heliolongitude of the associated flares and summarize related observations. Lower panels: *Dst* during these events.

Fig. 4 shows the Chicago/*IMP-8* event-integrated Fe fluences for the 1989 September-October events. Power laws give good fits to the fluences, at least above ∼80 MeV/nuc. Systematic uncertainties in the Chicago/*IMP-8* fluences are typically 10-20%, except for the 20 October shock enhancement. In this case, problems associated with very high rates in the electronic detectors left a factor-of-two systematic uncertainty in the Fe fluence.

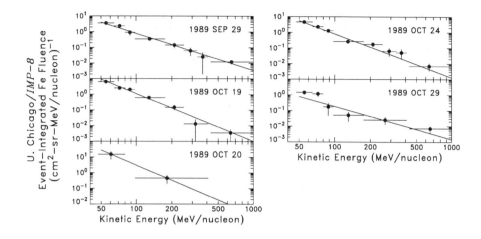

FIG. 4. Event-integrated Fe fluence for SEP events during the *LDEF* mission from the U. Chicago charged-particle telescope on *IMP-8*. Error bars are statistical only. Recent additional analysis (24) has systematically shifted these fluences, but with negligible effect on our <Q> measurement.

GEOMAGNETIC TRANSMISSION

The next step in comparing the Chicago/*IMP-8* and HIIS/*LDEF* Fe fluences was to determine the so-called "geomagnetic transmission function" (GTF), which specifies, as a function of magnetic rigidity, the fraction of the *LDEF* orbit over which an ion of given rigidity could have reached the HIIS detectors from interplanetary space. This GTF evaluation was based on ray tracings, which followed ions from randomly-chosen points along the *LDEF* orbit, backwards along their paths through a numerical model of the near-Earth magnetic field, to determine if the ions could have originated from interplanetary space. Our calculations utilized a state-of-the-art program (25) which incorporated both Earth's magnetic field (26) and fields generated by magnetospheric current systems (27). Details of the geomagnetic transmission calculations are given in (28,13).

One aspect of our geomagnetic transmission calculation deserves particular mention. During geomagnetically-quiet periods, ~300 MeV/nuc Fe ions with ionic charge Q < 14 could penetrate to a significant part of the *LDEF* orbit. During a large geomagnetic disturbance, in which enhanced magnetospheric current systems generate magnetic fields which are more effective in partially cancelling Earth's field, ~300 MeV/nuc Fe ions with ionic charge states as high as ~17 could just as easily reach *LDEF*. We accounted for the impact of these disturbances in our ray tracings and transmission calculations by adjusting parameters in the magnetospheric field model (27), to match the observed Kp and *Dst* variation. We also validated our techniques by applying them to the observed geomagnetic transmission of solar energetic protons during the 1989 October events, as determined from comparison of fluences from NOAA-10 (in low-Earth orbit) and GOES-7 (in geosynchronous orbit) (28).

Although we devoted considerable effort to validating our geomagnetic tranmission calculations and to understanding the impact of geomagnetic disturbances on our results, one should not be tempted to think that these complications produce a potential "loophole" in our conclusions about the Fe charge state. As shown in Fig. 3, the largest geomagnetic disturbance during the SEP events was $Dst = $ -268 nT, which occurred late on 21 October, when the high-energy particle fluxes had fallen to $\sim$1% of their peak values. We repeated our ray tracings and GTF calculations *as if this extreme geomagnetic disturbance had persisted throughout all of the SEP events*. But even this extreme calculation, which grossly overstates the potential impact of enhanced geomagnetic access on our results, shifted our best-fit <Q> value upward by only $\sim$10%.

RESULTS AND DISCUSSION

The interplanetary fluences from *IMP-8* were then convolved with the rigidity-dependent transmission functions to calculate the expected fluence at HIIS, leaving <Q> (the mean ionic charge *outside* the magnetosphere) as a free parameter. For these calculations, we neglected possible event-to-event variability and energy-dependence in the charge state distribution, making our <Q> value effectively an average over events and energy. The SEP curves in Fig. 1a show typical results. We believe that overall systematic errors in our fluences and calculations are small; but note that even a factor-of-two systematic change in the datapoints or curves would shift <Q> by less than one charge unit. Our best-fit result is <Q> = 14.2 $\pm$ 1.4, where the error comprises both statistical and systematic uncertainties, added in quadrature.

Because the HIIS observations span such a large energy range, our calculated spectrum – and hence, our best-fit <Q> – is sensitive to our assumptions about the *shape* of the charge state distribution. The SEP curves in Fig. 1a use flat distributions with rms widths of $\sim$2.5 charge units. (See (13) for details.) In fact, uncertainty about the shape of the charge state distribution is the single largest contributor to the systematic uncertainty in our result. Fig. 1b shows an attempt to obtain a better fit to the HIIS data by fine-tuning the charge state distribution while leaving <Q> unchanged (29). Neither a typical observed solar wind charge state distribution (30), nor a pure "coronal" distribution (as modeled by (7)) alone fits the spectrum. However, a combination of 10% solar wind and 90% coronal fits very well.

This result, if taken seriously, may provide a reasonable account of the origin of the SEP source plasma (29). The evolution of coronal charge state distributions into those found in the solar wind is presumably affected by the timescale on which the plasma escapes the inner corona. Plasma pushed ahead of a fast CME certainly leaves the corona much more quickly. Moreover, in the case of a very fast CME, a concave-inward shockfront develops while the CME is still in the corona (31). This geometry is perhaps amenable to some sort of "trapping" of plasma with a more corona-like charge-state distribution in the sheath region, between the shock and CME. (Such a charge-state distribution was indeed observed in the only direct observation of a sheath region at 1 AU reported to date (32).) This sheath region is generally believed to contain the primary SEP-source plasma (33). As the fast CME and shock move out into the interplanetary medium, ambient solar wind in front of the shock will also be accelerated, contributing roughly in the ratio of the relative sheath and solar wind densities. The SEPs will then show a mixture of coronal and solar wind charge states, provided that the effective "storage" time of the coronal plasma in the sheath region is at least comparable to the CME's transit time in the inner heliosphere. However, it has not been explained how this "storage" comes about. Alternatively, one might suppose that charge state distribution of the swept-up solar wind is somehow altered in the sheath region.

Fig. 5 compares our SEP Fe $<Q>$ measurement with results at lower energies. Although the most recent low-energy result from LICA (14) stands apart, the most impressive point here is the consistency – over three orders of magnitude in energy, for many different events, and from instruments employing very different detection and analysis techniques. These charge states are a fundamental feature of SEP production in gradual events; they cannot be ingored in any attempt to understand the underlying physical processes.

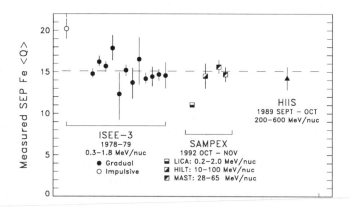

FIG. 5. Summary of SEP Fe $<Q>$ measurements from ISEE-3 (6), SAMPEX (14-16), and HIIS (13). The open circle is for impulsive events; all others are for gradual events. The dashed line shows a weighted average over all gradual-event measurements except LICA (14).

ACKNOWLEDGMENTS

We thank the Organizing Committee for the opportunity to present our results at the Workshop. This work has been supported by the Office of Naval Research, NASA contract W-18409 (NRL), and NASA grant NAG 5-706 (U. Chicago).

REFERENCES

1. D.V. Reames, Proc. Third SOHO Workshop, ed. A. Poland, Estes Park, CO, ESA SP-**373**, 107 (1994).
2. G. Kanbach et al., Astron. Astrophys. Suppl. Ser. **97**, 349 (1993).
3. L. G. Kocharov et al., Solar Phys **150** 267 (1994).
4. R.P. Lin, EOS Trans. AGU **40**, 457 (1995).
5. S. Kahler, Astrophys. J. , **428**, 837 (1994).
6. A. Luhn et al., Astrophys. J. , **317**, 951 (1987).
7. M. Arnaud & J. Raymond, Astrophys. J. , **398**, 394 (1992).
8. G. D. Reeves et al., J. Geophys. Res., **97**, 6219 (1992).
9. R. D. Belian et al., J. Geophys. Res., **97**, 16897 (1992).
10. H. H. Sauer, Proc. 23rd ICRC (Calgary), **3**, 254 (1993).
11. T. Mathews & D. Venkatesan, Nature, **345**, 600 (1990).
12. M. A. Shea et al., Proc. 23rd ICRC (Calgary), **3**, 846 (1993).
13. A. J. Tylka et al., Astrophys. J. Letters, **444**, L109 (1995).
14. G. M. Mason et al., Astrophys. J. **452**, 901 (1995).
15. M. Oetliker et al., Proc. 24th ICRC (Rome) 4, 470 (1995).
16. R. A. Leske et al., Astrophys. J. Letters, **452**, L149 (1995).
17. W.H. Kinard & G. D. Martin, Proc. First *LDEF* Post-Retrieval Symp., NASA CP-3134,1, 49 (1991).
18. R. Beaujean et al., Adv. Space Res., **17**, 167 (1996).
19. B. Wiegel et al., Adv. Space Res., **15**, (1)53 (1995).
20. A. J. Tylka et al., Astrophys. J. Letters, **438**, L83 (1995).
21. T. Kleis et al., Adv. Space Res., **17**, 163 (1996); Proc. 24th ICRC, 4, 481 (1995).
22. M. Garcia-Munoz et al., Astrophys. J. Letters, **201**, L145 (1975).
23. L. C. Gentile, Proc. 23rd ICRC (Calgary), **3**, 63 (1993).
24. W. F. Dietrich et al., in preparation (1996).
25. E. O. Flückiger et al., Proc. 22nd ICRC (Dublin), **3**, 648 (1991).
26. R. A. Langel et al., J. Geomag. Geoelectr., **43**, 1007 (1991).
27. N. A. Tsyganenko, N. A., Planet. Space Sci., **37**, 5 (1989).
28. P. R. Boberg et al., Geophys. Res. Letters, **22**, 1133 (1995).
29. P. R. Boberg et al., Proc. 24th ICRC (Rome), 4, 466 (1995).
30. R. von Steiger et al., Proc. Cosmic Winds and the Heliosphere, eds. J.R. Jokipii et al., (Tuscon: U. Arizona Press), (1995).
31. R. S. Steinoflson, Proc. 26th ESLAB Symp. (ESA SP-**346**), 51 (1992).
32. A. B. Galvin et al., J. Geophys. Res., **92**, 12069 (1987).
33. J. T. Gosling et al., J. Geophys. Res., **86**, 547 (1981).

High Energy Particles In Interplanetary Space on 11 June 1991

D. F. Smart and M. A. Shea

*Space Physics Division, Geophysics Directorate, Phillips Laboratory
29 Randolph Road, Hanscom AFB, Bedford, MA 01731-3010*

The solar cosmic ray Ground-Level Enhancement (GLE) observed on 11 June 1991 was mildly anisotropic with a velocity dispersive onset. The anisotropy determined by comparing the flux observed by "forward viewing" high latitude neutron monitors with the flux observed by "reverse viewing" high latitude neutron monitors had an approximate 2-to-1 ratio at the GLE maximum. The relativistic proton flux anisotropy persisted through most of the GLE suggesting an extended high energy injection of particles. Using improved modeling techniques we have fitted a shock acceleration spectrum to both the neutron monitor observations and high energy (350 to 550 MeV) spacecraft data. The shock acceleration spectrum used to fit the observations has a differential rigidity slope of -4.62 at 1 GV at the GLE maximum implying a shock compression ratio of 2.237.

INTRODUCTION

The source for the high energy solar particles studied in this paper is the solar activity associated with NOAA region 6659. The 11 June 1991 GLE is time-associated with the X12/3B solar flare at heliographic coordinates N31, W17 with an H-alpha onset at 0156 UT. The entire June 1991 episode of solar activity with its associated X-ray, gamma ray, and solar neutron emission, is described by other authors in this volume. This solar activity episode generated interplanetary shocks that propagated through the heliosphere. Six sudden commencement geomagnetic storm onsets were recorded at the earth between 4 and 12 June. The intense solar activity contributed to extreme cosmic ray modulation resulting in an historic minimum in the observed galactic cosmic ray intensity recorded on 13 June 1991. This extreme modulation effect strongly suggests that particle propagation conditions in the heliosphere were not quiescent. Also suggestive of the non-quiescent propagation conditions were the variations in the pre-event cosmic ray background which exceeded the variations expected from Poisson statistics.

Limitations of the Previous Analyses

Previously reported analyses of this event (1,2) were restricted to considering spectra represented by a power law in rigidity. These analyses used an asymptotic direction set (3) that describes quiescent magnetospheric conditions, but does not account for additional distortion in the asymptotic cones of acceptance generated by perturbed magnetic conditions. These earlier analyses also did not include the decrease in geomagnetic cutoff at mid-latitude stations which occurs during geomagnetic storms. Using only a spectral form of a power law in rigidity essentially forces a soft spectrum into the data fitting procedure for this event since the analysis is required to replicate the observations both at polar and mid-latitude stations. In the GLE on 11 June, the polar stations recorded increases of around 7% while the mid-latitude stations recorded small increases of the order of 1%.

In this analysis we have calculated cosmic ray trajectories and asymptotic directions appropriate for a disturbed magnetosphere (Kp=5) using the Flückiger-Kobel (4) trajectory code which employs the Tsyganenko (5) magnetospheric quantifications. When this magnetospheric model is used, the neutron monitors with the highest geomagnetic cutoff that recorded an increase during this event, Alma Ata, Kazatasan (~0.3% increase at the 18-NM-64 monitor at an altitude of ~3000 meters) and Rome, Italy, (a barely discernible increase by the 9-NM-64 monitor at sea level) were found to be responding to the solar particle spectrum down to about 5.5 GV, significantly below their 6.6 GV quiescent geomagnetic cutoff. We have also corrected for both the extreme modulation in the galactic cosmic ray background intensity that was present on 11 June 1991. We used the CHIME model (6) with a cosmic ray modulation factor of $\varphi = 1600$ to represent the cosmic ray spectrum for 11 June 1991 and to calculate the cosmic ray induced background counting rate in neutron monitors for this extreme modulation level.

METHOD OF DETERMINING HIGH ENERGY SOLAR PARTICLE SPECTRA FROM THE ANALYSIS OF NEUTRON MONITOR DATA

We have used our improved solar cosmic ray ground-level event technique (7) to determine the spectral characteristics and flux anisotropy for the 11 June 1991 GLE. The method is designed to reproduce the increase observed by the individual neutron monitors around the world. This analysis uses improved modeling techniques that account for the cosmic ray modulation and the cutoff depressions during a geomagnetic storm; it also utilizes an improved representation of the solar particle spectrum. A significant modification is the employment of the Ellison and Ramaty (8) shock acceleration spectrum to approximate the shape of the solar particle spectrum. This more flexible description of the spectrum avoids some of the severe constraints of the power law representation.

In describing the GLE modeling method it is necessary to explain the concept of asymptotic directions of approach and the asymptotic cone of acceptance. Charged particles of a specified energy (or rigidity) arriving at a detector from a

107

specific direction can be "mapped" through the geomagnetic field to a specific direction in space (9,10). The asymptotic direction of approach defines an allowed particle's direction in space prior to its interaction with the earth's magnetic field. The asymptotic cone of acceptance comprises all of the allowed asymptotic directions as a function of rigidity. In this analysis we have used the set of vertical asymptotic directions to approximate the asymptotic cone of acceptance. From the "geomagnetic optics" of high latitude neutron monitors, we can determine the orientation of the asymptotic cone of acceptance to the interplanetary magnetic field direction. From the individual asymptotic directions of approach we can determine the angle between each particle's velocity vector and the vector direction of the interplanetary magnetic field (or as a proxy, the direction of the maximum particle flux). From this information we can derive the particle flux anisotropy as a function of pitch angle.

In our advanced modeling technique, we account for both the level of the cosmic ray modulation during the solar cycle and the observed increase above the pre-event background at each neutron monitor station. Using the cosmic-ray trajectory-tracing technique the allowed cosmic ray particles (in the vertical direction) are calculated for each station. The results of these calculations are used to determine the asymptotic cone of acceptance for vertically incident particles for a specific station. Then the cosmic ray modulated pre-event background is calculated for each neutron monitor location by summing the response of the modulated cosmic ray spectrum (6) and the neutron monitor yield function (11) over all allowed rigidities from R_1 to 25 GV. (R_1 is the lowest allowed calculated rigidity.)

Next we model the increase above the pre-event background utilizing the functional form,

$$I = \sum_{R_1}^{\infty} J_\alpha(\alpha,R) \, S(R) \, G(\alpha) \, \Delta R \tag{1}$$

where I is the increase above pre-event background at the neutron monitor, R_1 is the cutoff rigidity typified by the lowest allowed calculated rigidity, $J_\alpha(\alpha,R)$ is the solar particle differential flux in the interplanetary medium at pitch angle α and rigidity R that is allowed through the asymptotic cone of acceptance, $S(R)$ is the neutron monitor specific yield as a function of rigidity, and $G(\alpha)$ is the anisotropic pitch angle distribution. In our modeling approach we sum the spectrum-yield response for each station from 0.7 GV, (or the lowest allowed rigidity) to 25 GV in varying intervals that typify each rigidity interval. Typical rigidity intervals are 0.1 GV. Larger intervals can be used when the asymptotic directions of approach are not changing rapidly as a function of rigidity and smaller intervals (as small as 0.01 GV) are used when there are rapid changes in direction. In our model we define pitch angle zero as the direction of the maximum particle flux which generally corresponds to the direction of the interplanetary magnetic field.

The calculated percentage increase is the additional increase above the pre-event background cosmic ray spectrum-yield at each station. If we have chosen the correct spectral and anisotropy parameters, then the calculations should reproduce the observations on a world-wide basis. Our numerical analysis consists of evaluating parameter choices of the solar particle spectrum, probable maximum flux source direction, and flux anisotropy until an acceptable solution is obtained. In our modeling, we also calculate the probable response to stations that did not observe the event since the verification of our simulation requires that we reproduce the response (or null response) observed at all neutron monitors on a world-wide basis.

THE GLE OF 11 JUNE 1991

An overall conceptual view of this small GLE can be obtained from Figure 1 which illustrates the increase observed at the Thule, Greenland 9-NM-64 neutron monitor. The disturbed interplanetary conditions present can be inferred from the fact that the cosmic ray background changes by about half the amplitude of the solar cosmic ray increase during the time of the GLE. In this analysis we have corrected for both the extreme modulation that was present on 11 June 1991 and for the cosmic ray intensity recovery in progress at the GLE onset. If a constant pre-event level were utilized throughout the GLE, then at the end of the event, the data from all high latitude stations would still be indicating a ~4% increase. To account for this background level change, we have adjusted the apparent percentage increase by 0.25 percent per hour from the pre-event normalization time of 00 UT.

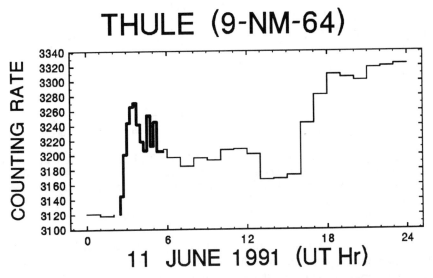

FIG. 1. The hourly averaged cosmic ray intensity and GLE recorded by the Thule, Greenland neutron monitor on 11 June 1991. The data points representing the small GLE are connected by the heavy solid line. Note the change in the cosmic ray background during the day.

This small, mildly anisotropic GLE on 11 June was observed by all high latitude neutron monitors. An impulsive onset was observed by the forward viewing stations (those having asymptotic directions of approach viewing into the solar particle flux propagating along the interplanetary magnetic field direction away from the sun) in the five-minute interval 0235-0240 UT. The onset for reverse viewing stations (those having asymptotic directions of approach viewing into the particle flux propagating along the interplanetary magnetic field back toward the sun) was after 0300 UT. At the time of the GLE maximum at about 0330 UT stations viewing in the probable forward direction such as Apatity in Russia, Oulu in Finland and Mawson in Antarctica recorded an increase of ~7 percent while stations viewing in the probable reverse direction such as Tixie Bay in Russia and Inuvik in Canada recorded an increase of ~3 percent as illustrated in Figure 2. The left panel of this figure shows the increase observed by stations viewing into the forward propagating flux as illustrated by the heavy line representing a statistically weighted composite of the increase observed at the Apatity, Russia (18-NM-64) neutron monitor and the Oulu, Finland (9-NM-64) neutron monitor. The light line shows the increase observed by the Tixie Bay, Russia (18-NM-64) neutron monitor was viewing into the reverse propagating particle flux. The right panel illustrates the pitch angle distribution required to generate the observed particle anisotropy at the GLE maximum. This form is similar to the exponential form derived by Beeck and Wibberenz (12).

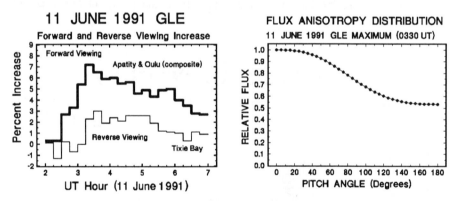

FIG. 2. The 11 June 1991 GLE. Left: relative increase observed by neutron monitors viewing into the forward flowing particle flux and the reverse flowing particle flux. (Fifteen-minute data averages are displayed to reduce statistical variations.) Right: solar particle flux pitch angle distribution necessary to produce the observed variation in the high latitude neutron monitors at the GLE maximum.

Unfortunately, there are no interplanetary magnetic field (IMF) measurements by earth-orbiting spacecraft on 11 June 1991. However, we can use the anisotropy of the observed increase at the onset time to approximate the probable IMF direction, at least to the proper octant. Figure 3 illustrates the orientation of the asymptotic viewing directions of selected high latitude neutron monitors with respect to the sun-earth line. At the time of the GLE maximum (0330 UT), the subsolar point was at 23° N, 127° E. From this figure we can also infer the orientation of these asymptotic cones of acceptance to the probable interplanetary magnetic field directions. The stations observing the largest increases (Apatity, Oulu and Sanae) indicate the direction of the maximum particle flux while stations with asymptotic cones of acceptance oriented in the opposite direction view the minimum particle flux.

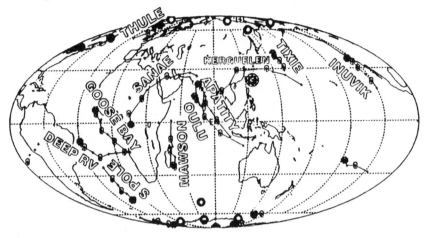

FIG. 3. Display of the asymptotic viewing directions responsible for 20 to 80 percent of the response of selected high latitude neutron monitors. At the time of the GLE maximum, 0330 UT, the subsolar point (indicated by the symbol ⊛) was at 23° N, 127° E.

The geomagnetic field was severely disturbed, and a geomagnetic storm was in progress. At the GLE onset the Dst was -96 nT and increasing toward the maximum of -140 nT which was observed at 06 UT. In our analysis of this event, the IMF direction did not appear to be stable during the GLE, but in our opinion, in this case this is not a serious impediment to our analysis.

We have used the shock acceleration spectral form developed by Ellison and Ramaty (8) as the basis for determining the incident proton flux. When the flux specified by the Ellison and Ramaty spectral form is plotted as a differential rigidity flux, the slope between the ultra-relativistic and low rigidity portion of the spectrum changes by one unit in index, the index changing as $1 - \beta^2$. As it happens, one GV is approximately mid way between the low rigidity and ultra-relativistic rigidity range and is a convenient reference point. For comparison purposes, we give the slope of a power law in rigidity that is tangent to the shock acceleration spectrum at 1 GV. At the GLE maximum the tangential differential rigidity slope at 1 GV is -4.62. At ultra-relativistic energies (approximated by 25 GV), the slope is -5.09. Solving for the shock compression ratio that would generate this spectral form yields a value of 2.237.

We can determine the magnitude of the particle flux in the steradian solid angle centered on the probable IMF direction and then obtain a 4π flux average by summing the flux anisotropy profile for all directions. From our analysis of this event, we find that the differential power law in rigidity tangent to the shock acceleration spectrum at 1 GV is:

$$J_{\parallel} = 4.23 \; P^{-4.62}; \qquad \text{and} \qquad J_{(avg)} = 3.19 \; P^{-4.62}. \qquad (2)$$

$J_{\parallel}$ is the flux in units of $(cm^2\text{-}s\text{-}ster\text{-}GV)^{-1}$ in the maximum flux direction (i.e. pitch angle of zero). $J_{(avg)}$ is the flux average over 4π steradians. This anisotropy averaged form is used to compare with spin averaged spacecraft data. We assume that the form of the spectra at the time of maximum, as derived from the neutron monitor data, can be extended to lower energies and that this time-of-maxima spectra may represent the near-sun form of the solar particle injection into space.

The Ellison-Ramaty spectral form in rigidity fitted to the GLE data has been numerically integrated and converted to energy so the results can be compared with other data. Table 1 gives this spectral information in a number of forms. The first three columns tabulate the values for the differential rigidity spectra, both for the anisotropic peak direction flux and the average flux obtained by a 4π steradian integral of the directional flux over the anisotropy function shown in figure 2. The center column of table 1 gives the average integral flux values obtained from integrating the differential flux values. The last 3 columns tabulate the values for the differential energy spectra, both for the anisotropic peak direction flux and the average flux.

TABLE 1. Derived Spectra for the 11 June 1993 GLE Maximum

Rigidity GV	Peak Flux Cm^{-2} S Sr GV	Average Flux Cm^{-2} S Sr GV	Average Flux Cm^{-2} S Sr	Energy MeV	Peak Flux Cm^{-2} S Sr MeV	Average Flux Cm^{-2} S Sr MeV
0.30	7.51E+2	5.66E+2	5.28E+1	46.8	2.47E+0	1.86E+0
0.40	2.29E+2	1.73E+2	2.09E+1	81.7	5.85E-1	4.41E-1
0.50	9.02E+1	6.80E+1	1.00E+1	125.	1.92E-1	1.45E-1
0.60	4.13E+1	3.12E+1	5.37E+0	175.	7.68E-2	5.79E-2
0.70	2.11E+1	1.59E+1	3.13E+0	232.	3.54E-2	2.67E-2
0.80	1.17E+1	8.79E+0	1.94E+0	295.	1.80E-2	1.36E-2
0.90	6.85E+0	5.16E+0	1.26E+0	362.	9.89E-3	7.46E-3
1.00	4.23E+0	3.19E+0	8.55E-1	433.	5.80E-3	4.37E-3
1.10	2.72E+0	2.05E+0	5.99E-1	507.	3.57E-3	2.70E-3
1.20	1.81E+0	1.37E+0	4.31E-1	585.	2.30E-3	1.73E-3
1.30	1.24E+0	9.36E-1	3.17E-1	665.	1.53E-3	1.15E-3
1.70	3.43E-1	2.59E-1	1.12E-1	1000.	3.92E-4	2.96E-4
2.00	1.55E-1	1.17E-1	5.92E-2	1270.	1.72E-4	1.29E-4
3.00	2.08E-2	1.57E-2	1.17E-2	2200.	2.18E-5	1.65E-5
4.00	4.92E-3	3.71E-3	3.66E-3	3170.	5.05E-6	3.81E-6
5.00	1.60E-3	1.20E-3	1.48E-3	4150.	1.62E-6	1.22E-6
6.00	6.34E-4	4.78E-4	7.03E-4	5130.	6.42E-7	4.84E-7
7.00	2.90E-4	2.19E-4	3.74E-4	6120.	2.93E-7	2.21E-7

Comparison of the 11 June 1991 GLE Spectra with Spacecraft Data.

In our analysis we have required that the spectrum of the more rigid particles >1 GV (>433 MeV) derived from the analysis of neutron monitor data be consistent with the spectrum derived from the GOES 6 HEPAD data and the GOES 7 high energy data. The spacecraft data show a velocity dispersive onset and time-of-maxima in the initial part of the event which is in agreement with the GLE data. The velocity dispersive flux maximum for energies >300 MeV to >30 MeV occurred between 0430 and 0500 UT, well after the 0330 UT GLE flux maximum. These data (indicated by the ■ symbol in Figure 4) are used to construct a time-of-maximum spectrum. After the ground-level event, it is our opinion that a spatial structure dominates the particle flux profiles at satellite energies.

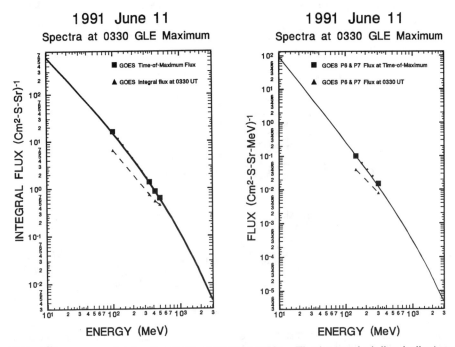

FIG. 4 Energy spectra for the GLE of 11 June 1991. The heavy dark line indicates the spectrum derived from neutron monitor data at the 0330 GLE maximum. Left: differential energy spectra. Right: integral energy spectra. The ■ symbol indicates the spacecraft measured time-of-maximum flux. The ▲ symbol indicates the spacecraft measured flux at 0330 UT.

We have taken the data from the GOES Space Environment Monitor as obtained from the National Geophysical Data Center (13) and applied the empirical correction provided by Sauer (14). The Ellison-Ramaty spectral form fitted to the GLE data has been numerically integrated so the results can be compared with the GOES data. The results of this comparison are shown in Figure

4. In this figure the heavy line indicates the spectrum that is derived from the high energy flux observed by the neutron monitors at the 0330 UT GLE maximum. We assume this also represents a time-of-maximum spectra that may represent the injection profile near the sun. The ■ symbol identifies the spacecraft time-of-maximum flux data at energies of >100, >355, >430, and >505 MeV. The ▲ symbol identifies the spacecraft observed fluxes at energies of 142 and 305 MeV at 0330 UT.

The GOES spacecraft particle flux data (13) for this event are shown in Figure 5. In our opinion these profiles display two maxima: a velocity dispersive flux maximum that occurs between 0330 UT and 0500 UT and a second, larger and non-velocity-dispersive flux maximum that occurs at about 14 UT during an extended period of increased magnetic activity when Dst exceeds -100 nT. It is our opinion that this strongly indicates an interplanetary source of the particles contributing to the second maximum. In view of this we are reluctant to take the integrated fluence observed by earth-orbiting satellites for this event and extrapolate it back to the sun to estimate the number of protons released from the acceleration site.

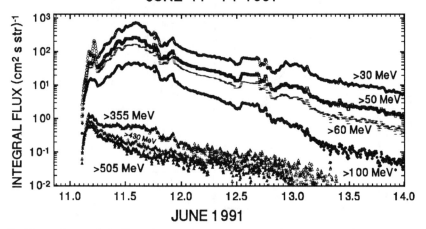

FIG. 5. The solar particle flux observed by the GOES spacecraft for the 11 June 1991 event. Note the non-velocity-dispersive maximum at about 14 UT which corresponds in time to the maximum of the geomagnetic storm.

ACKNOWLEDGMENT

We thank all the principal investigators who have contributed to the GLE data base. This data base is accessible to those who have the capability for remote network connections (15). Contact Gentile@PLH.AF.MIL, or AFGL::Gentile for access instruction. We wish to express special thanks to E. Eroshenko who provided data from the Russian neutron monitor network.

REFERENCES

1. Smart, D. F., Shea, M. A., and Gentile, L. C., 23rd Intl. Cosmic Ray Conf., (Calgary), **3**, 55 (1993).
2. Smart, D. F., Shea, M. A., and Gentile, L. C., in High-Energy Solar Phenomena - A New Era of Spacecraft Measurements, AIP Conference Proceeding 294, AIP Press, NY, NY., 1994, pp 222.
3. Gall, R., Orozco, A., Marin, C., Hurtado, A., and Vidargas, G., Tech. Rep., Instituto de Geofisica, Universidad Nacional Autonoma de Mexico (1982).
4. Flückiger, E. O. and Kobel, E., Geomag. Geoelectr., **42**, 1123 (1990).
5. Tsyganenko, A. N., Planet. Space. Sci., **37**, 5 (1898).
6. Chen, J., Chenette, D., Clark, M., Garcia-Munoz, M., Guzik, T. G., Pyle, K. R., Sang,Y., and Wefel, J. P., Adv. Space. Res., 14(10), 765 (1993).
7. Shea, M. A., and Smart, D. F., Space Sci. Rev., **32**, 251 (1982).
8. Ellison, D. and Ramaty, R., Astrophys. J., **298,** 400 (1985).
9. McCracken, K. G., J. Geophys. Res., **67**, 423 (1962).
10. McCracken, K. G., Rao, U. R., Fowler, B. C., Shea, M. A., and Smart, D. F. in Annals of the IQSY, **1**, The MIT Press, Cambridge, 1968, ch. 14, pp. 198.
11. Debrunner, H., Lockwood, J. A., and Flückiger, E. O., Preprint, 8th ECRC (1982).
12. Beeck, J. and Wibberenz, G., Astrophys. J., **311**, 437 (1986).
13. GOES Space Environment Monitor, National Geophysical Data Center, Boulder, Colorado (1992).
14. Sauer, H. Private communication, (1993).
15. Gentile, L. C., J. Geophys. Res., **98**, 107 (1993).

Neutron Decay Electrons after the Solar Flare of 1980 June 21

D. Ruffolo[1], W. Dröge[2], and B. Klecker[3]

[1] *Department of Physics, Chulalongkorn University, Bangkok 10330, Thailand*
[2] *Institut für Reine und Angewandte Kernphysik, Universität Kiel, Otto-Hahn-Platz 1, D-24118 Kiel, Germany*
[3] *Max Planck Institut für Extraterrestrische Physik, D-85740 Garching, Germany*

We have found evidence for fluxes of energetic electrons in interplanetary space on board the ISEE-3/ICE spacecraft which we interpret as the decay products of neutrons generated in a solar flare on 1980 June 21. The decay electrons arrived at the spacecraft shortly before the electrons from the flare and can be distinguished from the latter by their distinctive energy spectrum. The time profile of the decay electrons is in good agreement with the results from a simulation based on a scattering mean free path derived from a fit to the flare electron data. The comparison with simultaneously observed decay protons and a published direct measurement of high-energy neutrons places important constraints on the parent neutron spectrum.

INTRODUCTION

Previous studies have reported observations of interplanetary neutrons from solar flares by three methods: 1) direct detection of neutrons in space from flares on 1980 June 21 (1), 1982 June 3 (2), 1988 December 16 (3), 1991 June 9 (4), 1991 June 11 (5), and 1991 June 15 (6), 2) detection of their decay protons in space after flares on 1980 June 21, 1982 June 3, and 1984 April 25 [7-10], and 3) ground-based detection of neutrons from flares on 1982 June 3 [11-12], 1990 May 24 (13), 1991 March 22 (14), 1991 June 4, and 1991 June 6 [15-18]. These methods provide complementary information on the spectrum, angular distribution, and temporal distribution of escaping neutrons in different energy ranges, which can be compared with theoretical predictions (e.g., [19-21]) to constrain models of high-energy processes in solar flares.

Here we present observational evidence for a fourth type of detection based on decay electrons of solar flare neutrons on 1980 June 21. We also present detailed simulations of the injection and interplanetary transport of the decay electrons, which are used to fit those data. Preliminary results have been presented in (22).

As has been pointed out previously (23), solar neutrons of all energies yield a similar spectrum of decay electrons, so the decay electron intensity provides a measure of the total number of interplanetary neutrons, including those of ~ 1 MeV, which are not detected by other methods. There is a high flux of neutrons at these low energies, which propagate toward the hemisphere not obscured by the Sun and decay within $v_n \tau_n \sim 0.1$ AU, so with a reasonably good magnetic connection to the flare site, one can observe a significant flux of decay electrons with $E_e < 1$ MeV superimposed on the rising phase of the event.

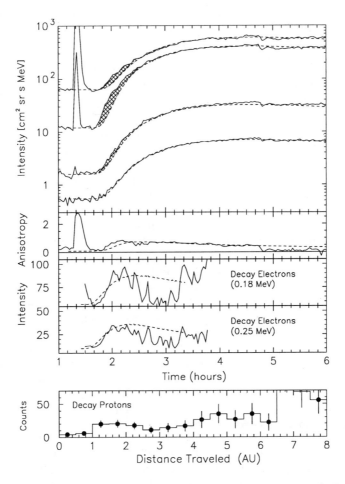

Fig. 1.–Electron intensities at 0.18, 0.25, 0.61, and 1.1 MeV (upper panel) and anisotropy of 0.18 MeV channel (second panel) of the 1980 June 21 solar event, observed on ISEE-3 (solid lines) and fits (dashed lines). Hatched areas indicate the excess flux attributed to neutron decay electrons. Middle two panels show the decay electrons and fits from a simulation. Lower panel shows protons as function of the distance traveled (see text for details).

OBSERVATIONS

The particle observations presented here were made with two instruments on board the *ISEE-3/ICE* spacecraft: the ULEWAT spectrometer (24), which measured the electron flux in the energy range of approximately $0.1 - 1$ MeV, and the University of Chicago MEH spectrometer (25), which measured protons from 27-147 MeV. Because no electron calibration was made with the ULEWAT spectrometer, a Monte Carlo simulation was performed to precisely determine its response to low energy electrons. During the time interval under consideration, *ISEE-3* was positioned at the Earth-Sun Lagrangian point well outside of the Earth's geomagnetic field.

Figure 1 (upper panel) shows electron fluxes at energies of $\approx 0.18, 0.25\ 0.61$, and 1.1 MeV, respectively, which were observed by *ISEE-3* on 1980 June 21 after a flare which occurred at 1:17 UT at N20 W88. The anisotropy of the lowest energy channel is shown in the second panel (no sectored data were available for the other channels). The spikes in the lowest two channels and the anisotropy lasting from $\approx 1:20$ UT to 1:30 UT are due to X-rays absorbed in the ULEWAT spectrometer. The electron event is characterized by a slow rise and late time of maximum despite a large and persisting anisotropy. Such a signature is indicative of a large interplanetary scattering mean free path (λ) in the vicinity of the observer and an extended injection of particles close to the Sun, or a short injection and strong scattering at small solar distances, or a combination of the latter two possibilities.

Fits to the intensity and anisotropy profiles (assuming the anisotropies of all four channels are similar) were performed using numerical solutions of the model of focused transport (26). A good fit can readily be obtained for the highest energy channel, and at late times for all four channels. However, among the many flares for which ULEWAT data have been analyzed in this manner, this flare is unique in that there is a peculiar excess at early times for the lower energy channels (hatched areas in Figure 1). This excess flux does not disappear for any reasonable combination of the fit parameters, and cannot be explained by contamination from energetic protons or γ-rays generated by them, or by variations in the observed interplanetary magnetic field.

We have investigated the hypothesis that this excess flux represents the detection of electrons from the decay of solar flare neutrons produced in the flare. Initial evidence is provided by the energy spectrum of the excess electrons (Figure 2), which is, within the uncertainty of the ULEWAT response functions (horizontal error bars in Figure 2) very similar to that expected for neutron decay electrons. To test this hypothesis in more detail, we have performed numerical simulations of the production and transport of such neutron decay electrons in interplanetary space.

Simulations

To model the production of electrons due to the decay of interplanetary neutrons, a Monte Carlo simulation was performed. For each of 5×10^7

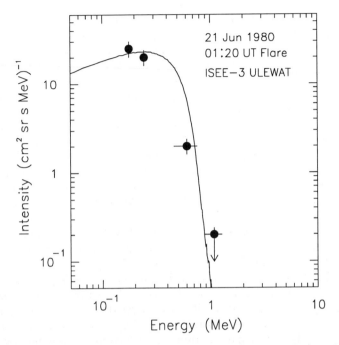

Fig. 2.–Energy spectrum of the excess electrons (filled circles) and theoretical prediction for decay electrons (solid line).

neutron decays, the decay electron was assigned a random energy, chosen according to the beta-decay energy distribution, and a random direction in the neutron rest frame. The electrons were then boosted into the fixed frame, for various neutron energies. A four-dimensional array stored the number of decay electrons per parent neutron for 5 electron momentum bins, 5 neutron energies, 4 magnetic field directions, and 25 pitch-angle bins. It was found that for higher neutron energies, the initial pitch angle distributions could be highly anisotropic in either the backward or forward sense, depending on the energy of the decay electrons.

Next, the injection of decay electrons into a section Δr of a flux tube subtending $\Delta\Omega$ from the Sun during an interval Δt was determined from the formula

$$\Delta N_e = \frac{dN_e}{dN_n} \left.\frac{dN_n}{dE_n d\Omega}\right|_{\text{at Sun}}$$
$$\cdot \frac{dE_n}{d\beta_n} \frac{d\beta_n}{dt} \, e^{-r/(\gamma_n \beta_n c\tau)} \frac{\Delta r}{\gamma_n \beta_n c\tau} \Delta\Omega\Delta t,$$

$$(1)$$

where dN_e/dN_n is the number of electrons per decaying neutron as determined from the Monte Carlo results, and $\beta_n = r/(ct)$, γ_n, and E_n are the appropriate values for neutrons arriving at a radius r after a time t.

Simulations of the interplanetary transport of decay electrons were performed using the finite-difference method of (10), as modified to include the effects of adiabatic deceleration and convection (27). The transport simulations were performed for electron momentum values of $0.2, 0.4, 0.6, 0.8$, and 1.0 MeV/c, $\Delta t = 1.0$ min, and assuming the same solar wind conditions, i.e., $V_{SW} = 290$ km/s, and $\lambda_\parallel$ as determined from the fits to the direct electrons. Even a simple fit, with a constant value of $\lambda_\parallel$ in each energy range and with no free parameters except the absolute normalization, yields a good qualitative fit to the electron excess, but the predicted onset of the flux is about 15 minutes too early. A better fit is obtained by assuming that λ_r is spatially constant, as recommended by (28), except within a postulated zone of enhanced scattering within 0.3 AU of the Sun. More details will be presented in a forthcoming paper (29).

The excess electron fluxes at 0.18 and 0.25 MeV (differences between total observed electrons and fits to direct electrons, plus the background intensity prior to the flare) are shown in panels 3 and 4 of Figure 1, together with the predictions for the decay electrons (dashed lines). There is good agreement between the two data sets until about 02:30 UT. After this time the difference fluxes do not give meaningful results any longer due to large, non-gaussian fluctuations in the electron counts rates caused by variations in the magnetic field. Given the above transport parameters, which are determined by fitting the flare electron fluxes at high energies and late times, the only free parameter of the fit to decay electron fluxes is the normalization, i.e., 3×10^{31} neutrons/sr (of all energies) *emitted toward the zenith*. The directional distribution and energy spectrum of the neutrons had no significant effect on the time profile or its normalization.

Discussion and Conclusions

Additional information about the production of neutrons in the 1980 June 21 flare can be gained from the observations of decay protons for that event (8). The bottom panel of Figure 1 shows the proton data from the MEH spectrometer, plotted in terms of the distance traveled, $s = v(t - t_{\text{flare}})$. The protons detected from $s = 1$ to 4 AU are believed to be mainly decay protons, because of their early arrival time and much harder spectrum. However, the statistical significance of the decay proton detection is marginal for this event, with only 51 proton counts (before the live time correction) and an uncertain contribution from direct protons. Based on simulations of the transport of neutron decay protons, we conclude that if all the counts were due to decay protons, the emission *toward the horizon* would be 2.6×10^{27} n/(MeV-sr) for $E = 27$ to 75 MeV and 1.1×10^{27} n/(MeV-sr) for $E = 75$ to 147 MeV. Given the possibility of a flux of quickly arriving direct protons, we take these to be upper limits of the neutron fluxes. The direct detection of neutrons from this event (1) indicated an integral flux $N(E > 50 \text{ MeV})$ of $\sim 3 \times 10^{28}$ n/sr *toward the horizon*, and a spectral index of 3 to 4. The upper limits to differential neutron fluxes that we derive from proton data are somewhat higher than those implied by the direct detection.

Each of these three observations of solar neutrons (using neutron decay electrons, neutron decay protons, and direct neutrons) imposes constraints on the parent neutron spectrum. Since typical theoretical results indicate that the neutron spectrum should be nearly energy-independent at low energies, with a steepening power law at higher energies, we have considered a spectrum of the form

$$\frac{dN}{dEd\Omega} = \frac{N_0}{1 + (E/E_0)^\delta} \qquad (2)$$

If we set $\delta = 4$, at the high end of the range of permissible power law indices for direct neutrons at high energies (1), and assume isotropic neutron emission, the measurements of $dN/d\Omega$ (>0 MeV) and $dN/d\Omega$ (>50 MeV) from direct neutrons imply that $N_0 = 360$ $n/$(MeV-sr) and $E_0 = 7.5$ MeV. The resulting spectrum is shown in Figure 3.

Note that these results for 1980 June 21 indicate that a rather steep power law persists down to an energy below ∼ 10 MeV. In contrast, results for the neutron flares of 1982 June 3 and 1984 April 25 have implied power law indices between 1 and 2 for $E = 27 - 147$ MeV (10), and for the latter flare there is a report (9) that the neutron flux actually declines with decreasing energy below about 30 MeV.

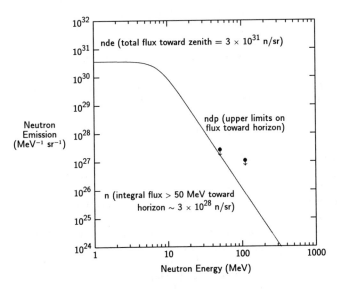

Fig. 3.–Possible neutron source spectrum for the 1980 June 21 flare that satisfies constraints from various observations (for details see text).

Finally, we note that decay electrons will usually be unobservable for flares at longitudes east of about 30° E, for which the inner portion of the magnetic field line connected to the detector lies within the "neutron shadow" (7), i.e., the volume beyond the horizon of the flare site. Since most decay electrons come from low energy neutrons ($\sim$ 1 MeV) which decay within $\sim$ 0.1 AU of the Sun, only a relatively few decay electrons are deposited on the portion of the field line that emerges from the neutron shadow. Because of this, we estimate that the peak flux of neutron decay electrons at *ISEE-3/ICE* was well below the background level for the two other flares for which neutron decay protons have been observed, namely those of 1982 June 3 (72°E) and 1984 April 25 (57°E relative to *ISEE-3/ICE*).

We thank Peter Meyer and Paul Evenson for kindly providing the proton data. DR thanks the Laboratory for Astrophysics and Space Research at the University of Chicago for remote access to their workstations. Thanks are also due to Bernd Heber for valuable assistance in preparing the figures.

REFERENCES

1. E. L. Chupp et al., Astrophys. J. **263**, L95 (1982).
2. E. L. Chupp et al., Astrophys. J. **318**, 913 (1987).
3. P. P. Dunphy, E. L. Chupp, and E. Rieger, Proc. 21st Int. Cosmic Ray Conf. **5**, 75 (1990).
4. J. Ryan et al., Proc. 23rd Int. Cosmic Ray Conf. **3**, 103 (1993).
5. G. Rank et al., in *High-Energy Solar Phenomena - A New Era of Spacecraft Measurements*, ed. J. M. Ryan and W. T. Vestrand, AIP Proc. **294**, 100 (1994).
6. H. Debrunner, J. A. Lockwood, J. M. Ryan, M. McConnell, V. Schönfelder, H. Aarts, K. Bennett, and C. Winkler, Proc. 23rd Int. Cosmic Ray Conf. **3**, 115 (1993).
7. P. Evenson, P. Meyer, and K. R. Pyle, Astrophys. J. **274**, 875 (1983).
8. P. Evenson, R. Kroeger, and P. Meyer, Proc. 19th Int. Cosmic Ray Conf. **4**, 130 (1985).
9. P. Evenson, R. Kroeger, P. Meyer, and D. Reames, Astrophys. J. Suppl. **73**, 273 (1990).
10. D. Ruffolo, Astrophys. J. **382**, 688 (1991).
11. H. Debrunner, E. Flückiger, E. L. Chupp, and D. J. Forrest, Proc. 18th Int. Cosmic Ray Conf. **4**, 75 (1983).
12. Yu. E. Efimov, G. E. Kocharov, and K. Kudela, Proc. 18th Int. Cosmic Ray Conf. **10**, 276 (1983).
13. M. A. Shea, D. F. Smart, and K. R. Pyle, Geophys. Res. Lett. **18**, 1655 (1991).
14. K. R. Pyle and J. A. Simpson, Proc. 22nd Int. Cosmic Ray Conf. **3**, 53 (1991).
15. K. Takahashi et al., Proc. 22nd Int. Cosmic Ray Conf. **3**, 37 (1991).
16. N. Chiba et al., Astropart. Phys. **1**, 27 (1992).
17. Y. Muraki et al., Astrophys. J. **400**, L75 (1992).
18. Y. Muraki et al., Proc. 24th Int. Cosmic Ray Conf. **4**, 175 (1995).
19. R. J. Murphy, C. D. Dermer, and R. Ramaty, Astrophys. J. Suppl. **63**, 721 (1987).
20. X.-M. Hua and R. E. Lingenfelter, Astrophys. J. **23**, 779 (1987).
21. V. G. Guglenko et al., Astrophys. J. Suppl. **73**, 209 (1990).

22. W. Dröge, D. Ruffolo, and B. Klecker, Proc. 24th Int. Cosmic Ray Conf. **4**, 183 (1995).

23. E. I. Daibog and V. G. Stolpovskii, Sov. Astron. Lett. **13**, 458 (1987).

24. D. Hovestadt et al., IEEE Trans. Geosci. Elect. **GE-16**, 166 (1978).

25. P. Meyer and P. Evenson, IEEE Trans. Geosci. Elect. **GE-16**, 180 (1978).

26. W. Schlüter, Ph.D. Thesis, University of Kiel (1985).

27. D. Ruffolo, Astrophys. J. **442**, 861 (1995).

28. I. D. Palmer, Rev. Geophys. Space Phys. **20**, 335 (1982).

29. W. Dröge, D. Ruffolo, and B. Klecker, Astrophys. J. Lett. (accepted for publication, 1996).

Longitudinal Extents of
Coronal/Interplanetary Shocks

H. V. Cane

Physics Department, University of Tasmania, Hobart, Australia and
Laboratory for High Energy Astrophysics, NASA/GSFC, Greenbelt, MD

Abstract. Whereas the majority of researchers accept the important role of shocks in accelerating particles at the Sun and in the heliosphere, there remain many details to be determined. In this paper the sizes of coronal shocks required by prompt solar particle events are compared with sizes of interplanetary shocks determined from in situ measurements. Energetic particle observations imply the existence of shocks extending at least 300° whereas interplanetary shocks at 1 AU extend at most about 180°. The observations can be understood if the longitudinal extents of shocks evolve as they propagate outwards from the Sun.

INTRODUCTION

A complete model to explain the observed particle profiles of large solar events has yet to be formulated. It will be a difficult task because it must include not only the evolution of the shock and the continuous interaction of the shock with the particles it produces, but also the constantly changing connection of the observer to different parts of the shock. Moreover there are the unknowns such as, for example, the possibility of separate coronal and interplanetary shocks. Furthermore the changing connection means one must also include longitudinal variations in the shock efficiency for particle acceleration.

One of the main reasons why shock acceleration is favoured for large proton events is that this process provides a straightforward explanation of how particles can gain rapid access to widespread regions of the corona and, subsequently, to interplanetary space (1,2,3,4). From the earliest observations it was realised that particles observed at Earth following an east limb flare were finding access to field lines some 150° from the supposed source region. It has proved impossible to find a mechanism by which particles diffuse such large distances across coronal field lines and at appropriately high speeds. In contrast it is certainly possible for a shock to cross field lines and then accelerate the particles directly onto distant field lines. Nevertheless it has recently been recognised (5,6) that there exists a discrepancy between the sizes of shocks required to account for widespread prompt particle acceleration and the sizes of interplanetary shocks inferred by in situ observations.

With single spacecraft observations the sizes of interplanetary shocks can only be determined if the source region can be established. It has been argued (7) that the only reliable technique for making associations is to use the presence of energetic particles. Since the low energy particles peak at shock passage and the particle event begins near the time of the solar event the particles provide the necessary link between source region and shock. Events extending beyond 50 MeV are almost always flare-associated and the high energy particles usually start within an hour or so of the beginning of the flare. Using this technique there has been only one shock seen at Earth that originated with a flare beyond either limb in the period from 1964 to the present. The shock of April 12, 1969 originated from just behind the east limb. Another extreme case was the shock of November 20, 1968 which originated from the west limb. Using multi-spacecraft observations some more extreme cases have been found. It was reported (8) that the November 1968 shock was detected at Pioneer 8 for which the flare was at W113°. Also reported (8) was that the April 1969 shock was detected at Pioneer 9 for which the flare was at E122°. These extreme values are still less by about 30-40 degrees than the extent of coronal shocks implied by east limb particle events. This discrepancy can be most readily explained if shocks decrease in longitudinal extent as a function of radial distance from the Sun.

THE OBSERVATIONS

The schematic in Figure 1 illustrates the problem raised by east limb events. If the shock is limited to about 180° in longitude as indicated by the observations discussed in the introduction then how do energetic particles gain rapid access to fieldlines 150° from the supposed source region? The figure suggests that particles should not be seen

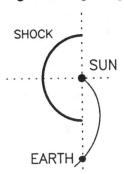

FIGURE 1. A schematic to illustrate how field line connection is not achieved until several days after the solar event if the shock extends only 180°.

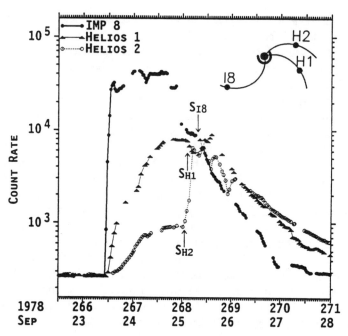

FIGURE 2. Particle count rates observed by the IMP 8 and Helios 1 and 2 guards for a period in September 1978. The locations of the spacecraft relative to the longitude of the associated flare (mid-point of the shock) are shown in the upper right of the figure. All three spacecraft saw an energetic shock at the times indicated.

until close to the time the shock reaches the Earth. This is expected to be several days after the solar event but this is not what is observed.

Using the multi-spacecraft combination of IMP 8 and Helios 1 and 2 we can examine particle profiles and onsets for poorly connected events. Figure 2 shows profiles at the three spacecraft for the event of September 1978. What is shown is the count-rate in the anticoincidence guards of the IMP 8 experiment (9) and the Helios experiments (10). The guards measure the rate for all particles capable of producing a signal, including electrons. Since we are only interested in the earliest onset time and the overall shape, the fact that the actual responses of the guards are not determined is not important. When compared with the differential intensities we find that the overall shape is not significantly different from a differential proton channel near 50 MeV. The advantage of the guard rate is the high counting statistics and the better coverage for some events as seen from the Helios spacecraft. In the top right of the figure is shown the locations of the spacecraft on September 23, 1978. For IMP 8 the event was at W50° whereas for the Helios spacecraft it was far east: E71° and E109° for Helios 1 and 2 respectively. Note that the profiles are very different. At IMP 8 the intensities were declining before the shock reached the spacecraft. (The times of passage of the shock past each spacecraft are indicated on the figure.) In contrast maximum

126

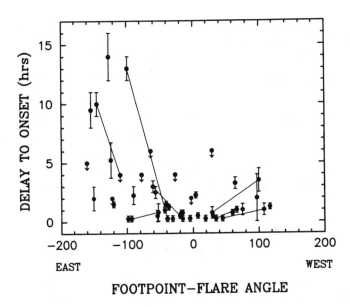

FIGURE 3. The delay times between the onset times of ~50 MeV protons and the time of maximum of the associated flare. Straight lines join some points which are the same event seen at multiple spacecraft.

intensities were not reached at the Helios spacecraft until after the shock passed. For an eastern event connection changes suddenly across the shock to a region closer to the nose of the shock because of wrapping of the upstream field lines around the CME. It should be noted that the profile seen at IMP 8 is not typical of a W50° event as, in general, shocks are not as extensive as the September 1978 shock . Most far western particle events rise rapidly to maximum intensity and then montonically decay (see Figure 4).

In Figure 2 it can be seen that particles appear at the Helios spacecraft within a few hours of the start of the event at IMP 8. At IMP 8, which was very well connected to the shock when it was low in the corona, the event commenced within about 0.5 hours of the associated Hα flare.

Figure 3 is a plot of delay to onset of ~50 MeV protons for all events for which protons of at least this energy were seen at all 3 spacecraft. The delay is plotted as a function of the angular separation between the associated flare and the footpoint of the observer's field line. For a solar wind speed of about 400 km/sec and an observer at 1 AU this angle is about 60° different from the distance between the flare and the observer's central meridian. Note that we find three cases for which Helios 2 saw particles associated with flares from ~20° behind the east limb as viewed by that spacecraft. The extreme case was that of September 19, 1977 with an angular separation between the flare and the spacecraft field line foot point of 178°. Unfortunately there was a data gap at the beginning of the event but it was near peak intensity at the end of the gap, 20 hours after the flare. Thus we conclude that the maximum delay for any event at 50 MeV is about half a day.

The asymmetry in Figure 3 is an artifact of the viewing conditions since we cannot identify flares for events from behind the west limb. Recall that the west limb is a much smaller angular distance from an observer's fieldline footpoint than the east limb. However there are a few events which are seen at both IMP 8 and Helios 1 which are most likely close to W180° for IMP 8. This is ascertained by assuming very good connection for Helios 1, based on the particle profile, and considering the separation between the two spacecraft. For these events the delay between the onset at Helios 1 and that at IMP 8 is less than 0.5 days.

DISCUSSION

The rapid onset of particles at angular distances far beyond those at which in situ shocks are observed is most easily explained if shocks shrink with time/radial distance. This shrinking probably is preceded by some expansion in order to explain the delays, of up to 14 hours, for the most distant events. A CME/shock traveling at 1000 km/sec reaches a distance of about 0.25 AU in 10 hours so the shrinking probably takes place quite distant from the Sun.

The observation that western events, for which no in situ shock is detected, have relatively smooth decays also suggests that shocks shrink. For a constant size shock one would expect a sudden turning-off when the observer, for which the event is western, gets onto a field line that never connected to the shock. In contrast, if shocks decrease in angular extent, there will still be some particles around at later times (because of scattering) which were generated when the observer's field line did connect to the shock in its early lifetime.

Another observation that supports the suggestion of evolving shock size is the existence of far eastern particle events for which no in situ shock is detected. By comparing a number of far eastern events the variations in the intensity profiles can be understood in terms of the evolution of the shock including its size. Figure 4 shows 50 MeV intensity profiles for three events seen at IMP 8. The time spans are such that the flare times line up vertically. The top event, which occurred in September 1977, shows a rapid onset and early time of maximum. No shock was detected at Earth. There is a local minimum in the profile at about the time for plasma, emitted close to the time of the flare, to convect to Earth. The middle event (associated flare on March 1, 1979) is somewhat delayed relative to flare time and has two intensity maximum on either side of the time of shock passage. The shock had a transit speed of 630 km/sec to Earth and was relatively weak as measured in situ. The bottom event (associated flare on August 18, 1979) is further delayed in its onset with maximum intensity after the passage of a relatively strong shock. The transit speed of this shock was 1030 km/sec.

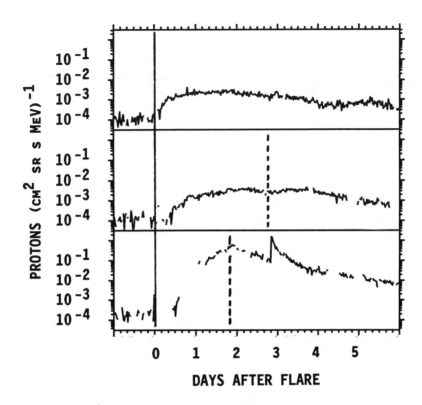

FIGURE 4. IMP 8 50 MeV profiles for three far eastern events which originated on September 7, 1977, March 1, 1979 and August 18, 1979. The solid vertical line indicates the times of the associated flares and the dashed lines indicate shock passages. The bottom event (August 1979) is the most intense and peaks behind the passage of a strong shock. (The sharply peaked feature near the end of day 2 is an additional western event with no local shock). The top event (September 1977) peaks early and there was no local shock.

A reasonable explanation for the differences between these profiles is that the angular extent of the shock size commences to decrease closest to the Sun for the top event and furthest from the Sun for the bottom event. Note that all three events were also seen at both Helios spacecraft. From the Helios observations it can be seen that the bottom (August 1979) event onsets very rapidly at these spacecraft, which were well connected, so the delay to onset at IMP 8 is unlikely to be related to evolution of the shock speed. It has been suggested (5) that for much less energetic events than the ones being discussed in this paper, differences between particle event profiles can result from evolution of the shock speed.

Finally, another possibility for explaining the difference in shock size between that required to explain widespread particles and the sizes observed for interplanetary shocks should be mentioned. This is that coronal shocks are not the progenitors of interplanetary shocks (11,12) and that the coronal shocks produce the earliest particles. Certainly there are some good arguments why the shocks which produce coronal Type II radio bursts are not driven by CMEs, which are widely accepted to be responsible for interplanetary shocks. However the rather uniform development of intensity profiles of particle events do not seem to support the idea of two shocks.

CONCLUSIONS

The rapid onset of particles at widely spread locations in the heliosphere, the relatively smooth decay of western and the existence of eastern events for which no shock is detected can be understood if shocks decrease in longitudinal extent as a function of radial distance from the Sun.

ACKNOWLEDGMENTS

I thank Ed Cliver for discussions and Gerd Wibberenz for some Helios data. Bill Erickson and Ed Cliver provided comments on the manuscript. Support at GSFC was provided by a contract with Universities Space Research Association.

REFERENCES

1. Palmer, I. D., and Smerd, S. F., *Solar Phys.* 26, 460-467 (1972).
2. Reinhard, R., and Wibberenz, G., *Solar Phys.* 36, 473-494 (1974).
3. Kahler, S. W., Hildner, E. and van Hollebeke, M. A.. I., *Solar Phys.* 57, 429-443 (1978).
4. Cliver, E.W., *Solar Phys.* 75, 341-345 (1982).
5. Cane, H. V., *Nuclear Phys. B (Proc. Suppl.)* 39A, 35-44 (1995).
6. Cliver, E. W., Kahler, S. W., Neidig, D. F., Cane, H. V., Richardson, I. G., Kallenrode, M.-B., and Wibberenz, G, in *Proc. Internat. Cosmic Ray Conf.* 4, 257-260 (1985).
7. Cane, H. V., *J. Geophys. Res.* 93, 1-6 (1988).
8. Pinter, S., in *Proceedings of COSPAR symposium B (STIP/1977),* 161- 188 (1977).
9. McGuire, R.E., von Rosenvinge, T. T., and Mcdonald, F. B, *Astrophys. J.* 301, 938- 948 (1986).
10. Kunow, H., Witte, M., Wibberenz, G., Hempe, H., Muller-Mellin, R., Green, G., Iwers, B., and Fuckner, J., *J. Geophys.* 42, 615- 631 (1977).
11. Wagner, W.J, and MacQueen, R.M, *Astron. Astrophys.* 120 136-138 (1983).
12. Cane, H. V, *Astron. Astrophys.* 140, 205-209 (1984).

Unusual Intensity-Time Profiles of Ground-Level Solar Proton Events

M. A. Shea and D. F. Smart

Space Physics Division
Geophysics Directorate/PL
29 Randolph Road, Hanscom AFB
Bedford, Massachusetts 01731-3010

Relativistic solar proton events have been prolific during the 22nd solar cycle. With improved measurement techniques and time resolutions several unusual characteristics have been observed in the intensity-time profiles of these events as recorded by ground-based neutron monitors. In particular several events exhibit an unusual initial anisotropic spike recorded by stations viewing into or through a narrow cone of asymptotic directions. A similar structure was observed during the 19th solar cycle associated with solar activity from the same heliolongitude range on the solar disk. Any solar particle acceleration and/or interplanetary propagation models must be able to explain these type of intensity-time profiles.

INTRODUCTION

The intensity-time profiles of solar proton events at any observation point in space are usually characterized by the location of the "parent" solar activity. From studies of the high energy solar proton events that occurred during the 19th through 21st solar cycles, the conventional wisdom was the following:

(a) Solar proton events from solar activity "well connected" to the earth were typified by a rapid rate of rise to maximum intensity followed by an exponential decay. "Well connected" implied that the associated solar flare activity was roughly between 50 degrees west heliolongitude to slightly behind the west limb of the sun as viewed from the earth.

(b) Solar proton events from solar activity near the central meridian of the sun or to the east of central meridian were typified by a "delay" in particle onset, a slow rate of rise in intensity, a lower maximum intensity than from "western" events, and a long decay rate to background.

The typical intensity-time profiles expected from these events is illustrated in Figure 1. The observations of the relativistic solar proton events during the 19th solar cycle were relatively crude with recording intervals of 10 to 60 minutes or more during most of the events. Although cosmic ray data were recorded at smaller intervals during the next two solar cycles, the intensity-time profiles observed during ground-level events were similar to those expected from previous observations. During the present (22nd) solar cycle we have observed several ground-level enhancements (GLE) that exhibit a "spike like" behavior during the initial portion of the event. The purpose of this paper is to identify some of these events in the expectation that theories on solar processes and/or interplanetary particle transport can be developed to explain this unusual behavior.

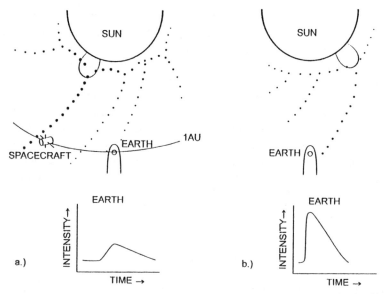

FIG. 1. Typical intensity time profiles associated with solar activity at (a) the central meridian and eastern hemisphere of the sun, and (b) the western hemisphere of the sun.

RELATIVISTIC SOLAR PROTON EVENTS IN THE 22ND SOLAR CYCLE

Solar cycle 22 rapidly rose from solar minimum conditions in September 1986 to a maximum in the smoothed sunspot number of 159 in July 1989 without the recording of a GLE during the rising portion of the cycle as had occurred during the previous four solar cycles. Nothing prepared the cosmic ray community for the unexpected plethora of ground-level solar proton events that occurred during solar maximum conditions, specifically the 15 events that occurred between July 1989 and November 1992. Seven of these events were during a four-month period from 25 July to 15 November 1989. This included the event on 29 September with the largest high energy peak flux intensity since 23 February 1956, and calculated to be the third largest event since 1942 (1).

With the increased time resolution available from many of the cosmic ray neutron monitors, the fine line structure recorded during several of these large ground-level enhancements offered an unusual insight into solar proton acceleration and propagation to the earth. Many of these events are so complex that they are still being intensively studied; some studies have resulted in more than one explanation (2,3).

With the use of the ground-level enhancement data base compiled at the Geophysics Directorate (4) we have examined some of the unusual time-intensity profiles recorded during the 22nd solar cycle in an attempt to identify any common characteristics that might be present. Any theories proposed for solar particle acceleration and propagation in the interplanetary medium (and through the earth's magnetosphere for ground-based stations) must be able to accommodate these unusual intensity-time increases.

EVENTS WITH UNUSUAL "SPIKE-LIKE" ONSETS

Event of 22 October 1989, 32° West Solar Activity

Probably the most spectacular and unexpected intensity-time profile for a ground-level enhancement occurred in conjunction with the 22 October 1989 event. The McMurdo, Antarctica neutron monitor recorded an onset between 1756-1758 UT with rapidly rising intensity to 193% ten minutes later. The intensity then decreased to less than 40% above background intensity ten minutes later. There was an additional slower decay to 27% after which the intensity again rose slightly to about 40% before decaying to background intensity several hours later. Figure 2 illustrates this complex intensity-time profile. Other stations recording similar structures during this event included South Pole and Calgary.

McMurdo (2 min. data) 22 October 1989

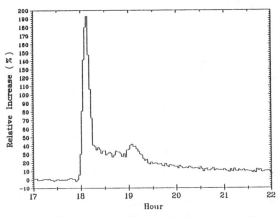

FIG. 2. Intensity-time profile for the GLE on 22 October 1989 as recorded by the McMurdo, Antarctica neutron monitor.

The initial pulse was recorded only by high latitude stations having narrow well-defined asymptotic cones of acceptance viewing directly into or through the relatively narrow beam of proton flux (5). Although stations viewing in the "reverse" direction observed a later onset time than McMurdo, implying perhaps "reflection" by an interplanetary disturbance beyond the earth, the presence of a small anisotropy at McMurdo during the "second" maximum (around 1900 UT) indicates particles are still propagating from the sunward direction. A comparison of the intensities recorded at McMurdo and Mawson is shown in Figure 3. Acceleration and propagation theories must be able to reconcile these unusual observations since the particles streaming from the sunward direction traverse the same conditions in the interplanetary medium over a period of one or two hours.

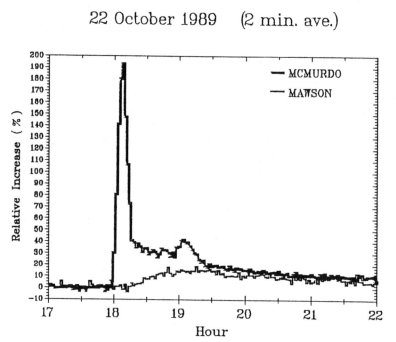

22 October 1989   (2 min. ave.)

FIG. 3. A comparison of the increases observed by the neutron monitors at McMurdo and Mawson, Antarctica during the 22 October 1989 GLE. Note that considerable anisotropy is still present around 1900 UT during the "second" maximum recorded by the McMurdo detector.

The associated solar flare was located at 32° West and was accompanied by a long duration soft X-ray event with magnitude X2.9. The Solar Maximum Mission spacecraft did not record a coronal mass ejection during this event because of operational procedures in progress at that time; however, the Mauna Loa solar observatory reported a coronal mass ejection (Burkepile, private communication). Type II radio emission was recorded; no world-wide sudden commencement geomagnetic disturbance was reported at the earth following this solar activity.

Event of 21 May 1990; 37° West Solar Activity

Four GLEs occurred in May 1990, with scientists concentrating their research on the 24 May event with its large solar neutron event precursor (6,7). Consequently the smaller events have not been intensively analyzed. One of these events, with maximum increases around 20-30%, occurred on 21 May with the associated solar activity at 37° West. Like the event on 22 October 1989, the intensity-time profiles for selected high latitude stations also recorded an initial "spike" followed by a more general enhancement. Figure 4 illustrates the increase detected by the Thule neutron monitor; a comparison of the increases detected by the Oulu and Inuvik neutron monitors is shown in Figure 5.

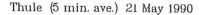

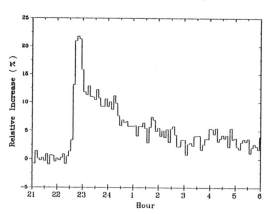

FIG. 4. Intensity-time profile for the GLE on 21 May 1990 as recorded by the Thule, Greenland, neutron monitor.

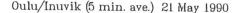

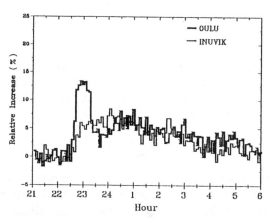

FIG. 5. A comparison of the increases observed by the neutron monitors at Oulu, Finland and Inuvik, Canada during the 21 May 1990 GLE.

The associated solar flare was accompanied by an impulsive short duration X-ray event with magnitude X5.5 (8). The Solar Maximum Mission spacecraft had re-entered the atmosphere late in 1989 so no spacecraft chronograph measurements are available. Type II radio emission was not recorded; no sudden commencement geomagnetic disturbance was reported at the earth following this solar activity.

Event of 15 November 1989; 28° West Solar Activity

Another considerably smaller event occurred on 15 November 1989 with ground-level enhancements between 5-15% at high latitude stations. Nevertheless, the same "spike like" characteristic is observed at Terre Adelie, Antarctica with somewhat similar profiles, albeit smaller intensities, at Oulu and Apatity. The increase at Terre Adelie is shown in Figure 6.

Terre Adelie (5 min. data) 15 November 1989

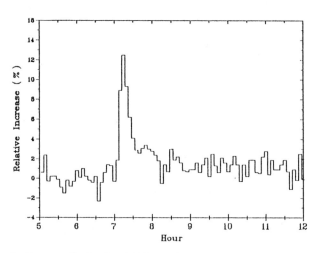

FIG. 6. Intensity-time profile for the GLE on 15 November 1989 as recorded by the Terre Adelie, Antarctica neutron monitor.

The associated solar flare was at 28°W and was accompanied by an impulsive short duration X-ray event with magnitude of X3.2 (8). Type II radio emission was recorded; a sudden commencement geomagnetic disturbance was reported at the earth following this solar activity.

Other Events

The only common thread between these three events is that the associated solar activity occurred in a longitude band between 28-37° West. Although we believe this may be fortuitous it behooves us to examine other GLEs since 1942 with associated solar flare activity between 28-37° West to see if similar intensity-time profiles are recorded. These events are the following:

17 July 1959	31°W	Extremely small event
15 November 1960	35°W	Large event
25 February 1969	37°W	
7 August 1972	37°W	
10 April 1981	36°W	Extremely small event

The ground-level enhancements associated with the events in July 1959 and April 1981 are extremely small, of the order of 1-2%, and any "spike" like profiles would not be easily discerned - if they are even statistically significant.

The event in February 1969 is associated with a maximum increase of 16% at Goose Bay. We have 5-minute data from only six high latitude neutron monitors, none of which display the characteristic "spike" described in the previous section. More small-time interval data from other high latitude stations would be necessary for additional analysis. This was a white light flare; no type II radio emission was reported. A sudden commencement geomagnetic storm was reported on 27 February; however, a major flare on 26 February with associated type II radio burst is in better time association with the sudden commencement than the flare associated with the GLE on 25 February.

The event on 7 August 1972 is even smaller with a maximum increase of 10% recorded at McMurdo. One interesting aspect of this event is the extremely rapid rate of increase observed at Inuvik in contrast to the slower increases observed by other neutron monitors. The asymptotic cone for Inuvik would have been viewing into the nominal interplanetary magnetic field direction at the time of the increase. Although some interesting fine structure is obvious at several stations, we need small-time interval data from key stations such as Calgary, Churchill, Kerguelen, Mawson, Resolute, Terre Adelie, and Tixie Bay for a more thorough analysis. Some of these stations have closed so these data, if recorded in smaller than one hour intervals, may be available only from private archives having been exchanged with other cosmic ray physicists.

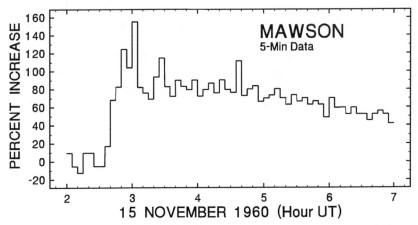

FIG. 7. Cosmic ray intensity profile recorded at Mawson, Antarctica during the onset phase of the GLE on 15 November 1960. The baseline level is 0100-0200 UT on 15 November. The standard deviation for the fine-minute background counting rate (as plotted) is 2%.

The only event from the previous solar cycles that occurred in this longitude band and for which we can positively identify the "spike" like characteristics was on 15 November 1960 with an initial spike of 124% followed by a second spike of 154% recorded by the Mawson, Antarctica neutron monitor. This increase is shown in Figure 7. At that time the Mawson monitor had a recording chart so that small time interval data could be retrieved and evaluated (9,10). Type II radio emission was reported associated with this event. No sudden commencement geomagnetic disturbance was recorded at the earth in an appropriate time interval following the solar flare; however, the geomagnetic field was extremely disturbed from a previous event and an additional sudden commencement might have been masked by the ongoing geomagnetic activity.

SUMMARY AND CONCLUSIONS

We have given examples where "spike-like" intensities have been recorded by neutron monitors during ground-level solar cosmic ray events. The stations which record these structures are typically polar stations having narrow asymptotic cones of acceptance presumably viewing in the direction of the initial particle increase. These unusual intensity-time profiles are evident for events when small-time interval data are available. It is possible that the events discussed in this paper may have occurred at the most opportune time and under the ideal interplanetary circumstances to record these features. In view of the fact that the initial anisotropy continues for an hour or more indicates that particles continue to enter the magnetosphere along a favorable propagation path for this initial time period. Other stations typically start to record a smaller, less structured, increase coincidental with or after the initial spike. All of these observations are dependent upon the viewing direction of the station and the amount of anisotropy present.

The identification of three of these "spike-like" structures during the present cycle indicates that these type of events may be more common than originally thought; the increased time resolution has resulted in our recognition of what would have been considered atypical profiles. The underlying physical circumstances leading to these profiles is not presently understood since, presumably, the solar protons traverse the same general regions of interplanetary space to reach their observation point. Any theories proposed for solar proton acceleration and/or propagation through the interplanetary medium must be able to explain these type of structures.

REFERENCES:

1. Smart, D.F. and Shea, M.A., *22nd International Cosmic Ray Conference, Contributed Papers*, **3**, Dublin Institute for Advanced Studies, Dublin, 101-104 (1991).
2. Shea, M.A., Smart, D.F., Wilson, M.D. and Flückiger, E.O., *Geophys. Res. Letters*, **18**, 829-832 (1991).
3. Stoker, P.H., Bieber, J.W. and Evenson, P., *24th International Cosmic Ray Conference, Contributed Papers*, **4**, 224-227 (1995).
4. Gentile, L.C., *J. Geophys. Res.*, **98**, 21,107-21,109 (1993).
5. Duldig, M.L., Cramp, J.L., Humble, J.E., Smart, D.F., Shea, M.A., Bieber, J.W., Evenson, P., Fenton, K.B., Fenton, A.G. and Bendorichio, M.B.M., *Proceedings of the Australian Astronomical Society*, **10** (3), 211-217 (1993).
6. Shea, M.A., Smart, D.F. and Pyle, K.R., *Geophys. Res. Letters*, **18**, 1655-1658 (1991).
7. Debrunner, H., Lockwood, J.A. and Ryan, J.M., *Astrophys. J.*, **409**, 822 (1993).
8. Kahler, S.W., Shea, M.A., Smart, D.F. and Cliver, E.W., *22nd International Cosmic Ray Conference, Contributed Papers*, **3**, Dublin Institute for Advanced Studies, Dublin, 21-24 (1991).
9. McCracken, K.G., *J. Geophys. Res.*, **67**, 435, 1962.
10. Shea, M.A., Cramp, J.L., Duldig, M.L., Smart, D.F., Humble, J.E., Fenton, A.G. and Fenton, K.B., *24th International Cosmic Ray Conference, Contributed Papers*, **4**, 208-211 (1995)

On the Formation of Relativistic Particle Fluxes in Extended Coronal Structures

L. I. Miroshnichenko* [1], J. Pérez-Peraza*, E. V. Vashenyuk**,
M. D. Rodríguez-Frías* [2], L. del Peral*[2] and A. Gallegos-Cruz***

*Instituto de Geofísica, U. N. A. M., 04510 - C. U., México D. F., MEXICO
**Polar Geophysical Institute of RAN, Apatity, 184200, RUSSIA
***Ciencias Básicas, UPIICSA, IPN, Té 950, Iztacalco 08400, México D. F.,
MEXICO

We analyze neutron monitor data of solar cosmic rays in order to obtain information about their sources. We use three methods for these
data analysis. As result, we obtain a set of evidences for two separate
solar cosmic rays sources that we call as prompt and delayed components. We attempt here to substantiate a two sources scenario for the
generation of both components. For the prompt component source, we
suggest regular acceleration in a neutral current sheet. For the delayed
one, we propose acceleration by magnetosonic wave turbulence.

INTRODUCTION

The first mention of the possible existence of two separate components for
relativistic solar cosmic rays was raised in (1). Independently, *Ramaty et al.* in
(2), investigating the relevant amount of all solar cosmic rays generated by the
June 3, 1982 flare, also approached towards a two components scenario. The
first population is accelerated during the flare flash phase and the particles
are trapped, low in the solar atmosphere, into closed magnetic structures.
The second component is generated high in the solar corona, at a second
acceleration phase, by a coronal shock wave belonging to the same flare.

Certain evidences of two relativistic protons ejections were found (3) for
the September 29, 1989 ground level event (GLE). Some peculiarities of this
GLE may be explained by a two separate acceleration sources model (4).

[1] Permanent address: IZMIRAN, Troistsk, Moscow Region, 142092, RUSSIA
[2] Permanent address: Depto. de Física, Universidad de Alcalá, 28871 Alcalá de
Henares, Madrid, SPAIN

OBSERVATIONAL DATA OF RELATIVISTIC SOLAR COSMIC RAYS

The study of solar cosmic rays at relativistic energies ($E > 500\ MeV$ for protons) provides an opportunity to obtain new information about acceleration processes in solar particle sources. In particular, it is usefull to clarify some features of the solar acelerators and estimate a number of important parameters of solar cosmic ray sources (upper energy limits of the acceleration mechanisms, the particle ejection time from the corona, etc.).

For the 54 GLEs registered since 1942, fourteen of them were recorded in the current 22nd solar cycle. Some of these GLEs display a set of peculiarities which seem to need interpretation under a new conceptual base.

The relativistic solar cosmic rays are detected at ground level by several neutron monitor stations around the world. We use three methods for the analysis and comparison:

1. Semiquantitative physical characteristics comparison of relativistic solar particles obtained by different neutron monitor during the same event.

2. vT_m technique, where v is the particle velocity and T_m is the maximum intensity time at 1 AU.

3. $T_{1/2}$ is the intensity-time profile half width.

Physical Characteristics of Relativistic Solar Particles

It is well known that the shape of the intensity-time profiles contains important information about the solar cosmic rays ejection time and their transport through the corona and the interplanetary space.

In several events, some neutron monitor stations, magnetically well connected with the source, are able to detect both, sharp and scattered peaks. In Fig. 1 we can observe that Apatity and Oulu neutron monitor stations located at the north hemisphere did not detect two peaks for the October 22, 1989 event, while the South Pole station did it. The reason is that South Pole station is located in the south hemisphere well connected with the flare site 27S32W.

The sharp peak particles show an anisotropic behaviour, while the smooth peak particles are fully isotropic distributed. This suggest us different sources for both components.

Fig.2. shows, for a given neutron monitor station, the dependence of the solar cosmic ray flux on the angular distance from the anisotropy axis ($\theta = 0°$) (boxes A and B in the upper left diagram) during the November 18, 1968 GLE; at the right there are the time profiles corresponding to the anisotropic component of box A and the scattered component in box B (5). The situation shown at Fig. 2. is typical for many GLEs, and in all these cases, one cannot

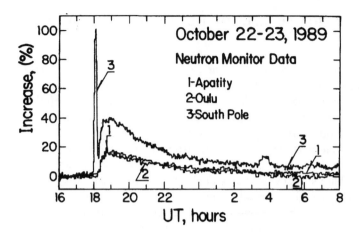

FIG. 1. Intensity-time profiles of the October 22, 1989 GLE obtained by the neutron monitors at three different stations: 1-Apatity, 2-Oulu and 3-South Pole.

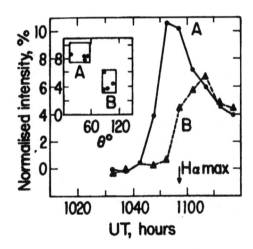

FIG. 2. Isotropic (A) and anisotropic (B) components of solar cosmic rays in the November 18, 1968 GRL. The angle $\theta = 0°$ corresponds to the average IMF direction [*Duggal et al.,* 1971].

Reprinted by permission of Kluwer Academic Publishers.

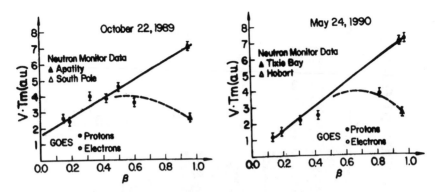

FIG. 3. vT_m-analysis for two GLEs observed in the 22nd solar cycle: October 22, 1989 and May 24, 1990.

always find any suitable shock or any other ideal reflecting boundary behind the Earth's orbit able to produce the scattered component, as it was often assumed (6). So, the scattered component is isotropically ejected from the corona and cannot be attributed to an isotropization during the transport in the interplanetary space.

The rigidity spectra of those two populations show different slopes in a single event (7). The particles of the sharp peaks show flatter spectra than those of the scattered peaks.(19)

vT_m Technique

This method was developed by *Reinhard and Wibberenz,* (8) and followed by *Van Hollebeke et al.,* (9) and *Ma Sung et al.,* (10), on basis to the following relations:

$$vT_n = A_n + B_n v \qquad (1)$$
$$vT_m = A_m + B_m v \qquad (2)$$

T_n : is the onset time;

T_m : is the maximum intensity time;

A_n : is the path of onset particles through IMF;

A_m : is the path of the main bulk particles through IMF;

B_n : is the time spent by onset particles in the corona;

B_m : is the time spent by the main bulk in the corona;

Fig. 3 demonstrate the results of vT_m-analysis for two GLEs observed in the 22nd solar cycle, namely, October 22, 1989 and May 24, 1990, respectively. It is seen, in particular, that the October 22, 1989 event had both components,

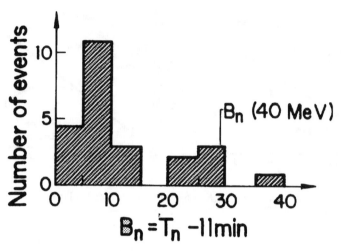

FIG. 4. Distribution of solar proton events number versus B_n parameter.

the prompt component being registered by NMs (above 500 MeV) as well as by proton detector on board the GOES satellite (above 200 MeV), meanwhile the delayed was registered in the entire range of SCR energies. The same is follows for the May 24, 1990 event, the delayed component being present also in relativistic electron population.

Conclusions of this analysis were that, if we take 11 min as the time spent by the first particles in the interplanetary space (11), we can obtain the spent time in the corona, B_n, from equation (1). As shown in Fig. 4 this parameter takes two different values for the time spent by energetic particles in the solar corona by the two different solar flare particles populations. The particle population of the sharp peak gives values for the B_n parameter between $0 - 15$ min and the particle population of the scattered peak gives B_n about 30 min in coincidence with the 29 min satellite data for the 40 MeV energy range (12). We also obtain two componets from the total path through the IMF of the main bulk particles (vT_m) analysis using equation (2). The particles of the sharp peak, have shorter paths ($vT_m \sim 2 - 3$ AU) than the particles of the scattered peak (with $vT_m \sim 7$ AU). This difference was also detected for protons by the GOES satellite in the October 22, 1989 GLE (13).

The Intensity-Time Profile Width

To derive essentially new information from GLE data it was suggested by *Vashenyuk et al.* (14) to use a specific parameter $T_{1/2}$. This parameter seems to be a measure of the time spent by the main bulk particles in the corona. The $T_{1/2}$ versus heliolongitude plot of the related flares for 42 GLEs points out that the dots concentrate in two zones, corresponding to two different heliolongitude distributions. The sharp component events have $T_{1/2} < 1$ hr and were produced at heliolongitudes between 20-130 W in the box region of

144

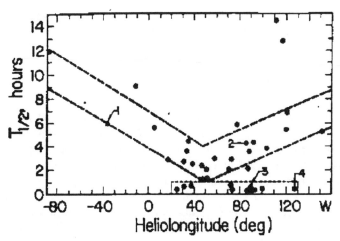

FIG. 5. Heliolongitude distribution of $T_{1/2}$, the intensity-time profiles half-width of GLEs.

Fig. 5. The scattered component events lie on a V-shape band with minimum at 50W, near the well connected line of IMF to the Earth and have similar behaviour to the same distribution for solar protons in the energy range 20-80 MeV (9). All of them, have $T_{1/2} > 1\ h$ and they were produced at heliolongitudes between 90 E - 150 W.

INTERPRETATION

Observational results considered above provide evidences of two components in relativistic solar protons events, that are apparently independent. According with previous investigations, we can state with more confidence that we deal with two components: a prompt component, which produces the sharp and anisotropic peak; and a delayed one, which produces the scattered and isotropic peak. However there are some stations that can only detect one of the components. The anisotropy characterizes the sources for both components. So, the prompt component must have an anisotropic source located in a region with open field lines (probably high in the corona), while the delayed component is associated with the magnetic bottle destroyed through plasma instabilities and reconnection of the disordered magnetic field lines.

The acceleration of the delayed component is carried out by the dissipation of local turbulence to a select number of particles able to undergo resonant interaction with the turbulence wave modes. When expanding, the flare-generated magnetic bottle gets in touch with the neighbouring magnetic arcade at heights $\sim 0.5 - 1\ R_{\odot}$ where a neutral current sheet may be formed due to magnetic reconnection between lines of opposite polarity (Fig. 6). Local particles in the non-adiabatic region of the neutral current sheet may be accelerated by the intense impulsive electric fields produced by the magnetic

145

merging process. According to *Pérez-Peraza et al.* (15) the energy spectrum of the accelerated particles in a neutral current sheet topology is:

$$N(E_k) = 1.47 \cdot 10^7 \left(\frac{nL^2}{BE_{k*}} \right) \left(\frac{E_k}{E_{k*}} \right)^{-1/4} \exp\left[-1.12 \left(\frac{E_k}{E_{k*}} \right)^{3/4} \right] \quad (3)$$

where $E_{k*} = 8.23 \cdot 10^{-3} B^2 (nL)^{2/3}$ MeV, n - plasma density and L - neutral current sheet length. In (16) *Pérez-Peraza et al.* demostrated that the source spectrum of the prompt component of three events, 23.2.1956, 7.12.1982 and 16.2.1984 may be adequately fitted by the relation (3) to the observational spectra provided the source parameters for the three events: $B = 30$, 20 and 20 G, $n = 2 \cdot 10^7$, $2 \cdot 10^6$ and $5 \cdot 10^6$ cm^{-3}, $L = 10^{10}$, $2 \cdot 10^9$ and $2 \cdot 10^{10}$ cm, respectively. These values correspond to generation altitudes $\geq 0.5R_\odot$ in the corona, and the accelerating electric field is in the range $\mathcal{E} = (U/c)B \sim 10^{-2} - 10^{-1}$ $V \cdot cm^{-1}$, where $U = v_A/18$, and v_A is the Alfvén velocity. The accelerated particles leaving the source undergo focusing in the diverging magnetic field of the corona and the IMF producing a major collimated component, though some fraction undergoes simultaneous azimutal drift ($\sim 0.7\, v_\perp$), which could be associated to the delayed arrival of particles generated in well connected events ($\sim 60° W$) as reported in (11).

According to the proposed scenario, the bulk of particles are generated in the flare volume or its vicinity (at coronal altitudes about $0.07 - 0.14$ $R_\odot$) and are ejected at the opening of a surrounding closed magnetic structure ("the magnetic bottle"). The acceleration of this component is carried out by the dissipation of local turbulence to a select number of particles able to undergo resonant interaction with the turbulence wave modes. In order to fulfill the resonant requirements (17) particles must have a relatively high initial energy, which is also necessary in order to overcome the Coulomb barrier during their "flight time" in the closed magnetic region. This entails the requirement of an injection mechanism to supply such kind of particles into the resonant stochastic process. Following (18) we assume monoenergetic injection, where all particles that are susceptible of paticipating to the acceleration resonant process with magnetosonic turbulence have an initial energy around E_0. In this case, according to (18), the source energy spectrum for a steady state situation is:

$$N(E_k) = \frac{q_0}{2} \left(\frac{a_f \alpha}{3} \right)^{-\frac{1}{2}} \left(\beta_0^{3/2} E_0 \right)^{-1} \left(\frac{\beta_0}{\beta} \right)^{\frac{1}{4}} \left(\frac{E}{E_0} \right)^{\frac{1}{2}} \exp\left[-\left(\frac{3a_f}{\alpha} \right)^{\frac{1}{2}} J_f \right]$$

$$(4)$$

where q_0 $(proton/MeV)$ is the injected flux, $\alpha(s^{-1})$ is the acceleration efficiency, E_0 is the injection energy, E_k = kinetic energy, $a_f = (\alpha/3)(\bar{F} + 3/\alpha\tau)$, τ is the mean confinement time of particles in the acceleration region, β is the particle velocity in terms of the light velocity, β_0 is the velocity cor-

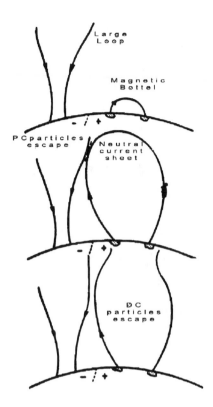

FIG. 6. Squematical drawing of the physical scenario proposed where the flare generated magnetic bottle expands and gets in touch with the extended coronal structure (large magnetic loop) and a neutral current sheet is formed.

responding to E_0, $\bar{F} = 0.5 \left[\beta^{-1} + 3\beta - 2\beta^3 + \beta_0^{-1} + 3\beta_0 - 2\beta_0^3 \right]$, and $J_f = \tan^{-1} \beta^{1/2} - \tan^{-1} \beta_0^{1/2} + 0.5 \ln \left[(1 + \beta^{1/2})(1 - \beta_0^{1/2})/(1 - \beta^{1/2})(1 + \beta_0^{1/2}) \right]$.

Calculated spectra with the equation (4) for the DC and observational spectra for the 29.9.89 and 22.10.89 events are shown on figs. 7.a) and 7.b), respectively. The best fits assuming monoenergetic injection at $E_0 = 1 \, MeV$ and $\tau \simeq 1 \, s$ are obtained in the range $\alpha = (0.03 - 0.037) \, s^{-1}$ for the first event and $\alpha = (0.034 - 0.065) \, s^{-1}$ for the second one. It should be noted that the fit was carried out without taking into account a possible interplanetary modulation of the observational spectra.

CONCLUSIONS

On basis to the observational data summarized in the beginning of this paper, we discuss here a qualitative scenario for a particular kind of solar events in which two relativistic components seem to proceed from two differ-

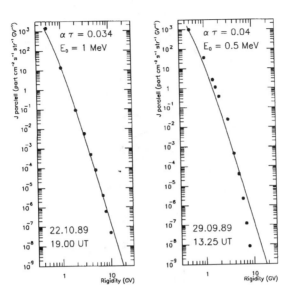

FIG. 7. Rigidity proton spectra of the delayed component of September 29, and October 22, 1989 events.

ent sources. One of them produces relativistic particles during the impulsive flare phase, deeply inside the corona, that we have designated as the delayed component; and another one operates later in the upper corona, where the conditions of particle escape are relatively easy, allowing particles to drift azimuthally through the corona and namely the prompt component because the first particle arriving to the Earth enviroment correspond to this population. Hence, during the development of the event both sources can contribute to a superposition of the observed fluxes.

The advantage of this scenario states on the fact that it does not need the assumption of continuous acceleration and/or prolonged trapping of particles to produce delayed particle arrival at the Earth's orbit. However, in order to build a model from such a scenario some of the hypothesis must be substantiated. In (16) it was shown that the energy spectra of the prompt component may be satisfactory reproduced assuming impulsive acceleration in a neutral current sheet. Here we have shown that the delayed component spectra may be satisfactory reproduced assuming stochastic acceleration by MHD turbulence. The source parameters for fitting the theoretical to the observational spectra turn to be within the order of the high and low coronal values, respectively. Similarly, the acceleration parameters range within the order of values inferred in other works on basis of the secondary radiation of flare emissions.

148

REFERENCES

1. Borovkov L. P. et al. Proc. 20th Int. Cosmic Ray Conf. **3**, 124 (1987).
2. Ramaty R. et al. Astrophys. J. **316**, L41 (1987).
3. Torsti J.J. et al. Proc. 21nd Int. Cosmic Ray Conf. **3**, 141 (1991).
4. Vashenyuk E. V. et al. Geomagnetism and Aeronomy **33**-5, 1 (1993).
5. Duggal S.P. et al. Solar Phys. **19**, 234 (1971).
6. Shea M.A. and Smart D.F. Space Sci. Rev. **32**, 251 (1982).
7. Pfotzer, G. Nuovo Cimento (Supp) **8-10**, 180 (1958).
8. Reinhard R. and Wibberenz G. Proc. 17th Int. Cosmic Ray Conf. **2**, 1372 (1973).
9. Van Hollebeke, M. A. I., Ma Sung, L. S. and McDonald, F. B. Solar Phys. **41**, 189 (1975).
10. Ma Sung, L. S., Van Hollebeke, M. A. I. and McDonald, F. B. Proc. 14th Int. Cosmic Ray Conf. **5**, 1767 (1975).
11. Cliver E.W. et al. Astrophys. J. **260**-1 362 (1982).
12. Bazilevskaya G.A. and Sladkova A.I. Geomagnetism and Aeronomy **26**, 187 (1986).
13. Solar Geophys. Data 1989, No. 542, pt. 1, p. 29; 1989, No. 543, pt. 1, p. 14; 1990, No. 550, pt. 1, p. 15.
14. Vashenyuk E.V. et al. Geomagnetism and Aeronomy **33**/5 1 (1993).
15. Pérez-Peraza J. et al. Adv. Space. Res. **18** 365 (1978); Proc. 15th Int. Cosmic Ray Conf. **5**, 23 (1977).
16. Pérez-Peraza J. et al. Geomagnetism and Aeronomy **32**-2 1 (1992).
17. Pérez-Peraza J. and Gallegos-Cruz A. Astrophys. J. (Supp.) **90**-2 669 (1994).
18. Gallegos-Cruz A. and Pérez-Peraza J. Astrophys. J. **446**-1 (1995).
19. Cramp J. L. et al. Proc. 23rd Int. Cosmic Ray Conf. **3**, 51 (1993).

GAMMA RAYS AND NEUTRONS

Thin Target γ-ray Line Production During The 1991 June 1 Flare

G. Trottet[1], C. Barat[2], R. Ramaty[3], N. Vilmer[1], J. P. Dezalay[2], A. Kuznetsov[4], N. Mandzhavidze[3,5], R. Sunyaev[4], R. Talon[2], O. Terekhov[4]

[1] *Observatoire de Paris, Section de Meudon, DASOP, URA 1756, F- 92195 Meudon, France*
[2] *Centre d'Etude Spatiale des Rayonnements, BP 4346, F-31029 Toulouse, France*
[3] *Laboratory for High Energy Astrophysics, Goddard Space Flight Center, Greenbelt MD 20771, USA*
[4] *Space Science Institute, Profsoyouznaya 84/32, 117810 Moscow, Russia*
[5] *Universities Space Research Association*

We present a time dependent analysis of the γ-ray line spectra recorded by the PHEBUS instrument on board GRANAT during the 1991 June 1 flare. For each studied spectrum we determine the electron bremsstrahlung continuum and the γ-ray line fluences $\Phi_{1.1-1.8}$ and $\Phi_{4.1-7.6}$, in the 1.1-1.8 MeV and 4.1-7.6 MeV bands respectively, as well as their ratio R= $\Phi_{1.1-1.8}$ /$\Phi_{4.1-7.6}$. We find that the power law index of the bremsstrahlung continuum in the 0.6-10 MeV band is 2.6 ± 0.1 and that it remains constant with time. The characteristic value of R, estimated for the spectrum accumulated over the whole event duration is about 4 but, surprisingly, R increases from a value of ~ 3 to a value of ~ 7 during the decay phase of the event. In order to interpret these findings we have compared the measured values of R with those obtained by using calculations of deexcitation γ-ray line spectra performed for different interacting particle spectra and compositions. We show that the high value of R and its large variation with time are inconsistent with a thick target γ-ray line source and that a thin target model is the appropriate one. This is consistent with an earlier study of the 1991 June 1 event which indicates that the observed nuclear line emission is produced in a coronal source. We finally argue that the increase of R with time strongly suggests an impulsive flare composition for the accelerated particles and the further enrichment of the heavy nuclei abundances as the flare progresses. For thin target interactions such an enrichment is most probably due to the acceleration process itself.

INTRODUCTION

Nuclear line emission in solar flares is thought to be produced in rather compact and dense ($> 10^{12}$ cm^{-3}) sources located between the Corona and

Photosphere [e.g. (1,2)]. On the other hand, the observations of the 1991 June 1 event reported in Barat et al. (3) showed that a substantial fraction of the γ-ray line emission is produced in, most probably extended, coronal sources of moderate density (some 10^{11} cm^{-3}). This coronal prompt γ-ray line emission is intense and comparable to that detected during the largest γ-ray line flares observed with the Gamma-Ray Spectrometer on board SMM. Such an observation was possible because: (i) the 1991 June 1 flare is among the most energetic flares observed so far (4); and (ii) the 1991 June 1 event is associated with an optical flare which occurred behind the limb so that the γ-ray emission from the thick target region, which was probably the most intense, is suppressed by occultation. This led Ramaty (5) to suggest that, during this flare, the observed γ-ray line emission was most likely produced in a thin target source.

In this study we have re-analyzed the observations of the 1991 June 1 event obtained with the PHEBUS instrument on board GRANAT in order to investigate which interaction model is consistent with the data. We have performed an analysis of the observed spectrum in successive intervals of time and we have obtained γ-ray line fluences $\Phi_{1.1-1.8}$ and $\Phi_{4.1-7.6}$, in the 1.1-1.8 MeV and 4.1-7.6 MeV bands respectively. We then compared the time evolution of the measured fluence ratio R= $\Phi_{1.1-1.8}$ /$\Phi_{4.1-7.6}$ with the values of R expected from calculated deexcitation γ-ray line spectra. We employ a nuclear deexcitation code (6) (updated by B. Kozlovsky, R. J. Murphy and R. Ramaty) which allows us to evaluate nuclear deexcitation gamma ray spectra for a broad range of parameters both for the ambient medium and the accelerated particles. The comparison is used to determine the nature of the interaction model which is appropriate for the 1991 June 1 flare, and to investigate the time behavior of the spectrum and composition of the accelerated particles during this flare.

DATA ANALYSIS

The PHEBUS instrument on the GRANAT spacecraft consists of 6 detectors with axes parallel to the cartesian coordinate system of GRANAT (7). Each detector is a cylindrical BGO scintillator surrounded by a plastic anticoincidence shield. In the present study of the 1991 June 1 γ-ray flare we have analyzed the records of one detector the effective area of which is $\sim$ 94 cm^2. Spectral data were obtained in the 0.12 - 90 MeV energy range, between 14:56:40 and 15:19:33 UT, with a time resolution of 1s and 4s below and above 10 MeV, respectively (3).

In a previous analysis of the 1991 June 1 event (3), it has been shown that the count rate spectrum consists of an electron bremsstrahlung continuum and a strong prompt γ-ray line structure above 1 MeV (most probably above 0.8 MeV). No significant 2.22 MeV neutron capture line was detected. The continuum spectrum was represented by a double power law with a break

154

energy at 0.40 ± 0.02 MeV during the main phase of the event and at 0.20 ± 0.02 MeV during the decay phase. As the present work focuses on the study of the prompt γ-ray line emission, the spectral analysis has been limited to the 0.57 - 10.25 MeV energy band. This energy domain has been chosen because: (i) with the exception of the α-α lines, it contains the whole γ-ray line spectrum; (ii) it does not include the delayed electron-positron annihilation line at 0.511 MeV, the study of which is outside the scope of this paper; and (iii) the continuum can be represented by a single power law which reduces the number of free parameters.

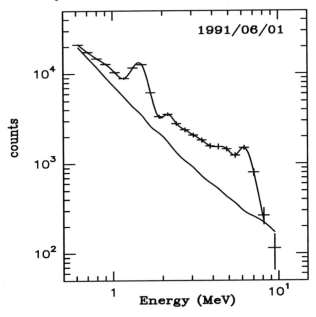

Fig. 1. Background subtracted total count spectrum of the 1991 June 1 flare recorded by PHEBUS. The solid curve drawn through the data points is the best-fit model; the fitted bremsstrahlung component is shown by the curve beneath the data points.

Figure 1 displays the background-subtracted count spectrum in the 0.57-10.25 MeV energy range accumulated in 23 energy channels from 14:57:30 to 15:19:33 UT which will be referred to as the total spectrum. Due to the moderate energy resolution of the BGO scintillators (9.8% FWHM at 1.63 MeV) and the small number of energy channels, the γ-ray line emission looks like a continuum emission with broad features in the 1-2 MeV and 3-8 MeV bands. The shape of the spectrum is not known in sufficient details neither to estimate the fluences of individual lines, as was done by Share and Murphy (8) for 19 γ-ray line flares observed with SMM, nor to separate the "narrow" and "broad" nuclear components, as was done by Murphy et al. (9). The goal of the present analysis is thus restricted to estimate the total prompt γ-ray line fluences $\Phi_{1.1-1.8}$ and $\Phi_{4.1-7.6}$ in the 1.1-1.8 MeV and 4.1-7.6 MeV energy bands [$\Phi_{1.1-1.8}$ and $\Phi_{4.1-7.6}$ are predominantly due to deexcitations in Ne, Mg, Si, S and Fe and in C, N, O respectively (6)].

A trial incident photon spectrum is constructed and convolved with the detector response function. The trial spectrum consists of:

- a bremsstrahlung continuum represented by a single power law $AE^{-\delta}$ in photon energy.

- a nuclear line component represented by a series of Gaussian functions with center energies placed at the mean energies of some (9 to 18, depending of the analyzed spectrum) of the instrumental channels and widths equal to the energy resolution (FWHM) of the detector at these energies.

The free parameters are thus A, δ and the amplitudes of the Gaussian functions. Though such a trial spectrum does not constitute a physical representation of the γ-ray line spectrum, it provides a convenient way to fit the bremsstrahlung and the nuclear components simultaneously and to get good estimates of γ-ray line fluences in wide energy bands. A χ^2-minimization algorithm is used to fit the data and find the free parameters. The fit is considered to be acceptable when a reduced χ^2 of ~1 is reached and when the uncertainties in all fitted parameters are smaller than the parameters themselves.

The results of the fitting procedure are illustrated in Figure 1 for the total spectrum accumulated from 14:57:30 till 15:19:33 UT (no line emission was observed before 14:57:30 UT). The solid curve drawn through the data points is the best fit model (reduced χ^2=1.17) and the fitted bremsstrahlung component is shown by the curve beneath the data points. In order to study the time evolution of $\Phi_{1.1-1.8}$ and $\Phi_{4.1-7.6}$ the fitting procedure is also applied to ten successive spectra accumulated over 128 s during the time interval corresponding to that of the total spectrum. According to the above criteria we got acceptable fits for 6 consecutive 128 s time intervals from 14:59:38 UT till 15:13:09 UT. The remaining 4 time intervals (one at the beginning of the event and the other 3 late in the decay phase), for which the fits are poor ($\chi^2 > 2$), will not be considered in the following.

RESULTS

The power law index, δ, of the bremsstrahlung continuum, the γ-ray line excess fluences $\Phi_{1.1-1.8}$, $\Phi_{4.1-7.6}$ and their ratio R=$\Phi_{1.1-1.8}/\Phi_{4.1-7.6}$ were estimated for the total spectrum and for each of the six spectra accumulated over 128 s for which good fits have been obtained. The results are:

- Within the uncertainties, the power law index, δ=2.6 $\pm$ 0.1, of the bremsstrahlung continuum in the 0.57-10.25 MeV band is found to be constant with time.

- For the total spectrum we get $\Phi_{1.1-1.8} = 379 \pm 27$ photons cm^{-2}, $\Phi_{4.1-7.6} = 89 \pm 7$ photons cm^{-2} and R= 4.26 ± 0.64. The excess fluence in the 3.7-7.6 MeV band, $\Phi_{3.7-7.6}$, is 101 ± 9 photons cm^{-2}. These values differ from those given in (3) who found $\Phi_{1.1-1.8} = 419 \pm 17$ photons cm^{-2} and $\Phi_{3.7-7.6} = 185 \pm 14$ photons cm^{-2}. Barat et al. (3) used mean photopeak efficiencies to convert count numbers into photon numbers while in the present study we have considered the detector response function which describes all interaction processes of incident photons in the BGO scintillator. Using the mean efficiencies given in (3), 0.58 ± 0.02 and 0.32 ± 0.02 in the 1.1-1.8 MeV and 3.7-7.6 MeV bands respectively, the present fit of the total spectrum leads to: $\Phi_{1.1-1.8} = 431 \pm 20$ photons cm^{-2} and $\Phi_{3.7-7.6} = 188 \pm 15$ photons cm^{-2}, in good agreement with the values derived in (3).

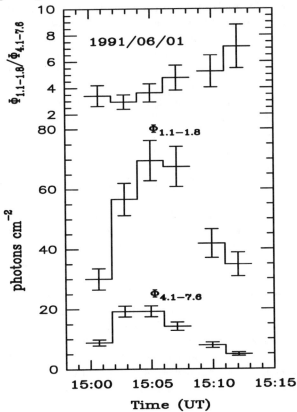

Fig. 2. Time evolution of the nuclear line fluences $\Phi_{1.1-1.8}$ and $\Phi_{4.1-7.6}$, in the 1.1-1.8 and 4.1-7.6 energy bands respectively and of their ratio R=$\Phi_{1.1-1.8}/\Phi_{4.1-7.6}$.

- Figure 2 displays the time evolution of $\Phi_{1.1-1.8}$, $\Phi_{4.1-7.6}$ and R. The missing point around 15:08 UT is due to a data gap between two triggers

157

of the PHEBUS burst mode. Figure 2 shows that $\Phi_{1.1-1.8}$ decays slightly slower than $\Phi_{4.1-7.6}$. This is reflected in the time evolution of R which increases after about 15:04 UT from a value of $\sim$ 3 to a value of $\sim$ 7 at the end of the analyzed time interval. This time behavior of R is statistically significant because the excess numbers of counts above the continuum, in the 1.1-1.8 MeV and 4.1-7.6 MeV energy bands, are everywhere larger than 17 σ and 5 σ respectively (here σ is the statistical error on the total count number in the considered energy band).

DISCUSSION

Calculations of the γ-ray line production have been carried out for both thick and thin target models (10). In the thick target model the accelerated particles lose energy while they produce the nuclear reactions. The heavier the projectile the larger is the energy loss. On the other hand, in the thin target model particles escape the interaction region before losing much energy. Consequently the contribution of the heavy accelerated nuclei to the γ-ray line production, relative to that of protons, α-particles and lighter nuclei, will be substantially larger for a thin than for a thick target model. Because the nuclear line emission in the 1.1-1.8 MeV band is predominantly due to deexcitations of Ne, Mg, Si, S and Fe while that in the 4.1-7.6 MeV band is mostly from C, N and O, the fluence ratio R = $\Phi_{1.1-1.8}/\Phi_{4.1-7.6}$ is expected to be larger for thin than for thick target interactions (5). In the following we show that the observed values of R indicate that the appropriate interaction model for the 1991 June 1 event is a thin target one.

The expected values of R have been computed by using the nuclear code of ref. (6). The source spectrum of accelerated particles has been taken for all species as a power law of index -s in kinetic energy per nucleon. The α-particle to protons ratio (α/p) is taken as 0.1 for the accelerated particles. We considered also other values of α/p (0.5 and 0.01) and found that this does not affect the conclusion discussed below. The ambient medium abundances are coronal (11) which are found to be consistent with spectroscopic obser-vations of γ-ray line flares (12–14). The enrichment of the coronal material in Mg/O, Si/O and Fe/O relative to the photosphere leads to a higher value of R. For the composition of the accelerated ions we followed the approach of ref. (12) where the calculations were done for both impulsive and gradual Solar Energetic Particles (SEP) compositions (15). The gradual composition is identical to the SEP derived coronal composition. The impulsive compo-sition is characterized by the enhancement of Ne/O, Mg/O, Si/O and S/O by a factor f_{Ne-S} and Fe/O by a factor f_{Fe} relative to their coronal values. The values of R determined for the 1991 June 1 event (vertical error bars) are shown in Figure 3 together with the computed ones for different values of s (horizontal lines). The two left panels are for the thin target model, the im-pulsive composition and different heavy element enhancement factors; f_{Ne-S}

=3, f_{Fe} =10 and f_{Ne-S} =10, f_{Fe} =20 represent respectively the mean and upper limit of the enhancement factors measured for impulsive SEP events (15). The upper right panel is for the thin target model and the gradual composition, while the lower right panel is for the thick target model and the maximal enhancement factors.

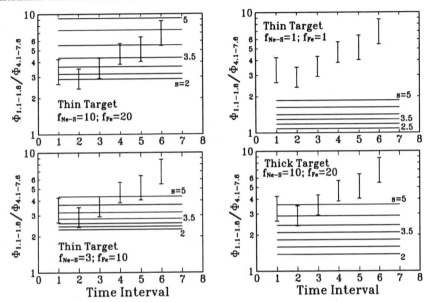

Fig. 3. Comparison between observed (vertical error bars) and computed (horizontal lines) values of R (see text).

The examination of Figure 3 leads to the following:

- Even if we considered upper limits for f_{Ne-S} and f_{Fe} , a thick target model (lower right panel) requires an extremely steep particle spectrum (s > 5) to account for the highest values of R ($\sim$ 7). Such steep spectra are unreasonable. For example, the analysis of 19 SMM γ-ray line flares (12) yielded values of s typically in the range from about 3.5 to 4.5. A thick target should thus be ruled out for the 1991 June 1 flare.

- On the other hand, a thin target model yields larger values of R, which for maximal enhancement factors (upper left panel), can account for even the highest observed R with reasonable values of s. The observed γ-ray line emission is most likely produced in the Corona (3), while the accelerated particles lose the bulk of their energy in the Chromosphere and the Photosphere.

- The gradual composition, even in the thin target model (upper right panel) is definitely inconsistent with the data. The common origin of the γ-ray producing particles and the SEP from impulsive flares was pointed out previously on the basis of both heavy element abundances and electron-to-proton ratios (16).

The two left panels of Figure 3 show that the time evolution of R reflects either a steepening of the particle spectrum with time, or an increase of the relative numbers of accelerated heavy ions in the course of the flare, or a combination of both effects. However, if the increase of R with time is due to a steepening of the particle spectrum alone this would lead to an unreasonably high energy content in the accelerated >1 MeV ions for the steepest spectra. Thus our observations strongly suggest that the increase of R with time is due, at least in part, to the continuous enhancement of the abundances of the interacting heavy ions, starting from abundances which are already enhanced relative to the gradual SEP composition. Such an evolution in the composition of the interacting particles is most likely due to the acceleration process itself.

CONCLUSION

Because it is associated with a behind the limb flare, the 1991 June 1 γ-ray line event constitutes a unique example where the production region of the observed nuclear line emission is located in the Corona (3). In this study we have performed a time-dependent analysis of the spectral data obtained with the PHEBUS instrument on board GRANAT during this flare. We have estimated the prompt nuclear line fluences $\Phi_{1.1-1.8}$ and $\Phi_{4.1-7.6}$ in the 1.1-1.8 MeV and 4.1-7.6 MeV bands, respectively, and their ratio R $=\Phi_{1.1-1.8}/\Phi_{4.1-7.6}$ for the event total spectrum and as a function of time during the rise, the maximum and the decay of the 0.57-10.25 MeV emission. For the total spectrum we got $\Phi_{1.1-1.8} = 379 \pm 27$ photons cm^{-2}, $\Phi_{4.1-7.6} = 89 \pm 7$ photons cm^{-2}, and R $= 4.26 \pm 0.64$. Furthermore we have found that $\Phi_{1.1-1.8}$ and $\Phi_{4.1-7.6}$ exhibit different time evolutions. $\Phi_{4.1-7.6}$ decays faster than $\Phi_{1.1-1.8}$ so that R increases from ~ 3 to ~ 7 during the decay phase of the event. We have computed the expected values of R from theory for both a thick and a thin target model, for different power law spectra and different compositions of the interacting particles. The comparison between the measured and computed values of R has led to the following conclusions:

- The appropriate interaction model for the 1991 June 1 flare is a thin target one. This demonstrates the earlier suggestion by Ramaty (5). For γ-ray line events associated with flares occurring on the visible disk, a thick target model seems to be more appropriate (12). In the absence of imaging observations, only when the thick target region is occulted, as it is the case for the 1991 June 1 event, can a thin target source be detected.

- The increase of R with time is most likely due to the combined effect of a steepening of the spectrum of the interacting particles with time and

160

of a relative enrichment of their composition in heavy nuclei (Ne, Mg, Si, S , Fe) as the flare progresses. If we assume that the acceleration region is located in the Corona, such a change in the composition is most likely due to the acceleration process itself.

In conclusion we have shown that the measurement of the γ-ray line fluence ratio R $= \Phi_{1.1-1.8}/\Phi_{4.1-7.6}$ as a function of time constitutes a simple and convenient spectroscopic diagnostic for thick versus thin target interactions in the γ-ray line production region, for the composition of the accelerated particles and for the acceleration process itself when a thin target model is appropriate.

ACKNOWLEDGEMENTS

One of us (GT) wishes to acknowledge Dr. R. Ramaty for inviting him to visit the Laboratory for High Energy Astrophysics (GFSC, Greenbelt), during spring 1994. The work of the French authors was supported by the Centre National d'Etudes Spatiales.

REFERENCES

1. E. L. Chupp, ARA&A, **22**, 359 (1984).
2. R. Ramaty and R. J. Murphy, Space Sci. Rev. **45**, 213 (1987).
3. C. Barat et al., ApJ, **425**, L109 (1994).
4. S. R. Kane, K. Hurley, J. M. McTiernan, M. Sommer, M. Boer, and M. Niel, ApJ, **446**, L47 (1995).
5. R. Ramaty, in The γ-ray Sky with COMPTON GRO and SIGMA, eds. M. Signore, P. Salati, and G. Vedrenne, (Dordrecht: KLUWER) 279 (1995).
6. R. Ramaty, B. Kozlovsky, and R. E. Lingenfelter, ApJ (Suppl.) **40**, 487 (1979).
7. C. Barat et al., in Proceedings of the Conference on Nuclear Spectroscopy of Astrophysical Sources, eds. G. H. Share and N. Gehrels, (New York: AIP), 395 (1988).
8. G. H. Share and R. J. Murphy, ApJ, **452**, 933 (1995).
9. R. J. Murphy, G. H. Share, J. R. Letaw, and D. J. Forrest, ApJ, **358**, 298 (1990).
10. R. Ramaty, in Physics of the Sun, (Dordrecht: REIDEL), 291 (1986).
11. D. V. Reames, Adv. Space Res. **15**, (7) 41 (1995).
12. R. Ramaty, N. Mandzhavidze, B. Kozlovsky, and R. J. Murphy, ApJ, **455**, L193 (1995).
13. R. J. Murphy, R. Ramaty, B. Kozlovsky, and D. V. Reames, ApJ **371**, 793 (1991).
14. R. Ramaty, N. Mandzhavidze, and B. Kozlovsky, this volume.
15. D. V. Reames, J-P. Meyer, and T. T. von Rosenvinge, ApJ (Suppl.) **90**, 649 (1994).
16. R. Ramaty, N. Mandzhavidze, B. Kozlovsky, and J. G. Skibo, Adv. Space Res. **13**, (9)275 (1993).

Gamma-Ray Line Measurements and Ambient Solar Abundances

Gerald H. Share, Ronald J. Murphy, Jeffrey G. Skibo[1]

E.O. Hulburt Center for Space Research, Code 7650, Naval Research Lab., Washington, D.C. 20375

We review measurements of ten narrow γ-ray lines in 19 X-class solar flares observed by the Solar Maximum Mission spectrometer from 1980 to 1989 (1) which showed that abundances of elements in the flare plasma are grouped with respect to their first ionization potentials (FIP), indicated that both the Ne/O and C/O line ratios are dependent on the spectral index of accelerated particles, and suggested that the range in low-FIP/high-FIP line ratios is similar to that in comparing coronal and photospheric compositions. We confirm this range in variability using the (Mg + Si + Fe)/Ne line ratio, which is not strongly dependent on the spectrum of accelerated ions. We also use measurements of the positronium continuum and annihilation line to show that γ-rays from flares with the lowest low-FIP to high-FIP ratios were produced deep in the chromosphere where the abundances should be close to photospheric.

INTRODUCTION

Gamma-ray line measurements provide information on the elemental abundances of ambient solar material and the accelerated particle composition and spectra in solar flares. In a recent publication (1) we applied an empirical technique (3) to study the relative intensities of 'narrow' γ-ray lines from proton and α-particle collisions with ambient material in 19 solar flares observed from 1980 to 1989 by the Solar Maximum Mission (SMM) Gamma Ray Spectrometer (GRS). We summarize this recent publication (1) in this Section and reflect on a recent interpretation of these data (2).

γ-Ray Line Measurements

The characteristics of the 19 flares are summarized Table 1. We provide the date, starting time of the spectral integration, the total accumulation time, the location on the Sun, the GOES X-ray classification, the optical importance and brightness, and the fluence (time-integrated flux) in the narrow nuclear

---

[1]NRL/NRC Postdoctoral Research Associate

TABLE 1. Parameters of 19 Flares Used in This Study. This table is reprinted courtesy of the Astrophysical Journal published by the Univ. of Chicago Press ©1995 (Amer. Ast. Soc.).

Flare	Date mm/dd/yy	Start (UT)	Accum. (s)	Location Lat Long	X-Ray Class	Optical	Narrow γ Lines γ cm^{-2}
1	04/10/81	16:46:13	524	N07 W36	X 2.3	2B	23.5 ± 3.1†
2	04/27/81	08:04:01	1916	N16 W90	X 5.5	1N	113.1 ± 6.2
3	06/03/82	11:42:27	1195	S09 E72	X 8.0	2B	28.6 ± 6.9†
4	07/09/82	07:35:24	327	N17 E73	X 9.8	3B	33.6 ± 3.4
5	11/26/82	02:29:21	393	S11 W87	X 4.5	2B	16.2 ± 2.7 †
6	12/07/82	23:38:29	2703	S19 W79	X 2.8	1B	147.9 ± 8.8
7	04/24/84	23:59:26	1097	S11 E45	X13.0	3B	55.3 ± 6.2 †
8	02/06/86	06:20:19	1228	S07 W02	X 1.7	2B	45.4 ± 4.5
9	12/16/88	08:28:50	3555	N27 E33	X 4.7	2B	219.6 ± 10.7
10	03/06/89	13:56:23	3515	N33 E71	X15.0	3B	293.8 ± 11.6
11	03/10/89	19:03:40	3341	N32 E22	X 4.5	3B	108.5 ± 7.5 †
12	03/17/89	17:31:22	835	N33 W61	X 6.5	2B	48.8 ± 4.6
13	05/03/89	03:42:49	1376	N28 E32	X 2.3	3B	24.3 ± 4.1
14	08/16/89	01:22:56	916	S15 W85	X20.0	2N	45.9 ± 3.9 †
15	08/17/89	00:47:52	2228	S17 W88	X 2.9	SN	54.4 ± 7.3
16	09/09/89	09:09:51	541	N17 E30	X 1.4	1B	17.0 ± 2.9
17	10/19/89	12:56:39	3260	S25 E09	X13.0	3B	179.7 ± 10.3 †
18	10/24/89	17:53:18	819	S29 W57	X 5.7	2N	44.7 ± 3.7 †
19	11/15/89	19:33:15	1016	N16 W27	X 1.8	2B	32.6 ± 4.4

†Indicates that the flare was only partially observed.

lines obtained by summing the fluences of the individual 'narrow' lines. Direct comparison with other parameters for these flares is not recommended because some flare data may have been lost due to earth occultation, saturation effects or bad data.

We note that 4 flares emitted higher fluences of prompt nuclear lines than did the well-studied flare on 1981 April 27. It is also noteworthy that the flare with the most prolific nuclear-line emission occurred on 1989 March 6. This is the same flare which exhibited episodes of intense electron-dominated emission (4). The 'narrow' nuclear component consists of ten line features (1). The dominant contributors to 8 strong and well-resolved lines are Fe (0.847 and 1.238 MeV), Mg (1.369 MeV), Ne (1.634 MeV), Si (1.778 MeV), C (4.439 MeV), O (4.439, 6.129, and ~7 MeV). Listed in Table 2 are the results of our fits for the 19 flares. Relative intensities are provided for these 8 lines as a percentage of the total prompt 'narrow' line fluence given in Table 1.

Implications for Ambient Abundances and Accelerated Spectra

Our analysis of these data indicated that the line fluences of low-FIP elements Mg, Si, and Fe are correlated in flares within the limited statistics of the

TABLE 2. Relative Line Fluxes for 19 Flares. This table is reprinted courtesy of the Astrophysical Journal published by the Univ. of Chicago Press ©1995 (Amer. Ast. Soc.).

Flare				Line Energy (MeV)				
	0.847	1.238	1.369	1.634	1.778	4.439	6.129	~ 7
1	8.7	7.7	18.0	27.6	9.8	9.5	13.4	4.6
	±4.7	±4.0	±5.4	±4.6	±4.4	±3.4	±2.1	±2.0
2	4.2	9.6	12.9	17.3	6.3	16.7	12.9	10.3
	±2.0	±1.8	±2.4	±2.1	±2.0	±1.7	±1.0	±0.9
3	6.5	3.9	15.9	28.3	7.7	33.2	7.8	6.1
	±8.0	±7.5	±9.0	±8.4	±7.9	±5.2	±2.8	±2.8
4	9.8	7.2	15.9	16.9	17.2	15.3	13.4	7.1
	±3.9	±3.1	±3.9	±3.5	±3.4	±2.9	±1.7	±1.6
5	16.7	8.6	13.3	12.0	6.4	6.6	11.5	14.5
	±6.4	±5.4	±6.9	±6.1	±5.8	±4.5	±2.7	±2.8
6	−0.9	8.0	14.8	23.9	8.6	18.0	14.7	8.7
	±2.1	±1.7	±2.1	±1.9	±1.7	±1.5	±0.9	±0.8
7	7.3	1.7	17.2	25.4	13.1	21.6	12.1	5.8
	±3.4	±3.1	±3.9	±3.6	±3.6	±2.8	±1.8	±1.7
8	6.1	3.1	20.5	14.4	3.6	14.3	16.6	7.4
	±3.7	±3.3	±4.2	±3.7	±3.7	±2.6	±1.6	±1.5
9	3.4	2.7	8.1	23.6	4.8	21.2	19.8	10.4
	±1.4	±1.3	±1.5	±1.5	±1.5	±1.2	±0.8	±0.7
10	6.4	4.3	15.2	21.8	9.3	16.6	13.6	9.0
	±1.3	±1.0	±1.2	±1.1	±1.1	±1.0	±0.6	±0.6
11	6.0	8.4	15.1	20.8	8.4	12.9	11.8	7.3
	±2.6	±2.3	±2.6	±2.5	±2.4	±2.0	±1.2	±1.2
12	−0.4	11.2	16.2	19.3	6.6	13.5	14.6	9.1
	±3.4	±2.9	±3.3	±3.2	±3.1	±2.7	±1.6	±1.5
13	6.6	−2.2	16.8	23.7	2.0	13.6	17.3	10.0
	±6.4	±5.9	±7.1	±6.5	±6.3	±5.1	±2.9	±2.8
14	6.3	4.2	13.2	25.0	10.3	13.4	13.9	8.5
	±3.0	±2.7	±3.0	±3.0	±2.8	±2.7	±1.7	±1.7
15	3.7	4.4	13.6	33.0	−0.4	21.2	15.0	6.1
	±5.3	±4.5	±4.9	±4.8	±4.5	±4.2	±2.5	±2.4
16	5.2	9.0	26.7	11.3	11.6	11.5	22.6	8.1
	±6.4	±5.9	±6.9	±6.4	±6.3	±5.2	±3.4	±3.0
17	2.7	1.3	10.7	23.2	8.3	20.1	17.6	8.5
	±1.5	±1.4	±1.6	±1.6	±1.6	±1.4	±0.9	±0.8
18	−4.0	7.7	7.4	22.5	10.3	17.8	18.9	9.2
	±3.2	±2.7	±3.0	±3.1	±2.9	±2.6	±1.6	±1.5
19	1.9	12.9	10.0	17.5	1.9	20.0	12.8	9.6
	±4.7	±4.4	±4.9	±4.6	±4.5	±4.2	±2.5	±2.6

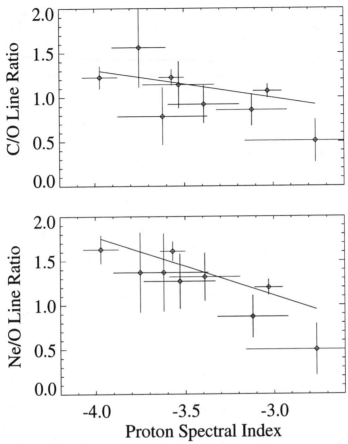

FIG. 1. Dependence of C/O and Ne/O line ratios on spectral index of accelerated particles.

measurements. On the other hand we found that the flare-to-flare C/O and Ne/O high-FIP line ratios exhibited significant scatter. This was explained by dependence of these ratios on the spectrum of interacting protons. The excited state of Ne is produced at lower impact energies than either C or O, and C is excited at lower energies than O. We would therefore expect to observe increases in the C/O and Ne/O line ratios with a steepening of the particle spectrum. This was found to be the case, as is shown in Figure 1, for 9 flares for which a measure of spectral hardness has been obtained using the 2.223 MeV to 4.4 MeV line ratios.

This suggested that the presence of a steep proton spectrum <10 MeV in the interaction region could produce an enhancement in the yield of 1.634 MeV γ-rays from Ne relative to other lines. This effect is expected from excitation cross sections when the α/p ratio in the accelerated particles is small.

Typical particle spectra from impulsive flares can be fitted with power laws having spectral indices of -3.5 or steeper above ~ 2 MeV/nucleon (5). This is consistent with spectral indices obtained by comparing the 2.223 MeV/4.439 MeV line ratio. We therefore suggested that such a steep proton spectrum in the interaction region might explain the anomalously high abundance of Ne found in a detailed analysis of the 1981 April 27 flare (6).

In contrast to the good correlation of line ratios within FIP groupings, we found that the low-FIP/high-FIP (specifically [Mg + Si + Fe]/[C + N +O]) line ratios were highly variable relative to the statistical uncertainties of the measurements. The probability that the ratios were distributed randomly about a mean was infinitesimal. We noted that these low-FIP to high-FIP line ratios varied by as much as a factor of about four and that this range is similar to the relative increase in low-FIP elemental abundances in going from the photosphere to the corona. This suggested to us that the composition of the ambient material in gamma-ray producing regions might vary from photospheric to coronal composition. Earlier analysis of the 1981 April 27 flare revealed a factor of three over abundance of low-FIP elements (6). The ratio of low-FIP/high-FIP line fluences for this flare is close to the average for all 19 flares. In contrast, the flare on 1988 December 16 had a low-FIP/high-FIP ratio at least a factor of two lower; thus we suggested that its ambient abundance was closer to that found in the photosphere.

Ramaty et al. (2) compared the SMM line measurements with detailed calculations of the γ-ray yield for varying ambient abundances, accelerated particle abundances and spectra, and interaction models. They studied the Mg/O and other abundance ratios and concluded that the mean composition of the plasma in which particles interact is coronal and that none of the flares exhibited a photospheric composition. When they used power-law spectra for the accelerated particles, as our analysis suggested, they found a Ne/O ratio of ~ 0.25 for a thick target model. This is still higher than the coronal ratio, 0.15, derived from solar energetic particles (SEP). Ramaty et al. also note that there are fluctuations in the Mg/O ratio significant at the 3σ level. The maximum range is about a factor of 4, the same factor as we derived using low-FIP/high-FIP line ratios; however, they point out that the lowest ratio is still above the photospheric value of 0.045.

In studying the Mg/O ratios in their paper we note that they are not symmetrically distributed about the coronal value; there seem to be a larger number with higher values. The highest of these may be a factor of two above the coronal value. If this effect is real then it implies that the γ-ray producing plasma can exhibit a stronger FIP effect than the SEP. Alternatively, it might suggest that the calculated Mg/O ratios derived from the γ-ray data may be systematically too high. If this were the case then the range of measured Mg/O abundance ratios might indeed reflect ambient compositions varying from photospheric (e.g. the 1988 December 16 flare) to coronal.

New γ-Ray Line Studies In the following Section we briefly summarize recent work we have done relating to flare-to-flare variability of the low-FIP/high

FIP line ratio and estimates of the density and temperature of the ambient material where the γ-rays are produced.

RECENT γ-RAY LINE STUDIES WITH SMM

We wish to determine whether there are significant flare-to-flare abundance variations in the ambient material onto which the accelerated ions impact. We previously noted that the $(Mg + Si + Fe)/(C + N + O)$ line ratio showed significant differences from flare to flare (10^{-10} probability that ratios were randomly distributed about a mean). Spectral variations of the accelerated ions could contribute to these differences, although we showed that this ratio did not appear to be correlated with the spectrum of accelerated particles (1). An alternative low-FIP/high FIP line ratio which should not have a significant dependence on the spectrum of accelerated particles is $(Mg + Si + Fe)/Ne$; this is true because the excitation cross sections have similar shapes (7). Even this ratio exhibited significant flare-to-flare variability as shown in Figure 5f in Ref. 1 (the probability rises to 6×10^{-4}, but this could be due to larger uncertainties due to reduced statistics). Ramaty et al. have also found that the Mg/O abundance ratio, even with its much larger uncertainties, appears to be variable at the 3σ level (2).

Much of the variability in the ratios is attributable to the flare which occurred on 1988 December 16. When this flare is excluded from the sample the significance of the variability in the line ratios is reduced, but is still apparent. For example the probabilities that the line ratios are randomly distributed about the mean rise from 10^{-10} to 10^{-5} for $(Mg + Si + Fe)/(C + N + O)$ and from 6×10^{-4} to 3×10^{-2} for $(Mg + Si + Fe)/Ne$. There is no residual statistical evidence for variability in the Mg/O abundance ratio, but the uncertainties are large.

Positronium Continuum and Elemental Abundances

We have searched for a physical reason for the variability in low-FIP/high-FIP line ratio and have identified an interesting correlation. One would expect that the low-FIP/high-FIP ratio would be smaller in more dense regions of the solar atmosphere, as it would reflect compositions closer to that found in the photosphere. An excellent measure of the atmospheric density is provided by the $3\gamma/2\gamma$ ratio from positron-electron annihilation. The 3γ continuum comes from annihilation of the triplet state of positronium while the 511 keV line comes from either free annihilation or annihilation from the singlet state. The triplet state is depleted at densities $> 10^{14} N_H$ cm^{-3} (8).

Seven of the nineteen flares exhibited a strong ($> 12\sigma$) 511 keV annihilation line. We have fitted spectra of these flares over the range from 300 to 750 keV with a photon model containing a power-law bremsstrahlung continuum, a positronium continuum, a nuclear spectrum based on the 1981 April

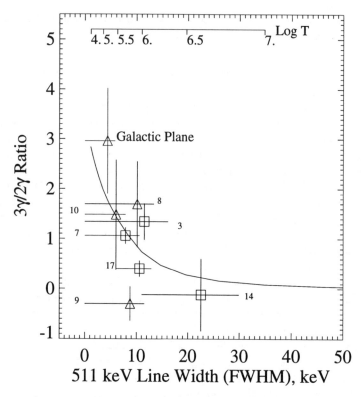

FIG. 2. Measurements of the $3\gamma/2\gamma$ positron annihilation ratio compared with the width of the 511 keV annihilation line for 7 strong flares. The curve shows the expected dependence on the temperature of the γ-ray producing plasma. The identifying numbers relate to the flares in Table 1.

27 flare, a Gaussian 511 keV line, and a broadened line feature from isotropic $\alpha - \alpha$ interactions. We have attempted to measure the width of the 511 keV line because it can provide a measure of the temperature in the region where positronium is formed. This is important because the probability for positronium formation falls rapidly at temperatures above $\sim 10^5$ K. Details of this analysis will be presented elsewhere. Here we summarize some preliminary findings as they relate to variation in abundance ratios.

Our ability to measure the width of the 511 keV annihilation line is limited by the instrumental resolution of the NaI detectors. We fit the data for the 7 flares to determine the best fit width as determined by minimizing χ^2. Plotted in Figure 2, are our measurements of the $3\gamma/2\gamma$ positron annihilation ratio versus width of the 511 keV line. A minimum of χ^2 was only found in four flares (designated with the open squares) and in only two of these were the widths measured at a significance exceeding 1σ. For the remaining three flares we were only able to measure the width at which χ^2 increased by unity, inferring a 1σ limit. These points are plotted as open triangles at 75% of the 1σ limits on the width. The plotted uncertainties range from 0 to $+1\sigma$ except for the two flares with $> 1\sigma$ significance widths. Also plotted for comparison is the width of the 511 keV line obtained from the SMM measurement of the Galactic spectrum (9); once again the plotted uncertainty is from 0 to $+1\sigma$ (~ 5.5 keV) and the data point is plotted at 75% of this value. This compares with a width of 2.6 ± 0.6 keV measured using a balloon-borne germanium detector (10).

The solid curve represents the calculated $3\gamma/2\gamma$ ratio as a function of 511 keV line width in hydrogen at densities $< 10^{13}$ cm^{-3} where quenching of the triplet state due to collisional breakup and spin flip can be ignored. We assume that the ambient medium is fully ionized hydrogen in which case the annihilation proceeds directly (11) or through radiative capture (12). At densities $< 10^{13}$ cm^{-3} the $3\gamma/2\gamma$ ratio depends only on temperature. We relate the 511 keV line width to temperature through the expression: FWHM $\simeq 1.1$ keV $\sqrt{T/10^4 K}$ (8). For densities $> 10^{13}$ cm^{-3} the corresponding curve would be lower. Detailed calculations of the $3\gamma/2\gamma$ ratio as a function of both density and temperature are in progress.

The SMM/GRS flare measurements are mostly consistent with the calculated curve, which relates to ambient material with temperatures ranging from $\sim 2 \times 10^5$ to 1×10^7 K and densities $< 10^{13}$ cm^{-3}. On the other hand, except for possibly two flares (14 and 17), the temperatures in the annihilation regions could all be lower than 10^5 K. This is due to the uncertainty in the measured widths of 511 keV line. Thus we could also attribute the low $3\gamma/2\gamma$ ratios to high densities. A local density $> 10^{14}$ cm^{-3} appears to be required to explain the lack of a positronium continuum for at least one of the flares (9). Not surprisingly this is the 1988 December 16 flare which had the lowest (Mg + Si + Fe)/Ne line ratio. This is suggestive that the annihilations took place low in the chromosphere.

It is important to determine whether the apparent variability in composition of the ambient material is generally related to density (and therefore location in the solar atmosphere). We plot the $3\gamma/2\gamma$ ratio versus low-FIP/ high-FIP line ratio in Figure 3. There is some evidence for a trend in the $3\gamma/2\gamma$ ratio versus (Mg + Si + Fe)/Ne line ratio; this is primarily due to the flares occurring on 1988 December 16 and on 1989 October 19. The flare on 1989 August 16 also had a very low $3\gamma/2\gamma$ ratio, but its 511 keV line was the broadest measured, suggesting that high temperature prevented formation of positronium. In future analyses we plan to improve the statistics in Figure 3 by using all the high-FIP lines instead of Ne. If the correlation improves then this would provide additional support for both variability in the composition of the ambient material and association with depth in the solar atmosphere.

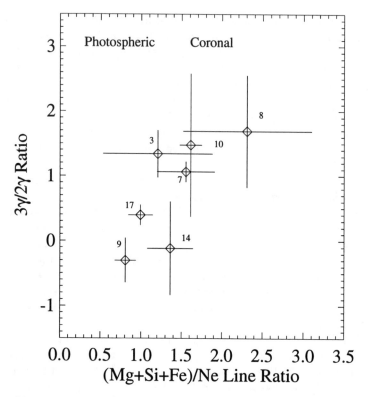

FIG. 3. Measurements of the $3\gamma/2\gamma$ positron annihilation ratio compared with the the (Mg + Si + Fe)/Ne line ratio for 7 strong flares. Two flares with low line ratios also have low $3\gamma/2\gamma$ ratios representative of high densities found deep in the chromosphere. The identifying numbers relate to the flares in Table 1.

DISCUSSION

We have shown that there is evidence for flare-to-flare variation in the line ratio of low FIP to high FIP elements in the ambient solar material which persists even after we correct for spectral effects. The range in variation is consistent with the range found in comparing coronal and photospheric abundances and suggests that ions accelerated in different flares may interact at significantly different depths. This conjecture appears to be confirmed because at least one, and possibly two of the flares, with low $(Mg + Si + Fe)/Ne$ (i.e. low-FIP/high-FIP) line ratios also have low $3\gamma/2\gamma$ positron annihilation ratios, suggesting that the γ-rays were produced at high densities. This suggests to us that γ-rays in flares may actually be produced in regions ranging from the upper photosphere to the corona.

We have noted that the Mg/O abundance ratios derived by Ramaty et al. (2) for most of the flares appear to be systematically higher than the coronal value and that one is as much as a factor of two higher. It seems unlikely that the material in which γ-rays are produced would exhibit a larger FIP effect than found in solar energetic particles. The Ne/O abundance ratio derived from γ-rays also appears to be higher than the accepted solar ratio. If the yield of γ-ray lines attributable to oxygen were overestimated by a factor of ~ 2 then the Ne/O ratio would approach the accepted value and the Mg/O ratios would range from coronal to photospheric compositions. This would be consistent with evidence presented here that a flare with one of the lowest Mg/O abundance ratios emitted γ-rays at densities near that of the photosphere.

Acknowledgment This work was supported under NASA contract DPR W-1832.

REFERENCES

1. G.H. Share and R.J. Murphy, Ap. J. **452**, 933 (1995).
2. R. Ramaty, et al., Ap. J. Letters **455**, L193 (1995).
3. R.J. Murphy, et al., Ap. J. **358**, 298 (1990).
4. E. Rieger, Ap. J. Supp. **90**, 645 (1994).
5. D.V. Reames, et al., Ap. J. **387**, 715 (1992).
6. R.J. Murphy, et al., Ap. J. **371**, 793 (1991).
7. R. Ramaty, et al., Ap. J. Supp. **40**, 487 (1979).
8. C.J. Crannell, et al., Ap. J. **210**, 582 (1976).
9. M.J. Harris, et al., Ap. J. **362**, 135 (1990).
10. N. Gehrels, et al., Ap. J. **375**, L13 (1991).
11. R.W. Bussard, et al., Ap. J. **228**, 928 (1979).
12. R.J. Gould, Ap. J. **344**, 232 (1989).

Solar Atmospheric Abundances from Gamma Ray Spectroscopy

Reuven Ramaty[1], Natalie Mandzhavidze[1,2] and Benzion Kozlovsky[3]

[1]*Laboratory for High Energy Astrophysics, Goddard Space Flight Center, Greenbelt MD 20771*
[2]*Universities Space Research Association*
[3]*School of Physics and Astronomy, Tel Aviv University, Ramat Aviv, Israel*

We used SMM gamma ray data from 19 solar flares to study ambient elemental abundances in the solar atmosphere. We found that the abundance ratios of low FIP (first ionization potential) to high FIP elements (Mg/O, Si/O, Fe/O) are enhanced relative to photospheric abundances and may vary around their respective coronal values. For the high FIP elements (C, O, Ne) we showed that: (i) The gamma ray data allows a good determination of the C to O abundance ratio; the data are consistent with a C/O which does not vary from flare to flare; and the best fit value is C/O=0.4. (ii) The derived value of Ne/O ($\sim$0.25) is higher than the coronal value of 0.15 obtained from solar energetic particle data and some EUV and X-ray observations of photospheric material. To avoid Ne/O higher than 0.3 a steep accelerated particle energy spectrum extending down to about 1 MeV/nucl is needed. This implies that a large fraction of the available flare energy is contained in accelerated ions.

INTRODUCTION

The interactions of flare accelerated protons and α particles with the ambient solar atmosphere produce narrow gamma ray lines whose intensities depend on the heavy element abundances at the gamma ray production site. Solar flare gamma ray spectroscopy can thus be used to determine abundances in the solar atmosphere. Strong narrow line emission at 4.44, 6.13, 1.63, 1.37, 1.78, and 0.85 MeV, resulting from deexcitations in ^{12}C, ^{16}O, ^{20}Ne, ^{24}Mg, ^{28}Si and ^{56}Fe, respectively, has been observed from many flares. An earlier analysis (1), carried out for the flare of 1981 April 27, showed that the abundances of Mg, Si and Fe relative to C and O are consistent with coronal abundances but enhanced in comparison with photospheric abundances. The elements of these two groups are distinguished by their first ionization potential (FIP). Mg, Si and Fe have low FIPs ($<$10 eV) while C and O have high FIPs ($>$11 eV). The enhancement of low FIP-to-high FIP element abundance ratios in the corona relative to the photosphere is well established from atomic spectroscopy and solar energetic particle (SEP) observations [e.g. (2,3)]. However, for Ne/O (both high FIP elements) the gamma ray analysis (1) yielded an abundance ratio that was more than a factor of 2 larger than the coronal Ne/O=0.15 derived from SEP data. The Ne abundance in the photosphere is not measured. However, the commonly adopted photospheric value is also 0.15 [e.g. (2,3)].

Solar Maximum Mission (SMM) gamma ray data, consisting of the above 6 lines and 4 additional weaker lines, have recently become available for 19 solar flares (4). It was shown that the ratios of line fluences from low FIP elements to those from high FIP elements vary from flare to flare by as much as a factor of 4, suggesting that the high FIP-to-low FIP element abundance ratios in the gamma ray production region also vary by about this factor (4). Such large abundance variations were derived from EUV and X-ray observations [e.g. (2,5)]. On the other hand, the line fluence ratios from elements in the same FIP group showed either no statistically significant variations or variations that could be interpreted as due to changes of the accelerated particle energy spectrum from flare to flare. This applied in particular to the 1.63 MeV ^{20}Ne and 6.13 MeV ^{16}O lines whose fluence ratios was found to correlate with spectral index, suggesting that in the gamma ray production region Ne/O does not vary from flare to flare. It was pointed out (4) that the strength of 1.63 MeV line may be the consequence of its low production threshold rather than an enhanced Ne/O, and that the high value of Ne/O obtained in (1) could have been the consequence of the Bessel function spectrum for the accelerated particles that was employed in that analysis; this spectrum bends over at low energies and hence cannot take full advantage of the low threshold.

In a previous paper (6) we analyzed these SMM data. In that paper, as well as in the present one, we employ a nuclear deexcitation code (7) (updated by B. Kozlovsky, R. J. Murphy and R. Ramaty) which allows us to evaluate nuclear deexcitation gamma ray spectra for a broad range of parameters both for the ambient medium and the accelerated particles. As 2.22 MeV neutron capture line fluences were provided for 9 of the 19 SMM flares (4), we also calculate fluences for this line using a neutron production code (8) (updated by B. Kozlovsky, R. E. Lingenfelter and R. Ramaty) and neutron-to-2.22 MeV photon conversion factors (9). In these calculations we again explore a broad range of parameter space, including the enriched abundances of accelerated ^{3}He and heavy elements that distinguish impulsive flares from gradual flares (3).

Our conclusions (6) were the following: (i) Mg/O in the gamma ray production region for all 19 flares is clearly enhanced relative to its photospheric value and shows some evidence for variability around a value which is consistent with the coronal Mg/O. (ii) Even by assuming an unbroken power law spectrum for the accelerated protons and α particles, Ne/O cannot be as low as 0.15; the available data are consistent with Ne/O=0.25. (iii) The implication of an unbroken power law energy spectrum extending down to 1 Mev/nucl (near the threshold for the ^{20}Ne line production) is that the energy contained in accelerated ions constitutes a major fraction of all the energy in flare accelerated particles.

In the present paper we describe our analysis in more detail. We present new calculations of neutron production and 2.22 MeV-to-4.44 MeV line fluence ratios and use the results to constrain the energy spectra of the accelerated ions. In addition to Mg/O we consider several other abundance ratios: C/O, Mg/Ne, Si/O and Fe/O to determine abundances and to further investigate the question of abundance variability.

We perform all of our calculations assuming a thick target interaction model (10). In our previous paper (6) we also considered a thin target model, but since the validity of a this model is quite doubtful, we do not use it in the present paper.

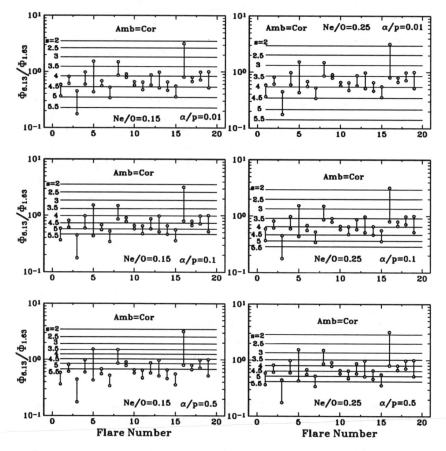

Fig. 1. O-to-Ne line ratios. The SMM data (4) (vertical bars) are plotted together with calculated line ratios (horizontal lines) for various power law spectral indexes.

ANALYSIS

We first consider the 6.13 MeV ^{16}O-to-1.63 MeV ^{20}Ne line fluence ratios, $\phi_{6.13}/\phi_{1.63}$. In Fig. 1 we compare the SMM data for the 19 flares with calculations; $\phi_{6.13}/\phi_{1.63}$ depends primarily on Ne/O, on the accelerated particle spectral index S, and on the α particle-to-proton ratio (α/p). We assume that the proton and α particle spectra are power laws with the same S. $\phi_{6.13}/\phi_{1.63}$ also depends on the Mg and Si abundances, but this dependence is quite weak (6). For the calculations shown in Fig. 1 we used coronal (11) values for Mg/O and Si/O. As we shall see below, the SMM data imply that both Mg/O and Si/O are more consistent with coronal than photospheric values.

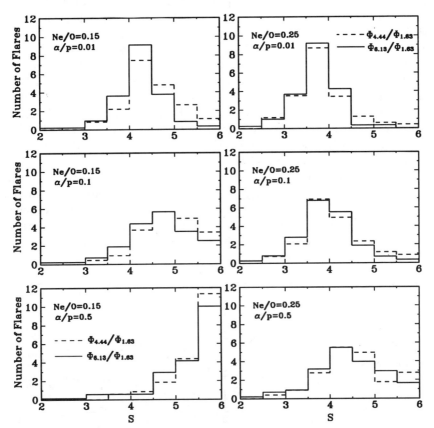

Fig. 2. Distribution of power law spectral indexes derived from the $\phi_{6.13}/\phi_{1.63}$ and $\phi_{4.44}/\phi_{1.63}$ data. Coronal abundances and C/O=0.43 were used in these calculations.

Because of the strong spectral and α/p dependencies, the observed values of $\phi_{6.13}/\phi_{1.63}$ alone cannot be used to determine Ne/O. Instead, we used the calculations of Fig. 1 to derive values of S for each flare, assuming the indicated values of α/p and Ne/O, and then constrained these parameters using observations of the 2.22 MeV and 4.44 MeV line fluences. Values of α/p ranging from about 0.01 to 0.5 are consistent with the broad variability seen in solar flare accelerated particle data (3,12). As already mentioned, both the coronal Ne/O (derived from energetic particle observations) and the commonly adopted photospheric Ne/O are about 0.15; however, as we shall see, the 2.22 MeV line data is more consistent with the higher assumed value, Ne/O=0.25.

We display the resultant spectral indexes by the distributions shown as solid histograms in Fig. 2. The dashed histograms were obtained by an analogous approach using the observed 4.44 MeV ^{12}C-to-1.63 MeV ^{20}Ne line fluence ratios, $\phi_{4.44}/\phi_{1.63}$; the calculated $\phi_{4.44}/\phi_{1.63}$ depends strongly on both Ne/C and Ne/O, but, similar to $\phi_{6.13}/\phi_{1.63}$, it is only weakly dependent on Mg/O and Si/O. As before, we have varied Ne/O, and fixed the Mg and Si abundances to that of O using coronal values (11) and we took C/O=0.43. The fact that the two distributions agree quite well for all choices of Ne/O and α/p provides evidence for the internal consistency of our approach. We carried out a similar analysis using the nominal coronal C/O=0.47 (11), but found that the agreement between the solid and dashed histograms in this case is not as good as for C/O=0.43. Below, by comparing the 4.44 MeV ^{12}C and 6.13 MeV ^{16}O line fluences, we show that the SMM data is consistent with C/O=0.43 but probably inconsistent with C/O=0.47.

Considering the results of Fig. 2, we see that the accelerated particle spectra derived from either the $\phi_{6.13}/\phi_{1.63}$ or the $\phi_{4.44}/\phi_{1.63}$ data are softer for the lower value of Ne/O, and soften with increasing α/p. This is the consequence of both the low threshold for the 1.63 MeV line production in (p, p') reactions and the important contributions of (α, α') reactions to all three (6.13, 4.44 and 1.63 MeV) lines at low energies.

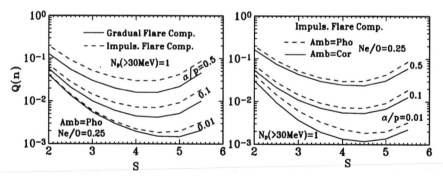

Fig. 3. Neutron yields showing the dependence on α/p and compositions.

We now consider the 2.22 MeV line which results from neutron capture in the photosphere. As opposed to the production of narrow line emission, which is due to only accelerated protons and α particles, neutron production also depends on the accelerated heavier nuclei. We considered both gradual and impulsive flare compositions for these particles. We took the gradual composition from ref. (11), which is identical to our assumed coronal composition. For the impulsive composition we increased Ne/O, Mg/O, Si/O and S/O by a factor of 3 and Fe/O by a factor of 10 relative to the gradual composition. These represent average heavy element abundance enhancements found in impulsive flares (3). For both the gradual and impulsive flare compositions we took He/O = 50. This value is bracketed by the gradual and impulsive flare values of 57 and 46 and does not exhibit much flare to flare variation (11). In addition, for the impulsive flare composition we assumed ^{3}He/^{4}He = 1. As before, for both the gradual and impulsive compositions we allowed α/p to vary between 0.01 and 0.5 and assumed power law spectra with the same S for all particle species.

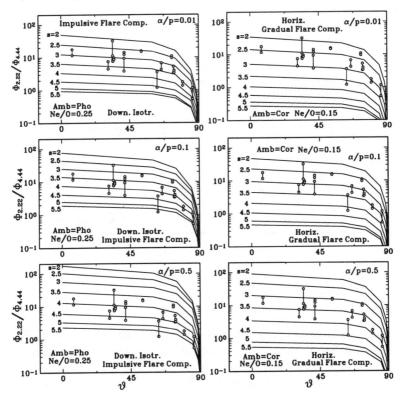

Fig. 4. Neutron capture-to-^{12}C deexcitation line ratios; θ is heliocentric angle.

We show the neutron yields in Fig. 3. The left panel is for an ambient medium with photospheric (13) composition, except for the recent update for Fe [see (14)], and accelerated particles with impulsive flare and gradual flare compositions and three values of α/p. We see that the neutron production depends strongly on α/p. The increased neutron production for the impulsive composition relative to that for the gradual composition is mostly due to the high value of ^{3}He/^{4}He and to a lesser extent due to the increased heavy element abundances. The right panel shows the dependence of the neutron production on the composition of the ambient medium. The lower yields for the coronal composition are due to the coronal He/H = 0.037 (11) which is lower than the nominal photospheric He/H=0.1 (13); even though the abundances of the low FIP heavy elements (Mg, Si, Fe) are higher in the corona than in the photosphere, this increase only slightly compensates the decreased neutron production owing to the lower He abundance. The present calculations employ the same neutron production cross sections as did our earlier calculations (15); however, they explore a broader range of parameter space. In particular, the dashed curves for $\alpha/p=0.5$ and the solid curve for $\alpha/p=0.1$ in the left panel are quite similar to the comp 2 and comp 1 neutron yields of ref. (15), respectively.

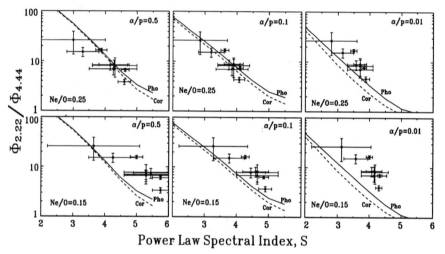

Fig. 5. Neutron capture-to-^{12}C deexcitation line fluence ratios for 9 flares (4). Both the data and calculations are corrected for heliocentric angle $\theta = 0°$ assuming a downward isotropic ion distribution.

In Fig. 4 we show 2.22 MeV-to-4.44 MeV line fluence ratios, $\phi_{2.22}/\phi_{4.44}$ as functions of heliocentric angle θ for various values of S. For each value of α/p the left and right panels correspond to maximal and minimal $\phi_{2.22}/\phi_{4.44}$. In the left (right) panel we used the impulsive (gradual) flare composition for the accelerated particles and the photospheric (coronal) composition for the ambient medium. Furthermore, in the left panel we used neutron-to-2.22 MeV photon conversion factors calculated for neutrons produced by ions with an isotropic distribution in the downward hemisphere (facing the photosphere), while in the right panel we use conversion factors appropriate for ions moving at 89° to a downward radius vector (9). The data in Fig. 4 are from ref. (4) for 9 flares (which are a subset of the 19 flares considered above), from ref. (16)

for 4 flares, and from ref. (17) for 2 flares. We see that the implied spectra become softer with increasing α/p, and those of the left panel are softer than those of the right panel. However, even the steeper spectra of the left panel are generally harder than those derived for the same α/p from the $\phi_{6.13}/\phi_{1.63}$ and $\phi_{4.44}/\phi_{1.63}$ data for Ne/O=0.15 (Figs. 1,2). This indicates that this value of Ne/O is probably too low.

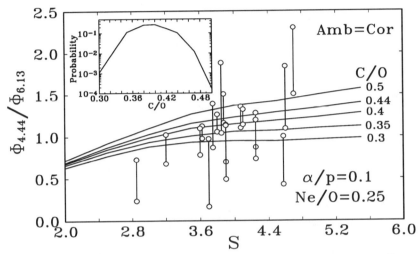

Fig. 6. ^{12}C and ^{16}O deexcitation line ratios. The insert gives the χ^2 derived probabilities corresponding to the fits of the various calculated curves to the data.

We note in addition that, when employing only $\phi_{2.22}/\phi_{4.44}$ data, the uncertainties in α/p, ^{3}He/^{4}He, heavy element abundances and flare geometry can lead to uncertainties as large as $\Delta S \simeq 1.5$ in the derivation of S (see Fig. 4).

In Fig. 5 we consider the 9 SMM flares for which $\phi_{2.22}/\phi_{4.44}$ data are available (4). The placement of the data on the horizontal axes represents the derived values of S from the Ne and O line ratios (Fig. 1). The placement of the data on the vertical axes represents the observed values of $\phi_{2.22}/\phi_{4.44}$ corrected to zero heliocentric angle for the derived spectral indexes using the downward isotropic neutron-to-2.22 MeV photon conversion factors. The values of $\phi_{2.22}/\phi_{4.44}$ given by the curves labeled 'Pho' are identical to those shown in the left panel of Fig. 4 for heliocentric angle $\theta = 0°$. As already pointed out, these values represent maxima of $\phi_{2.22}/\phi_{4.44}$. The dashed curves in Fig. 5 were obtained for the same parameters as the solid curves, except that the composition of the ambient medium was coronal. Both He/H (which affects the neutron production) and C/H (which affects the 4.44 MeV line emission) are lower in the corona (11) than in the photosphere (13). However, because the magnitude of the difference is larger for He/H, the dashed curves are lower than the solid curves. We see that even the highest calculated $\phi_{2.22}/\phi_{4.44}$ cannot account for the data if Ne/O=0.15. On the other hand, if

Ne/O=0.25, there is good agreement with the observations for α/p=0.1 and α/p=0.5. We thus exclude values of α/p much lower than 0.1 because these would require that Ne/O be larger than 0.25, which seems unlikely (2).

We now consider the $\phi_{4.44}/\phi_{6.13}$, $\phi_{1.37}/\phi_{6.13}$, $\phi_{1.37}/\phi_{1.63}$, $\phi_{1.78}/\phi_{6.13}$ and $\phi_{0.85}/\phi_{6.13}$ data, which give information on the ambient C/O, Mg/O, Ne/Mg, Si/O and Fe/O, respectively. For each flare we use the spectral index derived from $\phi_{6.13}/\phi_{1.63}$ (Fig. 1), limiting our considerations to Ne/O=0.25 and α/p=0.1. We have also tested cases with α/p=0.5 and found similar results.

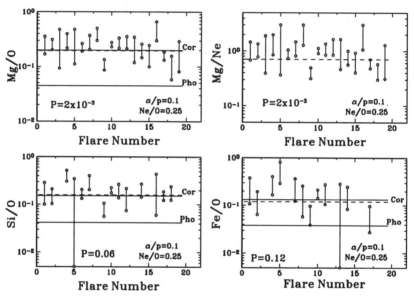

Fig. 7. Derived Mg/O, Mg/Ne, Si/O and Fe/O abundance ratios. The solid lines are coronal (11) and photospheric (13,14) abundances and the dashed lines are the best fitting abundances. For Mg/O and Mg/Ne there are data for all 19 flares; however, for Si/O and Fe/O the data are more limited.

In Fig. 6 we compare the $\phi_{4.44}/\phi_{6.13}$ data with calculations for various values of C/O. While for steep spectra $\phi_{4.44}/\phi_{6.13}$ is essentially proportional to C/O, for flatter spectra, because of the contribution of ^{16}O spallation, the dependence on C/O is weaker. We fitted the data to each one of the curves and we show the χ^2-derived probabilities in the insert. We calculated χ^2 by taking into account the uncertainties in both the $\phi_{4.44}/\phi_{6.13}$ data and S (as derived from Fig. 1) for each flare. We see that the curves for $0.35 \lesssim C/O \lesssim 0.44$ provide acceptable fits to the data, implying that C/O is consistent with a single value which does not vary from flare to flare. The best fit is given by C/O=0.4 (P=0.27). This value is consistent with the photospheric C/O=0.43±0.05 (13) and the SEP derived coronal C/O=0.43±0.05 (18). However, another SEP derived coronal C/O=0.47±0.01 (11) is marginally inconsistent with our results. But we point out that calculating the probabilities shown in Fig. 6 we

180

have included only statistical errors, without considering possible systematic errors. We also note that the recently updated photospheric C/O=0.48±0.1 (14) is higher than our gamma ray derived values, but taking into account the large uncertainty (14), there really is no discrepancy.

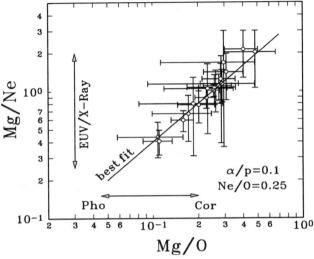

Fig. 8. Correlation between the derived Mg/Ne and Mg/O abundances. These abundances are identical to those shown in Fig. 7. The best fit corresponds to a value of Ne/O which is practically identical to the Ne/O=0.25 assumed in the calculations. The horizontal arrow indicates the photosphere-corona range for Mg/O (11,13) while the vertical arrow shows the Mg/Ne range determined from EUV and X-ray data (5).

In Fig. 7 we show derived values for Mg/O, Mg/Ne, Si/O and Fe/O. As opposed to C/O, for which both elements are have high FIP, these ratios are between low and high FIP elements. We obtained the results by using the spectral index S for each flare from the $\phi_{6.13}/\phi_{1.63}$ data (Fig. 1). Because Si spallation contributes to both the 1.37 MeV ^{24}Mg line and the 1.63 MeV ^{20}Ne line, for the derivation of Mg/O and Mg/Ne it is necessary to specify Mg/Si. We set this ratio equal to its coronal value (11) which is essentially the same as the photospheric value (13). The solid lines represent the corresponding coronal (11) and photospheric (13,14) abundances. Taking into account the uncertainties in both the data and S (from Fig. 1), we calculated the best fitting abundances by minimizing χ^2. The results are shown by the dashed lines and the corresponding probabilities are given in each panel. We see that for Mg/O, Si/O and Fe/O the best fitting abundances are either identical (for Mg/O) or very close to the respective coronal values. However, these best fitting abundances are inconsistent with photospheric abundances. Because of the uncertainty in the Ne abundance, we do not show the photospheric and coronal Mg/Ne.

Considering the results for Mg/O and Mg/Ne, the calculated probabilities exclude at about the 3σ level the hypothesis that these ratios do not vary from flare to flare. Indeed, these ratios are between low and high FIP elements and therefore are expected to vary (2,4,5). We find that Mg/O is confined to a range around the coronal value of 0.2 and does not get down to the photospheric value of 0.045. In contrast, the Mg/O determined from EUV and X-ray observations of flares and active regions varies from photospheric to coronal values (5). On the other hand, the range of variability of Mg/Ne seen in Fig. 7 is not inconsistent with the corresponding range obtained from EUV and X-ray data (5). The results of Fig. 7 for Si/O and Fe/O, because of the larger error bars, do not allow us to test the constancy, or lack of it, of the abundance ratios for these elements.

We have combined the derived Mg/Ne and Mg/O abundances in Fig. 8. The correlation between the two abundance ratios is very good ($\chi^2 = 0.74$ for 18 degrees of freedom) and precisely corresponds to Ne/O=0.25. While this result cannot be taken as a confirmation for our derived Ne/O, it does provide support for the internal consistency of the analysis as well as for the assumption that Ne/O does not vary from flare to flare.

CONCLUSIONS

We used solar flare gamma ray spectroscopy to derive abundances in the solar atmosphere. For the high FIP elements C and O we found that the SMM data is consistent with an abundance ratio $0.35 \lesssim$ C/O $\lesssim 0.44$ which shows no evidence for variability form flare to flare. Although this range seems to be lower than some of the recent solar C/O determinations, considering the uncertainties there is no real discrepancy. For another pair of high FIP elements, O and Ne, we find that the gamma ray data is in better agreement with Ne/O=0.25 than with the commonly adopted photospheric and coronal value of 0.15. Such a low Ne/O would require accelerated particle spectra which are too steep to produce sufficient neutrons to account for observations of the 2.22 MeV neutron capture line. In addition, the energy contained in ions with such steep spectra would be inconsistent with the overall flare energetics. Some EUV and X-ray observations also support a Ne/O higher than 0.15 (5,19,20).

For the low FIP elements Mg, Si and Fe we found the following: Both Mg/O and Mg/Ne show evidence for variability from flare to flare at about the 3σ level. However, Mg/O is confined to a range around the coronal value of 0.2 and does not go down to the photospheric value of 0.045. This is contrast to the variability of Mg/O obtained from EUV and X-ray data. On the other hand, Mg/Ne varies in a range similar to that obtained from EUV and X-ray observations. For Si/O and Fe/O the variations are also confined to a range around their respective coronal values; however, the error bars are too large to allow us to make a judgment concerning variability.

We also presented new calculations of neutron production and 2.22 MeV-to-4.44 MeV line fluence ratios for a broad range of parameters. This ratio can be used to determine the spectrum of the accelerated particles. However, we show that due to the uncertainties in the α/p ratio, ambient and accelerated particle compositions and flare geometry, there can be an uncertainty in the derived power law spectral indexes as large as 1.5.

REFERENCES

1. R. J. Murphy, R. Ramaty, B. Kozlovsky, and D. V. Reames, ApJ **371**, 793 (1991).
2. J-P. Meyer, in Origin and Evolution of the Elements, eds. N. Prantzos et al. (Cambridge), 26 (1992).
3. D. V. Reames, J-P. Meyer, and T. T. von Rosenvinge, ApJ (Suppl.) **90**, 649 (1994).
4. G. H. Share and R. J. Murphy, ApJ **452**, 933 (1995).
5. K. G. Widing and U. Feldman, ApJ **442**, 446 (1995).
6. R. Ramaty, N. Mandzhavidze, B. Kozlovsky, and R. J. Murphy, ApJ, **455**, L193 (1995).
7. R. Ramaty, B. Kozlovsky, and R. E. Lingenfelter, ApJ (Suppl.) **40**, 487 (1979).
8. R. J. Murphy, C. D. Dermer, and R. Ramaty, ApJ (Suppl.) **63**, 721 (1987).
9. X.-M. Hua and R. E. Lingenfelter, Solar Phys. **107**, 351 (1987).
10. R. Ramaty, in Physics of the Sun, (Reidel), 291 (1986).
11. D. V. Reames, Adv. Space Res. **15**, (7) 41 (1995).
12. J. E. Mazur, G. M. Mason, B. Klecker, and R. E. McGuire, ApJ **401**, 398 (1992).
13. E. Anders and N. Grevesse, Geochim. et Cosmochim. Acta **53**, 197 (1989).
14. N. Grevesse and A. Noels, in Origin and Evolution of the Elements, eds. N. Prantzos et al. (Cambridge), 26 (1992).
15. R. Ramaty, N. Mandzhavidze, B. Kozlovsky, and J. G. Skibo, Adv. Space Res. **13**, (9)275 (1993).
16. E. J. Schneid, private communication (1994).
17. M. Yoshimori, ApJ (Suppl.) **73**, 227 (1990).
18. T. L. Garrard and E. C. Stone, 23rd Internat. Cosmic Ray Conf. (Calgary) **3**, 384 (1993).
19. J. L. R. Saba and K. T. Strong, Adv. Space Res. **13**, (9)391 (1993).
20. J. T. Schmelz, ApJ 408, 373 (1993).

Abundance Study of the 4 June 1991 Solar Flare Using *CGRO*/OSSE Spectral Data

R.J. Murphy*, G.H. Share*, J.E. Grove*, W.N. Johnson*, R.L. Kinzer*, J.D. Kurfess*, M.S. Strickman*,G.V. Jung†

*Naval Research Laboratory, Washington, DC

†USRA, Naval Research Laboratory, Washington, DC

Spectral observations of the 4 June 1991 solar flare obtained with the Oriented Scintillation Spectrometer Experiment (OSSE) on board the *Compton Gamma Ray Observatory* (*CGRO*) are used to determine the interaction rate of the accelerated protons by fitting two-minute spectra with a narrow nuclear-deexcitation line template. The lines are produced primarily when accelerated protons interact with ambient solar material and thus reveal the proton interaction rate. The OSSE observations provide the only direct measurements of this emission during the intense portion of the 4 June flare. A subset of the data is then used to determine individual fluxes of 8 narrow γ-ray lines. These preliminary fluxes are compared to line fluxes determined for 19 flares obtained with the *Solar Maximum Mission* Gamma Ray Spectrometer (*SMM*/GRS). Ambient elements with low first-ionization potentials (FIP) appear to be enhanced relative to those with higher FIP (as compared to the photosphere) similar to the enhancement found previously for the 27 April 1981 *SMM* flare.

INTRODUCTION

Knowledge of the cosmic abundances of the elements is critical for testing theories of the early universe, stellar and galactic formation and dynamics, and nucleosynthesis. The Sun has been one of the primary sources of information on these cosmic abundances because its nearness has made possible detailed optical, UV and X-ray spectroscopic analyses of its atmospheric radiation. The compositions of various regions of the solar atmosphere have been studied using a variety of techniques (1) (2) revealing significant abundance variations. Based on solar energetic particle data and other data, it has been shown (3) that the coronal abundances of elements with low FIP are enhanced relative to those with higher FIP, as compared to photospheric abundances. Spectroscopic measurements of various regions of the solar atmosphere have also shown considerable variation in the low-FIP enhancement (2).

Line Energy	Emission Mechanism	Production Processes
1.238	$^{56}\text{Fe}^{*2.085} \rightarrow {}^{56}\text{Fe}^{*0.847}$	$^{56}\text{Fe}(p,p'){}^{56}\text{Fe}^{**}$
1.369	$^{24}\text{Mg}^{*1.369} \rightarrow \text{g.s.}$	$^{24}\text{Mg}(p,p'){}^{24}\text{Mg}^{*}$
1.634	$^{20}\text{Ne}^{*1.634} \rightarrow \text{g.s.}$	$^{20}\text{Ne}(p,p'){}^{20}\text{Ne}^{*}$
1.778	$^{28}\text{Si}^{*1.779} \rightarrow \text{g.s.}$	$^{28}\text{Si}(p,p'){}^{28}\text{Si}^{*}$
4.439	$^{12}\text{C}^{*4.439} \rightarrow \text{g.s.}$	$^{12}\text{C}(p,p'){}^{12}\text{C}^{*}$
6.129	$^{16}\text{O}^{*6.131} \rightarrow \text{g.s.}$	$^{16}\text{O}(p,p'){}^{16}\text{O}^{*}$

TABLE 1. Solar flare nuclear deexcitation lines >1 MeV.

Solar-flare γ-ray line spectra, such as observed with *CGRO*/OSSE, offer a powerful technique for studying elemental abundances of the solar atmosphere at the interaction site. Narrow nuclear deexcitation lines result primarily from reactions of accelerated protons and α particles with ambient gas. Broad nuclear deexcitation lines result from "inverse" reactions of accelerated C and heavier nuclei with ambient hydrogen. The most important lines from solar flares at energies >1 MeV are listed in Table 1. Two additional narrow features appear in solar flare γ-ray spectra at ~5.3 and ~7 MeV which are two complexes of lines resulting mostly from ^{16}O. The lifetimes of the excited states are very short ($<10^{-11}$ s) so a narrow γ-ray line time profile directly reflects the time profile of the proton interactions. The relative strengths of the narrow lines depend primarily on the ambient-gas abundances but also on the accelerated-particle energy spectrum. The excited nuclei can also be produced through other channels, such as spallation of heavier nuclei. Ambient elemental abundances can be obtained by comparing γ-ray spectra calculated from measured nuclear cross sections and kinematic calculations with observed spectra (5) (6) (4). But because most of these lines are dominated by deexcitation from a single element, some information about relative abundances can also be inferred directly from line fluxes.

Murphy et al. (4) performed the first determination of solar abundances using γ-ray spectroscopy. Gamma-ray data for the 27 April 1981 flare obtained with the *SMM*/GRS were used along with measured nuclear cross sections and kinematical calculations (5) (6). Their analysis suggested that the abundance ratios of ambient elements with low FIP (<10 eV, such as Mg, Si and Fe) to those with high FIP (>11 eV, such as C, N and O) were similar to the ratios found in the corona; i.e., about a factor of three higher than found in the photosphere.

Share and Murphy (7) measured the fluxes of 10 narrow γ-ray lines in 19 X-class solar flares observed by *SMM*/GRS. Flare-to-flare variations in line fluxes suggested that the abundances of elements in the flare plasma are grouped with respect to FIP: line fluxes from elements with similar FIP correlate well with one another. In contrast, the low-FIP to high-FIP line ratios were not consistent with a common value, varying by about a factor of four from flare to flare. The authors showed that there was little correlation of this ratio with the accelerated-particle spectral indices determined for a subset of the 19 flares, suggesting that the line flux variation is probably produced by a similar abundance variation rather than a spectral effect. The authors noted that the magnitude of this variation is similar to that of the low-FIP enhancement of the corona over the photosphere, implying that the flare abundances spanned a range from photospheric to coronal.

Ramaty et al. (8) used the *SMM* line fluxes (7) along with cross sections and kinematical calculations to determine the abundances for the *SMM* flares. They found that the flare-averaged composition of the plasma in which the particles interact is close to coronal and that the low FIP-to-high FIP ratio for individual flares was always higher than photospheric. The authors suggested that variability, if any, was less than that obtained with either atomic or γ-ray spectroscopy; however, this is in part due to limited statistics. The factor of four variability is still present in the best-fitting abundance values (e.g., for Mg/O).

In this paper, we extend the analysis of *SMM* solar-flare data to the 4 June 1991 X12+ flare observed with *CGRO*/OSSE. The intense flux from this flare resulted in a variety of instrumental problems but, using a combination of data from on- and off-pointed detectors, fluxes at >1 MeV have been determined throughout the observable emission period. The OSSE observations provide the only measurements capable of separating the ion- and electron-induced emissions during the intense portion. Other instruments were either saturated or lacked the required spectral capabilities. Preliminary results from the analysis of these data have been reported previously (9) (10) (11). We first determine the time profiles of the accelerated-proton interaction rate, as measured by the nuclear deexcitation lines, and the 2.223 MeV neutron-capture line. Using a subset of the data, we then show that the low FIP-to-high FIP line flux ratio is similar to that found for the 27 April 1981 *SMM* flare (7).

INSTRUMENT AND DATA DESCRIPTION

The OSSE instrument is described by Johnson et al. (12). OSSE consists of four large phoswich scintillation detectors surrounded by NaI annular shields. Each detector is mounted in an elevation-angle gimbal which allows independent 'scanning' through 192° of rotation. A tungsten collimator above the phoswich defines a $3.6° \times 11.5°$ FWHM field of view (FoV). During observations of the 4 June 1991 solar flare, two of the detectors maintained a fixed orientation with the Sun approximately centered in the FoV while the other two 'chopped'; i.e., alternately pointed on and 4.5° off the Sun at ~2-minute intervals. The Sun was therefore viewed by three of the detectors at all times while one of the detectors was off-pointed. In standard OSSE celestial observations, the off-pointed data is used for background estimation, but for the more intense portions of the 4 June flare the off-pointed observations provide the primary, non-saturated flare spectra (see discussion below). Background subtraction was performed via an algorithm using measured orbital, detector, and environmental parameters (12). Spectra were accumulated at ~8-sec intervals which were summed into ~2-minute intervals for this analysis. Analysis of the 8-sec data is in preparation.

In June of 1991, solar active region 6659 produced some of the largest GOES flare events ever recorded. As a result of the high probability for flare production, the Sun was declared an OSSE Target of Opportunity and on 4 June, AR 6659 produced an X12+ flare while OSSE was viewing the Sun. Observations were obtained during the rise, peak and decay of the event. The decay was interrupted by spacecraft night but observations were resumed at sunrise of the next several orbits. The Sun-pointed OSSE detectors suffered severely from saturation effects due to the intense flux of low-energy photons. The off-pointed detector, however, was largely protected by absorption of the low-energy photons in the shields and collimator. Therefore, only data from the one off-pointed detector are used when Sun-pointed detectors are saturated. Most of the flare photons <1 MeV in the off-pointed detector have been completely attenuated so the energy range addressed by the fitting procedure is limited to between 1 and 9 MeV. Sun-pointed detectors appear to be free of such saturation effects during the last 10 minutes of the first orbit and during orbits 2 and 3. These Sun-pointed data can be used for elemental abundance studies.

ANALYSIS TECHNIQUE

An assumed photon model was folded through the OSSE instrument response and compared to background-subtracted count spectra while the model parameters were adjusted until χ^2 was minimized. Flare spectra from 1 to 9 MeV are composed of narrow nuclear deexcitation lines, broad nuclear deexcitation lines, a narrow 2.223 MeV neutron-capture line and an electron bremsstrahlung continuum. The electron bremsstrahlung was represented by a power law with variable amplitude and index. The neutron-capture line was modeled with a Gaussian profile reflecting the instrumental width at 2.223 MeV with its amplitude allowed to vary. The broad lines and the weak, narrow lines unresolvable by NaI detectors that effectively blend into a continuum were represented with a single composite model composed of the five broad Gaussians used by Murphy et al. (13) in their 27 April 1981 flare analysis. The relative amplitudes of the broad Gaussians were held fixed at values derived from a new fit to the 27 April 1981 flare data using an improved response model for the *SMM* detector (7). This model component, representing both the broad deexcitation lines and the unresolved narrow deexcitation lines, is referred to simply as the *broad-line component.*

Two different approaches were used to model the narrow line component, depending on whether emission time profiles or abundance estimates were being determined. Determining emission time profiles through the peak of the flare required using unsaturated data from the off-pointed detector. Because of the poorer off-axis instrument response and the limited statistics of the data, the eight lines listed in Table 1 were combined into a single composite narrow-line model with their relative amplitudes held fixed at values derived from the new fit to the 27 April data mentioned above. The total amplitude of this composite narrow-line model, referred to as the *narrow-line component*, was varied during fitting. We note that using a narrow-line model generated with relative line fluxes derived for the 27 April flare in our analysis of the 4 June flare is reasonable for determining the proton interaction rate because the relative line fluxes in the two flares are not significantly different (see below).

For determining abundances, the improved statistics and instrument response of the Sun-pointed detectors are required. At present, this limits the usable data to those obtained during the last 10 minutes of the first orbit and during the second orbit. Two-minute accumulations from each of these two periods were summed separately and fit with a model composed of a power law, the 2.223 MeV neutron-capture line and the *broad-line component* as above, but with eight separate Gaussians replacing the *narrow-line component*.

RESULTS

Emission Time Profiles

Figure 1 shows a 2-minute spectrum from an off-pointed detector accumulated shortly after the peak of the flare. Only data >1 MeV are shown since photons at lower energy are attenuated. Also shown is the best fit obtained using the *narrow-line component* to model the narrow lines. The nuclei responsible for the various features are indicated. The model reproduces the line features present in the data reasonably well, implying that fitting with a *narrow-line component* based on the 27 April fit is adequate for determining the time profile of the proton interactions in the 4 June flare.

The time-dependent flux >1 MeV of the *narrow-line component* and the flux of the 2.223 MeV neutron-capture line throughout the three orbits of detectable emission are shown in Figures 2 and 3, respectively. The results for the first orbit when the Sun-pointed detectors were saturated are from fits to off-pointed detector data and the results from the second and third orbits are weighted means of fits to Sun-pointed detector data. During the third orbit these 2-minute results have been averaged into four intervals of seven 2-minute accumulations each to improve the statistics. Nuclear interactions and the resultant γ-ray emission are seen to continue after the peak of the flare for more than 110 minutes (until at least 5:37 UT, corresponding to the end of

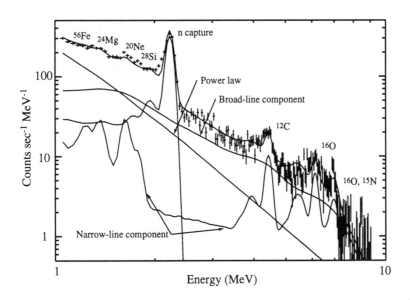

FIG. 1. OSSE 2-minute 1–9 MeV count spectrum accumulated from an off-pointed detector during the first orbit of the 4 June 1991 solar flare. Also shown is the model fit to the data using the *narrow-line component* with the components of the model indicated. The sources of the various line features are noted.

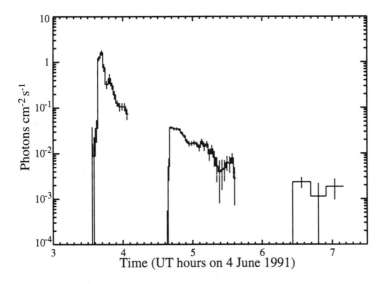

FIG. 2. The derived time profile of the narrow nuclear-line flux >1 MeV during the three orbits. Data are plotted at 2-minute resolution except for the third orbit.

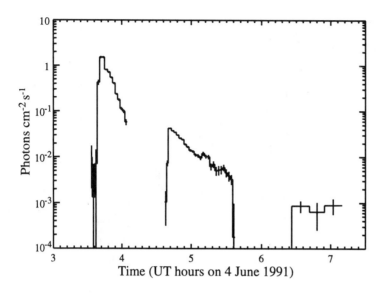

FIG. 3. The derived time profile of the 2.223 MeV neutron-capture line flux during the three orbits. Data are plotted at 2-minute resolution except for the third orbit.

the second orbit). Detection of flare photons into the third orbit is uncertain due to limited statistics and errors in background subtraction. Such extended high-energy emission is rare. The only similar example is that reported by EGRET (14) at >50 MeV from the 11 June 1991 flare.

The total, time-integrated fluxes (fluences) of the narrow lines >1 MeV and the 2.223 MeV line accumulated during the three orbits are 768 ± 42 and 931 ± 15 photons cm^{-2}, respectively. These values differ from those reported previously (10) primarily because the use of off-pointed detectors has allowed the inclusion of additional data at the flare peak. The narrow-line fluence is a factor of 2.5 higher than the largest fluence observed by *SMM*/GRS in its 10-year lifetime. A detailed analysis and discussion of the various emission components and their implications for accelerated-particle number and spectra is in preparation.

Abundance Study

Figure 4 shows a spectrum accumulated from one Sun-pointed detector during the last 10 minutes of the first orbit (3:53:58–4:04:37 UT). (Note: The feature at ~2.8 MeV is a calibration artifact.) In Panel a is shown the best fit obtained using the *narrow-line component*. In Panel b is shown the best fit obtained by replacing the *narrow-line component* with the eight individually-fit Gaussians. The reduced χ^2 has improved from 2.47 (d.o.f. = 200) to 1.35

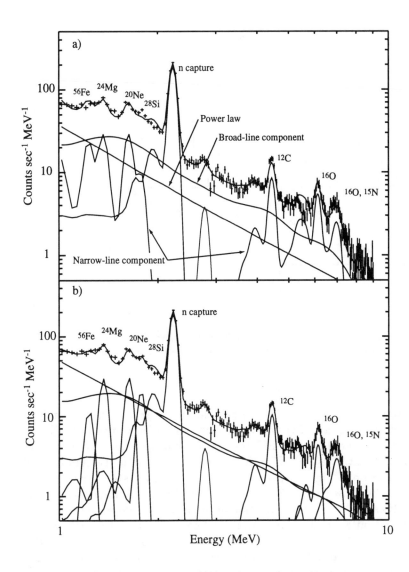

FIG. 4. OSSE 1–9 MeV count spectrum accumulated from one Sun-pointed detector late in the first orbit of the 4 June 1991 solar flare. In Panel a is shown the fit to the data using the *narrow-line component*. In Panel b is shown the fit using the eight narrow-line Gaussians.

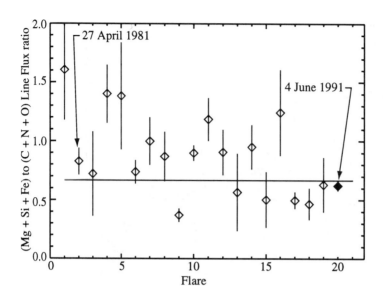

FIG. 5. Comparison of the lo-FIP (Mg+Si+Fe) to high-FIP (C+N+O) line flux ratio determined for the 4 June 1991 flare with the 19-flare *SMM* sample.

(d.o.f. = 193).

Similar to the *SMM* analysis (7), we have summed the line fluxes from low-FIP elements (Mg, Si and Fe) and those from high-FIP elements (C, N and O). We have compared the low FIP-to-high FIP ratio averaged over the last 10 minutes of orbit 1 and all of orbit 2 (0.63 ± 0.04) with the ratios found for the *SMM* sample. Figure 5 shows this 4 June average ratio along with the ratios for the 19 *SMM* flares. We note that even though the data used for the 4 June analysis represent only a fraction of the total flare data, the statistical uncertainty is quite small. The horizontal line represents the mean of this ratio for the 19 *SMM* flares and the value for 4 June is seen to be near this flare average. If the ratio determined for the 27 April 1981 flare (flare number 2) represents a factor of three enhancement of the corresponding abundances over the photospheric value as determined by Murphy et al. (4), then the value determined here for the 4 June flare represents about a factor of two enhancement.

The low FIP-to-high FIP line flux ratio for the the orbit-1 (last 10 minutes) and orbit-2 periods are 0.59 ± 0.05 and 0.74 ± 0.10, respectively. These two values are different, but only at the 17% level. If further analysis confirms that these values are indeed different, this would suggest that the corresponding abundances are changing with time, progressing from more photospheric values toward more coronal values. Such variability could possibly be due to time-dependent compositional changes at the flare site or possibly due to the flare site location changing with time, progressing from deeper in the chromosphere-photosphere toward the corona. Similarly, flare-to-flare variations in composition could reflect different heights of emission for each flare. Possible evidence for this is discussed in these Proceedings (15).

Following the individual line fluxes through the peak of the flare would better reveal any variation with time. This may require using Sun-pointed detector data since fits to off-pointed data are not sufficiently sensitive. The Sun-pointed data will have to be corrected for both spectral distortions and lost sensitivity and a technique is being developed to accomplish this. The results will be presented in a future publication.

REFERENCES

1. E. Anders, and N. Grevesse, Geochim. Cosmochim. Acta **53**, 197 (1989).
2. J. P. Meyer, in Origin and Evolution of the Elements, ed. N. Prantzos, E. Vangioni-Flam, and M. Cassé, (Cambridge:Cambridge Univ. Press), 26 (1993).
3. J. P. Meyer, Ap. J. Suppl. **57**, 151 (1985).
4. R. J. Murphy et al., Ap. J. **371**, 793 (1991).
5. R. Ramaty, B. Kozlovsky and R. E. Lingenfelter, Space Science Reviews **18**, 341 (1975).
6. R. Ramaty, B. Kozlovsky and R. E. Lingenfelter, Ap. J. Suppl. **40**, 487 (1979).
7. G. H. Share and R. J. Murphy, Ap. J. **452**,933 (1995).
8. R. Ramaty et al., Ap. J.— Letters **455**, L193 (1995).
9. R. J. Murphy et al., in Proceedings of the Compton Gamma Ray Observatory Symposium, ed. M. Friedlander, N. Gehrels and D. Macomb, (AIP:New York), 619 (1993).
10. R. J. Murphy et al., in Proceedings of the 23rd Internat. Cosmic Ray Conf. (Calgary) **3**, 99 (1993).
11. R. J. Murphy et al., in High Energy Solar Phenomena–A New Era of Spacecraft Measurements (Waterville), 99 (1994).
12. W. N. Johnson et al., Ap. J. Supp. **86**, 693 (1993).
13. R. J. Murphy et al., Ap. J. **358**, 298 (1990).
14. G. Kanbach et al., Astr. & Ap. Supp. **97**, 349 (1993).
15. G. H. Share, R. J. Murphy and J. G. Skibo, these Proceedings.

Spectral Evolution of an Intense Gamma-Ray Line Flare

Erich Rieger

Max-Planck-Institut für extraterrestrische Physik
D-85740 Garching, Germany

The flare of 16 December 1988 is one of the most intense gamma-ray line events that the Gamma-Ray Spectrometer (GRS) on SMM has recorded. It proceeded in several well separated bursts. By taking the fluences of selected energy bands, the spectrum of the primary particles can be determined. We find that it changes from burst to burst, suggesting even different acceleration mechanisms.

INTRODUCTION

In no other situation than during Solar flares the acceleration of charged particles can be explored in such great detail, because (a) events can be studied in their temporal history and (b) the Sun is near enough to investigate the phenomenon in a very wide energy range from X-rays to gamma-rays, where the accelerated particles leave their fingerprints most clearly. The flare of 16 December 1988 is one of the biggest gamma-ray line events recorded by the GRS on SMM (1). It shows emission also above 10 MeV and was, therefore, investigated for high energy neutron emission and for pion decay radiation (2) (3) (4). These papers deal primarily with a burst that exhibited high flux beyond 10 MeV photon energies. As the flare proceeded in several well separated bursts, the spectral evolution from burst to burst can be studied, which is the purpose of this paper.

OBSERVATIONS AND DISCUSSION

The X4.7/1B flare which occurred in NOAA Active Region 5278 at a heliographic position of N26E37 on 16 December 1988 (Preliminary Report and Forecast of Solar Geophysical Data, SESC PRF 694) was very much extended in time. Its temporal history is shown in Figure 1 at high energy X-rays and at selected gamma-ray energy bands. From measurements at low energy X-rays carried out with our small X-ray spectrometers (not shown in the Figure) we deduce that SMM came out of Earth ecclipse around 0829 UT, when the flare was already on. However, the fact that the flux of the 2.2 MeV neutron capture line (panel 3 of Figure 1) began to increase not before 0832 UT tells us that the flare got energetic after the night-day passage of SMM. Next satellite night begins around 0927:30 UT when the flare presumably faded away,

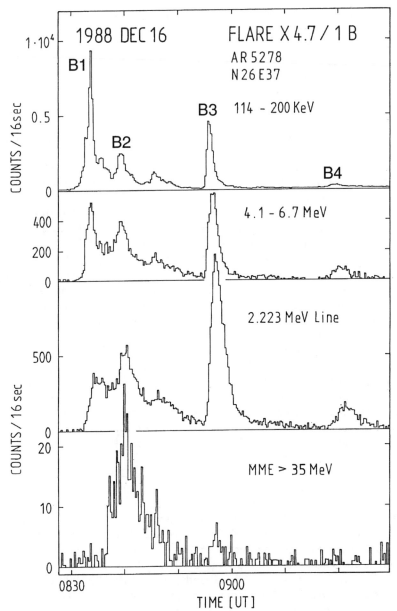

FIG. 1. Temporal history of the 16 December 1988 flare in different energy bands

195

giving almost one hour of uninterrupted observation. Over this time interval the flare evolves in 5 bursts, 4 of which are labelled. The burst around 0845 UT (best seen in panel 1) is considered a satellite of burst 2.

The energy range 4.1-6.7 MeV (panel 2) contains the strong nuclear de-excitation lines of Carbon and Oxygen. It is shown without reduction of a continuum from bremsstrahlung of high energy electrons. But in this energy range and for this special flare the continuum, which is determined at lower energies (0.3-1 MeV) and extrapolated to the nuclear energy range by assuming a power law (5), is mostly below 20total signal, so that the graph shown is a good representation of the nuclear excess radiation.

In panel 3 the flux of the 2.223 MeV neutron capture line is plotted. With an overall fluence of $610 \pm 30 \frac{\gamma}{cm^2}$ this event was one of the most prolific line flares that the GRS recorded (see also (1)).

In panel 4, named MME, events with energy losses above 35 MeV are shown which lead to a signal in the upper (NaJ) and lower (CsJ) part of the detector (6) (7) (8). As calculations have shown (9), these events are produced mainly by high energy gamma-ray photons. Signals due to high energy neutrons entering the spectrometer are suppressed by this method effectively. The measurement then is a superposition of a continuum resulting from bremsstrahlung of very high energy electrons and of pion decay photons. As shown by Alexander et al. (4), however, the bremsstrahlung component is negligible above 35 MeV.

The ratio of the neutron capture line fluence and the 4-7 MeV nuclear excess radiation fluence is a measure of the hardness of the primary proton spectrum at medium energies from about 10 to 100 MeV (10). We, therefore, see from an inspection of panel 2 and 3 of Figure 1 that the spectrum of the energetic particles must have hardened progressively from burst 1 to 3. Because of the delayed character of the 2.2 MeV neutron capture line, however, it is not sufficient to calculate the ratio of the fluences by taking the respective fluxes of certain time intervals. Following Prince et al. (11) we use the 4-7 MeV nuclear excess flux as the temporal injection profile of the energetic particles which is a measure of the production rate of the neutrons. The capture of these neutrons on Hydrogen and 3He and their decay determines the decay time of the 2.2 MeV line flux. The neutron production rate and the 2.2 MeV line decay time are varied until a sufficient agreement between the calculated and measured 2.2 MeV line flux versus time is obtained. The time intervals for the bursts are chosen as follows : (1) 0830:12 - 0835:56 UT; (2) 0835:56 - 0854:46 UT; (3) 0854:46 - 0901:03 UT and (4) 0917:10 - 0924:00 UT. The division between burst 1 and 2 is somewhat arbitrary. It was placed where the 4.1-6.7 MeV nuclear excess flux begins to raise again after burst 1.

The 4.1-6.7 MeV nuclear de-excitation fluxes, the 2.2 MeV/ 4-7 Mev fluence ratios and the 2.2 MeV decay time constants, calculated by the above mentioned method are listed in Table 1.

According to column 4 of the table, the flare as a whole is very hard, but the 2.2 MeV to 4-7 MeV ratio of burst 3 is the highest value observed so far for a flare or a burst within a flare. The parent particle spectrum must have been

TABLE 1.

Burst no.	Time interval [UT]	$\phi(4.1\text{-}6.7 \text{ MeV})$ γ/cm^2	$\dfrac{\phi_{2.2MeV}}{\phi_{4-7MeV}}$	τ_e [sec]
1	0830:12-0835:56	63 ± 6	1.4 ± 0.15	105 ± 10
2	0835:56-0854:46	125 ± 15	1.8 ± 0.16	80 ± 5
3	0854:46-0901:03	65 ± 4	3.1 ± 0.2	70 ± 5
4	0917:10-0924:00	14 ± 3	2.5 ± 0.22	90 ± 10

extremely hard. This, however, only pertains to the lower energies, because otherwise the pion decay flux above 35 MeV would have been tremendously high, which contrasts our measurements. The particle spectrum, therefore, must have steepened at higher energies considerably. Burst 2, on the other hand, because of the high photon flux above 35 MeV suggests a hard particle spectrum up to high energies. To assess this we use the Mixed Matrix Element (MME) emission above 35 MeV. The photon flux is obtained by taking the effective area published (12). For bursts 2 and 3 we get a flux of 9 ± 3 $\frac{\gamma}{cm^2}$ and 1.3 ± 0.6 $\frac{\gamma}{cm^2}$, respectively. For bursts 1 and 4 only 1sigma upper limits of 0.3 and 0.34 $\frac{\gamma}{cm^2}$, respectively, can be given. These values are normalized to the 4.1- 6.7 MeV fluences and, assuming that the signal is pion decay flux only, compared to yield calculations provided by Ramaty 1995 for a Bessel-function spectral shape and a power law (figure 2). The energetic particle spectral parameter is obtained by inserting the fluence ratios listed in column 4 of the table into new fluence ratio calculations (13) carried out for two different elemental compositions and two different directionalities of the primary particles, assuming a thick target situation.

It is evident that the energetic particle spectrum is compatible with a Bessel-function distribution for bursts 1 and 3, independent of the composition. The same holds for burst 4, whose upper limits are not listed in Figure 2. This calls for a stochastic acceleration mechanism.

Burst 2 is exceptional, because the measurements suggest a particle distribution resembling a power law. Some of the most intense events recorded by the GRS on SMM showed phases with enhanced pion decay flux, as well, which point to a hard particle spectrum. These are the flares of 3 June 1982 (7) (8), 24 April 1984 (14) and 6 March 1989 (4). But in all these cases the (gradual) burst appears only at photon energies beyond 10 MeV, not at lower energies, and is preceeded by an intense primary burst, suggesting a two-phase scenario. Also some of the giant flares of June 1991 show this time profile (15). Contrary to this temporal history, the second burst of the 16 December 1988 flare is seen simultaneously from X-rays to high energy gamma-rays as a distinct event. The power law like spectrum deduced for the energetic particles may therefore be explained by a shock wave acting as a primary accelerating agent, a situation which seems to be rather exceptional. On the other hand, due to missing spatial resolution, we cannot exclude that burst 1 has served as a preaccelerator for the particles.

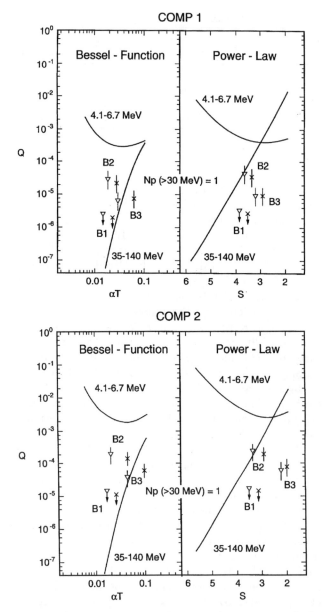

FIG. 2. Yield of the 4.1-6.7 MeV nuclear excess radiation and > 35 MeV flux resulting from pion decay versus the spectral parameters for a Bessel-function and a power law particle distribution (Ramaty, 1995, private communication) carried out for two different elemental compositions. The triangles and crosses mark the pion decay fluence normalized to the 4.1-6.7 MeV nuclear excess fluence of the bursts 1, 2 and 3 for, respectively, an isotropic downward and horizontal directionality of the primary particles.

CONCLUDING REMARKS

That a solar flare, which is extended in time, exhibits spectral variations is no surprise. But it is of interest to note how dramatic the changes are from burst to burst for this flare, even suggesting the action of different acceleration mechanisms. The present investigation once more demonstrates the importance to observe Solar flares with detectors sensitive in a wide energy range. Then insight into the phenomenon of particle acceleration can be gained which is out of the reach of particle detectors in space.

ACKNOWLEDGEMENTS

The author would like to thank Reuven Ramaty for photon yield calculations. Financial support by NASA to attend the meeting at Goddard Space Flight Center in August 1995 is acknowledged. This work was supported by the Bundesministerium fuer Forschung und Technologie under 010K017-ZA/WS/WRK0275:4 in West Germany.

REFERENCES

1. G.H. Share and R.J. Murphy, Astrophys. J. **452**, 933 (1995).
2. P.P. Dunphy and E.L. Chupp, Proc. 22nd Int. Cosmic Ray Conf. **3**, 65 (1991)
3. P.P. Dunphy and E.L. Chupp, Particle Acceleration in Cosmic Plasmas, eds. G.P. Zank and T.K. Gaisser (New York, AIP 1992), p. 253.
4. D. Alexander, P.P. Dunphy, and A.L. MacKinnon, Solar Phys. **151**, 147 (1994).
5. D.J. Forrest, Positron-Electron Pairs in Astrophysics, eds. M.L. Burns, A.K. Harding, and R. Ramaty (New York, AIP 1983), p.3.
6. D.J. Forrest et al., Solar Phys. **65**, 15 (1980).
7. D.J. Forrest et al., Proc. 19th Int. Cosmic Ray Conf. **4**, 146 (1985).
8. E.L. Chupp et al., Proc. 19th Int. Cosmic Ray Conf. **4**, 126 (1985).
9. J.F. Cooper et al., Proc. 19th Int. Cosmic Ray Conf. **5**, 474 (1985).
10. R.J. Murphy and R. Ramaty, Adv. Space Res. **4**, No7, 127 (1984).
11. T.A. Prince et al., Proc. 18th Int. Cosmic Ray Conf. **4**, 79 (1983).
12. The Gamma-Ray Spectrometer on the Solar Maximum Mission, March 11, 1991, Informal Comm. from SMM- Gamma-Ray Group, contact J. Gurman at SMM-Data Analysis Center, Goddard Space Flight Center, Greenbelt, Maryland.
13. R. Ramaty, N. Mandzhavidze, B. Kozlovsky, and J.G. Skibo, Adv. Space Res. **13**, No9, 275 (1993).
14. E. Rieger, Eruptive Solar Flares, eds. Z. Svestka, B.V. Jackson, and M.E. Machado, Lecture Notes in Physics **399**, p. 161 (1992).
15. E.J. Schneid et al., High-energy Solar Phenomena - A new Era of Spacecraft Measurements, eds. J.M. Ryan and W.T. Vestrand (New York, AIP 1993), p. 94.

COMPTEL Solar Flare Measurements

James M. Ryan and Mark M. McConnell

Space Science Center, University of New Hampshire, Durham, NH 03824

We review some of the highlights of the COMPTEL measurements of solar flares. These include images of the Sun in γ rays and neutrons. One of the important features of the COMPTEL instrument is its capability to measure weak fluxes of γ rays and neutrons in the extended phase of flares. These data complement the spectra taken with the COMPTEL burst spectrometer and the telescope during the impulsive phase of flares. We focus our attention on some of these general capabilities of the instrument and the latest results of two long-duration γ-ray flares, i.e., 11 and 15 June 1991.

INTRODUCTION

One of the original purposes of double-scatter (now more commonly known as Compton telescopes) was to measure the flux of neutrons arising from the earth's atmosphere as well as the flux of neutrons coming from the Sun. Either or both of these fluxes was postulated to be the origin of the energetic neutron-decay protons in the earth's radiation belts. A series of balloon flights with instruments of this type confirmed that the earth-albedo neutrons could populate the earth's radiation belts. Quiet-Sun measurements of the solar-neutron flux proved negative. Balloon-platform experiments have only short exposures to the Sun, so measuring the active-Sun neutron flux is consequently more difficult than with the same instrument on a spacecraft. However, with the launch of the Compton Gamma Ray Observatory, COMPTEL—a new generation double-scatter telescope—was now capable of fulfilling one of the original objectives of double-scatter telescopes, that of measuring the active-Sun emissions, including both γ rays and energetic neutrons. This review of COMPTEL solar data highlights these observations and measurements. A recent review of general high-energy solar-flare phenomena is provided by Hudson and Ryan (1).

Energetic electrons, protons and heavier ions are some of the significant products of the solar-flare process. They can account not only for emissions in generally quiet bands of the electromagnetic spectrum, i.e., microwaves, γ rays and

X rays, but can also account for a significant fraction of the flare energy. The solar-flare photon emission above 500 keV is particularly revealing. This part of the spectrum contains not only a host of nuclear lines but also the bremsstrahlung emission from relativistic electrons. Beyond 10 MeV lies the part of the spectrum containing radiation from ultrarelativistic electrons and π^0-decay photons at 68 MeV. The decay of charged π mesons yields electrons and positrons that in turn radiate. We consider these to be secondary radiations, much like 511 keV annihilation γ rays. COMPTEL is sensitive to much of this rich spectrum. In its bulk spectrometer mode it has measured solar γ rays from 600 keV to 10 MeV (2), while in its telescope mode it has measured γ rays from approximately 800 keV to 30 MeV (3). These measurements complement the concurrent measurements of BATSE (4) and OSSE (generally at lower energies) (5) and EGRET (generally at higher energies) (6,7). Unique to COMPTEL are the measurements of solar flare neutrons—a harkening to its heritage. These neutron measurements further complement the γ-ray measurements. Together they form a comprehensive set of solar measurements.

The data sample a wide range of the accelerated-particle population in a flare. Nuclear lines are produced by the energetic-proton spectrum in the range of 10–40 MeV, whereas the π^0-decay γ rays originate from the proton (or ion) spectrum above about 300 MeV. Neutrons in the range of 10–100 MeV, on the other hand, sample the proton spectrum at energies slightly higher than the neutrons themselves. The nuclear line at 2.223 MeV from the capture of free neutrons on hydrogen serves as an integral measure of the proton spectrum at all energies above the nuclear binding energy. It is useful as a crude measure of the total energy of the proton population but its origin for any given flare or any moment within a flare is uncertain. The free neutrons could be produced by protons that are also responsible for the nuclear lines but could also be produced by protons that are responsible for the π^0-decay γ rays—entirely different parts of the proton spectrum.

This review highlights some of the features and findings of the COMPTEL measurements of solar flares. We illustrate some of the technical properties of the instrument with measurements of so-called long-duration high-energy flares. Some of the more interesting phenomena are associated with these events. The remainder of the paper addresses these relatively rare events and the COMPTEL measurements of them. First, though, we briefly describe the COMPTEL instrument and its features that make it unique in solar-flare studies.

THE COMPTEL INSTRUMENT

The COMPTEL instrument is described by Schönfelder et al. (8). It is comprised of two detecting systems, D1 and D2. In the detection and measurements of photons, the γ rays Compton scatter in D1 and are ideally fully absorbed in D2. The locations of the interactions in the detector systems, the energy measurements and the Compton-scatter kinematics restrict the incident γ-ray direction to be on a circle in the sky that may pass through the location of the Sun. If it does, the measured energy is the full photon energy. A time-of-flight measurement between D1 and D2 serves to greatly improve the signal-to-noise ratio. This time-of-flight measurement is employed in the measurement of neutrons in the range of 10–100 MeV. Instead of Compton scattering in D1, the neutron elastically scatters off hydrogen in the scintillator material in D1. The scattered neutron—now with less energy—is detected in D2. The time-of-flight measurement is equivalent to measuring the scattered neutron energy. The sum of the two energies then represents the full neutron energy. Kinematics similar to the Compton scatter process also restrict the neutron incident direction to be on a circle in the sky, and this circle may also pass through the location of the Sun. There are many other complicating reactions, particularly for neutrons, that do not yield accurate energies or incident directions. The reader is referred to the instrument description by Schönfelder et al. (8) for more details.

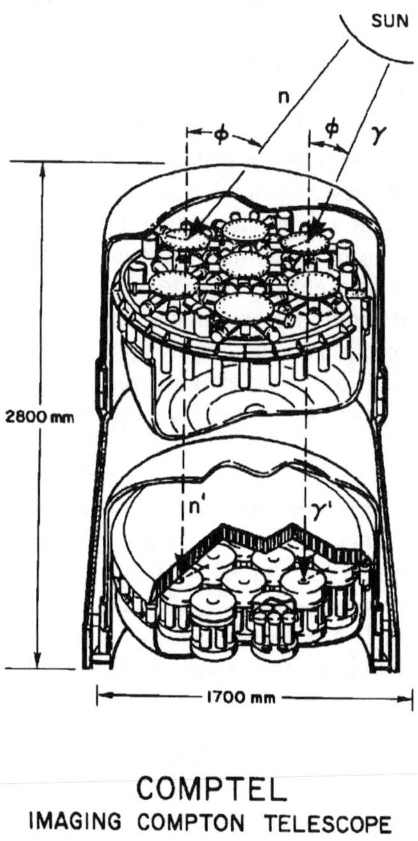

Figure 1. COMPTEL schematic.

OBSERVATIONS

γ-ray and Neutron Images

In general, the process of imaging greatly improves an instrument's signal-to-noise ratio, thereby enabling it to detect weaker signals and measure smaller fluxes. Such is the case with the COMPTEL. Exclusive of the science behind the flares that the Sun provides, the major feature of the COMPTEL solar-flare measurements is their sensitivity, best illustrated with the observations of the 15 June 1991 solar flare. The image displayed in Figure 2 is that of the Sun in the time following the

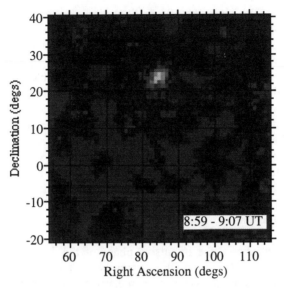

Figure 2. Gamma-ray image of the Sun from 0.8–8 MeV.

impulsive phase of a flare on 1991 June 15 (9). The image was obtained from only 9 minutes of data starting 30 minutes after the impulsive phase. Unfortunately, the resolution of the telescope does not permit resolving features smaller than ~ 1°—twice the diameter of the Sun. The Observatory missed the impulsive phase of the flare during an occulted part of the orbit. However, the lingering low-level MeV radiation was still detectable primarily because of the imaging properties of COMPTEL.

The Sun also emitted neutrons during this flare and these particles were detected and measured by COMPTEL concurrent with the γ-ray measurements. Since 10–100 MeV neutrons are subrelativistic these neutrons arose from earlier epochs of the flare. The γ-ray imaging procedures were applied to the neutron data resulting in a neutron image of the Sun (Figure 3). The neutron flux from this flare was 30× smaller than that of the 3 June 1982 flare (10). We will return to this flare because it exhibits what we now call long-duration γ-ray emission, that is γ-ray emission persisting for ~ 1 hour or more after the impulsive phase of a flare.

Neutron Measurements

The long-sought-after ability to perform neutron spectroscopy has yielded new information that is difficult or impossible to obtain with γ-ray measurements. Long-duration high-energy flares have been a poorly measured and understood phenomenon before the Compton Observatory. One of the simplest of such flares to observe and measure was that of 9 June 1991. Here, the impulsive phase was observed near the beginning of a sunlit portion of a spacecraft orbit, enabling uninterrupted γ-ray and neutron measurements for more than 40 minutes after the impulsive phase. The subrelativistic nature of the 10–100 MeV neutrons makes them arrive at the instrument mostly after the intense γ-ray flux has subsided. By measuring the energy of each registered neutron, we are able to trace the neutron back to its emission time at the Sun. A background-corrected and velocity-corrected intensity-time profile of these measured neutrons is shown in Figure 4. This 4 σ signal peaks well after the impulsive phase indicating that neutron emission is either delayed or of longer duration than that of the γ rays. This conclusion is supported by the EGRET detection of > 50 MeV γ-ray emission at this time or later (6). Whereas the flux of MeV photons is declining monotonically after the impulsive phase, the neutron flux stays at an elevated level for at least 5 minutes. The evidence points to a clear hardening of the proton spectrum over the course of 5 minutes. This phenomenon was observed again two days later with the flare on 11 June 1991.

As mentioned above, the impulsive phase of the 15 June 1991 flare was unobservable with the Compton Observatory because of earth eclipse and SAA transit. However, again because of the subrelativistic nature of the sub-100 MeV neutrons, the neutron measurements provide information about the flare high-energy emissions in the occulted and blacked-out time periods. Debrunner et al. (10) concluded that the neutron emission was consistent with the high-energy γ-ray measurements made with the GAMMA-1 instrument (11,12) that had more observing time on the Sun after the impulsive phase. The GAMMA-1 data show an exponentially declining flux of > 100 MeV γ rays after the impulsive phase, similar to what was measured by COMPTEL in γ rays (9).

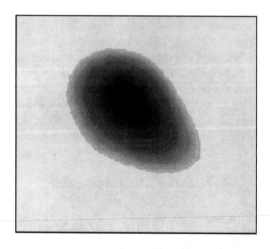

Figure 3. The Sun as it appears in neutron radiation. The image is ~ 10° in diameter and its asymmetrical shape is due to inaccuracies in the instrument point-spread function—the equivalent of optical distortion.

LONG-DURATION GAMMA-RAY FLARES

Because COMPTEL has different types and areas of detecting elements it is capable of performing measurements at different epochs of the flare that have greatly different emission rates. In particular, the burst spectrometer can handle greater fluxes without encountering dead time effects, while the telescope can measure weaker fluxes later in a flare. Such is the case for the 11 June 1991 long-duration, high-energy flare. In the impulsive phase of the flare the intense thermal X-ray flux and the limited telemetry of the telescope limited the number of registered photons. However, the burst spectrometer measured the 0.6–10 MeV spectrum every 12 s (2). Later in the flare after the flux subsided, no significant measurements in the nuclear-line region were possible with the burst spectrometer while the telescope was collecting data at its maximum rate.

The 11 June 1991 solar flare is a bellweather event in solar high-energy physics. Its γ-ray emission > 50 MeV persisted for more than 8 hours after the impulsive phase (13). Schneid et al. (6) also reported photon emission above 50 MeV minutes after the impulsive phase. The flare has attracted much attention since it tests our understanding of particle acceleration and transport. Opposing models of the particle behavior can be classified as to whether they revolve around the trapping of previously accelerated particles (14,15) or the continuous acceleration of particles (16,17). Distinguishing between these models is difficult with little more to go on than the γ-ray measurements.

COMPTEL measurements of the prolonged nuclear-line emission have been reported by Rank et al. (18) in these proceedings. The decay of the nuclear-line emission with respect to the emissions in the 2.223 MeV line and the 8–30 MeV band provides a clue into the nature of the particle populations late in this event.

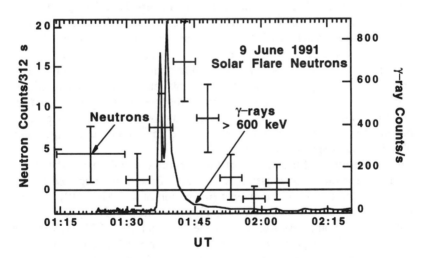

Figure 4. The velocity-corrected neutron emission from the 9 June 1991 flare.

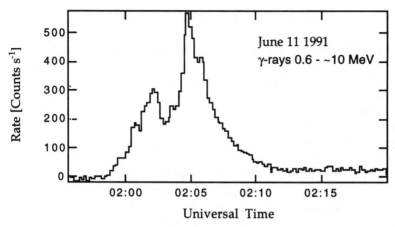

Figure 5. The background-corrected COMPTEL burst spectrometer count rate during the impulsive phase of the 11 June 1991 solar flare.

Rank et al. find that the behavior of these emission bands agree within statistics, supporting the hypothesis that the entire extended emission derives from protons, and that because of the constancy of the spectrum, the long lifetime of the proton population is due to prolonged acceleration rather than trapping. In the extended phase of the flare, the decay of all forms of γ-ray emission follows a double exponential decay—the first fast and the second much slower. The fast decay is on the order of 10 minutes and the slow decay is on the order of 100 minutes (13,18).

The situation differs during the *impulsive* phase of the 11 June 1991 flare. Figure 5 shows the count rate of the COMPTEL burst spectrometer. The burst spectrometer accumulated 256 channel spectra every 12 s during the impulsive phase. The flare consisted of an impulsive phase that included two large bursts starting at approximately 0158 UT and lasting for about 10 minutes. After about 0208 UT the flare as seen in γ rays showed little structure—only a simple exponential decay after a possible local minimum occurring at 0212 UT. This change in tempo suggests that some fundamental change in the nature of the flare was taking place at this time and this may be reflected in the spectrum and species of particles being accelerated and the photons that those particles would subsequently produce. We can investigate this putative change with the COMPTEL burst spectrometer data.

A simple but useful measure of the general spectral shape of the protons responsible for the γ-ray emission is the ratio of the time-integrated flux of the 2.223 MeV line to that of the nuclear-line region of 4–7 MeV (19). With some care this ratio can be constructed for different phases of the flare (20). The difficulty in this analysis stems from the fact that the 2.223 MeV deuterium formation line has a prolonged nature. The neutrons created by the energetic proton reactions require ~ 100 s (density and ^{3}He dependent) to capture on hydrogen, resulting in an exponential decay of the 2.223 MeV flux. The instantaneous 2.223 MeV flux is thus due to the integrated effect of all neutron production prior to that moment.

One can see in Fig. 5 that after 0210 UT the count rate does not return to background. This excess is statistically significant and is mostly in the 2.223 MeV line. We interpret this as the start of the extended emission phase. We can compare the 2.223 MeV/4–7 MeV flux ratio during the impulsive phase to that during the beginning of the extended phase to detect any change in the proton spectrum. Suleiman et al. (20) arbitrarily defined the interval 0158–0207 UT as the impulsive phase and the period 0210–0220 UT as the so-called tail phase or the beginning of the extended phase of the flare. The gap between these intervals was excluded to reduce 2.223 MeV cross-talk between the intervals. In the gap the 2.223 MeV flux appears to be declining exponentially in magnitude. Interpolating this decay allows one to estimate the 2.223 MeV flux lost from the impulsive phase interval and the excess flux introduced into the tail phase. (The corrections are both on the order of 30%. The lines in the 4–7 MeV range are prompt and one can directly account for all their emission by simply integrating the flux over time.) Doing this and combining with the time-integrated fluxes in the range of 4–7 MeV yields a 2.223 MeV/4–7 MeV flux ratio of 0.83±0.12 for the impulsive phase and 4.7±2.0 for the tail phase. The ratio in the tail phase could be larger than indicated by the error bars because so little flux is detected in the 4–7 MeV range. These ratios imply that the proton spectrum changes its shape significantly between the impulsive and tail phases of the flare. If the spectrum is of the form of a Bessel function, the hardness parameter αT changes from 0.02 to 0.12, or if the spectrum is power law in shape the spectral index changes from –3.6 to –2.3 (21). Stated either way, the change represents a marked hardening of the proton spectrum responsible for the γ-ray emission. The major evolution of the occurs sometime around or before 0210 UT when the proton spectrum hardens. Our value for the power-law index for the impulsive phase agrees with the value of –3.5 reported by Ramaty and Mandzhavidze (14), but our value for the tail phase represents a spectrum harder than what they concluded for the extended phase based solely on the high-energy spectrum hours after the impulsive phase. This disagreement in the extended or tail phases may not be significant because the two measures of the spectrum, i.e., the flux ratio and the shape of the photon spectrum > 100 MeV, span such a wide range of energy and different reaction channels that these simple measures may not be precise enough for such a comparison.

The 15 June 1991 solar flare in many ways resembles the 11 June 1991 flare. The two flares erupted in the same active region and both had prolonged high-energy photon radiation. Unfortunately, as mentioned above, the impulsive phase of the flare was unobservable to the Compton Observatory and the GAMMA-1 spacecraft as well. The similarity of the two flares may be more than just coincidental. Having originated in the same active region, they also may carry the signatures of the geometry and the topology of the region in their intensity-time profiles. Both models of prolonged γ-ray emission—trapping and continuous acceleration—are sensitive to the topology and dimensions of the magnetic structures containing the energetic particles. The time scales of the photon flux decay both follow the size of the active region, or more specifically, the size of the magnetic loops confining the particles. Rank et al. (18) have shown with the COMPTEL data that the time scales of the 2.223 MeV emissions have within uncertainties the same long and short exponential decays. This suggests that the long-duration flare scenario played out in two events four days apart. Although the impulsive phases may have been different, the particle populations, or more specifically protons, that produced the prolonged emission both resided in the same loops or, at least, loops of approximately the same dimensions.

CONCLUSIONS

Although of considerable importance and interest, our measurements of long-duration high-energy solar flares are largely based upon events stemming from a single active region in 1991. The specifics, as we saw, may depend on the details of the topology and dimensions of the particular region in which the flares occurred. Additional observations of flares in other active regions, supported by observations in other wavelength bands would broaden our knowledge base of this phenomenon. We must await the next solar maximum to study this further. The Compton Observatory and COMPTEL are both capable of operating and making good measurements in the next solar maximum. It is important that the mission be extended through the next maximum. Our experience in both solar physics and the operation of the instrument and some cooperation of the Sun will certainly yield data of the highest quality, thereby broadening our understanding of the enigmatic behavior of energetic particles in solar flares.

ACKNOWLEDGMENTS

We thank Raid Suleiman for help in preparing this paper. We also thank our COMPTEL colleagues and the staff at the Indiana University Cyclotron Facility for assistance in the neutron and gamma calibrations. This work was supported by NASA contract NAS5-26645.

REFERENCES

1. Hudson, H., and Ryan, J., *Annu. Rev. Astron. Astrophys.* **33**, 239-82 (1995).

2. Suleiman, R. et al., "COMPTEL'S Solar Flare Catalog," in *High-Energy Solar Phenomena—A New Era of Spacecraft Measurements*, **294**, New York: AIP, (1994), pp. 51-54.

3. McConnell, M., "An Overview of Solar Flare Results from COMPTEL," in *High-Energy Solar Phenomena—A New Era of Spacecraft Measurements*, AIP Conference Proceedings **294**, New York: AIP, (1994), pp. 21-25.

4. Schwartz, R. A. et al., "BATSE Flare Observations in Solar Cycle 22," in *The Compton Observatory Science Workshop*, NASA Conference Publications **3137**, NASA, (1992), pp. 457-468.

5. Murphy, R. J. et al., "OSSE Observations of Solar Flares," in *Compton Symposium*, AIP Conference Proceedings **280**, New York: AIP, (1993), pp. 619-630.

6. Schneid, E. J. et al., "EGRET Observations of Extended High Energy Emissions from the Nuclear Line Flares of June 1991," in *High-Energy Solar Phenomena—A New Era of Spacecraft Measurements*, AIP Conference Proceedings **294**, New York: AIP, (1994), pp. 94-99.

7. Dingus, B. L. et al., "EGRET Observation of the June 30 and July 2, 1991 Energetic Solar Flares," in *High-Energy Solar Phenomena—A New Era of Spacecraft Measurements*, AIP Conference Proceedings **294**, New York: AIP, (1994), pp. 177-179.

8. Schönfelder, V. et al., *Ap. J. Suppl.* **86**, 657-692 (1993).

9. McConnell, M. et al., *Adv. Sp. Res.* **13**, 245-248 (1993).

10. Debrunner, H. et al., "Neutrons from the 15 June 1991 solar flare," in *23rd International Cosmic Ray Conference*, **3**, (1993), pp. 115-118.

11. Akimov, V. V. et al., "Some Evidences of Prolonged Particle Acceleration in the High-Energy Flare of June 15, 1991," in *High-Energy Solar Phenomena—A New Era of Spacecraft Measurements*, **294**, New York: AIP, (1994), pp. 106-111.

12. Akimov, V. V. et al., "The high-energy gamma-ray flare of June 15, 1991: Some evidences of prolonged particle acceleration at the post-eruption phase," in *Proc. of Kofu Symposium*, **NRO Report #360**, Nobeyama Radio Observatory, (1994), pp. 371-374.

13. Kanbach, G. et al., *Astron. and Astrophys. Suppl.* **97**, 349-353 (1993).

14. Ramaty, R., and Mandzhavidze, N., "Theoretical Models for High Energy Solar Flare Emissions," in *High-Energy Solar Phenomena—A New Era of Spacecraft Measurements*, AIP Conference Proceedings **294**, New York: AIP, (1994), pp. 26-44.

15. Mandzhavidze, N., and Ramaty, R., *Ap. J.* **389**, 739-755 (1992).

16. Ryan, J. M., and Lee, M. A., *Ap. J.* **368**, 316-324 (1991).

17. Ryan, J. M., Bennett, E., and Lee, M. A., "Proton Acceleration in Long Duration Flares," in *Advances in Solar Physics*, **432**, Berlin: Springer-Verlag:, (1994), pp. 273-278.

18. Rank, G. et al., "Extended γ-Ray Emission in Solar Flares," in these proceedings, New York: AIP, (1996).

19. Hua, X.-M., and Lingenfelter, R. E., *Solar Phys.* **107**, 351-383 (1987).

20. Suleiman, R. M. et al., *Bull. Am. Astron. Soc.* **27**, 987 (1995).

21. Hua, X.-M., and Lingenfelter, R. E., *Ap. J.* **323**, 779-794 (1987).

Observations of Gamma-Ray Spectra and X-Ray Images of Solar Flares

M. Yoshimori, K. Morimoto, K. Suga, T. Matsuda and N. Saita

Department of Physics, Rikkyo University, Tokyo 171, Japan

Two gamma-ray flares were observed with a gamma-ray spectrometer and two X-ray imagers aboard *YOHKOH*. The gamma-ray flare on 27 October 1991 showed strong narrow deexcitation lines and two bright hard X-ray sources. We discuss a gamma-ray production site from comparison the gamma-ray(4.5 - 6.8 MeV) time profile with the hard X-ray (53 - 93 keV) time profile of each hard X-ray bright source. The gamma-ray lines are thought to be produced at one of the two hard X-ray sources. The gamma-ray flare on 3 December 1991 showed strong continuum extending about 8 MeV without detectable gamma-ray line. It implies that electrons were preferentially accelerated to relativistic energies. The rapid enhancement in hard X-ray (53 - 93 keV) emission was observed nearly simulateously with the large change in flaring loop structures. It suggests that the electrons were efficiently accelerated by strong electric field induced by the rapid variation in the magnetic field.

INTRODUCTION

YOHKOH observed several large flares in October - December 1991 with a gamma-ray spectrometer and two X-ray imagers in soft and hard X-ray energies. The simultaneous observations of gamma-ray spectroscopy and X-ray imaging contribute greatly to new findings in high-energy solar flare phenomena. Gamma-ray spectral observations, in particular, provide the most direct information on the highest-energy phenomena. The radiation in gamma-ray energies above 300 keV is emitted by relativistic electron bremsstrahlung, nuclear deexcitation, positron annihilation, neutron capture and neutral pion decay. The electron acceleration to relativistic energies is studied from the temporal and spectral analyses of bremsstrahlung continuum. The gamma-ray lines are produced from several nuclear reactions. A proton spectrum below 100 MeV is derived from a ratio of neutron capture line to deexcitation line fluences [1]. The electron and ion acceleration processes are investigated from the temporal and spectral analyses of the gamma-ray emission over wide energy ranges. Soft X-ray images reveal the spatial distribution of high-temperature coronal gas and the structure of the confining magnetic field. Hard X-ray images provide direct information on locations of nonthermal electron bremsstrahlung. From both the hard and soft X-ray images, we can obtain an essential clue to locate the hard X-ray sources on the flaring loop structure seen in soft X-ray images. The temporal evolutions of both X-ray images provide diagnostics for production, transfer and confinement of flare electrons.

In this paper, we discuss the following problems on the basis of *YOHKOH* gamma-ray spectra and soft and hard X-ray images: (1) Are gamma-rays produced at hard X-ray sources ? and (2) Does electron acceleration take place in association with large change of flaring loop structures ?

OBSERVATIONS

We present two gamma-ray flares observed with YOHKOH on October 27 and 3 December 1991 [2]. The two flares were detected with the gamma-ray spectrometer [3] and X-ray imagers in the soft X-ray [4] and hard X-ray [5] energies.

Flare on 27 October 1991

The gamma-ray flare started at 05:39:30 UT on 27 October 1991 and lasted for about 4 min. The GOES class and Hα importance were X6.1 and 3B, respectively. The location of flare was S13E15 (NOAA AR 6891). The *YOHKOH* observation was terminated at 05:41:30 UT before the flare was over. Hard X-ray (198 - 313 keV) and gamma-ray (4.5 - 12.6 MeV) count rate time profiles are shown in Fig. 1(a) and 1(b), respectively. The time resolutions of hard X-ray and gamma-ray time profiles are 1 and 2 s, respectively.

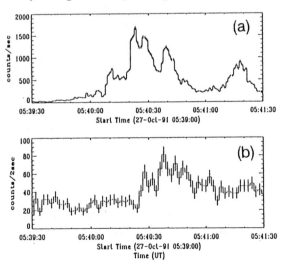

Fig. 1 Count rate time profiles of hard X-ray and gamma-ray emissions the for the flare on 27 October 1991. (a) 198 - 313 keV and (b) 4.5 - 12.6 MeV

The background-subtracted gamma-ray count spectrum observed in 05:40:18 - 05:41:26 UT is shown in Fig. 2. The bremsstrahlung continuum is dominant below 1 MeV, while nuclear deexcitation lines of C, O, Ne, Mg, Si and Fe and the neutron capture line at 2.223 MeV are dominant in the MeV region [2,6]. High-energy gamma-ray emission above 10 MeV was not detected with statistical significances. The hard X-ray images observed in the 53 - 93 keV band at four peak times P1 (05:40:23 UT), P2 (05:40:28 UT), P3 (05:40:38 UT) and P4 (05:41:17 UT) are shown in Fig. 3. Two bright sources A and B were observed at P1, P2, P3 and P4. It gives evidence to show that electrons accelerated above 100 keV precipitated to the two sources A and B at P1, P2, P3 and P4. Although a few additional weak hard X-ray sources were observed at P1, P2, P3 and P4, their X-ray intensities and locations varied with time. As an example, a three-dimensional intensity distribution of hard X-ray emission (53 - 93 keV) at P1 is shown in Fig. 4. We can see that the two sources A and B are much brighter than the other sources. The hard X-ray image overlaid with the soft X-ray image is shown in Fig. 5. These soft and hard X-ray images were observed at P1. The soft X-ray image did not significantly vary during the flare, as shown in Fig. 5.

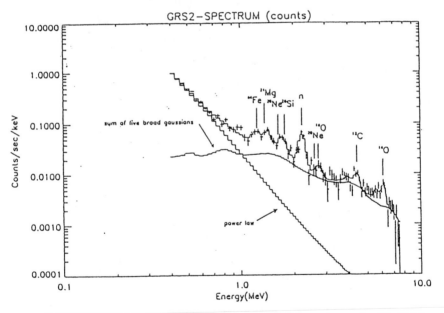

Fig. 2 Background-subtracted gamma-ray count spectrum in 05:40:18 - 05:41:26 UT for the flare on 27 October 1991.

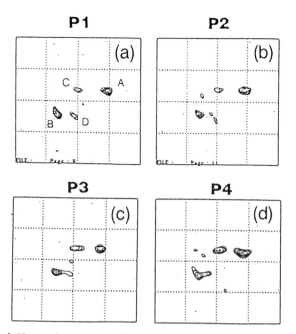

Fig. 3 Hard X-ray images in 53 - 93 keV at four peak times P1, P2, P3 and P4 for the flare on 27 October 1991. (a) 05:40:23 UT, (b) 05:40:28 UT, ; (c) 05:40:38 UT and (d) 05:40:38 UT.

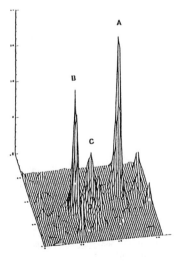

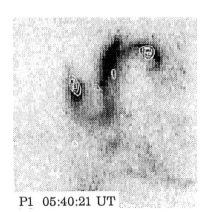

Fig. 4 Three-dimensional spational distribution of hard X-ray emission in 53 - 93 keV at P1 (05:40:23 UT) for the flare on 27 October 1991.

Fig. 5 Overlaid of hard X-ray images with soft X-ray images at P1 (05:40:23 UT) for the flare on 27 October 1991.

213

We can obtain the hard X-ray (53 - 93 keV) time profile for each hard X-ray bright source using the HXT data. It is very important to investigate whether the hard X-ray time profiles of two sources are similar. The hard X-ray count rate time profiles of two sources are shown in Fig. 6(a) and 6(b). The two time profiles are plotted with time resolution 1 s. Comparing the two time profiles, we can find a significant difference between two sources around 05:40:28 UT. The source A shows a strong peak at 05:40:28 UT, while there is not a corresponding peak for the source B. The hard X-ray emission of the source B decreases with time around 05:40:28 UT. It implies that timings of the electron precipitation are different between the two sources. We compare the gamma-ray (4.5 - 6.8 MeV) time profile with the hard X-ray (53 - 93 keV) time profile of each hard X-ray source. The gamma-ray (4.5 - 6.8 MeV) count rate time profile is shown in Fig. 6(c).

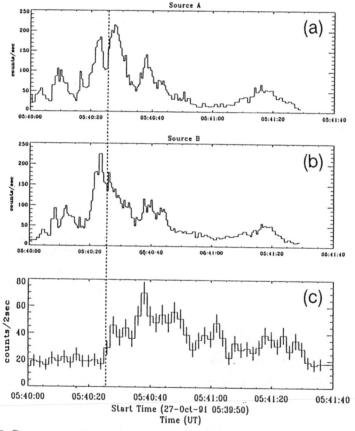

Fig. 6 Count rate time profiles of hard X-ray (53 - 93 keV) and gamma-ray emissions for the flare on 27 October 1991. (a) hard X-ray time profile for the source A, (b) hard X-ray time profile for the source B and (c) gamma-ray time profile in 4.5 - 6.8 MeV.

214

The time profile is plotted with time resolution 2 s. Most of the gamma-rays in the 4.5 - 6.8 MeV band are thought to result from C and O deexcitation lines at 4.438 and 6.129 MeV, respectively, as shown in Fig. 2. We can see that the gamma-ray emission started at 05:40:26 UT, reached the maximum and gradually decreased with time. The onset time of gamma-ray emission almost coincides with that of the hard X-ray emission of the source A. On the other hand, the source B does not show the hard X-ray increase associated with the gamma-ray emission. It suggests that the source A produced both hard X-rays and gamma-rays.

Flare on 3 December 1991

The gamma-ray flare started at 16:35:40 UT and lasted for about 2 min. The GOES class and Hα importance were X2.0 and 2B, respectively. The location of flare was N17E72 (NOAA AR 6952).The hard X-ray (198 - 313 keV) and gamma-ray (1.04 - 4.54 MeV) count rate time profiles are shown in Fig. 7(a) and 7(b), respectively. The two time profiles are plotted with time resolution 1 s. They are of impulsive characteristic and show similar temporal variations with no significant delay.

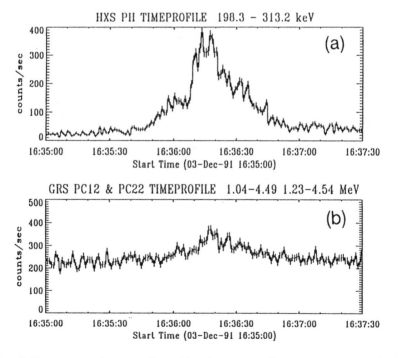

Fig. 7 Count rate time profiles of hard X-ray and gamma-ray emissions the for the flare on 3 December 1991. (a) 198 - 313 keV and (b) 1.04 - 4.54 MeV.

The background-subtracted gamma-ray count spectrum observed in 16:35:58 - 16:36:42 UT is shown in Fig. 8. The gamma-ray count spectrum is approximated by hard power-law without significant gamma-ray lines. It implies that most gamma-rays result from bremsstrahlung, suggesting that electrons were selectively accelerated to relativistic energies. This type of flare is named the electron-dominated event, which was reported from the SMM observation [7].

The hard X-ray emission increased rapidly around 16:35:40 UT, as shown in Fig. 7. We investigate whether the soft X-ray images changed simultaneously with the rapid increase of hard X-ray emission. The soft X-ray images before and after 16:35:34 UT are shown in Fig. 9(a) and 9(b), respectively. The soft X-ray image shows a large loop structure at 16:35:30 UT, as shown in Fig. 9(a), while it changed into a compact loop structure at 16:35:58 UT, as shown in Fig. 9(b). We can see that the rapid increase of hard X-ray emission is almost simultaneous with the large change of soft X-ray image. It suggests the possibility that the strong hard X-ray emission was caused by the large change of flaring loop structure.

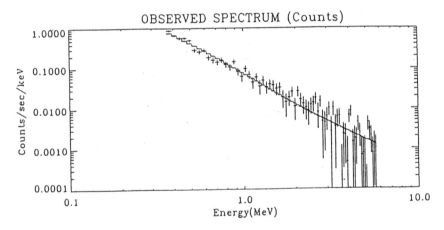

Fig. 8 Background-subtracted gamma-ray count spectrum in 16:35:58 - 16:36:42 UT for the flare on 3 December 1991.

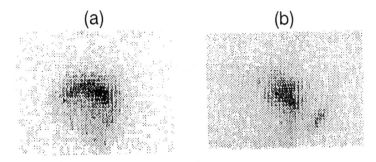

(a)   (b)

Fig. 9 Rapid changes in soft X-ray images for the flare on 3 December 1991.
(a) 16:35:34 UT and (b) 16:35:58 UT.

DISCUSSIONS

First we discuss the possible gamma-ray production site in the flare on 27
October 1991. It is well known that gamma-ray sources emit strong hard
X-rays. Since strong hard X-ray sources often show hard photon spectra extending to gamma-ray energies, the strong hard X-ray sources can be possible
gamma-ray sources. The flare on 27 October is the gamma-ray flare which
has two strong hard X-ray sources in the 53 - 93 keV band. The two hard
X-ray sources A and B are thought to be candidates for gamma- ray sources.
From comparison of the gamma-ray time profile in the 4.5 - 6.8 MeV band
with the hard X- ray time profile of each hard X-ray source, we can obtain
an essential key to infer the gamma-ray production site. As mentioned in
a previous section, the source A started the strong hard X-ray emission at
05:40:26 UT when the gamma-ray emission started, while the source B did
not show the increase of hard X-ray emission. In general, most of impulsive
flares (the flare on 27 October is the impulsive flare) are known to show no
significant delay of the gamma-ray emission with respect to the hard X-ray
emission. The observational fact that the onset time of hard X-ray emission
of the source A coincided with that of gamma-ray emission supports that
the gamma-rays were possibly produced at the source A around 05:40:26 UT.
That is, protons accelerated to above 10 MeV precipitated to the source A
and produced gamma-ray lines. Further, we can see the gamma-ray emission

at P3 and P4 when strong hard X-rays were observed from the two bright hard X-ray sources (See Fig. 6(a), 6(b) and 6(c)). Therefore, we can not infer the gamma-ray production site at P3 and P4.

Next, we discuss the possibility of electron acceleration associated with large change of flare loop structures. As previously mentioned, the rapid enhancement of hard X-ray emission took place simultaneously with the large change of flaring loop structures in the flare on 3 December 1991. Since the soft X-ray images reveal the structure of magnetic field which confines high-temperature plasma, the change suggests the possibility of restructuring of flare loops. The soft X-ray image observation [8] revealed evidence for the magnetic restructuring formed at the top of the flare loop. The magnetic restructuring plays an essential role in energy release. Rapid temporal variation of magnetic field can induce electric field which produce energetic particles. The acceleration rate is proportional to the rate of magnetic field variation. This type of acceleration process is the betatron acceleration which is more effective for electron acceleration [9]. If the large change of flaring loop structures represents the restructuring of magnetic field, the *YOHKOH* observation suggests the possibility of the betatron acceleration.

ACKNOWLEDGMENTS

We are grateful to the members of the *YOHKOH* HXT and SXT teams for their helpful discussions on the HXT and SXT data analyses.

REFERENCES

1. R. Ramaty and R. J. Murphy : Space Sci. Rev. **45** (1987) 213.
2. M. Yoshimori et al. : Astrophys. J. Suppl. **90** (1994) 693.
3. M. Yoshimori et al. : Solar Phys. **136** (1991) 69.
4. S. Tsuneta et al. : Solar Phys. **136** (1991) 37.
5. T. Kosugi et al. : Solar Phys. **136** (1991) 17.
6. M. Yoshimori et al. : Publ. Astron. Soc. Japan **44** (1992) L107.
7. E. Rieger and H. Marschhäuser : Max'91 Workshop No.3: Max'91/SMM Solar Flares: Observation and Theory, R.M.Winglee and A.L. Kiplinger, eds., p.61.
8. S.Tsuneta. : Fourth International Toki Conference on Plasma Physics and Controlled Nuclear Fusion, ESA SP-351, (1993) p.75.
9. S. Hayakawa et al. : Prog. Theor. Phys. Suppl. No.30 (1964) 86.

Extended γ-Ray Emission
in Solar Flares

G. Rank*[1], K. Bennett[†], H. Bloemen[‡], H. Debrunner[††],
J. Lockwood**, M. McConnell**, J. Ryan**, V. Schönfelder*
and R. Suleiman**

*Max Planck-Institut für extraterrestrische Physik,
85740 Garching, Germany
[†] Astrophysics Division, ESTEC, 2200 AG Noordwijk, The Netherlands
[‡] SRON-Utrecht, 3584 CA Utrecht, The Netherlands
[††] Physikalisches Institut der Universität Bern,
Sidlerstr. 5, 3012 Bern, Switzerland
**University of New Hampshire, Space Science Center,
Durham, NH 03824, USA

During the solar flare events on 11 and 15 June 1991, COMPTEL measured extended emission in the neutron capture line for about 5 hours after the impulsive phase. The time profiles can be described by a double exponential decay with decay constants on the order of 10 min for the fast and 200 min for the slow component. Within the statistical uncertainty both flares show the same long-term behaviour. The spectrum during the extended phase is significantly harder than during the impulsive phase and pions are not produced in significant numbers before the beginning of the extended emission. Our results with the measurements of others allow us to rule out long-term trapping of particles in non-turbulent loops to explain the extended emission of these two flares and our data favour models based on continued acceleration.

INTRODUCTION

Before the large solar flares in June 1991 occured, γ-ray emission from flares has been observed for no longer than about half an hour. For the 11 June flare EGRET detected emission > 50 MeV for more than 8 hours (1) and COMPTEL detected emission in the neutron capture line for about 5 hours (2) (3). The 15 June flare observations are not as detailed as those from 11 June. However, COMPTEL also measured emission of the neutron capture line for several hours for this flare. The GAMMA-1 instrument observed emission > 50 MeV (4) (5) for a different time period and reported significant emission during the following spacecraft orbit.

---

[1]present address: University of New Hampshire, Space Science Center, Durham, NH 03824

A model for the 11 June flare (6) based on the EGRET measurements explained the extended emission by particles that were accelerated during the impulsive phase and trapped in non-turbulent coronal loops. A different explanation for the emission during the extended phase of a flare (7) favours continuous acceleration of particles in turbulent magnetic loops for long time periods.

Our results for the 11 and 15 June 1991 flares allow us to investigate the question whether long-term trapping or continuous acceleration better describes the prolonged emission. We analyze the COMPTEL measurements and compare them to the results of the EGRET and GAMMA-1 instrument to distinguish between these types of processes.

OBSERVATIONS

The COMPTEL Instrument

The Compton telescope COMPTEL onboard CGRO is sensitive to γ-rays in the energy range from 0.75 to 30 MeV. This energy range contains the major nuclear deexcitation lines and the neutron capture line. Continuum emission in this range is produced by nuclear emission originating from inverse nuclear reactions (broad lines) and electron bremsstrahlung from both primary electrons accelerated directly in the flare and secondary electrons produced by pion decay.

The COMPTEL telescope uses a double scatter technique to get both spectral and directional information of the incoming γ-rays (8). In the ideal case the photons undergo a Compton scattering in one of seven detector modules of the upper detector plane and the scattered photon is subsequently fully absorbed in one of fourteen modules of the lower detector plane. Measuring the energy deposits in both detector planes yields the energy of the photon. It also provides the scatter angle of the Compton interaction in the upper detector and together with the event location in the modules constitutes the imaging capability of COMPTEL.

For solar flare observations the position of the source is known and the imaging properties of the instrument are used for suppressing the instrumental background and for excluding events that are not fully absorbed in the lower detector. Consequently the instrument offers high sensitivity and a nearly Gaussian response for an uncomplicated deconvolution of spectra. The instrumental background of COMPTEL is mostly due to cosmic ray interactions. The intensity of the incident cosmic ray flux varies periodically according to the spacecraft orbit. For a model background we use time intervals about 15 and 16 orbits before and after the flares where the orbit conditions at the time of the flare are reproduced.

All detectors of COMPTEL are shielded by anti-coincidence plastic scintillators. While these scintillators are essential for normal observations, they cause dead time problems for very intense flares. Due to the enormously high flux in soft X-rays during the impulsive phase, these anti-coincidence shields are triggered at a high rate. The dead time fraction can be calculated from housekeeping data. The statistics, however, remain poor under high dead time conditions. Therefore, the sensitivity of the telescope is degraded during the impulsive phase of the flare on 11 June. On 15 June the CGRO observation window allowed only measurements beginning about 40 min after the onset of the flare. At this time the soft X-ray flux has alrady faded and dead time was not a problem.

Extended Emission on 11 and 15 June 1991

The flare on 11 June 1991 was a X12/3B event starting at 0156 UT according to GOES-7. This flare began shortly after orbital sunrise for CGRO and could be observed during the whole orbital daylight period until about 0300 UT. The emission in the 2.2 MeV line was detected by COMPTEL for three more consecutive orbits.

To study the emission, the orbit was divided into twelve time intervals each with approximately equal statistics. Data for the subsequent orbits are time integrated. Each spectrum was corrected for background and dead time effects and was deconvolved using a SVD matrix inversion technique.

To estimate the bremsstrahlung of the primary electrons we use the data of the PHEBUS/GRANAT instrument, that measures photon energies down to 75 keV. The bremsstrahlung continuum as measured by PHEBUS can by described by a power law index of 2.55 ± 0.4 (9) (10). The hard X-ray time profile of PHEBUS in the energy range $310 - 540$ keV (9) is used to represent the time dependence of the bremsstrahlung emission. This bremsstrahlung model is necessary to determine the excess flux in the nuclear line region.

The impulsive phase consisted of several maxima, seen also in microwaves and hard X-rays. After the last peak of the impulsive phase at 0206 UT the prompt nuclear emission fades away and falls below the sensitivity limit at about 0209 UT. At 0213 UT the prompt emission returns. We define this to be the beginning of the extended emission.

The analysis reveals different emission characteristics during impulsive and extended phases: The fluence ratio of the 2.2 MeV line and the $4-7$ MeV range increases from 0.80 ± 0.12 to 1.24 ± 0.12. This indicates that the spectrum is significantly harder in the extended phase of the flare. Furthermore, the emission in the energy range $8 - 30$ MeV can be explained by the bremsstrahlung of the primary electrons during the impulsive phase. In the extended phase, however, the $8-30$ MeV flux exceeds the level expected from primary eletrons. We conclude that this additional component originates from secondary electrons of pionic origin. This observation shows that the major pion production does not begin during the impulsive phase, but is better associated with the extended processes. The time behaviour of prompt nuclear emission and emission at energies 8 − 30 MeV is correlated during the extended phase. From 0227 UT until the end of the first observation period at 0300 UT the 4−7 MeV and 8 − 30 MeV emission decay exponentially with constants of (6.4 ± 0.9) min for the nuclear emission and (7.0 ± 1.2) min for the 8 − 30 MeV emission.

The impulsive phase of the flare on 15 June 1991, also a X12/3B event, was not observable by CGRO. The observation window opened some 50 min after flare onset. Nevertheless, the γ-ray emission could still be detected by COMPTEL and the flux in the 2.2 MeV line was measured for several hours.

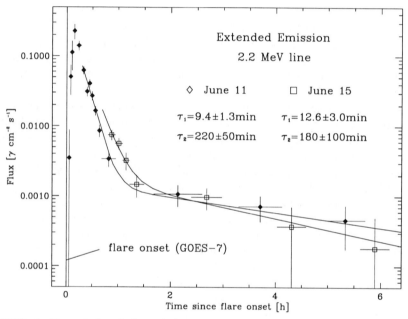

FIG. 1. Extended emission of the flares on 11 and 15 June 1991 in the neutron capture line as measured by COMPTEL.

The time profile of the 2.2 MeV line for both flares is given in figure 1. The origin of the time axis refers to the onset of the flares, i.e. 0156 UT on 11 June and 0810 UT on 15 June. The decay curves are fit by a double exponential decay. The decay constants for 11 June are (9.4 ± 1.3) min and (220 ± 50) min. For 15 June they are (12.6 ± 3.0) min and (180 ± 100) min. Both flares show a similar time profile. Their flux values are of the same order as are their time constants. The transition from the fast to the slow component occurs in the 15 June event about 20 min later than in the 11 June event.

DISCUSSION AND CONCLUSIONS

Our results are not in agreement with an explanation of the extended emission by long-term trapping of particles. The most important implication of a storage model in this context is, that we expect the trapped particles to suffer energy dependent losses (e.g. energy losses due to Coulomb collisions). Particles with MeV energies are removed more quickly than particles with GeV energies. Hence, different spectral components that originate from different energy ranges of the parent spectrum should have different decay times. In our case, the nuclear emission, produced mainly by $10 - 40$ MeV/nucl. particles, should decay about 30-times faster than the pionic emission, originating from particles with > 300 MeV/nucl. (6).

The COMPTEL measurements of the prompt nuclear emission (from protons with energies between $10 - 40$ MeV) and the excess in the $8 - 30$ MeV range from secondary electron bremsstrahlung from pion decay (requiring several 100 MeV protons) have similar temporal behaviours indicating a constant proton spectral shape. This does not agree with the predictions of a storage model. Thus, we conclude that particle trapping cannot play a dominant role in the extended phase of the 11 June flare up to 0300 UT.

For the flare on 11 June the long-term behaviour in the 2.2 MeV line can be compared to the emission > 50 MeV measured by EGRET. The time constants are 25 min and 255 min (1). However, these measurements are based on the spark chamber data only and are restricted to times after 0400 UT due to dead time effects. Therefore, the determination of the short time constant is not very reliable. By also using the TASC data the time profiles show a much steeper decay (11). This shows that both time constants measured by EGRET are consistent with the 2.2 MeV profile measured by COMPTEL. Since the neutrons responsible for the 2.2 MeV line are produced from lower energy protons than pions, their similar intensity-time profile further supports our conclusion that the parent proton spectrum was constant in shape during the extended phase.

The faster decaying component in the EGRET data was explained with primary electron bremsstrahlung (1). Our $8 - 30$ MeV measurement during the extended phase is above what we expect from primary electron bremsstrahlung. Furthermore, if we assume that this emission is due to primary electrons we expect different time evolution for the $8-30$ MeV emission and the EGRET > 50 MeV measurement and for the 2.2 MeV line or the nuclear emission between $4 - 7$ MeV (both of nuclear origin). Since we measure a correlated time behaviour, we conclude that the fast decay during the extended phase is dominated by pionic emission rather than primary electrons.

For the 15 June flare the scenario is not as complete, but the data have a similar interpretation. The COMPTEL data of the neutron capture line show the same temporal characteristics as the emission at > 50 MeV as measured by GAMMA-1. Recall that the COMPTEL measurements were made in a time interval after the GAMMA-1 observations. Also in the GAMMA-1 data there

223

is a signal in the following orbit that indicates that there exists an additional slow component declining with a time constant > 100 min. As with the 11 June flare data, these similar time profiles in two different spectral components argue against long-term trapping.

Our data favour an explanation by continuous or episodal acceleration (7). In these models the plasma turbulence does not inhibit the establishment of long time scales but serves to accelerate a new particle population.

Both flares are characterized by at least two components showing different exponential decay times. This may be interpreted in terms of several large scale loops of different sizes. In a highly turbulent environment the particle transport is dictated by diffusion. If the diffusion constant does not vary within the loops, the decay constants are proportional to the square of the loop length. An impulsive loop of 10^9 cm and extended phase loops of $(0.5-1) \cdot 10^{10}$ cm are in the proper ratio to explain the measured time behaviour. Indeed, these are typical loop length for the extended phase that were observed for the 15 June flare in H_α (12). Moreover, both flares show similar time profiles. It appears that the magnetical loop structure on large scales remains stable for several days and the same regions offer the most efficient sites for particle acceleration in both flares.

REFERENCES

1. G. Kanbach et al., Astroph. J. Suppl. **97**, 349 (1993).
2. J. M. Ryan et al., in The 1st Compton Symposium, St. Louis, MO, Ed. M. Friedlander, N. Gehrels and D. J. Macomb, AIP Proc. **280**, 631 (1993).
3. G. Rank et al., in The 1st Compton Symposium, St. Louis, MO, Ed. M. Friedlander, N. Gehrels and D. J. Macomb, AIP Proc. **280**, 661 (1993).
4. V. V. Akimov et al., in 22nd Internat. Cosmic Ray Conf., Dublin, **3**, 73 (1991)
5. G. E. Kocharov et al., in High-Energy Solar Phenomena — A New Era of Spacecraft Measurements, Waterville Valley, NH, Ed. J. M. Ryan and W. T. Vestrand, AIP Proc. **294**, 45 (1994).
6. R. Ramaty and N. Mandzhavidze, in High-Energy Solar Phenomena — A New Era of Spacecraft Measurements, Waterville Valley, NH, Ed. J. M. Ryan and W. T. Vestrand, AIP Proc. **294**, 26 (1994).
7. J. M. Ryan and M. A. Lee, Astroph. J. **368**, 316 (1991).
8. V. Schönfelder et al., Astroph. J. Suppl. **86**, 657 (1993).
9. G. Trottet, in High-Energy Solar Phenomena – A New Era of Spacecraft Measurements, Waterville Valley, NH, Ed. J. M. Ryan and W. T. Vestrand, AIP Proc. **294**, 3 (1994).
10. G. Trottet, private communication (1995).
11. E. J. Schneid et al., in High-Energy Solar Phenomena – A New Era of Spacecraft Measurements, Waterville Valley, NH, Ed. J. M. Ryan and W. T. Vestrand, AIP Proc., **294**, 94 (1994).
12. L. G. Kocharov, Solar Ph. **150**, 267 (1994).

Pion Decay and Nuclear Line Emissions from the 1991 June 11 Flare

Natalie Mandzhavidze[1,2], Reuven Ramaty[1], David L. Bertsch[1]
and Edward J. Schneid[3]

[1] *Laboratory for High Energy Astrophysics, Goddard Space Flight Center,
Greenbelt MD 20771*
[2] *Universities Space Research Association*
[3] *Northrop Grumman, Bethpage, NY 11714*

We reexamined the issue of continuous acceleration vs. trapping in the 1991 June 11 flare using a much broader data set than was available previously. We consider updated EGRET spark chamber data, high energy continuum and nuclear line data from EGRET/TASC, and 2.22 MeV line data from COMPTEL covering an extended time period. We find that the data indicate the existence of at least three distinct emission phases characterized by changes in the ion spectrum during transitions from phase to phase. By combining the 2.22 MeV and 4.44 MeV line fluxes with the pion decay emission flux in the first two phases, we show that ion spectrum hardened during the transition from the first to the second phase. We derive the ion spectrum in the third phase from a detailed spectral analysis of the EGRET spark chamber data and show that this spectrum is consistent with the 2.22 MeV line-to-pion decay flux ratio in this phase. The ion spectrum in the third phase is softer than that in the second phase. Concerning variability within the phases, we find that the ion spectrum probably remained constant during the second and third phases. This implies that the hitherto developed ion transport models are not appropriate for explaining the extended emission observed from the June 11 flare. We discuss a different scenario in which the ions are trapped in the low density coronal portions of loops but produce the gamma rays in the denser subcoronal interaction regions; this model of episodal acceleration and subsequent trapping could be consistent with the data.

I. INTRODUCTION

Gamma ray observations from several X-class flares in June 1991 provided evidence for the interaction of high energy ions in the solar atmosphere for long periods of time. The most striking example is the June 11 flare from which 50 MeV to 2 GeV gamma rays were observed with the EGRET spark chamber on the Compton Gamma Ray Observatory (CGRO) for 8 hours after the impulsive phase of the flare (1). Gamma rays up to GeV energies were also observed for about 2 hours with GAMMA-1 from the June 15 flare (2,3).

Such high energy gamma ray emission is most likely pion decay radiation produced by ions of energies up to several GeV interacting with the ambient solar atmosphere. The ions could either be accelerated in the impulsive phase and subsequently trapped at the Sun, be accelerated continuously over the duration of the emission, or be accelerated episodically and subsequently trapped between the acceleration episodes.

We have previously developed a transport model for high energy ions in solar flare loops which allowed us to calculate the time dependence, energy spectrum and angular distribution of the pion decay radiation produced by accelerated ions injected impulsively at the top of the loop (4). We then applied this model to the EGRET spark chamber observations of the June 11 flare and showed that the nuclear line emission, which is expected to accompany the high energy photons, should decay faster than the pion decay emission (5). On the basis of subsequent observations of nuclear lines (6–8), which provided some indication that the pion decay and 2.22 MeV line emissions had similar time profiles, we concluded that at least during the first 3 hours of the June 11 flare, pure trapping cannot account for the observations (9). An argument against trapping, based on the similarity of the combined pion decay–nuclear line time profile and microwave time profiles, was also presented for the June 15 flare (3,10).

In the present paper we reexamine this conclusion using new 2.22 MeV and 4.44 MeV nuclear line data and 150–210 MeV continuum data from EGRET/TASC (11), and 2.22 MeV line data from COMPTEL on CGRO (12). These data, together with microwave data (13) provide a much more detailed description of the time profiles of the various gamma ray emissions than was hitherto available. In particular, interpreting the 150–210 MeV continuum emission as pion decay emission, we are able to consider the time profile of the pion decay emission for the entire 8 hours starting from the impulsive phase of the flare; the EGRET spark chamber could not observe the impulsive phase because of the saturation of its anticoincidence dome. We present these data in §2. We find that the data indicate the existence of at least three distinct emission phases characterized by changes in the ion spectrum during transitions from phase to phase, with the spectrum probably remaining constant during the second and third phases. In §3 we determine power law spectral indexes for the ions in the three phases using 2.22 MeV and 4.44 MeV line, and pion decay flux ratios. We also carry out a full spectral analysis of the gamma rays observed with the EGRET spark chamber during a portion of the third phase. In this analysis, in addition to power laws we also consider the ion spectrum predicted by shock acceleration (14) with the spectral index which provides the best fit to the ground based neutron monitor observations of the interplanetary charged particles (GLE) from the June 11 flare (15). For simplicity, we carry out the calculations in an isotropic thick target model using various abundances for the accelerated particles [see (16,17) for more details]. In §4 we use the results to reexamine the issue of trapping vs. continuous acceleration and show that modifications to our

226

transport model would allow the long term trapping of ions in the second and third phases.

II. DATA

We present the various gamma ray fluxes in Fig. 1 together with the 17 GHz flux density (13).

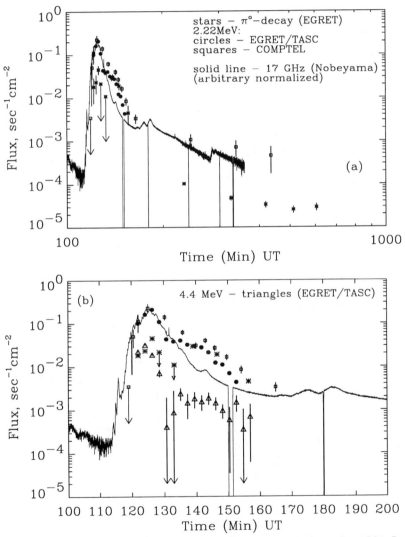

Fig. 1. Time dependence of the various emissions during the 1991 June 11 flare. The notation of panel (a) also refers to panel (b).

Panel (a) shows the data on an extended time scale covering 6 consecutive

CGRO orbits, while panel (b) shows the data on a more expanded time scale for only the first orbit. The closed circles and triangles represent, respectively, the 2.22 MeV and 4.44 MeV line fluxes observed with EGRET/TASC [see (11) for more details]. The open squares represent the 2.22 MeV line flux observed with COMPTEL (12). The 2.22 MeV COMPTEL data extend to much later times than do the 2.22 MeV EGRET/TASC data (Fig. 1a); during the first orbit (Fig. 1b), when the two data sets overlap, there is agreement between the fluxes at some times but differences as large as a factor of 2 can also be seen.

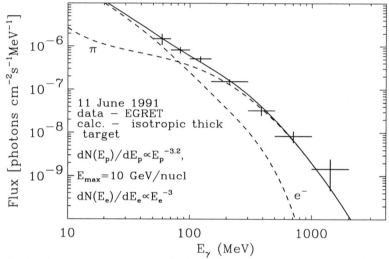

Fig. 2. Gamma ray spectrum during the 3.5 to 6 hour UT period on June 11 measured with the EGRET spark chamber. The calculated spectra are the best fitting combination of bremsstrahlung and pion decay emission.

The stars represent our estimated total π^o decay gamma ray fluxes. During the first orbit we based our estimate on the 150–210 MeV fluxes observed with EGRET/TASC. We assumed that all the observed counts in this highest energy channel are due to photons from π^o decay. Our calculations show that the ratio of the total π^o decay gamma ray flux to the 150–210 MeV flux is practically independent of the assumed particle energy spectrum and is equal to ≈ 6.3. There is practically no contribution from secondary $e^\pm$ bremsstrahlung and annihilation in flight in this channel. There might be, however, contributions from neutrons and primary electron bremsstrahlung. We have calculated the neutron flux that is expected to accompany the observed gamma ray emissions and found that the predicted time profile is quite different from the 150–210 MeV EGRET/TASC time profile; in particular the neutron emission peaks at about 130 min UT, a time at which the 150–210 MeV EGRET/TASC flux is minimal, showing only upper limits (Fig. 1b). We thus believe that the contribution to the 150–210 MeV EGRET/TASC

228

data from neutrons is not very significant. On the other hand, to estimate the contribution of the primary electron bremsstrahlung a full spectral analysis of the EGRET/TASC data is required which is beyond the scope of this paper. However, by using the highest energy channel we have minimized this contribution.

During the next 5 orbits the π^o decay gamma ray flux is based on a new analysis of the EGRET spark chamber data (11). Using the results of this analysis we fitted the 50–2000 MeV gamma ray spectrum observed from 3.5 to 6 hours UT (second and third orbits) with theoretical spectra which include both pion decay radiation and primary electron bremsstrahlung. In Fig. 2 we show the calculated spectra which provide the best fit to these data. The calculations were performed using an isotropic thick target model and assuming power law energy spectra for the accelerated particles with various ion and electron spectral indexes. The spectra shown in the figure correspond to the indexes which minimize χ^2. This fitting allows us to determine the total π^o decay gamma ray emission which is the quantity plotted in Fig. 1a. As there is no evidence for the variability of the observed gamma ray spectrum over the entire period out to the sixth orbit (11), we used the same 50–2000 MeV to total π^o decay gamma ray flux ratio for orbits 4, 5, and 6 as that calculated for the 3.5 to 6 hour UT period.

III. ANALYSIS

Considering the fluxes shown in Fig. 1, we can distinguish three major emission phases. The first and second phases are separated by a transition at about 130 min UT when both the 4.44 MeV line and pion decay emission show only upper limits (Fig. 1b). The 2.22 MeV line is still detectable during this time because of its delayed nature. The onset of a 'bump' in the 2.22 MeV line flux is another indication of this transition. The transition between the second and third phases occurs around 170 min UT. The evidence for the third phase is provided by: (i) the change of slope in the time profiles of the 2.22 MeV line, pion decay and 17 GHz microwave data; (ii) the two peaks in the microwave data at 175 and 182 min UT indicating additional particle acceleration which may have started already at 165 min UT.

During the first phase the 2.22 MeV line flux exceeds the pion decay flux, while in the second phase these two emissions have almost equal fluxes (compare the closed circles and open squares with the stars before and after 130 min UT in Fig. 1b). This immediately implies that the ion spectrum is harder during the second phase than during the first phase. These two phases, however, are not manifest in the microwave emission, for which the 'bump' between 130 and 150 min UT seen in the nuclear emissions is absent (Fig. 1b). In the third phase the pion decay flux is again lower than the 2.22 MeV line flux, indicating a softening of the spectrum.

In Fig. 3 we compare the 2.22-to-4.44 MeV line fluence ratio calculated for various power law spectral indexes s with the data for the two phases. As this ratio is strongly dependent on the accelerated α particle to proton ratio (17), we show calculations for three values of α/p. The range from about 0.01 to 0.5 is consistent with the broad variability of α/p seen in solar flare accelerated particle data (18). For the composition of the accelerated particles we used the same impulsive flare composition as used in our previous paper (17). The composition of the accelerated particles from impulsive flares is enriched in ^{3}He and heavy ions relative to the particle composition from gradual flares (18). We found (16,17) that the enhanced neutron production due to these enrichments is needed to account for the SMM observations of narrow gamma ray lines (19). For the ambient medium we assumed that the composition is photospheric; as we have shown previously (17) the coronal composition (20) leads to very similar results. We used neutron-to-2.22 MeV photon conversion factors appropriate for ions moving at 89° to a downward radius vector (21); this angular distribution provides a good approximation to an isotropic ion distribution; 35° is the heliocentric angle of the 1991 June 11 flare.

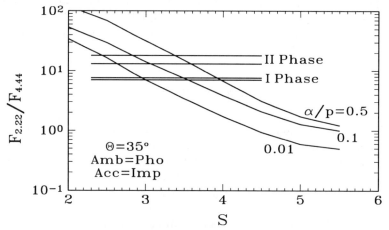

Fig. 3. 2.22 MeV-to-4.44 MeV line flux ratios. The horizontal lines are the measured (EGRET/TASC) ratios for the first and second phases; the calculated ratios are discussed in the text.

In Fig. 4 we show the π^o gamma ray to 4.44 MeV line fluence ratio. As we can see, this ratio is much less dependent on α/p. As opposed to the previous ratio, which probes the 10–100 MeV/nucl range, this ratio is sensitive to energies up to about a GeV. By comparing the calculations with the data we obtain $3.1 < s < 3.3$ for the first phase and $2.6 < s < 2.9$ for the second phase, showing that indeed the spectrum is harder in the second phase.

Considering the results of Fig. 3, we see that for the first phase the spectrum implied by $F(\gamma_\pi^o)/F_{4.44}$ is harder than the spectrum implied by $F_{2.22}/F_{4.44}$ for $\alpha/p=0.1$ or 0.5, but softer than that implied by $\alpha/p=0.01$. Since it is unlikely that the accelerated particle spectra flatten at high energies, either α/p is less than 0.1 or there is a significant non-pionic contribution to the 150–210 MeV EGRET/TASC data. Such a contribution would lower $F(\gamma_\pi^\circ)/F_{4.44}$ and hence lead to a softer spectrum. On the other hand it is unlikely that α/p is as low as 0.01 since in our previous analysis (17) we found that this is not consistent with the narrow gamma ray line data from flares. For second phase, the spectrum implied by $F(\gamma_\pi^\circ)/F_{4.44}$ is consistent with that derived from $F_{2.22}/F_{4.44}$ for $\alpha/p \lesssim 0.1$, suggesting an unbroken power law spectrum from about 10 MeV to 1 GeV.

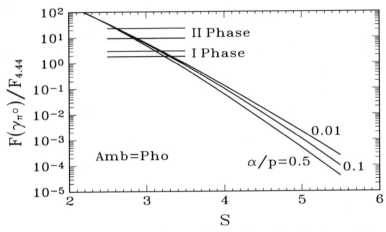

Fig. 4. π° decay-to-4.44 MeV line flux ratios. The horizontal lines are the measured (EGRET/TASC) ratios for the first and second phases; the calculated ratios are discussed in the text.

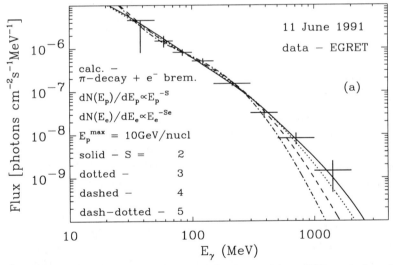

Fig. 5. Gamma ray spectrum during the 3.5 to 6 hour UT period on June 11. The curves show the effects of variations of the ion power law spectral index.

For the third phase, as already mentioned, we have carried out a full spectral analysis of the EGRET spark chamber data for the 3.5 to 6 hour UT period. The results are shown in Figs. 2, 5, 6 and 7. In Figs. 2, 5 and 6 we used power law ion spectra extending up to a high energy cutoff E_p^{max}. In Fig. 5 E_p^{max} is fixed at 10 GeV/nucl and the power law spectral index s is varied. For each s we select the primary electron spectral index s_e which provides the best fit to the data by minimizing χ^2. The overall best fit is provided by $s=3.2$ and $s_e=3$ for which the spectrum is shown in Fig. 2. For values of s greater than about 4 the calculated spectra become too steep at high energies. In Fig. 6 s is fixed and we varied E_p^{max}. We see that to account for the highest energy data point E_p^{max} has to be greater than about 8 GeV/nucl. While not shown in the figure, we found that increasing E_p^{max} above 10 GeV/nucl does not affect the calculated spectrum up to 2 GeV.

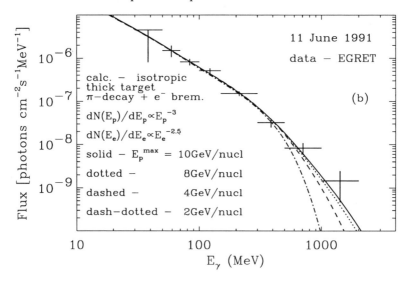

Fig. 6. Gamma ray spectrum during the 3.5 to 6 hour UT period on June 11. The curves show the effects of variations of the high energy cutoff.

In Fig. 7 we used the ion spectrum predicted by shock acceleration (14) with a spectral index which provides the best fit to the GLE observations of the June 11 flare (15). As before, we varied s_e to obtain the best fit for this ion spectrum. We set the high energy cutoff at 10 GeV/nucl. The GLE data seem to indicate a high energy cutoff at about 5 GeV/nucl (15). However, because of the steepness of the observed shock acceleration spectrum at high energies, the gamma ray spectra up to 2 GeV are practically identical for $E_p^{max}=5$ and 10 GeV/nucl. We see that calculated spectrum in Fig. 7 provides a reasonable fit to the data, except for the highest energy point. This may indicate that the spectrum of the interacting particles is harder than that of the interplanetary particles from the flare. This conclusion would be even stronger if we compared the interplanetary particles with the interacting particles in phase two, since we found that the spectrum of the interacting particles during this phase was harder than that during phase three.

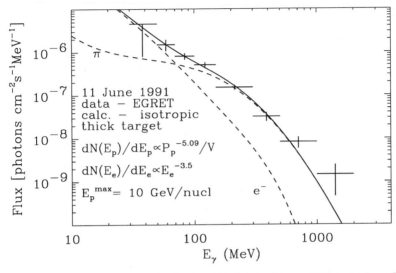

Fig. 7. Gamma ray spectrum during the 3.5 to 6 hour UT period on June 11. The calculations are for the shock acceleration spectrum which best fits the GLE observations of the June 11 flare (15).

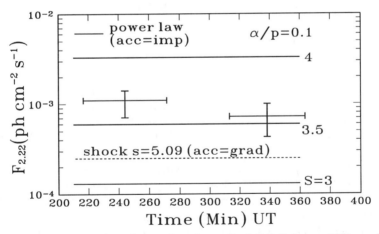

Fig. 8. 2.22 MeV line fluxes for the period 3.5–6 hours UT on June 11. Data points: COMPTEL (12); horizontal lines: average fluxes calculated by normalizing to the EGRET spark chamber data.

In Fig. 8 we compare the 2.22 MeV line flux observed with COMPTEL (12) with 2.22 MeV line fluxes obtained by normalizing the calculations to the EGRET spark chamber data in the 3.5 to 6 hour UT interval. The solid lines are for power law ion spectra for which, as in Fig. 3, we used the impulsive flare composition for the accelerated particles. The dashed line is for the assumed shock acceleration spectrum. In this case we used the gradual flare composition because it is unlikely that shock acceleration will produce the impulsive flare enhancements of ^{3}He and heavy nuclei. For all the calculations in this figure we used the same neutron-to-2.22 MeV photon conversion factors as we did for the calculation in Fig. 3. We see that an ion spectrum with $s \sim 3.5$ or slightly higher is consistent with the 2.22 MeV line flux. Such a spectrum is also consistent with the pion decay gamma ray spectrum (Fig. 5). The shock acceleration spectrum and a power law spectrum with $s=3.2$ (not shown in Fig. 8) underproduce the 2.22 MeV line flux. However, the following two effects can increase the 2.22 MeV-to-pion decay flux ratio: (i) an increase of α/p above the assumed value of 0.1 at least at low energies; (ii) an ion angular distribution pointing preferentially towards the photosphere which will enhance the neutron-to-2.22 MeV photon conversion and at the same time decrease the number of pion decay photons moving towards the observer. On the other hand, the assumption of the gradual flare composition for the accelerated particles in the case of the power laws would have decreased the calculated 2.22 MeV line flux; this could lead to a discrepancy between the ion spectrum determined from the spectral analysis and the 2.22 MeV line-to-pion decay emission flux ratio.

IV. DISCUSSION AND CONCLUSIONS

Considering the available data and calculations, it is clear that there were three major acceleration episodes in the June 11 flare giving rise to the inter-acting particles in the three phases that we have identified. The ion energy spectra in the three phases are different. The spectrum in the second phase is flatter than that in both the first and third phases. The data in the first phase do not allow us to study the possible variability of the spectrum within this phase. However, during the second and third phases the fact that the 2.22 MeV line-to-pion decay flux ratio is practically constant suggests that the spectrum within each phase does not vary with time. This is in conflict with the prediction of trapping models in which the particles are removed from the trap due to Coulomb and nuclear interactions in the trapping volume itself. Such models predict that the ion spectrum should harden with time leading to a faster decay of the 2.22 MeV line flux relative to the pion decay flux.

The loop model that we developed previously (4) made the same prediction. The model consisted of a coronal part in which both the magnitude of the magnetic field and the gas density were constant and two subcoronal portions in which the magnetic field and ambient gas density increased with increasing depth towards the photosphere. We took into account the magnetic mirroring of the ions in the subcoronal portions (but not in the coronal portion where the field magnitude was constant), their pitch angle scattering due to plasma turbulence in the coronal portion, and their energy loss and removal due to Coulomb and nuclear interactions throughout the loop. Particles injected at the top of the loop were trapped between mirror points in the subcoronal regions, which, depending on the initial particle pitch angle, could lie fairly deep in the atmosphere. Thus, except for particles with pitch angles in the loss cones, the trapped particles interacted in the trapping volume causing the hardening of the ion spectrum with time and the faster decay of the 2.22 MeV line.

On the other hand if there is a strong convergence of the magnetic field lines also in the coronal portions of the loops, particles could be stored in a region of sufficiently low density, so that their spectrum remains constant in time. To trap protons of energies greater than 30 MeV for 5 hours (the duration of the observed overlap of the 2.22 MeV and pion decay fluxes in the third phase), the density in the trapping volume must be lower than 10^9 cm^{-3}. This value is reasonable for the coronal portion of moderately large loops. The gamma rays in this case will not be produced in the trapping volume; they will be produced after the precipitation of the ions into the subcoronal interaction regions. The trapping region will thus provide to the interaction region a source of particles with spectrum which remains constant in time, as required by the observed 2.22 MeV line and pion decay data. As the magnetic fields are expected to further converge in the subcoronal regions, the structure of the field and the depth profile of the ambient density must be such that all the precipitating particles, including those at the highest energies, will interact or lose their energy before they reach their mirror points in the interaction region. Otherwise, they will be reflected back into the trapping volume (with the reflection being more probable the higher particle energy) causing the energy spectrum of the trapped particles to flatten with time. However, if the particles interact before they reach their mirror points, their angular distribution will be forward peaked along the magnetic field lines. If the magnetic field is normal to the photosphere, as was assumed in the previous ion transport models (4,5,22), the pion decay radiation from flares near disk center (e.g. the June 11 flare) will show a strong cutoff at high energies due to the forward beaming of the emitted radiation (23). As this cutoff is in conflict with the EGRET spark chamber data [see fig. 10 of ref. (23)], our proposed model must further assume that there is a broad distribution of the magnetic field directions relative to the normal to the photosphere. Detailed calculations of gamma ray production in such models have not yet been carried out.

The available data for the June 11 flare may thus be consistent with particle trapping between episodes of acceleration. The scenario that we proposed will naturally lead to interacting particles with a constant spectrum. On the other hand, it is questionable whether continuous acceleration can produce particles with a time independent spectrum over hours. Furthermore, the argument based on the similarity of the gamma ray and microwave time profiles in favor of continuous acceleration for the June 15 flare (3,10) would not hold for the June 11 flare for which the microwave time profile and the gamma ray time profiles are quite different (Fig. 1).

We wish to acknowledge Paul Evenson for discussions on the trapping scenario outlined above.

REFERENCES

1. G. Kanbach et al., A&A (Suppl.) **97**, 349 (1993).
2. V. V. Akimov et al., 22nd Internat. Cosmic Ray Conf. Papers **3**, 73 (1991).
3. V. V. Akimov et al., 1993, 23rd Internat. Cosmic Ray Conf. Papers **3**, 111 (1993).
4. N. Mandzhavidze and R. Ramaty, ApJ **389**, 739 (1992).
5. N. Mandzhavidze and R. Ramaty, ApJ **396**, L111 (1992).
6. J. Ryan et al., in Compton Gamma Ray Observatory (AIP: NY) 631 (1993).
7. R. J. Murphy et al., in Compton Gamma Ray Observatory (AIP: NY) 619 (1993).
8. E. J. Schneid, AAS Winter Meeting, Crystal City, Virginia (1994).
9. R. Ramaty and N. Mandzhavidze, in Proc. of the Kofu Symp., eds. S. Enome and T. Hirayama, Nobeyama Radio Observatory Rept. No. 360, 275 (1994).
10. G. E. Kocharov et al., 23rd Internat. Cosmic Ray Conf. Papers **3**, 107 (1993).
11. D. L. Bertsch et al., ApJ, in preparation (1996).
12. G. Rank et al., this volume (1996).
13. S. Enome & H. Nakajima, private communication (1993).
14. D. C. Ellison & R. Ramaty, ApJ **298**, 400 (1985).
15. D. F. Smart and M. A. Shea, this volume.
16. R. Ramaty, N. Mandzhavidze, B. Kozlovsky, and R. J. Murphy, ApJ, **455**, L193 (1995).
17. R. Ramaty, N. Mandzhavidze, and B. Kozlovsky, this volume.
18. D. V. Reames, J-P. Meyer, and T. T. von Rosenvinge, ApJ (Suppl.) **90**, 649 (1994).
19. G. H. Share and R. J. Murphy, ApJ **452**, 933 (1995).
20. D. V. Reames, Adv. Space Res. **15**, (7) 41 (1995).
21. X.-M. Hua and R. E. Lingenfelter, Solar Phys. **107**, 351 (1987).
22. X.-M. Hua, R. Ramaty, and R. E. Lingenfelter, ApJ **341**, 516 (1989).
23. R. Ramaty & N. Mandzhavidze, in High Energy Solar Phenomena—A New Era of Spacecraft Measurements, (AIP: NY), 26 (1994).

Origin of the High Energy Gamma-Ray Emission
in the March 26,1991 Solar Flare

Viktoria Kurt[*], V.V. Akimov[†], N.G. Leikov[†]

[*]*Institute of Nuclear Physics, Moscow State University*
Moscow,119899
[†]*Russian Space Research Institute, Moscow, 117810, Russia*

The solar flare on March 26,1991 presents a unique case when high
energy (up to 300 MeV) gamma radiation was registered in both, impulsive and delayed, phases of the flare. The radiation in the delayed
phase has been attributed to neutral pions decay (1) analogous to the
high energy gamma-ray emission at the late stages of the solar flares
on June 11 and June 15,1991 (2), (3), (4). On the contrary, spectra
of the emission in the impulsive phase of the March 26 flare definitely
indicate a bremsstrahlung origin of this emission. From the position of
the flare close to the center of the disc we conclude that the high energy
gamma-rays could be radiated only by moving upward electrons. We
compare time profiles of the gamma-ray and the microwave emissions
and show that the high and the low energy electrons responsible for
these emissions were accelerated in the same acts. We put forward
arguments in favour of an acceleration of the electrons in the upper
chromosphere or in the transition layer.

INTRODUCTION

The data on the high energy gamma-ray (GR) emission from solar flares
obtained with the EGRET(CGRO) and GAMMA-1 gamma-telescopes made
solar phisicists to revise their views on the acceleration mechanisms working
in solar flares. Lasting for many hours high energy GR emission of π^0 decay
nature from the flares on June 11 and June 15 (2), (3) stimulated search for
the origin of the long duration GR emission in the late phase of solar flares (5),
(6), (7), (8) (9) . Unfortunately for both these flares the gamma-telescopes
being in the night part of their orbits could not observe an impulsive phase of
the flare. The only flare registerd from its very beginning by the GAMMA-1
telescope was the one on March 26,1991 (4). It was shown that this flare also
exibits a feature of the proton acceleration in the late phase which coincides
in time with the delayed component of the microwave (MW) emission. More
detailed analysis based on unfortunately very limited set of data for this flare
is presented in this paper.

DEVELOPMENT OF THE FLARE ACCORDING TO H_α AND MW DATA

The March 26, 1991 flare occured in the region NOAA AR 6555. The central meridian passage date for this region was March 24.6, 1991.

A description of H_α filtergrams available for this region is compiled from (10). The region was dominated by two large negative polarity sunspots ($H \sim 2400\ G$). The positive polarity spot was insignificant in size and strength. An inverted polarity bipole was observed between the main sunspots. The dominant negative polarity fluxes clearly resulted in a large magnetic flux disbalance in the active region, which was remarkably flare-productive (150 flares according to SGD).

At 14:25 UT on March 26, 1991 there was observed a filament eruption associated with an H_α SF/(M1.1) subflare in the negative polarity subarea. At 20:11-20:24 UT an H_α subflare of SF importance took place in the positive polarity subarea, simultaneousely with a MW enhancement (20:10.8 -20:14.8-20:20.0 UT, $f_{max}=35$ sfu at 9.4 GHz) and soft X-ray (C3.3) preflare heating. Meanwile the filament was undergoing restructuring and eventually reformed along the neutral line at 20:24 UT. At this moment an impulsive and energetically large 4B/X4.7 two ribbon flare occurred with one of the ribbons situated at the site of the filament reformation. A position of the flare was S28, W23. The H_α emission maximum was observed at 20:27-20:29 UT. White light emission was also registered (SGD).

The associated radioemission consisted of MW burst ($T_{ons}=20:25:05$ UT) and intensive Type III burst ($T_{ons}=20:28:00$ UT). The Type II and Type IY bursts were not reported (SGD).

The large scale time behaviour of the highest available frequency 15.4 GHz flux density is depicted in the top panel of Figure 1. The peak flux density is 16000 sfu at 20:29:05 UT. A peak frequency spectrum at 20:29:05 UT was compiled from SGD. In spite of the lack of data for higher frequencies a flattening of the spectrum allows to consider 10-15 GHz frequency range as a spectral maximum.

The last significant peak f=700 sfu of the MW flux density curve started at 20:33:20 UT. A new source of H_α emission appeared at this time slightly aside the impulsive phase ribbons area. We defined this peak as "delayed enhancement" and all the time interval after 20:32 UT as "delayed phase of the flare."

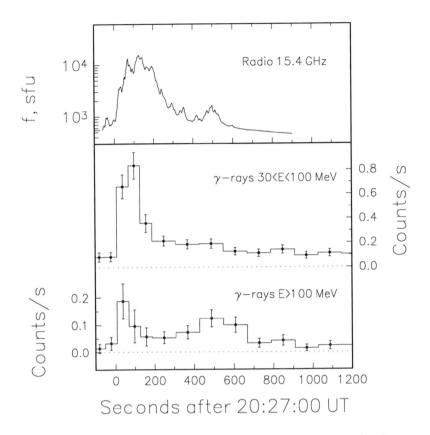

FIG. 1. MW and GR time profiles of the March 26,1991 solar flare.

GR MEASUREMENTS

The only data from this flare were provided by the GAMMA-1 high energy gamma-telescope (4). The telescope began observation of the Sun at 20:25:05 UT. The large scale time profiles in energy ranges of photons 30 MeV < E < 100 MeV and E > 100 MeV are shown in the two bottom panels of Figure 1. The statistics available allowed to perform spectral maximum likelihood analysis up to 300 MeV for two time intervals of the GR impulsive phase: first 11 s corresponding to the most bright outburst and subsequent 73 s of a "trailer" (2). The deconvolved spectra are depicted in Figure 2. The continuous lines present best approximations by a logarithmic parabola. The vertical bars show a 68% confidence area evaluated by multiple bootstrap

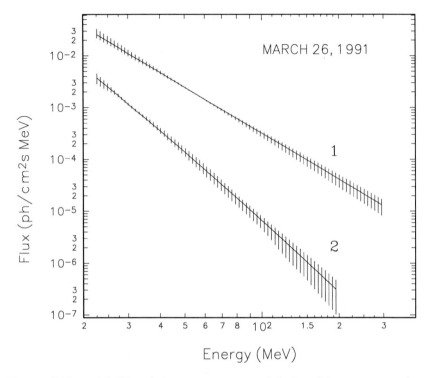

FIG. 2. Differential GR emission energy spectra of the impulsive component (curve 1, 20:27:56-20:28:07 UT) and trailer (curve 2, 20:28:07-20:29:20 UT). Shaded area contains 68 % of bootstrap test solutions.

sampling of the original data. The second power terms of the best approximations by the logarithmic parabola appeared to be very small indicating good agreement of the data with simple power law spectra. The best power law approximations correspond to exponents 2.9 ± 0.2 and 4.1 ± 0.4 for the first and the second time intervals respectively. The power law shapes of the spectra indicate an electron bremsstrahlung origin of the gamma-ray emission in the impulsive phase.

For the delayed phase scarce statistics does not allow detailed deconvolution of the energy spectrum but in terms of spectral ratios the delayed phase (see the bottom panel of Figure 1) is significant at the confidence level of 4 σ . This increase characterized by very hard spectrum was attributed to decay of

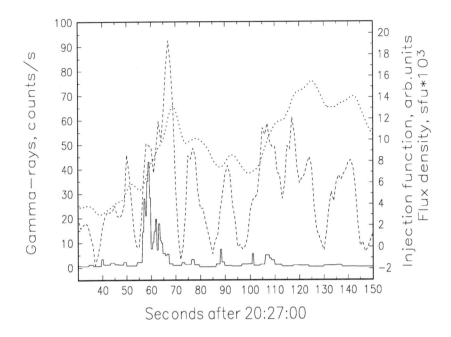

FIG. 3. MW (dotted line) and GR (solid line) time profiles and the low energy electron injection function I(t) (dashed line) in the impulsive phase of the flare.

neutral pions born in interactions of high energy ions accelerated in corona during the magnetic field relaxation (1).

COMPARISON OF GR AND MW EMISSION

Detailed time structure of 15.4 GHz emission in the period of the impulsive GR emission is shown in Figure 3 by the dotted line.

It attracts ones attention that the flux decays have very close characteristic time ($\tau = 13 - 27\ s$) as if the electrons are injected into the trap in shorter pulses and precipitate at more or less stable conditions. If we take into account only Coulomb collisions of 1 MeV electons we find plasma density $\sim 10^{11} cm^{-3}$ for these decay times. It is obviously an upper limit of the emission region plasma density because the electrons can leave the trap also by scattering on the plasma turbulence or by drift across the trap.

Relation between the MW flux f(t) and the function I(t) of injection of electrons into the trap can be expressed by the differential equation:

$$\frac{df(t)}{dt} = I(t) - \frac{f(t)}{\tau} \tag{1}$$

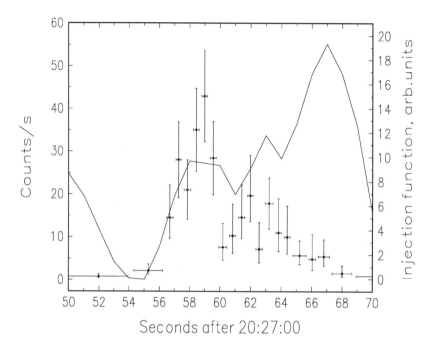

FIG. 4. Zoom of the GR time profile (points) and the electrons injection function I(t) in the period of the most intense GR.

The injection function I(t) numerically derived from the shown in Figure 3 by the dotted line function f(t) is depicted in the same Figure by the dashed line. In our calculations we used a step-vise $\tau(t)$ dependence found in the intervals of f(t) decays. The time profile of the high energy E >(30 MeV) GR counts rate is presented in Figure 3 by the solid line. The small increases of the GR counts rate around 75, 88, 100 and 107 s are statistically significant.

Figure 4 shows a zoom of the injection function and the GR counts rate time profiles in the period of the most impulsive GR flux.

The similarity of the injection function and the GR time profiles and the exact coincidance of the leading edges of their increase makes one to conclude that

- *MeV electrons responsible for MW and hundreds MeV electrons responsible for high energy GR are accelerated in the same acts;*
- *the acceleration acts have duration not longer than several seconds;*

- ratio of the low energy electrons injection function and the observed gamma-ray flux is very different in different acceleration acts;

- the GR emission seems to arise in the beginning of groups of the acceleration acts.

DISCUSSION

H_α measurements showed that the powerful impulsive white-light flare on March 26, 1991 presented a classical case of a two-ribbon flare associated with a filament movement and reformation and with changes in the strength and orientation of transversal magnatic field in the distinguishing features of the An important for us particularity of this flare is its position close to the center of the Sun disk. An opening of a bremsstrahlung radiation cone is of the order of $m_e c^2/E_e$, that is about 0.3^o for 100 MeV electron. Multiple scattering of 100 MeV electrons is also very small - about 10° in one radiation length (r.l.) corresponding to 63 $g * cm^{-2}$ of hydrogen. It means that such electons lose energy by Coulomb collisions faster than scatter by large angle. These two factors make it practically impossible to see in this flare the high energy bremsstrahlung from electrons moving downward or even horizontally at mirror points of a trap.

Calculations of bremsstrahlung emissivity angular distributions presented in (11) show that the flux of the gamma-rays with energy 100 MeV is by about 11 orders lower at 90° than at 0° to the direction of the electron beam. Even if we could see some radiation at large angles to the direction of electrons the GR spectrum would be much softer than that of the accelerated electrons (the same calculations in (11) give steepening of the GR spectrum at 100 MeV by $\Delta\gamma = 1.8$ for the bremsstrahlung radiation perpendicular to the electron beam direction). But it is not a case in this flare because the exponent of the impulsive phase GR spectrum $\gamma \sim 3$ corresponds to the most hard spectra of relativistic electrons observed in space after powerful solar flares (12). The very sharp GR flux decays also witness against trapping of the radiating electrons.

Thus, we have to accept that the high energy gamma-rays were emitted by moving upward electrons. What number of these electrons is needed to provide the observed fluence of GR? If we suppose that the electrons move upward from the level of matter density 10^{12} cm^{-3} the column density along the line of sight equals to about $2*10^{-4}$ cm^{-2} that is $3*10^{-6}$ r.l. of hydrogen. This value is very sensitive to the altitude of the injection region because of high gradient of matter density in the transition region. Accepting the target thikness $3*10^{-6}$ r.l. and supposing that at the Sun the electrons are uniformly distributed in 1π solid angle we find $N_e(> 30$ MeV) about 10^{33} to account for the GR fluence in the main 11s peak. With the energy spectrum exponent $\gamma = 3$ these electrons contain total energy about 10^{29} erg which is large but for flares of this importance corresponds to only $10^{-3} - 10^{-4}$ of total energy release (13).

Let's compare these estimates with that we would get in the case of downward moving electrons. To yield the same GR fluence we have to compensate at least 11 orders of the suppression of the bremsstrahlung radiation at 90°. The only way to do it is to increase by the same factor the product of the number of electrons and the column density. The number of electrons can be increased not more than by 2 orders not to contradict to the total energy release of such flares. The column density can not be increased more than by 5 orders, otherwise the 100 MeV electrons would lose significant fraction of their energy in Coulomb collisions. Another restriction on the column density comes from the fast variation of the GR flux in the impulsive phase: to pass 0.1 r.l. in 1 second the relativistic electron would have to move in hydrogen with density exceeding $10^{14} cm^{-3}$. So, the model with downward moving electrons disagrees with the observational data by many orders.

One can suppose some very special conditions in this flare which allow to observe the high energy GR from electrons, accelerated at the top of magnetic loops and trapped: very oblique, nearly horizontal, magnetic loops; very low, situated in dense matter, mirror points, etc. It can not be excluded, but we do not see anything exceptional in this flare and it is hard to believe that the very first flare registered by the GAMMA-1 telescope happened to be so special.

To us the most simple and straightforward explanation of the upward directed electron flux is an acceleration of the electrons close to footpoints of magnetic loops in the chromosphere or in the transition layer. In the framework of this model it is reasonable to suppose that the observed strong variation of the ratio of the GR flux to the MW injection function is caused not just by changes of the electron energy spectrum but rather by variation of the acceleration altitude resulting in the bremsstrahlung efficiency variation.

It is interesting to compare the number of the high energy electrons responsible for the GR emission with the number of the low energy electrons producing the MW emission. For this purpose we used the gyrosynchrotron emission approximations given in (14). These approximations, as the authors show, are valid for continuous power law spectra of electrons with slopes ranging from 3 to 7, harmonic numbers 10 to 100, and viewing angles from 20° to 80° degrees. Recent calculations by the authors of (15) with a use of a new gyrosynchrotron code confirmed the validity of these approximations within the enumerated restrictions.

In our calculations we used the electron spectrum slope $\gamma = 3$ in accordance with the spectrum of the high energy GR emission in the most bright 11 s outburst; the viewing angle 80° corresponding to the MW emission by the electrons trapped at the top of the magnetic loop situated close to the center of the Sun disc; the optical thickness equal to 1 corresponding to the peak MW frequency, and the emitting region size $10^{18} cm^2$. The latter figure was derived from the given above estimate of the plasma density in the MW emitting region $10^{11} cm^{-3}$ which corresponds to the loop height about 10^{10} cm. We remind that this estimate is an upper limit of the density.

Under this conditions we found the magnetic field strength $B \sim 110$ G, the harmonic number $n \sim 50$ and the number of electrons $N_e (> 30 keV) \sim 10^{36}$

for the 15.4 GHz MW flux density 10^4 sfu observed at the end of the 11 s GR outburst. If we extrapolate N_e ($> 30\mathrm{MeV}$)=10^{33} derived from the high energy GR data to lower energies with differential spectrum slop $\gamma = 3$ we find N_e (> 30 keV)=10^{39} which is much higher than the number found from the MW emission. This discrepancy can not be removed by any plausible variations of the parameters used in the estimation of the number of the low energy electrons responsible for the obsreved MW flux. Thus, we have to suppose that either the electron spectrum flattens significantly below 30 MeV or the trapping efficieny is very low. The latter seems to be natural if the injection of the electrons into the trap takes place close to mirror points.

All the featurs of this impulsive flare seem to be compartable with the models of the particles acceleration resulting from magnetic field reconnection during the loop-loop coalescence or the filament-loop interaction. Although discussion of these prosesses is out of scope of this paper we suppose that our results provide an opportunity to join already developed current loop interaction models for powerful flares (16), (17) with intrinsic high energy particle acceleration characteristics.

Acknowledgements. The authors are grateful to Dr. G. Deuel for providing the MW data and to Dr. I. Chertok for useful discussions. V.V.A. and N.G.L. are indebted to RFFR for support by GrantN94-0203316-a.

REFERENCES

1. V. Akimov et al., AIP Conf. Proc. **294**, 130 (1994).
2. N. Leikov.et al., Astr and Ap. Suppl. **93**, 345 (1993).
3. G. Kanbach et al., Astr and Ap. Suppl. **97**, 349 (1993).
4. V. Akimov et al., 22-nd ICRC, Dublin Proc. **3**, 73 (1991).
5. V. Akimov et al., AIP Conf. Proc. **294**, 106 (1994).
6. G. Kocharov et al., AIP Conf. Proc. **294**, 45 (1994).
7. N. Mandzhavidze and R. Ramaty, Ap. J. **396**, L111 (1992).
8. N. Mandzhavidze et al., 23-nd ICRC, Culgary Proc. **3**, 119 (1993).
9. V. Akimov et al., 23-nd ICRC, Culgary Proc. **3**, 111 (1993).
10. A. Ambashta, M. Hagyard and E.West, Sol. Phys. **148**, 277 (1993).
11. D.Dermer, R. Ramaty, Ap. J. **301**, 962 (1986).
12. D. Moses et al., Ap. J. **346**, 523 (1989).
13. V. Kurt Basic plasma processes on the Sun, **IAU**, 409 (1990).
14. G. Dulk, K. Marsh, Ap. J. **259**, 350 (1982).
15. R. Ramaty et al., Ap. J. **436**, 941 (1994).
16. J. Szao, B. Charglishvili, J. Sakai, Sol. Phys. **146**, 331 (1973).
17. J. Sakai, Ap.J. Suppl. **73**, 321 (1990).

The 1990 May 24 Solar Flare and Cosmic Ray Event

Leon Kocharov[1], Gennadi Kovaltsov[2], Jarmo Torsti[1], Ilya Usoskin[2], Harold Zirin[3], Antti Anttila[1] and Rami Vainio[1]

[1] *Space Research Laboratory, University of Turku, Turku SF-20520, Finland*
[2] *A.F.Ioffe Physical-Technical Institute, St.Petersburg 194021, Russia*
[3] *Big Bear Solar Observatory, Caltech, Pasadena CA 91125, U.S.A.*

We have analyzed data on solar protons, neutrons, electrons, gamma-ray, optical and microwave emissions for the 1990 May 24 solar flare. Taking into account high energy neutron and gamma-ray observations, we have suggested two neutron injections occurred during the flare. These two injections are called *f*- (first) and *s*- (second). Two components of interacting protons correspondingly existed to produce these neutrons at the Sun. The flare gave also a rise to solar cosmic ray event, which was detected by the neutron monitor network and GOES satellites. Two components of protons were observed in the interplanetary medium (*p*- (prompt) and *d*- (delayed) components). A possible spectrum of the *s*-component of interacting protons coincided with injection spectrum of *p*-component of interplanetary protons. For this reason, *s*- and *p*- components of protons may be considered as different portions of a single population of accelerated particles in the solar corona. The net result is that three proton components (*f*-, *p/s*- and *d*-) were accelerated during flare process developing from the Sun to the interplanetary medium.

INTRODUCTION

The May 24 1990 flare has been observed in many high energy bands and this provided a great deal of information on the high energy particles (1-8). In particular the flare was a source of high energy particle flux detected by the neutron monitor network (2, 5). The 1990 May 24 increase of neutron monitor counting rate had two distinct peaks. The characteristics of the second increase (at ~21:11UT) were usual for a solar proton event. The first increase (at ~20:51 UT) of short duration was detected only by monitors on the day side of the Earth, and this increase was less for larger air mass along the line of sight to the Sun. All these circumstances allowed Shea et al. (2) to ascribe the first increase to the arrival of solar neutrons. The neutron origin of this increase was reliably shown by Kovaltsov et al. (9) who noted that the response of neutron monitors to solar neutrons does not follow a simple

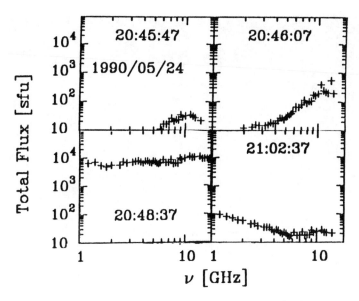

FIG. 1. Time-sequence of the total power spectra of microwave emission as obtained at the Owens Valley Radio Observatory (Caltech) during observations of the 1990 May 24 solar flare.

exponential dependence on the air mass along the line of sight to the Sun because of contribution of neutrons scattered through large angles. Neglect of this fact might be responsible for misinterpretation of the event. Recently we have examined the properties of the 1990 May 24 flare source region based on integrated analysis of gamma-ray, high energy neutron, optical and microwave data (10). In the present paper, we supplement this study with an analysis of solar cosmic ray event and comparison of spectra of escaping and interacting protons.

NEUTRAL EMISSIONS

The flare of 1990 May 24 (X9.3/1B; N33 W78) occurred during 20:46-21:45UT in the active region NOAA 6063. This flare was a source of the most powerful high-energy solar neutron event ever detected by ground based neutron monitors (2). Earlier, we obtained the parameters of neutron injection from the Sun during this event (10-12). Keeping in mind the time profile of 57.5-110 MeV gamma-rays (4), we suggested that the first (f-) injection of high-energy neutrons took place at the beginning of the impulsive phase of the flare with a maximum at 20:48:30 UT and with e-folding time $\approx$20 s. On the other hand, the 2.2 MeV gamma-ray line emission had an extremely long decay time (6) which was a signature of the second (s-) production of

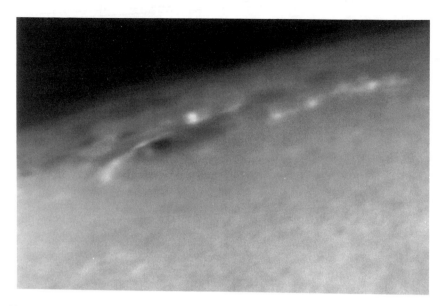

FIG. 2. HeD3-line picture of the active region NOAA 6063 at 20:46:56 UT on 1990 May 24 (Big Bear Solar Observatory).

neutrons. In what follows, we use designations f- and s- for those interacting protons which produce neutrons and gamma-ray emissions at the Sun.

In Fig. 1 is shown the time-sequence of the observed total power spectra of microwave emission during this flare (13). Note that the intensity of high frequency emission rose first (at ~20:45:47 UT) and had short decay time, while the intensity of lower frequency emission rose later and decayed longer. This may be another signature of two sources of emissions in the flare. After the beginning of the microwave burst, the Moreton wave (MHD shock wave) was observed at Big Bear Solar Observatory. The wave was seen near the flare site for the first time at ~20:47:50 UT (10). The start of the Moreton wave was followed shortly by the hardening of high-energy (15.9-110 MeV) gamma-ray spectrum. This hardening was observed at 20:48:20-20:48:40 UT with GRANAT/Phebus detectors and considered as a signature of relativistic proton acceleration (4). As a result of an integrated analysis of high-energy, microwave and optical emissions of the flare, it has been concluded that high-energy neutrons were produced by relativistic protons trapped in magnetic loops of two scales (10). Height of loops were found to be $\approx 2 \times 10^9$ cm and $\approx 2 \times 10^{10}$ cm, respectively for f- and s- components. As an illustration, in Fig. 2 is shown a HeD3-line picture of the flare site obtained at Big Bear Solar Observatory at 20:46:56 UT. It is believed that the HeD3-line brightening is due to heating of solar chromosphere by energetic particles (14). Note that the brightening and decay of the HeD3-line emission correlated with time profiles of both 14.0 GHz and 15.9-110 MeV emissions of the flare. Hence, one can

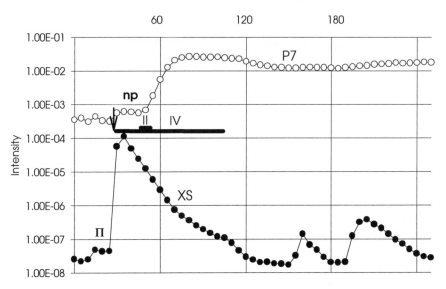

FIG. 3. The 5-min-average counting rates in proton (P7) and 0.5-4 Å X-ray (XS) channels at the beginning of the 1990 May 24-26 event as detected by GOES 6. The proton intensity is given in units of counts/(cm^2 sec sr MeV). The X-ray intensity is given in units of Watts/m^2. The proton precursor is mared with 'np'. The soft X-ray precursor is marked with 'II'. The arrow shows the onset of Moreton wave. Bars illustrate time of II and IV type radio bursts.

consider two chains of bright points clearly seen in Fig. 2 as footpoints of a magnetic arcade where particles of the *f*-component were accelerated and trapped.

HIGH ENERGY PROTONS IN THE INTERPLANETARY MEDIUM

The arrival of high-energy solar neutrons was followed by arrival of protons. During May 24 and May 25, two main maxima in high-energy proton channels were observed by GOES 6/7 satellites (GOES Data distribution disk, 1992). The first (prompt or *p*-) increase in GOES channel P7 (nominal band is 110-500 MeV) was observed during 21:05-22:30 UT of May 24 (Fig. 3). The onset of the second (delayed or *d*-) maximum was delayed for about two hours (Fig. 4). In addition, a small precursor was seen in GOES P7 channel during 20:49-21:05 UT. Solar neutrons produce neutron decay protons in the interplanetary medium. These secondary protons could produce such a precursor. This possibility has been recently studied in ref. (15). It is concluded that only the latest portion of the precursor (20:54-21:05 UT) may be explained as caused by neutron decay protons. The rest of the precursor was probably caused by high energy gamma-rays.

In order to obtain the parameters of proton injection from the Sun we use the model of focused transport of particles in the interplanetary medium (e.g., (16)). We assume a pitch angle scattering of particles being independent of the angle. Spatial, temporal and rigidity dependencies of the mean free path are taken into account in a phenomenological manner. We adopt the mean free path increasing with distance z from the Sun (along the magnetic field line): $\lambda \sim z^n$ at $z \gg 0.1$ AU. A power law decay of P7-channel counting rate was observed during the latest phase of the event (Fig. 4). This decay may be fitted with $n=1.0-1.4$ and $\lambda \sim 0.2$ AU for 1 GV proton near the Earth. Using this circumstance and basing on the results of previous studies (17), we adopted the value $n=1.4$. We also adopted the dependence of the mean free path on particle rigidity, R, as $\lambda \sim R^k$, where $k=1/3$ (see (15, 18) for more details, discussion and references). Let us consider a magnitude of the proton mean free path during the event. A very anisotropic angular distribution of high-energy protons was detected by the world neutron monitor network during the prompt peak (5). By means of a comparison of the observed pitch angle distribution with Monte Carlo simulations of focused transport we obtained the parallel mean free path of 1 GV proton near the Earth to be as large as 1.8 AU at the beginning of the event. According to neutron monitor observations, the anisotropy of high-energy proton flux abruptly decreased in one hour after the onset of the event. In order to fit the data, we proposed that the mean free path of 1 GV proton near the Earth correspondingly decreased down to 0.2 AU and remained constant thereafter. Note that, when analyzing GOES data, we make a direct fit of 'uncorrected' counting rates taking into account 'secondary' channels coming from the passive shield of the detector (for details see (15, 19)). A prolonged and anisotropic flux of protons was observed in this event, leading us to search for an extended injection of particles into the interplanetary medium. We considered two models for the injection time profiles: (i) The traditional coronal diffusion model (20). (ii) The exponential injection model. None of certain models of coronal transport is speculated in the later case. The time profile of the injection is proposed to be of the form: $q(t, E) = \{N(E)/(\Omega \tau_0)\} exp\{\Phi(t)\}$, where $N(E)$ is the number of particles with energy E injected into the interplanetary medium at the Earth-connected magnetic field line, Ω is a solid angle of the injection, τ_0 is the normalization time, $\Phi(t)$ is a piecewise linear fitting function. In the case of the prompt component, a source of accelerated particles was situated at $z_0=0.006$ AU. In the case of the delayed proton injection, we proposed a source of the delayed component moving from the Sun with the average velocity of $V_s=1200$ km/s to arrive to the Earth in 40 hours.

The results of the fitting of GOES counting rates are presented in Fig. 4. The exponential injection model fits better the rise of P7-channel counting rate because of proposed exponential shape of the injection increase. The decay of p-component injection was fitted with e-folding time $\tau=40$ min. The spectrum of p-component protons at the source is shown in Fig. 5. This

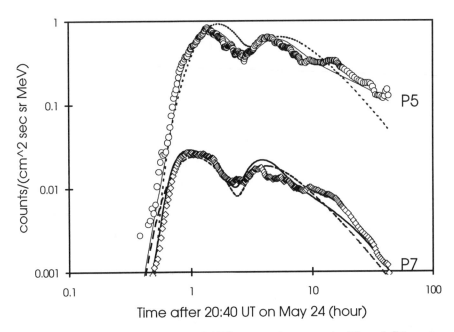

FIG. 4. Illustration of fitting to GOES 6 counting rates in P5 and P7 proton channels (nominal energy band 39-82 MeV and 110-500 MeV respectively). Solid lines correspond to the exponential injection model. Dashed lines correspond to the coronal diffusion model.

spectrum can be considered as constant in time. On the contrary, the spectrum of d-component protons was found to vary in time. The hardest spectrum of d-component protons was produced at ∼300 min. after the flare onset. This spectrum may be fitted by Ellison-Ramaty type function (21): $N(E) \sim (E^2 + 2EMc^2)^{-\gamma/2} exp(-E/E_0)$, where $\gamma = (\sigma+2)/(\sigma-1)$ with the shock compression ratio σ=2.7 and $E_0 \approx$350 MeV.

COMPARISON OF INTERACTING AND INTERPLANETARY PARTICLES

We have considered a spectrum of protons interacting in the solar atmosphere. These protons produce high-energy neutrons and gamma-rays in the flare site. An analysis was carried out by means of a comparison of a calculated response of the Climax neutron monitor to this neutron event with the observed one. Production of neutrons in the solar atmosphere was calculated by means of the technique described in ref. (22). We used for the analysis a power law shape of the primary proton spectrum, $N(E) \sim E^{-S}$ with cutoff energy ϵ. The decay time of neutron production was taken as T_f=20 sec and T_s=200 sec for f- and s- components, respectively (11). We obtained that the

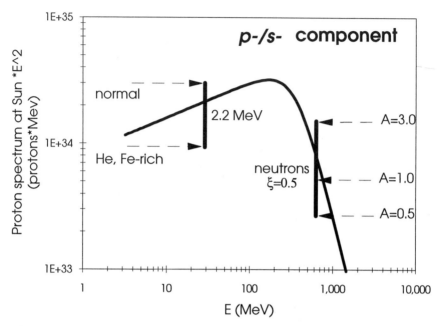

FIG. 5. The energy spectrum at the Sun, $N(E) \times E^2$, of the prompt (p-) component of interplanetary protons as derived from GOES and neutron monitor data for the 1990 May 24 solar cosmic ray event (for the exponential injection model). The same spectrum is valid for the s-component of interacting protons which is responsible for production of the 2.2 MeV gamma-ray line and high-energy neutron emissions. Vertical bars represent derived numbers of interacting protons which produce these secondaries at the Sun (theoretical parameters A and ξ are defined in ref. (15)).

proton spectral index for the first proton component was S_f=3-3.5, while protons of second component were of softer spectrum: S_s=4.5-6. A finite value of the cutoff energy was found to be essential only for the first component: $\epsilon_f \leq 2.5$ GeV. The above-stated results are correct for protons with energy above several hundreds MeV. When extrapolating the obtained spectrum of high-energy protons down to lower energies, one can calculate the corresponding total number of $\geq$30 MeV protons. On the other hand, the number of the primary protons with energy 30-100 MeV can be estimated from observed fluence of 2.2 MeV gamma-ray line emission. It turned out that, in the case of the s-component, none of single power law spectra of interacting protons (in entire 30-2000 MeV energy band) can fit data on solar neutrons and 2.2 MeV gamma-rays simultaneously (for more details see ref. (15)).

As an alternative to a single power law spectrum of the s-component of interacting protons, a broken power law spectrum may be considered. Such a spectrum has been obtained for the prompt (p-) component of protons in the interplanetary medium (Fig. 5). The 2.2 MeV bar in Fig. 5 shows the number of interacting 30 MeV protons as estimated from the observed fluence of the 2.2 MeV line emission. We considered various compositions of accelerated particles (23) and adopted $S_s = 1.8$ which corresponds to the index of the p-component spectrum in the energy band of 30-100 MeV. The number of 600 MeV protons (see neutrons bar in Fig. 5) was calculated from Climax neutron monitor response with $S_s = 4.5$ which corresponds to the index of the p-component spectrum at ≥ 600 MeV, and with varied values of anisotropy of high-energy neutron production. It is seen that the p-component spectrum fits both high-energy neutron and 2.2 MeV gamma-ray line observations. Thus, a possible spectrum of the s-component of interacting protons coincides with the spectrum of p-component protons in the interplanetary medium. Besides, total numbers of both the components of particles agree surprisingly well (at reasonable value of escape solid angle $\Omega = 2$ sr).

Let us compare the number of relativistic electrons in the interplanetary medium with the number of electrons in the flare site. The number of electrons injected at the root of the interplanetary magnetic field line was deduced from GOES data as $N_e^{IPM}(\geq 100$ keV$) \approx 3 \times 10^{36}$ for the prompt injection component with $\Omega = 2$ sr and differential spectral index being equal to 3. On the other hand, the total number of electrons in a giant magnetic loop at the Sun, as deduced from microwave data (13), was $N_e^{\mu}(\geq 100$ keV$) \approx 3 \times 10^{36}$ at the time of maximum flux (20:49 UT). At last, the number of interacting electrons was derived from continuum gamma-ray spectrum as $N_e^{\gamma}(\geq 100$ keV$) \approx 1 \times 10^{36}$ (10). One can see that the numbers of high-energy electrons generating electromagnetic emissions during first minutes of the flare and the total number of escaping electrons coincided rather well.

DISCUSSION

The 1990 May 24 flare was an event with clearly seen Moreton wave; that is, a shock wave manifests itself through a fast and bright front of Hα emission that propagates away from the flare site. The value of the wave linear speed was ≈ 1600 km/s. The shock started 40 - 50 sec prior to the time of the maximum of gamma-ray emission and about 10 minutes prior to the time of the first injection of protons into the interplanetary medium. We suggested that this shock wave initiated acceleration processes of all the proton components discussed in the present paper. Let us estimate the size of acceleration regions responsible for the prompt (p-) and delayed (d-) components of SCR. By the time of the first p-proton injection into the interplanetary medium, the shock wave had passed the distance $\leq 10^{11}$ cm. Note that this is just a characteristic scale of the longitudinal distribution of source flares for prompt SCR events on the solar disk (24, 25). Thus, we can conclude that acceleration and (or) injection of the p-component of particles took place at the distance ≤ 1 solar radius from the flare site. For the d-component, by the same reasoning, one can find a scale of the acceleration region of about 10 solar radii.

As it was shown above, the spectrum of the p-component of escaping protons

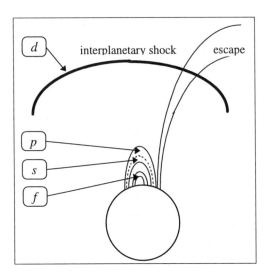

FIG. 6. Illustration of a multicomponent nature of proton production during the 1990 May 24 solar flare and cosmic ray event. Letters denote regions of acceleration or trapping of first (f-) and second (s-) components of interacting protons, prompt (p-) and delayed (d-) components of protons in the interplanetary medium.

agreed well with the spectrum of the s-component protons which generated the 2.2 MeV gamma-ray line emission and high-energy neutrons. With this noteworthy coincidence, protons of both the s- and p- components may be considered as accelerated at the beginning of the flare during 20:48-20:50 UT. However, the p-component of escaping protons was released ≥ 10 minutes later with e-folding time of the injection being $\tau=40$min. Such a late and prolonged injection of protons into the interplanetary medium may be explained with a leakage of particles from a giant magnetic loop. It was proposed in (10) that protons of the s-component were trapped in the highest loop which had still survived after the impulsive energy release. In this case, some neighboring loops may be unstable with respect to the 'flute' instability (26). Expansion and opening of such a trap may explain the observed onset of proton injection into the interplanetary medium. When proposing the leakage velocity to be equal to Alfven velocity and taking into account the value of τ, one can estimate the loop height as $(3-4) \times 10^{10}$ cm, which is roughly twice that proposed for trapping of s-component of interacting protons. Note that the decay time of the 2.2 MeV gamma-ray line intensity increased with time (6). Recent analysis of Mexico neutron monitor data also revealed a rise of the decay time of the neutron injection function with time (12). The existence of two or more giant loops of $(2-4) \times 10^{10}$ cm height can explain such a phenomenon. Expansion of a giant loop where neutrons were produced during the post-impulsive phase of the flare could lead to such a rise as well. Under the later assumption, the rate of the loop size rise can be estimated as

$\sim 3 \times 10^7$ cm/sec.

The scheme of the most likely scenario of the 1990 May 24 flare is illustrated in Fig. 6. We propose a very strong initial energy release which resulted in occupation of a number of magnetic loops by accelerated particles. Explosion of the highest of those loops could be a source of the prompt component of particles in the interplanetary medium. In addition, several hours after the impulsive phase of the flare, the delayed SCR component was presumably accelerated by interplanetary shock wave at the distance of tens of solar radii from the Sun.

Acknowledgements. We thank Frances Tang for providing frames from BBSO. L.K. was partially supported by NASA COMPTON GRO Cycle 4 Guest Investigator Program. The Academy of Finland is thanked for financial support for the Space Research Laboratory of the University of Turku.

REFERENCES

1. K. R Pyle, M. A. Shea and D. F. Smart, Proc. 22nd Int. Cosmis Ray Conf., Dublin **7**, 57 (1991)
2. M. A. Shea, D. F. Smart and K. R. Pyle, Geophys. Res. Lett. **18**, 1655 (1991).
3. F. Pelaez, P. Mandrou, M. Niel et al., Solar Phys. **140**, 121 (1992).
4. G. Talon, G. Trottet, N. Vilmer et al., Solar Phys. **147**, 137 (1993).
5. H. Debrunner, J. A. Lockwood and J. M. Ryan, Astrophys. J. **409**, 822 (1993).
6. O. V. Terekhov et al., Pis'ma v Astron. J. **19**, No. 3, 163 (1993).
7. G. E. Kocharov, L. G. Kocharov, G. A. Kovaltsov et al., Proc. 23rd Internat. Cosmic Ray Conf., Calgary, **3**, 107 (1993).
8. H. Debrunner, J. A. Lockwood and J. M. Ryan, Astrophys. J. **387**, L51 (1992).
9. G. A. Kovaltsov, Yu. E. Efimov, L. G. Kocharov, Solar.Phys. **144**, 195 (1993).
10. L. G. Kocharov, J. W. Lee, H. Zirin et al., Solar Phys. **155**, 149 (1994).
11. G. A. Kovaltsov et al., Astronomy Letters **20**, 762 (1994).
12. G. A. Kovaltsov et al., Proc. 24th Internat. Cosmic Ray Conf. **3**, 155 (1995).
13. J. W. Lee, D. E. Gary, H. Zirin, Solar Phys. **152**, 409 (1994).
14. H. Zirin, Astrophysics of the Sun, Cambridge: Cambr. Univ., 1988, p. 358.
15. J. Torsti, L. G. Kocharov, R. Vainio, G. A. Kovaltsov et al., Solar Physics (to be published), Papers I and II (1996).
16. I. N. Toptyghin, Cosmic Rays in Interplanetary Magnetic Fields, Moscow: Nauka, 1983, p. 119.
17. M.-B. Kallenrode, J. Geophys. Res. **98**, 19 037 (1993).
18. J.W. Bieber, W.H. Matthaeus, C.W.Smith et al., Astrophys. J. **420**, 294 (1994).
19. R. Vainio, J. Torsti, L. G. Kocharov et al., Proc. 24th Internat. Cosmic Ray Conf. **3**, 131 (1995).
20. G. C. Reid, J. Geophys. Res. **69**, 2659 (1964).
21. D. C. Ellison and R. Ramaty, Astrophys. J. **298**, 400 (1985).
22. V. G. Gueglenko et al., Solar Phys. **125**, 91 (1990).
23. R. Ramaty, N. Mandzhavidze et al., Adv. Space Res. **13(9)**, 275 (1993).
24. L. P. Borovkov et al., Proc. 20 Int. Cosmic Ray Conf., Moscow, **3**, 124 (1987).
25. M.-B. Kallenrode et al., Astrophys. J. **391**, 370 (1992).
26. V. V. Zaitsev and A. V. Stepanov, Solar Phys. **99**, 313 (1985).

Solar Neutrons on May 24th, 1990

Y. Muraki* and S. Shibata[†]

*Solar Terrestrial Environment Laboratory,
Nagoya university, Nagoya 464-01, Japan
[†]Engineering Science Laboratory, College of Engineering,
Chubu University, Kasugai 487, Japan

The solar neutron event on May 24th 1990 has been analysed with use of Shibata's response function. It has been found from those solar neutron data that ions are accelerated within one minute with power index $\gamma=2.5$.

OPEN NEUTRON CHANNEL

Recently, due to the efforts of solar physicists throughout the world, a lot of important information has been obtained on the acceleration mechanism of electrons, especially through X-ray, radio, and gamma-ray channels. However, for the mechanism of ion acceleration, only the emitted neutrons, together with neutral pions, gamma-ray lines, give us direct information from the solar surface. Gamma-rays may also be produced by the bremsstrahlung of electrons, so the gamma-ray signals do not always represent pure information on the acceleration of ions. Moreover, gamma-ray lines are produced by the interactions of low-energy ions with the solar atmosphere and, in the case of the 2.2 MeV line, for example, the emission of gamma-rays is delayed by the time required for the accelerated ions to slow to the thermal energy (deceleration time) to form the deuteron. Only solar neutrons bring us unique information on the mechanism of ion acceleration to high energies.

Studies of solar neutrons are still rather sparse. The first detection of solar neutrons was made by the SMM satellite (1), then, neutron detectors located at high mountain altitudes succeeded in observing solar neutrons in the June 3rd flare of 1982 (2,3). That was the entirety of reliable data in the previous solar cycle (cycle 21). In this solar cycle (cycle 22), some data have been added on solar neutrons in association with the gigantic flares of May 24th 1990, Mar. 22nd 1991, June 4th 1991 and June 6th 1991. In the CGRO satellite, COMPTEL detectors also observed solar neutrons. However, the number of solar neutron events is still not large.

One interesting report has been made by Russian groups on the basis of the Γ-1 detector on board the French-Russian satellite (4–6). They found an interesting event on June 15th 1991, using this detector, in which the time profile of gamma-rays shows the same time behavior as the radio intensity. The gamma-rays observed by the Γ-1 detector cover the energy range above

50 MeV and CGRO/COMPTEL detector recorded photons in the 0.8-10 MeV energy range with the same intensity-time profile. It is natural to consider that the high-energy photons are produced by highly accelerated ions while the radio signals are emitted by the accelerated electrons. Can the Sun equalize the acceleration of both particles so skillfully ? This is our strong motivation for investigating May 24th 1990 event.

It might be possible to speculate a theory that electrons and ions are accelerated at the same time. Then the acceleration does not depend on the particle's identity, but rather on its velocity. Therefore, if we examine the maximum energy, it just depends upon the mass difference, m_e/m_p. Radio signals in the 5-9 GHz region are emitted from electrons with a few MeV by synchrotron radiation, and gamma-rays are emitted by the interaction between few-GeV protons and the solar atmosphere. This can be considered as evidence that an acceleration mechanism described above occurs at the solar surface. However, if we consider this in detail, we may easily find a controversial point in this simple assumption. What kind ? When we look at the data obtained by the Nobeyama radiometer at three different frequencies, 17, 35, and 80 GHz, for the June 1991 flares, the time profile of the three bands differ from each other, as seems more natural to us. A very close similarity in time the profiles between 9.1 GHz , 5.2 GHz and the Γ-1 photon data on June 15th 1991 seems miraculous.

Another doubt arises about the above simple assumption (although the simplest is usually the best) in that electrons can emit photons by synchrotron radiation. If the magnetic field is as weak as 10 gauss, the lifetime of an electron about 10 MeV is around 4×10^3 minutes, when the field strength is around 1 gauss, the lifetime can be prolonged to as much as 4×10^5 minutes. The observed slope of the photon and radio data on June 15th decays with a mean lifetime of about 11-20 minutes. This suggests the possibility that all these emissions are made by electrons, then it is very clear why the time profiles of each wavelength should resembles each other so closely.

Recently we have noticed other problems with low energy gamma-ray lines. The production time profile of gamma-ray lines does not necessarily reflect the acceleration time of charged particles, one to one. This process was treated by Wang and Ramaty more than 20 years ago by Monte Carlo calculations (7). The data used for nuclear interaction cross-sections at low energies has not changed since that time, so the results of their paper are essentially correct. They have already pointed out that there is a delay of the gamma-ray line emission from the acceleration time. According to them, the delay time is of the order of one minute. The experimental data of the flares of June 15th 1991 indicate about a three minutes delay. Probably the two-minute difference comes from the acceleration time of the ions (8). However, we must still consider other alternatives for the two-minute difference. We believe that there still remains a possibility that this difference can be explained by the deceleration time. If this is correct, an interpretation for the long-duration photon tails of the CGRO/COMPTEL data on June 11th and 15th of 1991 can be made with an assumption that ions are accelerated instantaneously but are trapped in magnetic loops (9). The photons are created by the precipitation of the ions into the solar atmosphere. The data are shown with a logarithmic scale, not linear scale, so we must be careful not to forget this in considering the above possibility.

Thus, we can make a simple hypothesis for the mechanism of ion acceleration to the multi GeV region, namely, that the ions are accelerated very rapidly, say within a few seconds. Based on this working hypothesis, we now try to re-analysis data from the neutron GLE event of May 24th, 1990. The remainder of this paper gives the results of our data analysis of this noble event.

EVENT, DATA ANALYSIS AND RESULTS

First we would like to review the May 24th, 1990 event. This large solar flare was observed using various detectors located at the ground level throughout the world and in spacecraft. One remarkable feature of this event may be found in the double peaked structure observed by the neutron monitor at Mt. Climax. Shea, Smart and Pyle (10) interpreted the first peak as being

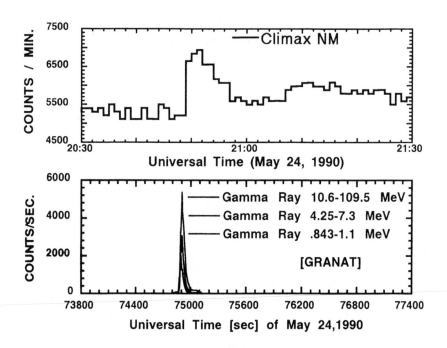

FIG. 1. Mt. Climax Solar neutron data on May 24th, 1990 are presented together with gamma-ray data taken by GRANAT/PHEBUS detector as a function of photon energies.

258

produced by solar neutrons and the second by protons. The event was reinterpreted by Debrunner *et al.* (11) as being two stages of proton acceleration, since they originally believed that the first peak was too large to be a neutron increase, but these authors later found that including the effects of scattering on the attenuation mean free path allowed the first peak to be interpreted as a neutron signal (12). It is now accepted that the first strong enhancement of the event was produced by a powerful beam of solar neutrons, and the second by a flux of directly-accelerated protons. This event can be called the largest observed GLE event produced by neutrons.

The time profile of the 2.7 GHz radio signal indicates that a majority of electrons were accelerated within 70 seconds (13). Later data were published on gamma-rays which were observed by the Russian Granat satellite in the energy ranges 0.843-1.1 MeV, 4.25-7.3 MeV and 10.6-109.5 MeV. These data are shown in Fig. 1 together with the neutron monitor data. All the gamma-ray data indicate that gamma-rays are produced within 40-60 seconds (14). Another interesting feature was reported by Leon Kocharov *et al.* (8) on the 2.2 MeV line gamma-rays which were seen by the GRANAT/PHEBUS detector. The 2.2 MeV gamma-rays decayed with a characteristic decay time 240 seconds. The 2.2 MeV time profile clearly deviated from that of gamma-rays with 57.5-110 MeV. Leon Kocharov *et al.* have concluded that two acceleration phases must occur in the short time in this flare. The present authors have regarded the delay of the 2.2 MeV gamma-rays relative to the high-energy gamma-rays as only a reflection of the deceleration time; it is not considered to represent an acceleration time of solar particles in this paper.

We have assumed that the ions are accelerated within one minute, i.e., at the same time as electron acceleration, and we have obtained an expected neutron flux at Mt. Climax as a function of the energy spectral index γ. The result of a χ-squared fit to the data is shown in Fig. 2. Fig. 2 indicates that $\gamma=2.5$ gives the best result. In Fig. 3, we show the difference between the fitted curve and the data. The deviation from the experimental data is very small and no room is left to introduce a neutron-decay-proton (NDP) effect here, as was proposed by Leon Kocharov *et al.* (15). The muon telescope data at Mt. Norikura has indicated a small enhancement before a large rise, but we are now examining this rise to see whether it was produced by NDPs or just by simple large angle scattering of the solar neutrons in the upper atmosphere (16).

The data can be reproduced very well by a simple power law. The power index turns out to be 2.5 ± 0.2. This implies that particles are produced with a very hard spectrum. Our result is consistent with a picture in which charged particles are accelerated instantaneously at the solar surface in this event. It is not possible to determine whether ions are accelerated within one second or one minute. However, the experimental result is consistent with a picture in which the ions are accelerated along with the electrons, within one minute.

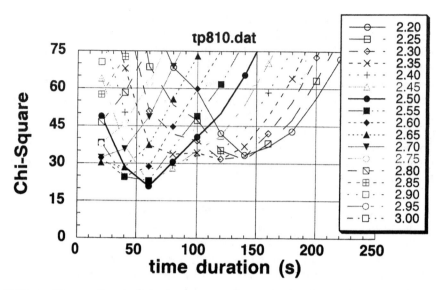

FIG. 2. The results of χ^2-fit are plotted as a function of the acceleration time of ions in the solar surface. To the power index $\gamma = 2.5$ and acceleration time $\tau = 60$ seconds, solar neutron data give the minimum χ^2 value.

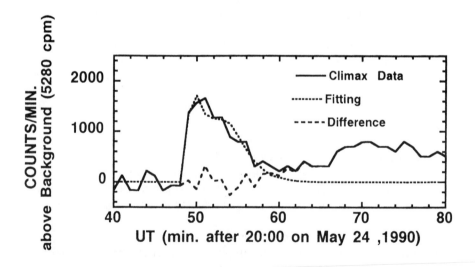

FIG. 3. The time profile of the expected neutron flux at Mt. Climax is given together with the observed data. Present fit can even reproduce a small shoulder at 20:52 UT.

REVIEW, DISCUSSION AND SUMMARY

In this section, we first review solar neutron events from the point of view that solar neutrons are produced INSTANTANEOUSLY by the nuclear interaction of accelerated ions with the solar atmosphere. However this hypothesis can not be applicable to other acceleration mechanisms such as that suggested by Kahler *et al.* (17).

The first neutron event was detected by the detectors on the SMM mission on June 21st, 1980. The event can be naturally understood as an initial strong gamma-ray peak, followed by an enhancement caused by neutrons. The event is consistent with a hypothesis that solar neutrons are produced during the same short time interval as the gamma-rays.

However, the famous event detected on June 3rd 1982, includes a controversial feature in the data themselves. Chupp *et al.* (2) give the interpretation that ions were accelerated, NOT instantaneously BUT CONTINUOUSLY, within a certain duration of time. The evidence arises from two data sets, data taken by the neutron monitor at Jungfraujoch station, and gamma-neutron data detected by the SMM GRS detector (Fig. 4).

At that time, however, only one absorption curve was available for the attenuation of solar neutrons in the atmosphere (18). According to that attenuation curve for neutrons in the atmosphere, neutrons with energies less than 200 MeV are absorbed and will not penetrate to the mountain detectors. Neutrons with an energy of 200 MeV arrive at the earth only 6 minutes later than light from the flare, so that it was natural to interpret the enhancement between 11:50 UT and 11:55 UT on June 3rd in terms of continuous acceleration, lasting at least 5 minutes.

However, Shibata (19) has given a new attenuation curve for solar neutrons. The essential difference between Debrunner-Fluckiger and Shibata curves will be found at the cut-off energy of neutrons by the atmosphere. Shibata predicts that neutrons with energy less than 100 MeV still arrive at ground level. This difference comes from the model of nucleus-neutron collisions. Debrunner and Fluckiger treated the nuclear interaction of neutrons with air nuclei as a superposition of individual neutron-neutron or proton-neutron collisions, while Shibata treated the nucleus as a whole (nuclear effect).

Therefore, the long tail in the time profile of the neutron monitor in the June 3rd 1982 event cannot be taken as evidence for continuous acceleration. Only the SMM HE-matrix event (25 MeV) remains as possible evidence for continuous acceleration of solar protons, however, this detector was never calibrated by an accelerator neutron beam. It must be checked the instrument's ability to distinguish neutrons from photons. The GRS spectrometer has about 0.4 nuclear mean free paths and 2.9 radiation lengths. Photons will be converted into electron pairs and those electrons behave like protons converted from a neutron in the detector. The pattern recognition might not completely distinguish the two. When incident energy of neutrons increases, neutrons can produce many fragment particles from NaI nucleus. The authors

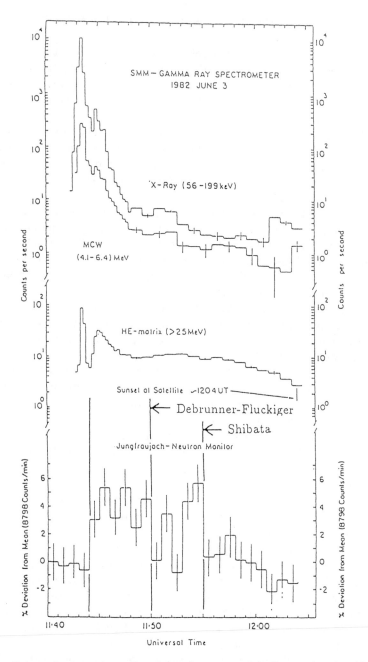

FIG. 4. June 3rd 1982 event taken from tha paper of ApJ **318**, 913 (1987). The neutron time spread is shown by arrows for the Shibata's function and Debrunner-Fluckiger function in the figure (bottom). The middle time profile of the figure indicates high energy photon and neutron combined data.

plan to make copy of the GRS detector in Japan and calibrate it with an accelerator beam. If the second peak of the HE matrix channel of GRS on the SMM satellite are proven to be created by neutrons, then we can say immediately that the June 3rd 1982 event has been produced by the instantaneous acceleration of charged particles.

Next, we would like to mention the June 4th event of 1991. Originally, the authors wrote a paper suggesting that those neutrons, together with the electrons, were made within 20 seconds, based on CGRO/BATSE time profile (20,21). Since, we have published our data that is fit by a decay time 14.9 seconds, and the CGRO/OSSE group has published a paper of the time profiles on 2.2 and 4.4 MeV gamma-ray lines (22). Although the highest points were not detected, due to saturation of the detector, the time profile can be fit quite well by a decay time of 330 seconds. Struminsky *et al.* also fit the Riken neutron monitor data at Mt. Norikura to this time profile (23). A long time scale describes the neutron monitor data rather well, however, the neutron monitor also responds to protons, and at that time the proton background was also rapidly increasing due to a shock from the June 1st flare and the separation of the neutron signal from this background was difficult.

Thus, our present conclusion is as follows: it might be still too early to conclude which acceleration mechanism does work at the solar surface, however, particles are accelerated almost instantaneously and they are trapped in magnetic loops. It is important to pursue the following things in the future to understand the nature of the particle acceleration mechanism, i.e., [1] checking experimentally the ability of the SMM GRS detector to resolve neutrons from photons, [2] checking the efficiency of neutron monitors using accelerator beams, [3] Monte Carlo studies of photon production mechanisms (especially for gamma-ray line production), and [4], the most important, collecting more data on solar neutrons with new detectors which have the ability to determine neutron energies. It will be very exciting to know the true acceleration mechanism which operates at the solar surface.

The authors thank to Prof. Don Reames who has read this draft carefully.

REFERENCES

1. E. L. Chupp *et al.*, Astrophys. J. **263**, L95 (1982).
2. E. L. Chupp *et al.*, Astrophys. J. **318**, 913 (1987).
3. Y. E. Efimov, G. E. Kocharov and K. Kudela, Proc. 18th Int. Cosmic Ray Conf. **10**, 276 (1983).
4. V. V. Akimov *et al.*, Proc. 22nd Int. Cosmic Ray Conf. **3**, 73 (1991).
5. N. G. Leikov *et al.*, Astron. Astrophys. Suppl. Ser. **97**, 345 (1993).
6. L. G. Kocharov *et al.*, Solar Phys. **150**, 261 (1994).
7. H. T. Wang and R. Ramaty, Solar Phys. **36**, 129 (1974).
8. L. G. Kocharov *et al.*, Solar Phys. **155**, 149 (1994).
9. N. Mandzhvidze and R. Ramaty, Astrophys. J. **396**, L111 (1992).
10. M. A. Shea, D. F. Smart and K. R. Pyle, Geophys. Res. Lett. **18**, 1655 (1991).

11. H. Debrunner, J. A. Lockwood and J. M. Ryan, Astrophys. J. **387**, L51 (1992).
12. H. Debrunner, J. A. Lockwood and J. M. Ryan, Astrophys. J. **409**, 822 (1993).
13. Solar-Geophys. Data **550** Part 1, 54 (1990).
14. F. Pelaez *et al.*, Solar Phys. **140**, 121 (1992).
15. L. G. Kocharov *et al.*, Proc. 24th Int. Cosmic Ray Conf. **4**, 163 (1995).
16. D. F. Smart, M. A. Shea and K. O'Brien, Proc. 24th Int. Cosmic Ray Conf. **4**, 179 (1995).
17. S. Kahler *et al.*, Astrophys. J. **302**, 504 (1986).
18. H. E. Debrunner *et al.*, Proc. 18th Int. Cosmic Ray Conf. **4**, 75 (1983).
19. S. Shibata, J. Geophys. Res. **99**, 6651 (1994).
20. R. A. Schwartz *et al.*, NASA Conf. Publ. **3137**, 457 (1992).
21. Y. Muraki *et al.*, Astrophys. J. **400**, L75 (1992).
22. R. Murphy *et al.*, Proc. 23rd Int. Cosmic Ray Conf. **3**, 99 (1993).
23. A. Struminsky, M. Matsuoka and K. Takahashi, Astrophys. J. **429**, 400 (1994).

X-RAYS

Solar Flare Energy Release
and Particle Acceleration
as Revealed by Yohkoh HXT

Takeo Kosugi[1]

National Astronomical Observatory
Mitaka, Tokyo 181, Japan

Solar flare observations with the Hard X-ray Telescope (HXT) on board *Yohkoh* are briefly reviewed with an attention to the impulsive phase. Hard X-rays in the impulsive phase are typically emitted from three sources, namely, two "footpoint sources" and a "loop-top source", with the former pair usually predominating over the latter. The double footpoint sources vary their fluxes simultaneously to each other with time lags less than a few tenths of a second, and are located in or near the chromosphere at magnetically conjugate footpoints, i.e., at the two ends of a flaring loop seen in soft X-rays. The loop-top source is a coronal source, located at an altitude of more than 10^4 km above the photosphere. At least in some cases, this source is located well above the apex of the corresponding soft X-ray loop; this source may be better named "above-the-loop-top source". Implications of these and related observations are discussed for revealing the site(s) and mechanism(s) of magnetic energy release, particle acceleration, and energy transport in solar flares.

INTRODUCTION

The impulsive phase is an interval during which solar flare energy release takes place most violently; electrons and ions are accelerated up to $\sim$ MeV and $\sim$ GeV/n, respectively, and super-hot plasmas with temperatures exceeding 3×10^7 K are created. The total energy released during a few to several minutes of this phase amounts to $\sim 10^{32}$ erg for a most intense flare, and a large portion of it is converted into kinetic energies of individual particles, at least as the first step. The energy released originates, with no doubt, from coronal magnetic fields, since no other sources can supply such a big amount. The problem to be solved now is, therefore, in what configuration of magnetic fields, and how, the energy is released and particles are accelerated.

Hard X-ray (HXR) imaging is one of the most relevant, though not unique, to observationally answer this problem. The emission in this regime is directly

[1]also at Nobeyama Radio Observatory of NAOJ, Minami-maki, Minami-saku, Nagano 384-13, Japan

produced by collisions of high-energy electrons with ions (bremsstrahlung). Thus, with an appropriate hard X-ray imaging instrument, we expect to obtain information on the spatial distribution and temporal behavior of energetic electrons, one of key clues to unveil the flare enigma.

The Hard X-ray Telescope (HXT) on board *Yohkoh* was so designed as to achieve this goal (1), and is actually approaching this goal as will be discussed in this review. The capabilities of HXT include (i) simultaneous imaging in four energy bands in the 14 – 93 keV range, (ii) angular resolution as high as ~ 5 arcsec with a wide field of view covering the whole Sun, (iii) basic temporal resolution of 0.5 s, and (iv) high sensitivity with a total geometrical aperture of ~ 60 cm^2 (1,2). With these capabilities, HXT has observed about 1000 flares since its start of routine observations in October, 1991, among which more than two thirds are intense enough for synthesizing images (3). Preliminary but comprehensive reviews of solar flare observations with HXT were given elsewhere, e.g. in (4,5); for individual work see references therein.

In this paper we will summarize typical characteristics of two types of hard X-ray sources which appear in the impulsive phase, namely, "double footpoint sources" and "loop-top impulsive source", and discuss their implications on the site(s) and mechanism(s) of magnetic energy release, particle acceleration, and energy transport in solar flares.

DOUBLE FOOTPOINT HXR SOURCES
— EVIDENCE FOR ELECTRON PRECIPITATION —

Even in the era of the first hard X-ray imaging experiments made by *SMM* and *Hinotori* more than a decade ago, it was recognized that some impulsive flares are characterized by the double-source structure, and it was claimed that this is evidence for nonthermal electron precipitation towards the two ends of a single flaring loop (6,7). Although this claim is now confirmed as discussed below, there was no firm observational basis at that moment which unambiguously supports this interpretation; the X-ray energy range used for the observations was below ~ 30 keV where contamination from thermal emission is not negligible, and the temporal resolution was not sufficient as to confirm the expected simultaneity between the two sources (8). In addition the double-source structure was observed only rarely; the number of flares with the double-source structure is small, probably due to the limited spatial resolution then available.

The HXT observations have confirmed that the double-source structure is one of the fundamental characteristics of impulsive flares. First, Sakao et al. (9) have revealed, in the case of the X-class flare of 15 November, 1991, that two sources are located at both sides of a magnetic neutral line. The double-source structure is most pronounced at higher X-ray energies ($\geq$ 30 keV) at the times of individual peaks, whereas it is not so clear at lower X-ray energies ($\leq$ 20 keV) or in between peaks. In this event the separation between the

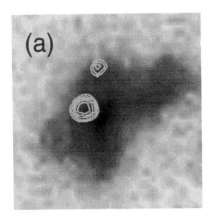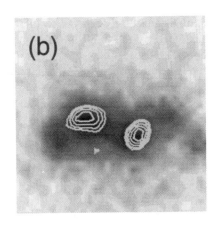

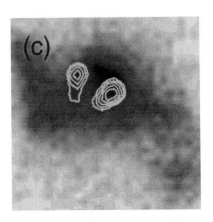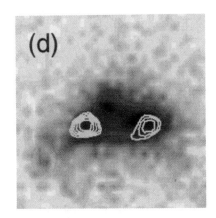

FIG. 1. Examples of double-footpoint HXR sources (M2-band; 32.7–52.7 keV; in contours) overlaid upon soft X-ray images taken almost simultaneously with SXT (in grey scale). (a) 7 February, 1992, 11:54 UT flare at S21W53; (b) 22 April, 1993, 14:08 UT flare at N11E04; (c) 27 January, 1994, 05:09 UT flare at N11W65; and (d) 14 August, 1994, 17:36 UT flare at S12W08. The solar north is to the top, west to the right. The field of view of each map covers 78×78 arcsec. Contour levels are 18, 25, 35, 50, and 71 % of the maximum brightness. Accuracy of overlay is estimated to be within $\sim 1\%$ arcsec.

double sources is observed to increase with time. At the same time the angle, sustained by the line connecting the double sources and the magnetic neutral line, increases. These are suggestive of a multiple loop system flaring successively from lower-lying, strongly-sheared loops to higher-lying, less-sheared ones with rising energy release site (10,11).

Subsequent analyses by Sakao (10,12) of a few dozens of impulsive flares have confirmed the following:

a. Statistics. The double-source structure is predominant in ~ 40 % of the events examined, and is better seen at higher X-ray energies (≥ 30 keV). The remainder show single- or multiple-source structure.

b. Source location w.r.t. magnetic neutral line. Double sources are located at a low altitude ($\leq 10^4$ km) at both sides of a magnetic neutral line. In some cases, they are cospatial with white-light brightenings (9,13). Usually we see in low-energy (≤ 20 keV) hard X-ray or soft X-ray images a loop-like structure connecting the two hard X-ray sources.

c. Separation versus simultaneity. Double sources are typically apart from each other by a linear dimension of ~ 20 arcsec or along the loop by $\sim$ a few $\times 10^4$ km. On the other hand, double sources almost simultaneously vary in intensity with time lags less than a few tenths of a second.

d. Intensity and spectral hardness. The brighter source of a pair tends to appear at a location where the photospheric magnetic field is the weaker, irrespective of the sense of the magnetic field. Recently Li et al. (14) has independently confirmed this result. Also the brighter source tends to show a harder hard X-ray spectrum than the other source of a pair.

The observational fact (d) above is easily understood as resulting from behaviors of *charged* particles that undergo magnetic mirroring in an asymmetric magnetic loop. Towards the weaker fields the particles precipitate without strong mirroring, and hence yield the brighter source. In this bright source the energy spectrum of particles becomes harder due to longer collisional energy-loss lifetimes for higher-energy particles, resulting in the harder hard X-ray spectrum ("thick-target" effect (15)). Thus, this observation is one of conclusive pieces of evidence that charged particles precipitating downwards are the primary agent which finally yields the hard X-ray emission; the possibility that the primary agent is the high-velocity, electrically-neutral beam, which is supposed to produce ~ 100-keV electrons *in situ* at the hard X-ray sources (16), can be rejected. [Note that this interpretation is consistent with the fact (b). Also it is noteworthy here that the asymmetry of magnetic loop may explain single-source cases (cf. fact (a)); hard X-rays from a region with quite strong magnetic fields might be too weak to observe. On the other hand, more than one loops may be involved in multiple-source cases. In this regard, note that the double-source structure is not necessarily suggestive of a simple, single loop flaring; see the case of the 15 November, 1991 flare already discussed.]

The observational fact (c), then, suggests with nearly 100-% confidence that the charged particles are electrons; the small, if any, time lag over the loop length cannot be explained by sub-MeV ions due to their low velocities unless the acceleration would take place exactly at the mid point of the flaring loop. Ions with energies $\geq$ MeV as the primary agent are also rejected from the energetics consideration. Thus, we conclude from the above observational facts combined together that electrons are accelerated in the corona, maybe near the apex of a loop or a system of loops, and precipitate towards the two footpoints with preference towards the weaker magnetic field footpoint.

ABOVE-THE-LOOP-TOP HXR SOURCE
— EVIDENCE FOR MAGNETIC RECONNECTION —

In the impulsive phase of several flares occurring near the solar limb, Masuda et al. (17–20) have found, in addition to the double footpoint sources, an isolated, blob-like source ("loop-top impulsive source") at an altitude of more than 10^4 km above the photosphere. This source looks isolated from the double footpoint sources in the X-ray energy range above ~ 20 keV. Surprisingly, at least in some cases, the source is located well above the apex of the corresponding soft X-ray loop; this source may be better named "above-the-loop-top source". Although this hard X-ray source is weak in comparison with the double footpoint sources by about an order of magnitude, it varies its intensity similarly to the footpoint sources, i.e., impulsively, as far as the effective temporal resolution of several to ten seconds (limited by poor photon statistics) is concerned. It shows a relatively hard spectrum; if the emission is assumed to be of thermal origin, we get the temperature $T \sim 200$ MK and the emission measure $EM \sim 10^{44} - 10^{45}$ cm^{-3}.

The existence of an impulsive hard X-ray source above the soft X-ray flaring loop is of crucial importance. With no doubt some energetic process, including particle acceleration, is in progress *outside* the bright soft X-ray loop. Most plausibly this process is a part of magnetic field disruption which is supposed to take place in the impulsive phase and release magnetic energy. In fact, soft and hard X-ray images combined together lead us to speculate a magnetic field geometry with a vertical current sheet, sandwiched by anti-parallel magnetic fields, overlying the loop structure.

Such a "loop-with-a-cusp" structure appears frequently when the solar corona dynamically changes its shape; the Soft X-ray Telescope (SXT; (21)) on board *Yohkoh* has found such structures in various spatial scales from $\sim 10^5$ km to $\geq 10^6$ km, i.e., from long-duration flares (22) to large-scale arcade formation (23). Tsuneta (24,25) discusses in detail why this "loop-with-a-cusp" structure can be regarded as evidence for magnetic reconnection; here we only mention that the outer edge of the cusp shows the highest temperature (while the bright loop itself is relatively cool) and also that the whole structure increases its height and expands with time. In this context it is reasonable to hypothesize the "above-the-loop-top" hard X-ray source as evidence for similar, but more violent, magnetic reconnection taking place in a smaller spatial

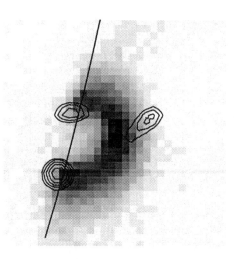

FIG. 2. "Above-the-loop-top" HXR source seen over the limb (13 January, 1992 flare at 17:29 UT). A hard X-ray image (M2-band; 32.7–52.7 keV; in contours) is overlaid upon the corresponding soft X-ray image (Be filter; in grey scale). The west solar limb is denoted by solid line. The others are the same as in figure 1.

scale in impulsive flares.

Let us speculate further how the "above-the-loop-top" source (plus double footpoint sources) can be interpreted on the basis of this reconnection hypothesis. Reconnection involves merging of anti-parallel magnetic fields at the X-point of the current sheet and inevitably ejects bidirectional plasma outflows. The downward plasma outflow, ejected from the X-point at a speed close to the local Alfvén velocity, collides with an underlying closed magnetic loop and forms a shock. Electrons may be energized in the X-point and/or in the shock. In either case, the hard X-ray source represents a site where energetic electrons easily accumulate, i.e., individual electrons somehow stay much longer in this source region than their free-transit times; otherwise the loop-top source would not look like an isolated bright blob nor so bright as observed. [If we would arbitrary introduce an ambient density high enough to explain the hard X-ray intensity of this source, such a trap would not be required (26). This is not the case, however; the ambient density determined from SXT observations is much lower.] The intensity ratio of this source to the footpoint sources suggests that electrons mirror back and forth in this source over the integrated column density of the order of one tenth of the column density that they pass before completely losing their energies. Tangled magnetic fields expected to be present in the shock may provide such a partial trap, which not only traps electrons for a while but also enables them to escape before completely losing their whole energies. Then energetic electrons

272

further stream down along the magnetic fields (without emitting detectable hard X-rays due to low ambient plasma densities along the path) to the double footpoint sources, where their energies are consumed in the thick target to heat and "evaporate" the dense chromospheric material. Bright soft X-ray loops appear as a byproduct resulting from this chromospheric evaporation.

So far we have not yet been successful in constructing any elaborated quantitative models based upon this speculative scenario. We have tried, however, to estimate from our observations several key parameters involved in this scenario, such as velocity and density of the plasma outflow, flux of accelerated electrons, ambient plasma densities in the loop-top source as well as along the path towards the footpoints, and so on. We have found a set of parameters which does not contain any trivial internal inconsistency (17,18).

A few consequences of our scenario are worthwhile mentioning here. Contrary to the widely-accepted classification scheme of solar flares, our view claims that there is no basic difference between confined (compact, short-lived, impulsive) flares and eruptive (two-ribbon, long-lived, gradual) flares. The presence of a vertical current sheet requires a rising plasmoid that stretches anti-parallel magnetic fields upwards, and the X-type reconnection requires, in addition to the downward outflow, an upward outflow which may collide with the plasmoid. In fact, Shibata et al. (27) have found faint upward ejecta seen in soft X-rays in association with impulsive flares for all the events analyzed by Masuda (17). Thus it is certain that some eruptive process is involved even in impulsive flares. Recently an interesting, new trial has been made to locate the acceleration site. Aschwanden and his colleagues (28,29) have examined energy-dependent peak delays of hard X-rays and interpreted them in terms of time-of-flight distances of electrons along the path from the acceleration site to the thick-target hard X-ray emission site. From a comparison with *Yohkoh* HXT and SXT images, they have concluded that the acceleration site is far away beyond the apex of the soft X-ray loop, maybe even above the above-the-loop-top HXR source.

CONCLUDING REMARKS

The two new observations, as well as the small step for constructing a quantitative model, mentioned at the end of the preceding section, are all encouraging, though far away from convincing. We expect that further, quantitative studies to be made straightforwardly along this line, will cast a new insight, both observationally and theoretically, and unveil the flare enigma.

In this paper we have tried to clarify implications of some of *Yohkoh* HXT observations without hesitating to oversimplify the problem and present a cartoon from a unified view. No attempt has been made to give an unbiased review for the wide variety of *Yohkoh* science.

Reuven Ramaty is acknowledged for a wonderful organization of and inviting the author to the HESP Workshop. The author is also thankful to his colleagues in the HXT experiment, especially to Taro Sakao and Satoshi Masuda. The *Yohkoh* satellite is a Japanese national project with international collaboration with U.S. and U.K. ISAS, NASA, and SERC have been continuously supporting its operation.

REFERENCES

1. T. Kosugi, K. Makishima, T. Murakami, et al., Solar Phys. **136**, 17 (1991).
2. T. Kosugi, T. Sakao, S. Masuda, et al., Publ. Astron. Soc. Japan **44**, L45 (1992).
3. T. Kosugi, M. Sawa, T. Sakao, et al., "The Yohkoh HXT Databook" (published from NAOJ; available upon request from the author) (1995).
4. T. Kosugi, in J. Linsky and S. Serio (eds.), "Physics of Solar and Stellar Coronae" (Kluwer Academic Publ.), pp.131–138 (1993).
5. T. Kosugi, in S. Enome and T. Hirayama (eds.), "Proc. of Kofu Symposium", NRO Report No. **360**, pp.11–18 (1994).
6. P. Hoyng, M. E. Machado, A. Duijveman, et al., ApJ. Lett. **244**, L153 (1981).
7. A. Duijveman, P. Hoyng, and M. E. Machado, Solar Phys. **81**, 137 (1982).
8. A. L. MacKinnon, J. C. Brown, and J. Hayward, Solar Phys. **99**, 231 (1985).
9. T. Sakao, T. Kosugi, S. Masuda, et al., Publ. Astron. Soc. Japan **44**, L83 (1992).
10. T. Sakao, Ph.D. thesis (University of Tokyo) (1994).
11. T. Sakao, T. Kosugi, S. Masuda, et al., in Y. Uchida et al. (eds.), "X-ray Solar Physics from Yohkoh" (Universal Academy Press: Tokyo), pp.91–94 (1994).
12. T. Sakao, T. Kosugi, S. Masuda, et al., in S. Enome and T. Hirayama (eds.), "Proc. of Kofu Symposium", NRO Report No. **360**, pp.169–172 (1994).
13. H. S. Hudson, L. W. Acton, T. Hirayama, and Y. Uchida, Publ. Astron. Soc. Japan **44**, L77 (1992).
14. J. Li, T. R. Metcalf, R. C. Canfield, et al., in these proceedings (1996).
15. J. C. Brown, Solar Phys. **18**, 489 (1971).
16. G. M. Simnett, Sp. Sci. Rev. **73**, 387 (1995).
17. S. Masuda, Ph.D. thesis (University of Tokyo) (1994).
18. S. Masuda, T. Kosugi, H. Hara, et al., Nature **371**, 495 (1994).
19. S. Masuda, in S. Enome and T. Hirayama (eds.), "Proc. of Kofu Symposium", NRO Report No. **360**, pp.209–212 (1994).
20. S. Masuda, T. Kosugi, H. Hara, et al., Publ. Astron. Soc. Japan **47**, 677 (1995).
21. S. Tsuneta, L. Acton, M. Bruner, et al., Solar Phys. **136**, 37 (1991).
22. S. Tsuneta, H. Hara, T. Shimizu, et al., Publ. Astron. Soc. Japan **44**, L63 (1992).
23. S. Tsuneta, T. Takahashi, L. W. Acton, et al., Publ. Astron. Soc. Japan **44**, L211 (1992).
24. S. Tsuneta, in H. Zirin, G. Ai, and H. Wang (eds.), "The Magnetic and Velocity Fields of Solar Active Regions", IAU Colloq. No. **141**, AIP Conf. Series No. **46**, pp. 239–248 (1993).
25. S. Tsuneta, ApJ., in press (1996).
26. M. S. Wheatland and D. B. Melrose, Solar Phys. **185**, 283 (1995).
27. K. Shibata, S. Masuda, M. Shimojo, et al., ApJ. Lett. **451**, L83 (1995).
28. M. J. Aschwanden, in these proceedings (1996).
29. M. J. Aschwanden, H. Hudson, T. Kosugi, and R. A. Schwartz, ApJ., in press (1996).

Reconnection Dynamics in Cusp-Shaped Flare Loops

T.G. Forbes

Institute for the Study of Earth, Oceans, and Space
University of New Hampshire, Durham, New Hampshire 03824

The soft X-ray telescope on *Yohkoh* has observed plasma structures at the top of flare loops which are suggestive of reconnection jets. However, these structures are relatively cool compared to the surrounding plasma, and their location within the dense flare loops is not consistent with the location expected of a reconnection jet. Numerical simulations of field line reconnection in a radiative plasma suggest that the observed structures are instead condensations which are formed below a reconnection jet lying at higher altitude. In the simulations the condensation is caused by the increase in density downstream of the fast-mode shock which terminates the reconnection jet. The increased density locally enhances the radiative cooling and causes the top of the loop to cool faster than the legs.

INTRODUCTION

During the last few years high resolution images obtained from the Hard X-ray Telescope (HXT) and the Soft X-ray Telescope (SXT) on *Yohkoh* show several features that are consistent with a reconnection site in the corona. These features include: *i*, a hard X-ray source located above the soft X-ray loops (1-3); *ii*, cusp structures suggestive of either an *x*-type or a *y*-type neutral line (4-6); *iii*, bright features at the top of the soft X-ray loops (7-9); and *iv*, high temperature plasma along the field lines mapping to the tip of the cusp (5, 10).

Some of the cusp-shaped loops observed by *Yohkoh* have a linear, trunk-like feature which extends from the top of the cusp all the way down to the inner most arch of the flare loop system. An example of such a trunk feature is shown in Figure 1 for a flare which occurred on 1992 February 21. This flare was a long duration event which began at 02:30 UT, and it occurred in association with a coronal mass ejection. The flare was located 15° away from the limb (heliographic coordinates E75 N8), and it produced a large scale system of loops with legs already well already separated at the start of the flare. This flare did not have an impulsive phase as such, but it did produce hard X-rays for about 30 minutes after onset (11). Throughout all but the very late phase,

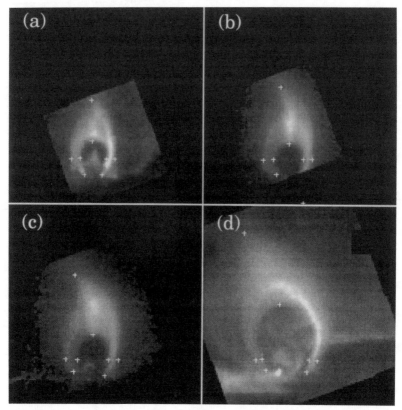

FIG. 1. SXT images of the flare loop system of 1992 February 21 at four different times: (a)3:12 UT, (b) 4:11 UT, (c) 5:02 UT, and (d) 12:19 UT. The crosses show the location of the points used to determine the widths and heights of the inner and outer edges of the loop system.

the flare loop system exhibited a trunk-like feature which moved slowly upward with time as shown in Figure 1.

The hottest regions in the loop system do not lie in the trunk feature but along the edges of the cusp formed by the outer most loop (10). Along these edges the temperature is 1.6×10^7 K, and density inferred from the emission measure assuming a line of sight of 5×10^4 km is 4×10^9 cm^{-3}. The temperatures in the trunk feature itself range from 9×10^6 K at the top to 6×10^6 K at the bottom. For the same line of sight as above, the observed emission measure leads to a density of 5×10^9 cm^{-3} at the top of the trunk to one of 2.5×10^{10} cm^{-3} at the bottom. Thus the trunk feature is both cooler and denser than the plasma surrounding it.

Figure 2 shows the evolution of the loop system over a period of about 10 hours. The location of the inner and outer edges of the legs of the loop are shown in Figure 2a, while the corresponding loop heights are shown in Figure 2b. These locations are determined systematically from the intersection of contours of the maximum intensity gradient with lines which are tangent and normal to the solar surface as shown by the crosses in Figure 1. Further details on the method used to determine these locations are in Forbes and Acton (12).

RECONNECTION MODEL OF FLARE LOOPS

At the present time, reconnection is thought to be the only mechanism which can account for the loop motions shown in Figures 1 and 2. Although some alternate mechanisms have been suggested in the past, these mechanisms

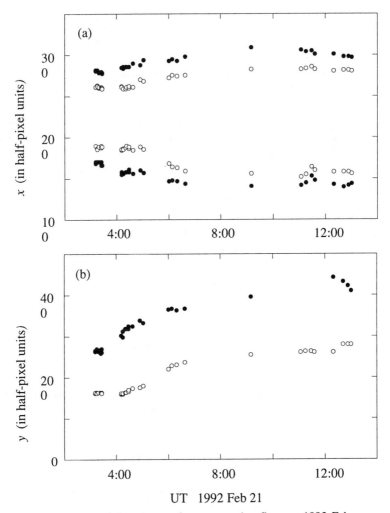

FIG 2. Evolution of flare loops after an eruptive flare on 1992 February 21. The locations (x) of the inner and outer edges of the loop legs at the chromosphere are shown in (a), while the corresponding heights (y) are shown in (b). The unit of distance is the SXT full resolution half-pixel unit (hpu) equal to 1.25" or, equivalently, 890 km.

277

have become less tenable as observations have improved. For example, simple expansion of the loops due to outward motion of the plasma from the flare site has been ruled out by Doppler-shift measurements of Hα loops. The Hα loops appear when the plasma in the X-ray loops becomes thermally unstable, and consequently these cooler loops lie just below the X-ray loops. Determination of Doppler shifts in the Hα line show that the plasma in the Hα loops flows downward at speeds of a 100-500 km/s during the time that the loops appear to be expanding (13). Thus the loop motions are not due to mass motions of the plasma, but rather to the continual propagation of an energy source onto new field lines.

Another non-reconnection model of the loop motions is the one proposed by Hudson (14). In this model the motion of the flare loops is due to the progressive thermalization by collisions of nonthermal energetic particles. The nonthermal particles are all injected at flare onset, and they are assumed to have a density which decreases with height such that the time required for them to thermalize by collision increases with height. The apparent upward motion of the thermal X-ray and Hα loops is then explained as a wave of thermalization propagating upwards in the corona. As Hudson himself noted, his model predicts a diffuse outer edge for the thermal X-ray loops and for the flare ribbons which is not consistent with observations (15). A comparison by Hiei, Hundhausen, and Sime (16) of Yohkoh SXT observations with white light coronagraph images has made this model even less likely. Hudson's model requires high density loops (about 100 times denser than the corona) to be formed over an extensive region (from the surface to heights in excess of 10^5 km) at flare onset. The coronagraph observations, which measure the total electron density independently of whether the electrons are thermal or nonthermal, show that such an extended, high density region is not present above the thermal X-ray loops.

Although one can argue that the loop and ribbon motions are strong evidence of reconnection (17), they are, nevertheless, indirect evidence. In principle, X-ray observations of sufficient spatial resolution and sensitivity can provide direct evidence. However, the ability to determine whether or not there is a reconnection site in the corona depends very much on theoretical expectations of what such a site should look like. One such proposal is shown schematically in Figure 3. This figure is a modified version of an earlier proposal based upon the ideas of Carmichael (18), Sturrock (19), and Hirayama (20) and upon various simulations of reconnection (21-23), evaporation (24-28), and condensation (29). According to this scenario, flare loops are created by chromospheric evaporation on field lines mapping to slow-mode shocks in the vicinity of the neutral line (21). These slow shocks are similar to those proposed originally by Petschek (30) except that the conduction of heat along the field lines causes them to dissociate into isothermal shocks and conduction fronts as shown in the figure. The shocks annihilate the magnetic field in the plasma flowing through them, and the thermal energy which is thus liberated is conducted along the field to the chromosphere. This in turn drives an upward flow

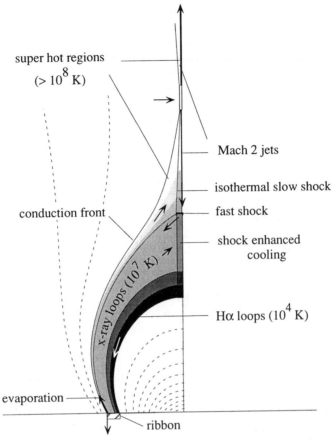

super hot regions
$(> 10^8 \text{ K})$

Mach 2 jets

isothermal slow shock

conduction front

fast shock

shock enhanced
cooling

x-ray loops (10^7 K)

Hα loops (10^4 K)

evaporation

ribbon

Fig. 3. Schematic diagram of a flare loop system formed by reconnection in the supermagnetosonic regime. This regime is most
likely to occur in the early phase of the flare when the reconnecting
fields are strong. Solid lines indicate boundaries between various
plasma regions while dashed lines indicate magnetic field lines.

of dense, heated plasma back towards the shocks, and compresses the lower
regions of the chromosphere downward.

In order for strong slow shocks to form on the field lines below the neutral
line, the outflow from it must be supermagnetosonic with respect to the fast-
mode wave speed. If the magnetic fields are sufficiently strong, the outflow
from the neutral line produces two supermagnetosonic jets – one directed upward and the other downward. Because of the obstacle presented by the closed
field lines attached to the photosphere, the lower jet terminates at a fast-mode
shock after traveling a short distance (21). Below this termination shock the
flow is deflected along the field, and only weak, field aligned slow-mode
shocks are present. Consequently, the magnetic energy released below the
termination shock is relatively small.

When the magnetic field is relatively strong, the fast-mode Mach number of the jets is about two, but as the field decreases, the Mach number decreases and the jets eventually become submagnetosonic (21). The transition occurs when the plasma β in the X-ray loops exceeds $(3 - \gamma)/\gamma$ where γ is the ratio of specific heats (21, 31). For $\gamma = 5/3$ this give $\beta = 4/5$ which for typical loop parameters corresponds to a field strength of a few Gauss. When the fast-mode Mach number becomes less than one, the lower jet disappears and almost all of the outflow from the neutral line is directed upwards. Only a relatively slow flow, aligned along the separatrices, remains below the neutral line as shown in Figure 4, and most of the plasma flowing into the reconnection region is deflected upwards. An upward reconnection jet is still present, but it is located at the upper end of a current sheet which may extend far out into the corona. Slow shocks are still present below the neutral line at the lower tip of the current sheet, but they are aligned along the separatrices and are much weaker than the Petschek-type shocks. Due to the weaker field and shocks, the loop temperature and density are reduced, and thermal condensation is less likely to occur.

An important difference between Figure 3 and earlier published versions (21, 22) is that the inner edges of the ribbons and loops are not determined by the location of the termination shock, but rather by the cooling time of the plasma as proposed by Moore et al. (32) and Švestka et al. (33). An individual field line travels through the jet in about one minute, but it takes hours for this same field line to pass from the outer to the inner edge of a flare ribbon. During the time that a field line is within the jet, its footpoint has traveled only a very small fraction of the total ribbon width.

The mapping of Figure 3 implies that the trunk feature seen in the SXT images cannot be a supermagnetosonic reconnection jet along its entire length. If a jet is present, it would have to be confined to the upper most part of the feature. Below the supermagnetosonic jet there is a thin layer of submagnetosonic flow which tends to spread the plasma out along the loops. In the absence of a cooling process, the plasma below this thin layer is essentially static (31).

Švestka et al. (33) argue convincingly that the ribbon width, W say, is approximately determined by

$$W = V_R \tau_{cr} \tag{1}$$

where V_R is the ribbon velocity and τ_{cr} is the cooling time due to conduction and radiation acting together. Švestka et al. (33) calculate τ_{cr} from the observed ribbon width and ribbon velocity and show that this value is consistent with the cooling times inferred from the emission measure and the thermal temperature. For the 29 July 1973 flare observed by *Skylab*, they find $\tau_{cr} \approx 15$ minutes ($n \approx 10^{11}$ cm^{-3}) shortly after flare onset, and $\tau_{cr} \approx 240$ minutes ($n \approx 10^{10}$ cm^{-3}) 10 hours after onset. Cooling by thermal conduction is important for the hottest loops, but radiative cooling becomes significant as the loops cool. Averaged over all the loops, the radiative and conduction losses were roughly comparable (within a factor of two) as predicted by Hirayama (20).

The thickness, W_j, of the region of the ribbon which maps to the jet, is roughly

$$W_j = (V_R/V_j) W \qquad (2)$$

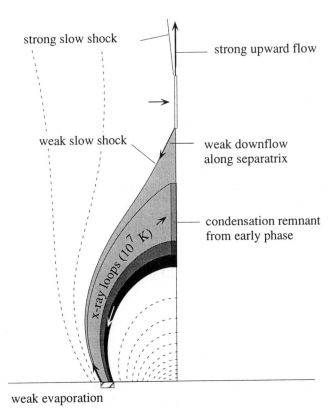

Fig. 4. Schematic diagram of a flare loop system formed by reconnection in the submagnetosonic regime. This regime is most likely to occur when the reconnecting fields are weak.

where V_j is the speed of the plasma flow in the jet. According to the reconnection-evaporation model, this speed is approximately the Alfvén speed in the dense loops which is about a factor of 10 smaller than the Alfvén speed in the corona (22). Assuming $V_j = 500$ km/s, $V_R = 10$ km/s, and $W = 2 \times 10^4$ km, gives $W_j = 400$ km which is less than an arc second. Because the value of W_j is of the same order as the thin ($\approx 1"$) region of red shifted chromospheric plasma observed by Švestka, Martin, and Kopp (15), it is much more likely that it is this red-shifted region which maps to the jet rather than the entire width of the ribbon.

One possible explanation for the lower part of the trunk feature in Figure 1 is that it is created by the onset of the thermal instability which produces the cool Hα loops lower down. A very similar feature has been found in numerical simulations which include radiation but not conduction (23). In these simulations the termination shock triggers a thermal instability in the downstream region, and as the reconnection site travels upward, the condensing region forms a trunk-like feature on the static loops.

If the reconnecting field is too weak to create the downward jet, then no condensation occurs, and the trunk feature will not form. Since the magnetic field decreases with height, we expect the jet to disappear when the reconnection site reaches a sufficient height. At this height the condensation process will stop, and newly formed loops will not contain the trunk feature as Figure 4 illustrates. Here the downward jet of Figure 3 is replaced by a weak bifurcated flow along the field lines mapping from the tip of the current sheet to the chromosphere. Because of the weaker fields, the evaporation process is greatly reduced and the plasma density in the loops becomes too low to trigger a thermal condensation. The condensation remaining on the loops at lower altitude is a remnant from the earlier supermagnetosonic phase. Eventually, as the older loops continue to cool and disappear, the trunk feature will vanish. This scenario accounts for the gradual disappearance of the trunk feature in the event of 1992 February 21 shown in Figure 1. By the time of the last image (Figure 1d) the trunk feature is no longer present.

CONCLUSIONS

Some SXT images of flare loops show a trunk-like feature which is imbedded within the loop and which extends vertically downward from a cusp at the top of the loop system. This feature has sometimes been interpreted as a supermagnetosonic reconnection jet (10). However, this interpretation encounters two difficulties. First, the trunk-like feature is several times cooler than the rest of the loop, and second, the feature extends across loops mapping from the outer to the inner edges of the flare ribbons. If the feature were a reconnection jet it should have a relatively high temperature, and the loops on which it lies should map to a thin (≈ 1 ") region lying at the outer edge of the flare ribbons (see Figure 3). Numerical simulations of the reconnection process in a radiative plasma like the corona suggest that the trunk-like feature is a thermal condensation which is triggered by a downward directed reconnection jet lying at higher altitude (21, 23). In the simulations the condensation is triggered by the density compression caused by the termination shock at the tip of a supermagnetosonic jet (about Mach 2 with respect to the fast-mode wave speed).

A few SXT images (see Figure 7 in the paper by Forbes and Acton) show a faint, spike-like feature which extends upwards from the cusp region and lies above the flare loops. This spike-like feature is relatively hot and has a low density, and it might possibly be a reconnection jet. Alternatively, the spike-like feature might simply be a static current sheet extending upwards from the cusp. In the absence of velocity measurements, it is difficult to distinguish a static current sheet from a jet. In the numerical simulations, the jet lies between a pair of currents which are slow-mode, Petschek shocks radiating from the diffusion region current sheet at the x-line. Since the angle between the shocks is only a few degrees, the two separate current sheets produce essentially the same field configuration as a single sheet.

According to both theory and simulations, a downward reconnection jet is most likely to occur when the reconnecting fields are stronger than a few Gauss (31). Thus a jet is most likely to be present in the early phase of an eruptive flare when the reconnecting fields are relatively strong. Such a jet may play a role in the production of energetic particles during both the impulsive and gradual phases. The high speed flow in the jet may be an important source of turbulence for the production of energetic particles by stochastic processes, or, alternatively, the shock structures associated with the jet may provide a site for a diffusive particle acceleration mechanism. Parallel electric fields generated by reconnection in three-dimensions are yet another possibility (34).

This work was supported by NASA grants NAGW-3463 (Solar SR&T) and NAG5-1479 (Space Physics Theory Program) to the University of New Hampshire.

REFERENCES

1. T. Sakao, et al., Publ. Astron. Soc. Japan, **44**, L83 (1992).
2. S. Masuda, T. Kosugi, H. Hara, S. Tsuneta, and Y. Ogawara, Nature, **371**, 495 (1994).
3. R.D. Bentley, et al., Astrophys. J., **421**, L55 (1994).
4. L.W. Acton, et al., Publ. Astron. Soc. Japan, **44**, L71 (1992).
5. S. Tsuneta, in: The Magnetic and Velocity Fields of Solar Active Regions, H. Zirin, G. Ai, and H. Wang, Editors (Astron. Soc. Pacific, San Francisco, 1993) p. 239.
6. G.A. Doschek, K.T. Strong, and S. Tsuneta, Astrophys. J., **440**, 370 (1995).
7. W.C. Feldman, et al., Geophys. Res. Lett., **11**, 599 (1984).
8. S. Tsuneta, H. Hara, T. Shimizu, L.W. Acton, K.T. Strong, H.S. Hudson, and Y. Ogawara, Publ. Astron. Soc. Japan, **44**, L63 (1992).
9. J.M. McTiernan, et al., Astrophys. J., **416**, L91 (1993).
10. S. Tsuneta, Astrophys. J., in press (1996).
11. H.S. Hudson, et al., in: X-Ray Solar Physics from Yohkoh, Y. Uchida, T. Watanabe, K. Shibata, and H.S. Hudson, Editors (Universal Academy Press, Tokyo, 1994) p. 143.
12. T.G. Forbes and L.W. Acton, Astrophys. J., in press (1996).
13. B. Schmieder, T.G. Forbes, J.M. Malherbe, and M.E. Machado, Astrophys. J., **317**, 956 (1987).
14. H.S. Hudson, in: 19th International Cosmic Ray Conference, C.J. Jones, J. Adams, and G.M. Mason, Editors (NASA CP2376, Washington, D.C., 1985) p. 58.

15. Z. Svestka, S.F. Martin, and R.A. Kopp, in: IAU Symp. 91, Solar and Interplanetary Dynamics, M. Dryer and E. Tandberg-Hanssen, Editors (Reidel, Dordrecht, 1980) p. 217.
16. E. Hiei, A.J. Hundhausen, and D.G. Sime, Geophys. Res. Lett., **20**, 2785 (1993).
17. R.A. Kopp and G.W. Pneuman, Solar Phys., **50**, 85 (1976).
18. H. Carmichael, in: AAS-NASA Symposium on the Physics of Solar Flares, W.N. Hess, Editor (NASA, SP-50, 1964) p. 451.
19. P.A. Sturrock, in: Structure and Development of Solar Active Regions, K. Kiepenheuer, Editor (IAU, Paris, 1968) p. 471.
20. T. Hirayama, Solar Phys., **34**, 323 (1974).
21. T.G. Forbes and J.M. Malherbe, Astrophys. J., **302**, L67 (1986).
22. T.G. Forbes, J.M. Malherbe, and E.R. Priest, Solar Phys., **120**, 258 (1989).
23. T.G. Forbes and J.M. Malherbe, Solar Phys., **135**, 361 (1991).
24. F. Nagai, Solar Phys., **68**, 351 (1980).
25. C.-C. Cheng, Astrophys. J., **265**, 1090 (1983).
26. G.A. Doschek, C.C. Cheng, E.S. Oran, J.P. Boris, and J.T. Mariska, Astrophys. J., **265**, 1103 (1983).
27. R. Pallavicini, G. Peres, S. Serio, G. Vaiana, L. Acton, J. Leibacher, and R. Rosner, Astrophys. J., **270**, 27 (1983).
28. G.H. Fisher, in: Radiation Hydrodynamics in Stars and Compact Objects, D. Mihalas and K.-H. Winkler, Editors (Springer-Verlag, New York, 1986) p. 53.
29. S.K. Antiochos and P.A. Sturrock, Astrophys. J., **254**, 343 (1982).
30. H.E. Petschek, in: The Physics of Solar Flares, W.N. Hess, Editor (NASA, SP-50, 1964) p. 425.
31. T.G. Forbes, Astrophys. J., **305**, 553 (1986).
32. R.L. Moore, et al., in: Solar Flares, P.A. Sturrock, Editor (Colorado Assoc. Univ. Press, Boulder, 1980) p. 341.
33. Z. Svestka, H.W. Dodson-Prince, S.F. Martin, O.C. Mohler, R.L. Moore, J.T. Nolte, and R.D. Petrasso, Solar Phys., **78**, 271 (1982).
34. E.R. Priest and T.G. Forbes, J. Geophys. Res., **97**, 1521 (1992).

Yohkoh OBSERVATIONS OF FLARES WITH SUPERHOT PROPERTIES

H.S. Hudson[*] and N. Nitta[†]

[*]*Institute for Astronomy, University of Hawaii*
[†]*Lockheed Palo Alto Research Laboratory*

Solar flares, almost as their defining property, fill coronal magnetic flux tubes with hot plasma. When the temperature of a significant fraction of this plasma exceeds about 30×10^6 K, we call the event "superhot", following the initial observation of the hard X-ray continuum of such an event by Lin et al. (11). The Yohkoh observations include many examples of similar events, of which three have been published thus far. This paper reports a survey of the Yohkoh observations, based mainly on the hard X-ray spectra obtained by the HXT instrument. While comprehensive conclusions will not be possible until the survey includes the Yohkoh imaging observations, we make tentative suggestions here about the nature of flares with superhot properties.

INTRODUCTION

Pioneering observations with high spectral resolution (11) showed that the hard X-ray continuum of a solar flare may contain an apparently isothermal component at a temperature considerably exceeding that of the associated soft X-ray source. This hottest component of thermal flare plasma may be that part most closely related to the heating mechanism, but further study has been difficult because of the absence of definitive high-resolution hard X-ray spectra. The superhot component could essentially form a bridge between the inherently non-thermal energy release of a solar flare, and its thermal consequences. Tsuneta et al. (19) suggest, for example, that flares with spectral properties similar to that of Lin et al. may result from intrinsically thermal energy release, rather than the usual non-thermal energy release involving strong particle acceleration.

The Yohkoh data (17) provide our most comprehensive observations yet of superhot sources, since they combine hard X-ray continuum spectroscopy and imaging, soft X-ray imaging, and soft X-ray emission-line spectroscopy, including spectroscopy in the FeXXVI emission lines. Figure 1 shows the thermal responses of the different instruments on board Yohkoh.

FIG. 1. Responses of the different detectors on board *Yohkoh*, as functions of temperature. The calculations assume an isothermal source with Mewe atomic-physics tabulations (12), with assumptions about abundances and ionization equilibrium. The dotted lines refer to emission at the formation temperatures of the resonance lines. The HXT responses are normalized to the values at 10^8 K.

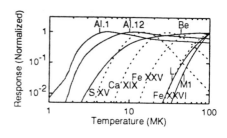

Thus far only three superhot events observed by *Yohkoh* have been described in the literature (Appendix B). This paper begins a broader survey, starting with a list of FeXXVI events compiled by Pike *et al.* (15). The FeXXVI lines require high temperatures to excite them. According to Figure 1, the low-temperature branch of the FeXXVI contribution function almost exactly matches that of the HXT low-energy channel. The Pike *et al.* list of 75 events is therefore a good starting point, although we note that instrumental effects restrict the numbers. In any case, the *Yohkoh* data show that FeXXVI emission or "Type A" behavior (18) is relatively common. We study these events using the hard X-ray spectral evolution seen in the four broad-band energy channels of the *Yohkoh*/HXT instrument. This paper represents a progress report for a fuller survey, which would include the hard and soft X-ray images from *Yohkoh*.

In this paper we use certain properties of the hard X-ray spectral evolution to show the superhot nature of an event: high apparent temperatures (above 30×10^6 K), smooth time profiles, and the characteristic pattern of joint variation of temperature and emission measure seen in thermal sources at lower temperatures (7). These conditions do not unambiguously establish the thermal nature of a source, but the smooth time profile and cooling pattern are not found in impulsive-phase non-thermal sources (4), and the cooling pattern is not found in gradual coronal sources (6).

BACKGROUND

The appearance of hot plasma in the solar corona is nowadays almost the defining property of a solar flare. Soft X-rays and Hα emission occur in a one-to-one relationship, except for limb flares where occultation confuses the relationship; the effective temperatures of flare X-ray plasmas distinguish them clearly from ordinary active-region loops. Soft X-ray emission from a

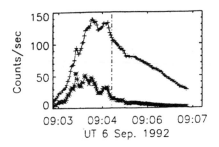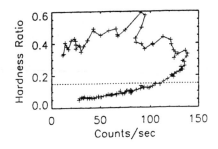

FIG. 2. A superhot flare observed by *Yohkoh*/HXT on 6 September 1992 (Pike #52). Left, time profiles of the two lowest energy channels of HXT, 14-23 keV and 23-33 keV. Right, the hardness ratio of these two channels plotted against the rate in the low channel. After the clearly-defined impulsive phase, the spectrum radically softens. the time profile becomes smooth, and the flux decays smoothly. After the dotted line (left) and below it (right) the apparent temperature drops below about 35×10^6 K.

flare occurs in a set of one or more magnetic flux tubes and develops late in the flare process. The Hα emission tends to occur at the chromospheric footpoints of the magnetic loops. This picture is oversimplified but adequately represents the important part of the phenomenology.

The bulk of the emission measure typically falls in the range 10-20 × 10^6 K, although lower peak temperatures occur in flare-like structures outside active regions *e.g.* (1). The temperature of the hot flare plasma typically reaches a maximum before the emission measure (or the flux) does, *e.g.* (5). This pattern of spectral evolution clearly distinguishes thermal X-ray sources from the non-thermal (hard) impulsive bursts, which tend to show a symmetrical soft-hard-soft spectral evolution (8).

Similar behavior (hot first, then bright) also characterizes the superhot component (11). Figure 2 shows a *Yohkoh* flare with superhot characteristics. In this case HXT photometry in the 14-33 keV range showed temperatures above 30×10^6 K, while GOES temperatures did not exceed 16×10^6 K. The ratio of emission measures of the two "components", although this is unlikely to be the correct way to describe the broad distribution of differential emission measure, is about 10 in this case.

We normally allow for the possibility of a distribution of temperatures in a given observation via the concept of the differential emission measure (for a recent discussion in the context of moving plasmas, see (13). There are two distinct ways in which a distributed emission measure can occur. First, in an optically-thin medium a given observational line of sight may include radiation from physically distinct plasmas at different temperatures. Second, the plasma itself may be homogeneous but have a non-Maxwellian particle velocity distribution. This is a transient situation, expected to prevail during heating. The time scale for thermalization can be estimated from Coulomb collision theory (20):

$$\tau_{ee} = 5.5 \times 10^6 T_6^{1.5}/n \quad sec,$$

where T_6 represents the minority electron temperature in MK, n the ambient density, cm^{-3}, and τ_{ee} the electron energy-exchange time with an assumed background component at a much lower temperature. For $T_6 = 40$, τ_{ee} ranges from 20 sec to 2 msec for densities in the range 10^8-$10^{12} cm^{-3}$. These times may be long enough to permit a non-Maxwellian electron distribution function to exist, especially if the particle heating occurs in a low-density domain and on the rapid time scales recently discovered in solar flares (2).

Yohkoh/HXT HARD X-RAY PHOTOMETRY

The Yohkoh/HXT instrument is designed to image hard X-ray sources (9). It also has provided the best broad-band photometry to date. The photometry is good because the imaging function of HXT dictated the use of many (64) small, independent detectors, each with a peak exposed area of about 2 cm^2. Thus the counting rate does not become large even in a major event, and saturation problems such as pulse pile-up are minimized. The total detector area is large, for photometric purposes ~55 cm^2, to provide good photon statistics for image reconstruction. In addition, imaging requires careful detector gain stabilization (gain is defined here as d(PH)/d(hν), where PH represents the pulse height in the detector electronics, and hν the photon energy). With these advantages HXT can do precise photometry over a wide dynamic range. There is normally a telemetry-imposed restriction on the number of energy channels of four (Low, M1, M2, High: nominally 14-23-33-53-93 keV) in the normal mode of operation.

Appendix A describes the modeled response of the four HXT energy channels to thermal sources (cf. Fig. 1). It is important to note that the HXT entrance windows are thin enough to permit substantial response even at "normal" flare temperatures of 15-20 $\times$ 10^{6}K, so it is not surprising that most flares show thermal characteristics at least in the 14-23 keV channel.

The recognition of a thermal source from HXT data alone is inherently ambiguous, since the two-channel comparison that determines a temperature could as easily be used to fit a power-law distribution. It is not enough to note that a thermal spectrum of 30 $\times$ 10^6 K corresponds to an implausibly soft non-thermal spectra (the equivalent power-law index $\gamma \sim h\nu/kT$, so that for HXT energies we would have $\gamma \sim$ 8-10). There is no theoretical reason not to have

TABLE 1. *Yohkoh*/HXT spectra of flares in the Pike *et al.* list

Property	Number of events
Number with superhot properties	~46/64
"Good" events	33
Number with >5% "superhot" emission measure	13/33
Number with $T_{M1/Low} > 30$ MK	26/33

such distributions even though they do not occur at higher energies in normal impulsive-phase spectra (4). Therefore in identifying superhot plasmas we also make use of the spectral time evolution. The normal pattern of impulsive-phase hard X-ray spectral evolution is the soft-hard-soft variation (8). As noted above, thermal plasmas may tend instead to show a pattern of cooling with time, or if the heating phase is included, then an open loop pattern.

SURVEY FOR SUPERHOT PROPERTIES

The preliminary survey presented here starts from the Pike *et al.* list of FeXXVI flares observed by *Yohkoh* (15). This list includes the three prominent examples described in Appendix B, but we note that this does not represent a complete survey because of detector saturation and other effects.

For each of the events in the Pike *et al.* list, we have examined (i) the thermal fits to HXT photometry (*i.e.*, non-imaging), (ii) the HXT and GOES light curves, and (iii) the joint variation of HXT spectral hardness ratio with respect to flux. Table 1 presents the results of estimates of the thermal parameters, which were carried out in an interactive manner for background selection and time range. We find that most events in the Pike *et al.* list do exhibit superhot characteristics, as inferred from the HXT photometry. We find many cases (*e.g.*, the one shown in Figure 2) in which the two HXT temperatures match reasonably well, and some in which they do not. In the latter cases the temperature derived from the M2/M1 ratio, at a representative energy of 33 keV, is normally higher than the temperature derived from the M1/Low ratio, at a representative energy of 23 keV.

The *Yohkoh*/HXT data show events similar in spectral behavior to the original Lin *et al.* event, namely the appearance of isothermality over the range of photon energies detected. We interpret this to mean that, in such events, a truly Maxwellian particle energy distribution dominates, and that the apparent differential emission measure (*i.e.*, that resulting from line-of-sight averaging) has a well-defined upper-limit temperature. These events tend strongly to show time variations of thermal types, in the sense of the loop seen in Figure 2.

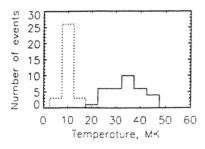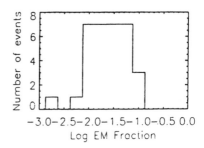

FIG. 3. Results of the survey of hard X-ray spectral properties of the Pike *et al.* list (15) of FeXXVI flares observed by *Yohkoh*. Left, the distribution of temperatures (solid line, the HXT M1/Low hardness-ratio temperature; dotted line, the simultaneously observed temperature from GOES; right, the distribution of ratios of emission measures between the hard X-ray source and the GOES sources. Both of these distributions are for 33 of the 75 Pike events matching the criteria described in the text.

CONCLUSIONS

From the *Yohkoh* observations, we find abundant evidence that superhot temperatures occur. The Pike *et al.* list of 75 members, a lower limit, represents a substantial fraction (on the order of 10%) of all the flares that triggered *Yohkoh* flare mode during the interval. Do the events showing this phenomenon constitute a separate *class* of flares, as it appeared from *Hinotori* (18)? Do the observations even suggest the existence of a physically distinct *component*, as suggested by Lin *et al.*? The answers to these two questions are "probably not", even though the superhot phenomenon is observationally distinctive when it occurs. In two of the three events already published (Table 1, Appendix A), the superhot characteristics appear in a physically distinct structure.

The present work deals mainly with a survey of the morphology of the hard X-ray spectral evolution in the Pike *et al.* list of 75 FeXXVI events. A future more comprehensive work will include image properties, which are available both from SXT and from HXT. We summarize our present results as follows:

- The occurrence of FeXXVI emission lines (15) is a good guide to the presence of superhot characteristics in the hard X-ray spectrum.

- The superhot sources behave like normal hot plasmas in solar flares, except for having higher temperatures.

- The differential emission measure of the superhot sources may abruptly drop at a relatively low temperature, say 30×10^6K, but in some cases it clearly extends much higher.

These observations are consistent with the notion that, at least sometimes, the superhot characteristics represent the thermalization of a plasma whose electron distribution function has been driven out of equilibrium. In other cases, the superhot source may be relaxed but seen together with normal flare sources (line-of-sight effect), thus giving the appearance of a broad differential emission-measure distribution. The survey described here could not easily distinguish these two alternatives, but the future survey using images may be able to help further. There is always the possibility that these superhot properties result actually from non-thermal particles mimicking thermal behavior. The nature of the particle distribution function is difficult to determine from remote-sensing observations.

Acknowledgements. NASA supported this work under contract NAS 8-37334. The *Yohkoh* satellite is a project of the Institute of Space and Astronautical Sciences of Japan. We thank A. Sterling for comments on the manuscript.

APPENDIX A. *Yohkoh*/HXT THERMAL RESPONSE

Figure A1 (left) summarizes the spectral response of the HXT instrument, based upon detailed calculations performed by M. Inda-Koide. These calculations include the detailed material properties of the detectors, entrance windows, and housings, plus allowance for detector resolution spreading and K-photon escape. Each of the 64 subcollimators views a different part of any given source, due to the modulation collimators that provide the imaging response. Accordingly each will have a different total counting rate, and the measurement of total hard X-ray flux at any given instant may have an uncertainty on the order of $1/\sqrt{64}$ or about 10%.

We have taken these probability functions and convolved them with model thermal spectra (12) for isothermal sources at different temperatures. From the convolved counting rates, the ratio of counts in any adjacent pair of energy channels determines a temperature and an emission measure. Figure A1 (right) shows these "hardness ratios" for the three independent pairs of adjacent channels. Note that the thermal hardness ratios are much smaller than the hardness ratios rsulting from power-law spectra: HXT was designed to have hardness ratios of approximately unity at number spectral index $\gamma = 2$.

APPENDIX B. THREE *Yohkoh* SUPERHOT EVENTS

Three *Yohkoh* events with superhot properties have already been presented in the literature (Table 2). The published analyses include image morphology. We describe these briefly here.

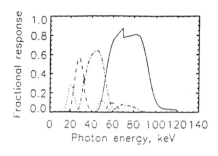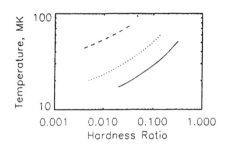

FIG. A1. Left, the responses of the four HXT broad-band channels (Low, 14-23 keV, dots; M1, 23-33 keV, dashes; M2, 33-53 keV, dot-dash; and High, 53-93 keV, solid). Right, the variation of hardness ratios with temperature: Med1/Low, solid; Med2/Med1, dots; High/Med2, dashes.

TABLE 2. Three previously reported superhot events

Date	T_{HXT} [a]	Spatial Pattern	Reference
16 Dec 1991	35	Legs	(3)
6 Feb 1992	35	Separate	(10), (16)
21 Aug 1992	40	Separate	(14)

[a]MED1/LOW channel ratio, effective energy ~23 keV, evaluated at end of impulsive phase.

16 December 1991, Pike #10 (3). The light curves show a distinct second phase, during which the HXT and Fe XXVI temperatures matched. The HXT images showed a loop with bright footpoints, but there unfortunately were no SXT data. Culhane *et al.* suggested that the superhot component in this case could be identified with the legs of the loop, filled with the heated and upward-flowing material ablating from the chromosphere.

6 February 1992, Pike #25 (10); (16). The superhot signatures occurs later in time by several minutes than a more normal flaring loop. The HXT images clearly define the superhot location to be a separate structure. The SXT image sequence gives the appearance of gradual filling from one leg during the source evolution.

21 August 1992. Pike #47 (14). The superhot source appears in a separate loop following a more normal event, in a pattern possibly similar to that of the 6 Feb. 1992 event. The temperatures determined by SXT filter photometry clearly define the superhot location to be a separate longer loop structure in this case.

REFERENCES

1. D. Alexander, K.L. Harvey, H.S. Hudson, J.T. Hoeksema, and X. Zhao, submitted to *Solar Wind 8* (1995).
2. M.J. Aschwanden, A.O. Benz, and R.A. Schwartz, Ap. J. **417**, 790 (1993).
3. J.L. Culhane, A.T. Phillips, M. Inda-Koide, T. Kosugi, A. Fludra, H. Kurokawa, K. Makishima, C.D. Pike, T. Sakao, T. Sakurai, G.A. Doschek, and R.D. Bentley, Solar Phys. **153**, 307 (1984).
4. B.R. Dennis, Solar Phys. **100**, 465 (1986).
5. D. Horan, Solar Phys. **21**, 188 (1971).
6. H.S. Hudson, Ap.J.**224**, 235 (1978).
7. J. Jakimiec, B. Sylwester, J. Sylwester, S. Serio, G. Peres, and F. Reale, Astron. Astrophys. **253**, 269 (1992).
8. S.R. Kane and K.A. Anderson, Ap. J. **162**, 1003 (1970).
9. T. Kosugi, K. Makishima, M. Inda, T. Murakami, and T. Dotani, Solar Phys.
10. T. Kosugi, T. Sakao, S. Masuda, H. Hara, T. Shimizu, and H.S. Hudson, Proc. Kofu Symposium, 127 (1994).
11. R.P. Lin, R.A. Schwartz, R.M. Pelling, and K.C. Hurley, Ap. J. **251**, L109 (1981).
12. R. Mewe, J.R. Lemen, and G.H.J. Van den Oord, Astron. Astrophys. **65**, 511 (1986).
13. E.K. Newton, A.G. Emslie, and J.T. Mariska, Ap. J. **447**, 915 (1995).
14. N. Nitta and K. Yaji, submitted to Ap. J. (1995).
15. C.D. Pike, K.J.H. Phillips, J. Lang, A. Sterling, T. Watanabe, E. Hiei, J.L. Culhane, M. Cornille, and J. Dubau, submitted to *Ap. J.* (1995).
16. A. Sterling, Proc. Kofu Symposium, 131 (1994).
17. Z. Svestka, and Y. Uchida, "The *Yohkoh* (Solar-A) Mission" (Dordrecht: Reidel) (Solar Phys. **136**, 1991).
18. K. Tanaka, in P.B. Byrne and M. Rodonò (eds), Activity in Red-Dwarf Stars (Dordrecht: Reidel), p. 307 (1983).
19. S. Tsuneta, N. Nitta, K. Ohki, T. Takakura, K. Tanaka, K. Makishima, T. Murakami, M. Oda, and Y. Ogawara, Ap. J. **284**, 827 (1984).
20. B.A. Trubnikov, 1965, *Revs. Plasma Physics* **1**, 105.

Conditions for Energetic Flares

Nariaki Nitta

Lockheed Palo Alto Research Laboratory
O/91-30, B/252
3251 Hanover Street
Palo Alto, California 94304

Hard X-ray emission is commonly observed in association with solar flares and it is believed to signify interactions of accelerated electrons with the ambient ions. As a first step to understand what conditions are conducive to acceleration of electrons in solar flares, we have analyzed *Yohkoh* X-ray images for a total of four-dozen events whose hard X-ray spectra extend at least to 100 keV. The morphology of soft X-ray emission relative to hard X-ray sources shows a wide variety, but it is usually confined in a compact loop, which is sometimes topped with a cusp that is much more diffuse. While eruptive behaviors are often seen in the soft X-ray images, it is hard to decide whether they trigger or result from the nonthermal processes. We give a case study which indicates that double hard X-ray sources do not always come from conjugate footpoints of a loop.

INTRODUCTION

We empirically know that most solar flares include an impulsive phase, producing hard X-ray emission whose intensity varies with time scales from subseconds to minutes. In such "common" flares, the classic thick-target model (1) is often applied. Although this idea has been generally supported by recent observations (10), we have long known that some flares have considerably lower ratios of hard and soft X-ray fluxes than others, suggesting a more complex reality. Moreover, the model is not meant to answer the fundamental question of how to accelerate electrons.

In recent years it has become fashionable to explain solar flares in terms of reconnection of initially open field lines, following a long duration event that showed a cusp shaped geometry in soft X-rays (13). Shibata *et al.* (11) has pushed the reconnection hypothesis in such a field geometry to compact flares and claimed that X-ray ejecta are detected around the time of the hard X-ray intensity peak. Comparing the observations with MHD simulations, these authors argue that such ejecta permit field lines to reconnect, resulting in closed loops that appear bright in soft X-rays. The causal relation may be observationally marginal, however, given the limited number of the large-field-of-view images that are essential to their identification of the ejecta. In addition, it is unclear how nonthermal processes can be incorporated quantitatively in MHD models.

Thus it is important to extract observationally the conditions that must be satisfied for accelerating electrons. As a first step, we have compared the soft and hard X-ray images from the *Yohkoh* spacecraft (12) for those events that produced strong hard X-ray signals. Such events have been analyzed in detail by Sakao (10) but mostly in hard X-rays only. With the soft X-ray data, we can also find preflare conditions and locate hard X-ray sources in terms of the structures that represent coronal magnetic field. While we are preparing for a full paper on our analysis to be presented elsewhere, some important results are given in this report. We also report on a controversial flare that is part of our sample.

EVENT SELECTION

Our study is based on data from the Hard X-ray Telescope (HXT) and the Soft X-ray Telescope (SXT) on board *Yohkoh*. We have selected 48 flares for analysis in a similar manner as (10), but a little differently. Our criteria require the flare (i) to have significant counts ($\gtrsim 5$ counts/s/subcollimator in a light curve averaged over 10 data points) in the HXT H band (53-93 keV), (ii) to be observed by SXT from the beginning, and (iii) to have its hard X-ray peak observed by HXT. The number of our flares overlapping with (10) is 21. According to data from the *Yohkoh* hard X-ray spectrometer, only six of our flares show significant counts above 500 keV (the last one being the 1992 January 26 flare (5)). Therefore the flares we study are generally not unusually energetic.

ANALYSIS

For each flare, we have synthesized hard X-ray images in the M2 band (33–53 keV) around the hardest peak using both the standard Maximum Entropy method and the Pixon method (8) to see their consistency. They are coaligned with the corresponding SXT images with an accuracy of $\sim 1''$ (6). The SXT images have been examined to keep track of how the soft X-ray sources evolve, and to see if there are signatures of eruption or ejection. We have also produced soft X-ray light curves at the hard X-ray sources to look for impulsive variations in synchronous with hard X-rays (5). Using filter ratios, we have obtained temperature maps at various times to see if a hot region is formed. We have summed 5–10 images in each filter to improve the counting statistics. In order to measure the extension of soft X-ray emission at the time of the hard X-ray peak, we take the number of the first brightest pixels of the SXT image (in Be filter), in such a way that flux from those pixels reaches 50 % of total flux from the entire image. We denote such a number as N_{50}.

295

TABLE 1. Number of events having various characteristics

Hard X-ray Appearance	Eruption	Impulsive SXR	Extended	Hot Region
Single (16)	12	4	2	5
Double (25)	16	12	6	11
Multiple or Extended (7)	3	3	4	7

We summarize the results in Table 1. We come to a similar result as (10) regarding the appearance of hard X-ray sources for the overlapping events. More than half of the events look double.

Concerning the existence of eruptive features, we have included everything however or whenever it may appear. In fact, some flares show jet-like motions, probably X-ray counterparts to Hα surges (2), whereas others show disturbances in a wide area or expanding loops. We have been able to identify eruptions from disk flares also. It is likely that the direction of eruptions affects their visibility more than the background emission. Only in a few flares have we seen eruptions preceding the impulsive phase; these flares have comparatively gradual light curves in hard X-rays.

Detection of the impulsive soft X-ray variation is often difficult either because the soft X-ray image at the hard X-ray peak is saturated or flux has already built up at the apparent hard X-ray source locations. We include all the flares that show at least one area of soft X-ray impulsive variations. Here we call a flare extended if N_{50} (see the above definition) exceeds 100. Our flares are generally not extended. We regard a flare as having a hot region if several adjacent pixels persistently have temperature above 18 MK in the filter ratio Be/Al.12. We have not included those cases where only a few pixels at the edge show high temperatures, since they may represent artificial effects such as slight misalignments. It is probably easier to see high temperatures from a flare of large dimension.

FLARE ON 1991 DECEMBER 3

Here we present a case study to illustrate how challenging it is to interpret the data. This flare, located close to the limb (N17E78), is one of the most energetic ones that has been observed by *Yohkoh* and its spectrum in the γ-ray range is dominated by continuum rather than lines (9). The light curves are shown in Figure 1. The 50–90 keV hard X-rays exhibit spiky variations that are typical in impulsive flares. Indeed, the hard X-ray images show double sources with close temporal correlations (10). We are inclined to an interpretation that these hard X-ray sources represent two footpoints of a loop.

A quick look at the soft X-ray images gives an impression that the flare is compact and it is hard to identify a loop or loops. On large-field-of-view images an eruption is perceived only in the form of an expanding closed loop

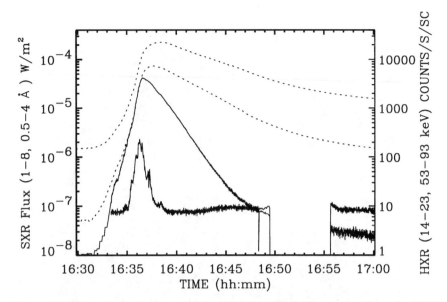

FIG. 1. Light curves for the 1991 December 3 flare. Solid lines come from the HXT L and H bands. Dotted lines represent the GOES flux

lasting 8 min after the strong hard X-ray emission. Then the outer diffuse structure starts to be affected. Thus we were about to conclude that this flare has essentially a closed nature like other double footpoint flares (4). However, a close look at the images in the Be filter with different scaling has revealed a small but growing loop that appears to be topped with a cusp structure (See Figure 2). The footpoint separation at the hard X-ray peak (Figure 2(b)) is only $\sim 12''$; it increases to $\sim 20''$ after the hard X-ray burst is over.

Comparing the X-ray images with the white-light image (Figure 2(d)), both footpoints of the X-ray loop are anchored on the western side of a sunspot. That means that we have difficulty regarding the north-eastern hard X-ray source as coming from a loop footpoint. Figure 2 (c) suggests, instead, that the northeastern source comes from the top of a loop, which probably represents a compact core as is commonly seen (3). But since the loop is compact, there is the possibility that the hard X-ray source may be located above the loop as in the famous 1992 January 13 event (7).

FURTHER COMMENTS

From the example of the last section, wee see that even if the hard X-ray sources appear double and their temporal variations are similar, they may not necessarily come from conjugate footpoints of a loop. In order to understand the magnetic field configuration around the hard X-ray sources, it it useful

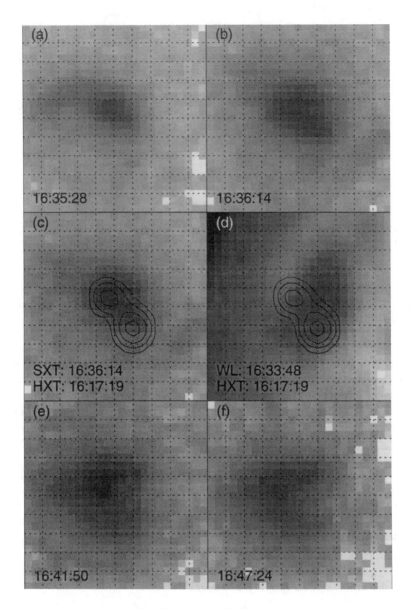

FIG. 2. (a), (b), (e), and (f) Sequence of SXT images in Be filter. The color is inverted. In (d) an SXT white-light image is shown, in which a sun spot is seen. The contours in (c) and (d) represent a hard X-ray (33–53 keV) image at the peak. The field of view is $61.5'' \times 61.5''$.

to study time sequences of soft X-ray images. The example suggests that there may be flares in which the loop and cusp are too small to be clearly indentified by the SXT. However, such a configuration may not occur in other flares, since we also know at least a few nice examples of hard X-ray sources located at conjugate footpoints of a loop (4).

In connection with limited resolution of the present instrumentation, Sakao (10) indeed pointed out a possibility that HXT sometimes cannot resolve two footpoints close by, resulting in the apearance of a single source. This hypothesis may be applicable to one or both hard X-ray sources when they appear double.

Acknowledgements. NASA supported this work under contract NAS 8-37334. We are thankful to H. Hudson, K. Shibata and M. Aschwanden for fruitful discussions.

REFERENCES

1. J. C. Brown, Solar Phys.,**18**, 489 (1971).
2. R. C. Canfield, K. P. Reardon, K. D. Leka, K. Shibata, T. Yokoyama, and M. Shimojo, ApJ, in press, (1996).
3. U. Feldman, J. F. Seely, G. A. Doschek, K. T. Strong, L. W. Acton, Y. Uchida, and S. Tsuneta, ApJ, **424**, 444 (1994).
4. N. Gopalswamy, J.-P. Raulin, M. R. Kundu, N. Nitta, J. R. Lemen, R. Herrmann, D. Zarro, and T. Kosugi, ApJ, **455**, 715 (1995).
5. H. S. Hudson, K. T. Strong, B. R. Dennis, D. Zarro, M. Inda, T. Kosugi, and T. Sakao, ApJL, **422**, L25 (1994).
6. S. Masuda, PhD Thesis, University of Tokyo (1994).
7. S. Masuda, T. Kosugi, H. Hara, S. Tsuneta, and Y. Ogawara, Nature, **371**, 495 (1994).
8. T. R. Metcalf, H. S. Hudson, T. Kosugi, R. C. Puetter, and R. K. Piña, submitted to ApJ (1995).
9. K. Morimoto, PhD Thesis, Rikkyo University (1996).
10. T. Sakao, PhD Thesis, University of Tokyo (1994).
11. K. Shibata, S. Masuda, M. Shimojo, H. Hara, T. Yokoyama, S. Tsuneta, T. Kosugi, and Y. Ogawara, ApJL, **451**, L83 (1995).
12. Z. Svestka, and Y. Uchida, "The *Yohkoh* (Solar-A) Mission" (Dordrecht: Reidel) (Solar Phys. **136**, 1991).
13. S. Tsuneta, H. Hara, T. Shimizu, L. W. Acton, K. T. Strong, H. S. Hudson, and Y. Ogawara, Publ.Astr.Soc.Jap., 44, L66 (1992).

Hard X-Ray Timing

Markus J. Aschwanden

University of Maryland, Astronomy Department, College Park, MD 20742
internet: markus@astro.umd.edu

High-time resolution (64 ms) hard X-ray (HXR) data from *BATSE/ CGRO* allow us to study the energy-dependent timing of acceleration, propagation, energy loss, and trapping of $\gtrsim 20$ keV HXR-emitting electrons during solar flares. In many flares two different HXR flux components can be distinguished: (1) the fine structure of a HXR time profile (containing sequences of subsecond pulses) exhibits delays of $\approx 10 - 100$ ms for the low-energy electrons, while (2) the unmodulated HXR time profile (a smooth lower envelope to the fine structure) shows a delay of opposite sign and much larger magnitude, of typically 1-10 s. We model the timing of various acceleration mechanisms and find that the delay of the HXR pulses is dominated by time-of-flight differences rather than by acceleration time scales, while the timing of the unmodulated HXR flux is governed by trapping and collisional time scales.

INTRODUCTION

Many flares show double footpoint hard X-ray emission, where the time histories of the two footpoint sources seem to be synchronized with an accuracy of $\lesssim 100$ ms (16). This synchronization suggests that the acceleration site of electrons that produce thick-target emission at the footpoints is located near the top of flare loops. Recently, HXR sources have also been discovered near or above flare looptops (12), (13). The HXR looptop emission is always fainter than the bright HXR footpoint emission, but its time history shows a striking similarity and simultaneity with the footpoint emission, and thus, seems to be modulated by the same non-thermal electrons that modulate the footpoint emission (10). This indicates a topological link that places the acceleration site near or above the flare looptop. Recently, another constraint on the timing of HXR-producing electrons was established, that makes use of the energy-dependent time delays observed in high-time resolution data from the *Burst and Transient Source Experiment (BATSE)* (8) onboard the *Compton Gamma Ray Observatory (CGRO)*. For the 1992 January 13 flare, which became the prototype of flares with looptop HXR emission, an electron time-of-flight path length was inferred that corresponds to an altitude of $h = 44,000 \pm 6000$ km for the acceleration site, which is significantly higher than the apex of the bright SXR flare loop (with a mean height of $h = 12,500$ km), or the HXR source in the cusp region of the flare loop (in a height of $h = 22,000$ km)

(1). Additional statistics of flare loop geometries observed with *Yohkoh/HXT* and electron time-of-flight path lengths inferred from *BATSE/CGRO* confirm this initial result, that the timing of HXR pulses is consistent with an acceleration site located significantly above the SXR flare loop (2). The timing of HXR emission seems to provide powerful constraints on the inference of the electron kinematics. In this paper we present one of the most accurate HXR delay measurements available sofar (from the 1991 Dec 15 flare) and test various timing models that constrain the kinematics of electron acceleration and propagation.

ELECTRON KINEMATICS

We can measure time delays τ_{ij} between HXR fluxes in different detector channels i and j (with characteristic energies of ε_i and ε_j),

$$\tau_{ij} = t_i^X(\varepsilon_i) - t_j^X(\varepsilon_j) , \tag{1}$$

by means of cross-correlation, to determine the relative timing t_i^X of a selected time structure. Because an electron with kinetic energy E can produce hard X-ray photons over a large range of lower energies ($\varepsilon \leq E$) by bremsstrahlung, the relation of electron energies E to photon energies ε represents a convolution with the bremsstrahlung cross-section $\sigma(\varepsilon, E)$ and also with the instrumental response function $R(\varepsilon)$. We simulated this time-dependent, spectral convolution for electron injection pulses over a wide range of spectral indices γ and cutoff energies E_0 of the electron injection spectrum and derived numerical values for the energy conversion,

$$E = \varepsilon \cdot q_E(\varepsilon, \gamma, E_0) \tag{2}$$

that allows one to invert the timing of HXR energies $\varepsilon_i(t)$ into the timing of electron energies $E_i(t)$ (4). For typical HXR spectra (with a power-law slope of $\gamma \approx 4$) the conversion factor is $q_E \approx 2.0$ for the thick-target model.

The kinetic electron energy E is related to the relativistic Lorentz factor γ by

$$E(\gamma) = m_e c^2 (\gamma - 1) , \tag{3}$$

with $m_e c^2 = 511$ keV, and the velocity v is defined by the relativistic relation

$$v(\gamma) = c\sqrt{1 - \gamma^{-2}} . \tag{4}$$

The observed timing of HXR time structures $t^X(E)$ can be affected by a number of processes, such as the acceleration time $t^{acc}(E)$, the time-of-flight (TOF) or propagation time $t^{prop}(E)$ [which may be prolonged by some trapping time $t^{trap}(E)$], and the energy loss time $t^{loss}(E)$,

$$t^X(E) = t^{acc}(E) + t^{prop}(E) + t^{loss}(E) , \tag{5}$$

which all have different functional dependences on the energy E, and thus may serve as useful diagnostic of the dominant timing process.

Propagation time scales

The probably most important effect in the timing of HXRs is the propagation time of electrons between the acceleration site and the energy loss site, often located at the chromospheric footpoints of flare loops in the thick-target model. If we assume that electrons are simultaneously accelerated, the timing of HXRs is essentially determined by the time-of-flight difference of electrons with different speeds, i.e.

$$\tau_{ij} \approx t^{prop}(E_i) - t^{prop}(E_j) = \frac{l}{v_i} - \frac{l}{v_j} = \frac{l}{c}\left(\frac{1}{\sqrt{1-\gamma_i^{-2}}} - \frac{1}{\sqrt{1-\gamma_j^{-2}}}\right) . \quad (6)$$

Implicitely it is assumed in Eq.6 that acceleration times and energy loss times are much smaller than the free-flight propagation times, $(t^{acc} \ll t^{prop}, t^{loss} \ll t^{prop})$, and can be neglected in the general timing relation (Eq.5).

We apply first this approximation $(t^X \approx t^{prop})$ to the observed HXR delays τ_{ij} measured between the channels $i = 3$ and $j = 4 - 10$ in the 1991 Dec 15, 1832 UT, flare (Fig.1). The results of these delays and their uncertainties, determined with a high-precision cross-correlation technique described in (3), are shown in Fig.2 for the HXR energy range of $40 - 400$ keV. The energy range of the HXR-producing electrons (with $E \approx 2.0\ \varepsilon$) is $E \approx 80 - 800$ keV. The fit of the kinematic model specified in Eq.6 has only one free parameter, the time-of-flight distance l, and is found here to be $l = 29,000 \pm 1,800$ km. Note, that the TOF distance l is constrained with an accuracy of 6% from the 7 independent delay measurements (being one of the most accurate HXR delay measurements available at the time of writing). The 1-parameter model is fully consistent with the data (with $\chi^2_{red} = 0.59$) within the uncertainties of the Poisson noise and does not require additional parameters. The same model was also found to fit the functional form of observed HXR delays in a large number of other flares (1), (2). However, we will test in the following whether the same data could also be consistent with other kinematic models that include timing effects of the acceleration process.

For comparing the so evaluated electron TOF distance l with actual flare loop geometries, one has also to take into account projection effects, e.g. to correct for the pitch-angle of the spiral-like Larmor motion of electrons or the helical twist of magnetic field lines. These projection factors are estimated to be of order $q_\alpha \approx 0.64$ for the pitch-angle effect and $q_{twist} = 0.85$ for the helical twist (1), yielding a projected electron propagation path length L of

$$L = l \cdot q_\alpha \cdot q_{twist} , \quad (7)$$

with a value of $L = l \cdot 0.54 = 16,000 \pm 1,000$ km. This projected path length L is the relevant parameter to be compared with flare loop geometries (projected into the loop plane) to localize the acceleration site [see examples in (2)].

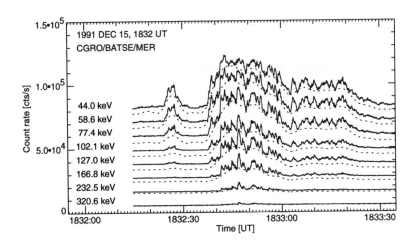

FIG. 1. HXR observations of the 1991 Dec 15, 1832 UT flare, recorded with *BATSE/CGRO* in the *Medium Energy Resolution (MER)* mode that contains 16 energy channels. The channels #3-10 are shown, with the low energy edges indicated on the left. The channels are incrementally shifted and a lower envelope is indicated, computed from a Fourier filter with a cutoff at a time scale of 4 s. The fine structure (with the envelope subtracted) are cross-correlated in the delay measurements shown in Fig.2.

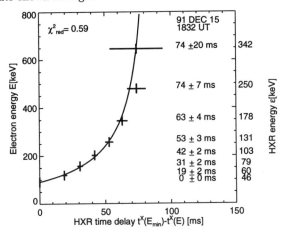

FIG. 2. HXT time delay measurements between channel #3 and #4-10 from the data shown in Fig.1. The delays and statistical uncertainties are indicated on the right. The HXR energies on the right axis represent the medians of the count spectra in each channel, while the electron energies on the left axis are computed from the conversion described in Eq.2. The solid line represents a fit of the 1-parameter model given in Eq.6, yielding a TOF distance of $l = 29,000 \pm 1,800$ km, with $\chi^2_{red} = 0.59$.

303

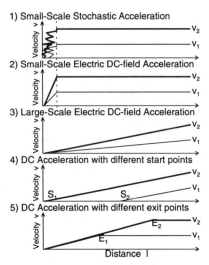

1) Small-Scale Stochastic Acceleration

2) Small-Scale Electric DC-field Acceleration

3) Large-Scale Electric DC-field Acceleration

4) DC Acceleration with different start points

5) DC Acceleration with different exit points

FIG. 3. Five different models for the timing of electron acceleration and propagation. The velocity changes of a low-energy (v_1) and a high-energy electron (v_2) are shown along a 1-dimensional path from the beginning of acceleration (left side) to the thick-target site (right side). Model 1 and 2 characterize small-scale acceleration processes, while Model 3-5 depict large-scale acceleration situations. Model 4 and 5 illustrate different start (S_1, S_2) and exit positions (E_1, E_2) for the accelerated electrons.

Acceleration time scales

In this Section we discuss how finite acceleration times affect the timing of HXR emission, or the arrival times of electrons at the thick-target site. Basically one can distinguish between two opposite scenarios: small-scale and large-scale acceleration processes.

In small-scale acceleration processes, one can generally neglect acceleration time scales compared with propagation time scales, $t^{acc} \ll t^{prop}$, supposed that the acceleration path is small compared with the free-flight propagation path. Such situations are depicted in Model 1 and 2 (Fig.3). The HXR timing can then be described with Eq.6, which was found to be fully consistent with the data for the flare shown in Fig.2.

Stochastic Acceleration: Model 1 illustrates a stochastic acceleration process as it can occur in coronal regions with enhanced wave turbulence (or similarly in shock fronts). However, a small spatial scale of the acceleration region (compared with the TOF propagation distance) does not necessarily imply that the acceleration time is also much smaller than the propagation time. In the case of diffusive stochastic acceleration the particles can be bounced around in a turbulence region significantly longer than the travel time through this region. For instance, LaRosa et al. (11) estimate the bulk energization time of electrons in a reconnection-driven MHD-turbulent cascade to $t^{acc}(E=20 \text{ keV}) \approx 300$ ms, which is comparable with the propagation time in-

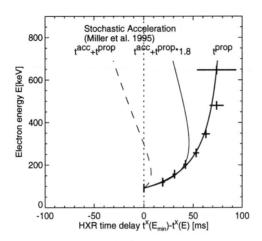

FIG. 4. Fit of the stochastic acceleration model of Miller et al. (1996) to the same data as shown in Fig.2. Adding the predicted acceleration time to the propagation time (thick line) yields negative delays (dashed line). Adjusting the propagation time by a factor of 1.8 (thin line) yields an acceptable fit in the 100-200 keV range, but not at higher energies.

ferred in our flare (Figs.1-2), $t^{prop}(E = 20keV) = l/v = 29,000\text{km}/(0.27c) = 360$ ms. More specifically, Miller et al. (14) estimate acceleration times of 70 ms to energize electrons from ≈ 5 to 50 keV, or about 180 ms to 511 keV. They specify an energy dependence of

$$t^{acc}(E) = [(\frac{E}{m_e c^2})^{1/6} - 0.48] \cdot 350 \text{ ms}, \quad (E > 5 \text{ keV}) \tag{8}$$

to energize electrons by gyroresonant interactions with fast mode waves in an MHD-turbulent cascade. We fit this model to the 1991 Dec 15 flare and show the expected HXR timing in Fig.4. First we add the acceleration time to the same propagation time inferred in Fig.2 (based on a TOF distance of $l = 29,000$ km). The expected HXR delay (dashed curve in Fig.2) becomes negative above 200 keV, meaning that the high-energy electrons arrive later than the low-energy electrons at the chromosphere due to the longer acceleration time. If we perform a fit of the combined expression $t^X = t^{acc} + t^{prop}$ (using Eq.6 and Eq.8) we find that the data can be reasonably fitted in the 100-200 keV range with a 1.8 times larger TOF distance (to compensate for the acceleration time), but the HXR delay decreases above 200 keV significantly below the measured values. Thus, the acceleration time scale specified in Miller et al. (14) cannot fit the observed delays over the entire energy range of 80-800 keV for this specific flare.

Electric DC-field Acceleration: Acceleration mechanisms employing DC electric fields have been studied by various researchers, e.g. (9), (17), (7), but there is no detailed comparison of the predicted timing with observations.

The models 2-5 shown in Fig.3 depict various scenarios where the acceleration and propagation time scales have different weighting, depending on the spatial location and extent of the DC field. Because the free-flight path of electrons is complementary to the acceleration path length in unidirectional DC fields, in a more direct fashion than in the case of stochastic acceleration, the resulting HXR timing provides a crucial test between different models.

The simplest case is given in Model 2 (Fig.3), where the spatial extent of the DC field is small compared with the TOF distance, and necessarily also implies that the acceleration time is small compared with the free-flight time ($t^{acc} \ll t^{prop}$), and thus can be neglected. The HXR timing can then adequately be described with Eq.6, which fits the data satisfactorily for the 1991 Dec 15 flare (Fig.2).

Another simple approach is to assume that an electric field extends over the entire flare loop and that electrons are accelerated from one end of the loop to the other. Assuming a constant electric field, the electrons would end up with a monoenergetic spectrum and coincident timing, and thus cannot explain the observed delays (Fig.2). A first variant is to assume that a flare loop consists of a number of current channels with different electric fields. In this scenario electrons are accelerated in separate channels with different electric fields $\mathcal{E}$, obeying the force equation

$$F = m_e \ a = \frac{d}{dt}(\gamma m_e v) = e \ \mathcal{E} = \frac{E}{l} \ , \tag{9}$$

where E represents the kinetic energy of an electron gained by the electric field $\mathcal{E}$ over a distance l. The acceleration a can then be expressed as function of the kinetic electron energy E or Lorentz factor γ (using Eq.3) by

$$a(\gamma) = \frac{E}{m_e \ l} = \frac{c^2}{l}(\gamma - 1) \ . \tag{10}$$

The acceleration time t^{acc} as function of the energy can then be derived by integrating the force equation (Eq.9) and inserting Eq.4,

$$t^{acc}(\gamma) = \frac{c}{a(\gamma)}\sqrt{\gamma^2 - 1} \ . \tag{11}$$

Inserting the acceleration $a(\gamma)$ from Eq.10 into Eq.11 yields then

$$t^{acc}(\gamma) = \frac{l}{c}\sqrt{\frac{\gamma + 1}{\gamma - 1}} \ . \tag{12}$$

In Model 3 (Fig.3), the HXR timing is entirely determined by this energy dependence on the acceleration process, corresponding to the approximation $t^X \approx t^{acc}$ in the general timing equation (Eq.5). We fit this model to the observed HXR timing of the 1991 Dec 15 flare and show the results in Fig.5 (left). Interestingly, this model shows a very similar energy dependence as

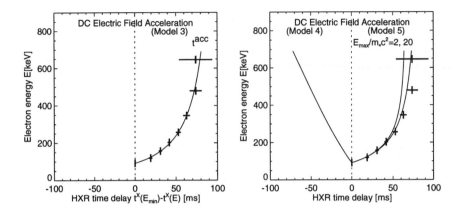

FIG. 5. Fit of the DC electric field Model 3 (left) and Model 4 and 5 (right) for the same data as shown in Fig.2. With respect to the TOF distance $l^{prop} = 29,000$ km fitted in Fig.2, the best fit of Model 3 (left) yields an acceleration distance of $l = l^{prop} \cdot 0.44$, Model 4 is shown for $l = l^{prop}$, and Model 5 yields $l = l^{prop} \cdot 1.1$ (for $E_{max}/m_e c^2 = 2$) and $l = l^{prop} \cdot 1.0$ (for $E_{max}/m_e c^2 = 20$).

the TOF propagation model (shown in Fig.2) and thus fits the data equally well. The inferred acceleration path length is a factor of 0.44 shorter than the path length in the TOF propagation model, because the average electron speed is about half of the final speed applied in the propagation model (being exactly half in the nonrelativistic limit). Thus, the two models cannot be distinguished from the timing alone, but the inferred distance scale is a factor of ≈ 2 different.

In Model 4 and 5 (Fig.3) we investigate two further variants of electric DC field acceleration, where the accelerated electrons are allowed to enter into (Model 4) or exit from (Model 5) an electric field channel at different locations. In both models the resulting electron energy is proportional to the acceleration path length, assuming a constant (mean) electric field $\mathcal{E}$ in all current channels. Model 4 is a natural situation in the sense that all electrons in a current channel experience acceleration once the electric field is turned on. In this scenario, the acceleration path length of each electron is defined by the distance between its start position and the loop footpoint. Defining the acceleration a by the maximum electron energy $E_{max} = m_e c^2 (\gamma_{max} - 1)$ obtained by the electric field $\mathcal{E}$ over the loop length l, (or an electron energy E gained over a proportionally smaller distance x^{acc}),

$$a = \frac{E_{max}}{m_e \, l} = \frac{E}{m_e \, x^{acc}} \; , \tag{13}$$

we find the following timing for electrons in Model 4:

$$t^{acc}(\gamma) = \frac{c}{a} \sqrt{\gamma^2 - 1} = \frac{l}{c} \frac{\sqrt{\gamma^2 - 1}}{(\gamma_{max} - 1)} \; . \tag{14}$$

307

The resulting timing in Model 4 is shown in Fig.5 (right) for the same TOF distance $l = 29,000$ km used in Fig.2 and for $\gamma_{max} - 1 = 2$ (or $E_{max} \approx 1$ MeV). The high-energy electrons arrive later at the thick-target site than the low-energy electrons for every parameter combination, and thus, cannot fit the data. Therefore, Model 4 can clearly be rejected for all flares where the high-energy electrons arrive first.

In Model 5 the electrons are allowed to exit a current channel with an accelerating electric field at an arbitrary location. Because electron spectra have always negative slopes, this means that more electrons leave the current channel after a short distance than after longer distances. This model may mimic a realistic situation when the current channels are relatively thin [for most extreme aspect ratios see (7)] so that electrons exit a current channel by cross-field drifts. The length of the acceleration path can then be determined from the final electron energy E using Eq.13,

$$x^{acc}(\gamma) = l\frac{E}{E_{max}} = l\frac{(\gamma - 1)}{(\gamma_{max} - 1)} \tag{15}$$

The resulting timing of electrons arriving at the thick-target site is then composed of the sum of the acceleration time t^{acc} over the acceleration path length x^{acc} and the free-flight propagation time t^{prop} over the remaining path length $l - x^{acc}$,

$$t^X(\gamma) = t^{acc}(\gamma) + t^{prop}(\gamma) = \frac{c}{a(\gamma)}\sqrt{\gamma^2 - 1} + \frac{l - x^{acc}(\gamma)}{v(\gamma)}$$

$$= \frac{l}{c}[\frac{\sqrt{\gamma^2 - 1}}{(\gamma_{max} - 1)} + \frac{1 - \frac{(\gamma - 1)}{(\gamma_{max} - 1)}}{\sqrt{1 - \gamma^{-2}}}] . \tag{16}$$

The fit of the timing Model 5 is also shown in Fig.5 (right) for two different parameter combinations ($E_{max} = 1$ MeV and 10 MeV). The essential result is that Model 5 fits the data the better the smaller the acceleration time is relative to the propagation time, a situation that approaches asymptotically Model 2 for high electric field strengths. Consequently, the best fit is consistent with a small-scale acceleration region like in Model 2.

Energy loss time scales

In the general timing relation (Eq.5) we have a third term that has been neglected in the previous considerations: the energy loss time $t^{loss}(E)$. The energy loss time for electrons that interact via Coulomb collisions in a plasma with electron density n_e is given by, e.g. Book (6),

$$t^{loss}(E) = \frac{E_{eV}^{3/2}}{2.91 \cdot 10^{-6} \ln \Lambda \, n_e} , \tag{17}$$

with the Coulomb logarithm $ln\Lambda = 23 - \ln(n_e^{1/2}T_{eV}^{-3/2})$. From this collisional time scale we see immediately that, in the absence of other timing effects, the resulting HXR bremsstrahlung should be delayed for high-energy electrons. Since $\approx 70\%$ of the flares reveal an opposite timing (5), (3), or an even higher fraction for the timing of HXR fine structure, the energy loss time can obviously not be the dominant timing factor in Eq.5. From this fact we conclude that the electron density in most of the flare loops is not sufficiently high $(n_e \gtrsim 10^{12}$ cm$^{-3})$ to warrant significant energy loss inside the coronal part of the flare loop, so that that the principal energy loss for $\gtrsim 25$ keV occurs in the chromospheric footpoints. This is confirmed by the frequent observations of HXR footpoint emission seen by *Yohkoh/HXT* (16). Because the transit time through a 2500-km chromosphere is generally much smaller than the propagation times through the flare loop (e.g. $l = 29,000$ km in the 1991 Dec 15 flare here), the energy loss time (t^{loss}) can readily be neglected in the HXR timing of flares with footpoint emission.

Trapping time scales

For the timing of trapping effects, similar arguments apply as for the energy loss times, because the collisional deflection time, an upper limit of particle trapping in a mirror loop, is closely related to the energy loss time. Trapped electrons are expected to be scattered into the loss-cone near the mirror points latest after the collisional deflection time, where they precipitate to the footpoints and produce HXR emission at arrival. During the 1992 Jan 13 Masuda flare, an electron density of $n_e \lesssim 2 \cdot 10^{11}$ cm^{-3} was inferred from *Yohkoh/SXT*, for which a collisional deflection time of 2-5 s is expected for 50-150 keV electrons. Such a delay was indeed observed for the unmodulated HXR flux during this flare (1). Because pitch-angle scattering in the trap randomizes the orbits of injected electrons after a few bounce times, the timing of the rapidly varying injection function becomes smeared out and the escaping precipitating electron flux exhibits a smooth time profile. The time structure of the HXR flux can therefore be used to separate the directly precipitating from the trapped electron flux, i.e. by filtering the fine structure from the smoothly varying lower envelope of the HXR time profile (as indicated in Fig.1).

CONCLUSIONS

We analyzed the energy-dependent HXR timing in the 1991 Dec 15 flare, which provides one of the most accurate timinig measurements todate, with uncertainties of a few ms over the range of 80-800 keV in electron energies. We found that the HXR timing can accurately be fit with a 1-parameter model that includes only electron time-of-flight differences between the acceleration site and the HXR emission site. We find that acceleration models with homogeneous, large-scale electric DC fields (Model 4 and 5) cannot explain the observed timing, unless for very high fields that accelerate electrons over short distances only (Model 5 in the limit of $E_{max} \gg m_e c^2$). The only candidate of large-scale electric DC fields that can accomodate for the observed timing is the inhomogeneous case (Model 3), where different current channels have different electric field strengths. We tested also a model with stochastic acceleration (Model 1) based on time scales computed by Miller et al. (14) and found that it can only be reconciled with the data if the acceleration time scales are significantly reduced.

In summary, propagation time differences clearly dominate the timing of HXR fine structure, while acceleration time scales seem to represent second-order effects at most. Trapping effects and collisional time scales, on the other hand, account for the timing of the smoothly-varying HXR flux. The detailed matching of the electron timing to the observed HXR delays found here (Fig.2) rules out protons or ions as cause of the rapidly varying HXR flux, although commensurate ion and electron abundances were identified above 1 MeV (15).

Acknowledgements: I thank Richard Schwartz and the *BATSE* PI team of G.Fishman for providing high-quality *BATSE* data products. The work was supported by NASA grant NAG-5-2352.

REFERENCES

1. Aschwanden,M.J., Hudson,H., Kosugi,T., and Schwartz,R.A., *Ap.J.*, **464**, (June 20 issue) in press (1996).
2. Aschwanden,M.J., Wills,M.J., Hudson,H., Kosugi,T., and Schwartz,R.A., *Ap.J.*, submitted (1996).
3. Aschwanden,M.J. and Schwartz,R.A., *Ap.J.* **455**, 699-714 (1995).
4. Aschwanden,M.J. and Schwartz,R.A., *Ap.J.*, **464**, (June 20 issue) in press (1996).
5. Aschwanden,M.J., Schwartz,R.A., and Alt,D.M., *Ap.J.* **447**, 923-935 (1995).
6. Book,D. 1980, *NRL Plasma Formulae*, Washington/DC:NRL, 1980.
7. Emslie,A.G. and Henoux,J.-C., *Ap.J.* **446**, 371-376 (1985).
8. Fishman,G.J. et al. 1989, in Proc. GRO Workshop, ed. W.N.Johnson, Greenbelt:GSFC, p.2-39.
9. Holman,G.D., 1985, *Ap.J.* **293**, 584-594 (1985).
10. Hudson,H.S. and Ryan,J., *Annu.Rev.Astron.Astrophys.* **33**, 239-282 (1995).
11. LaRosa,T.N., Moore,R.L, Miller,J.A., and Shore,S.N., *Ap.J.*, in press, (1995).
12. Masuda,S., Ph.D. Thesis, National Astronomical Observatory, University of Tokyo, Mitaka/Tokyo, (1994).
13. Masuda,S., Kosugi,T., Hara,H., Tsuneta,S., and Ogawara,Y. *Nature* **371**, No.6497, 495-497, (1994).
14. Miller,J.A., LaRosa,T.N. and Moore,R.L. *Ap.J.*, April 10 issue, in press (1995).
15. Ramaty,R., Mandzhavidze,N., Kozlovsky,B., Murphy,R.J., *Ap.J.Lett.*, in press (1995).
16. T. Sakao,T., Ph.D. Thesis, National Astronomical Observatory, University of Tokyo, Mitaka/Tokyo, (1994).
17. S. Tsuneta, *Ap.J.* **290**, 353-358 (1985).

Subsecond Time Variations in Solar Flares around 100 keV: Diagnostics of Electron Acceleration

N. Vilmer*, G. Trottet*, H. Verhagen*,
C. Barat**, R. Talon**, J.P. Dezalay**,
R. Sunyaev***, O. Terekhov***, A. Kuznetsov***

*Observatoire de Paris, Section de Meudon, DASOP, URA 1756,
F- 92195 Meudon, France
**Centre d'Etude Spatiale des Rayonnements, BP 4346, F-31029 Toulouse, France
***Space Science Institute, Profsoyouznaya 84/32, 117810 Moscow, Russia

The present paper discusses the results of a systematic study of subsecond time variations performed on the PHEBUS observations of more than one hundred solar bursts detected around 100 keV. The analysis shows that: 1)a large percentage (73%) of hard X-ray bursts observed around 100 keV presents time features with rise times in the 100 ms- 1 s range, 2)these time features are less systematically observed in long duration events, 3)there is a continuous distribution (during a single burst or for all the bursts of the sample) of the values of the different timescales derived from the analysis. The results are discussed with respect of the electron acceleration timescales in solar flares as well as in the context of the "statistical flare" scenario.

INTRODUCTION

The study of rapidly time varying emissions from energetic electrons has great potential to infer the timescales of non- thermal energy release in solar flares. From the analysis of TD1A observations, Van Beek et al (1) and de Jager and de Jonge (2) proposed that hard X-ray bursts be decomposed in a series of short lived spikes of duration ranging from a few seconds to a few tens of seconds ("Elementary Flare Bursts") representing the fundamental timescales of energy release. With the increased sensitivity of hard X-ray experiments, a systematic search of faster time variations has been performed. At photon energies $\geq$ 30 keV Kiplinger et al. (3) found that about 10% of the bursts show significant structures with time scales of a few tens to a few hundreds of milliseconds, suggesting that the "basic" fragments of non-thermal energy release occur on these timescales. However, Brown et al (4) pointed out that the fastest transients (50 ms) reported were not inconsistent with extreme statistical fluctuations in the counting rates, thus leading to a real limit of the order of a few hundred of milliseconds for the fastest time structures.

More recently, temporal fluctuations with duration ranging from 200 to 600 ms (FWHM) at energies above 25 keV have been reported through the analysis of some events observed with the even more sensitive BATSE experiment (5). Aschwanden et al (6) have performed a systematic study on 640 flares observed with the burst trigger mode of BATSE and have found a continuous distribution of the duration of X-ray pulses in the range 0.3 to 3 seconds in the 25 - 100 keV energy range. Finally, at energies around and above 100 keV where a clearer signature of non-thermal emission is expected subsecond time variations (sometimes claimed to be as short as tens milliseconds) have been reported for a few events (7, 8, 9, 10). The preliminary results of a more systematic search for rapid time structures at energies around 100 keV performed on the observations of the PHEBUS experiment aboard GRANAT were presented in Vilmer et al (11). The results based on one third of the total PHEBUS solar data base showed that the majority of these bursts exhibited significant fast time structures on time scales less than 500 ms and that there was no unique value of fast characteristic times but rather a distribution of values within a burst or from one flare to the other. The present work extends the previous study to the complete PHEBUS data base of solar bursts recorded between January 1990 and July 1993 (111 events).

DATA ANALYSIS

The PHEBUS experiment consists of 6 independent BGO detectors measuring solar and cosmic hard X-ray/gamma-ray bursts in the energy range $\simeq$ 0.10-100 MeV (12, 13). The present study uses the high time resolution integral count-rate recorded for one sunward detector in the $\simeq$ 0.10-1.6 MeV energy range for 86.6s at the onset of the burst mode. The time profile is acquired with a basic time accumulation $\leq$ 31.25 ms. For short duration events, this time profile may represent the whole burst while for longer events it covers only the first part of the emission (usually the rising phase of the event).

The analysis is performed on the 111 solar bursts recorded by PHEBUS between January 1990 and July 1993. As in previous works (10, 11, 13), the sample of events has been divided into two classes, thereafter referred to as impulsive and extended bursts according to their total duration at energies around 100 keV (smaller or larger than 5 minutes.) The limit of five minute duration is somewhat arbitrary, although it has been found on a limited sample that these bursts tend to exhibit different temporal behaviors.

The existence of intrinsic statistical fluctuations of the count-rates renders the identification of hard X-ray time structures difficult. There is also no unique way of defining these structures and the present definition is the one already used in (13) which focuses on the rise times of features encountered in the time profiles. A variation in time of the count-rate is considered as significant if it corresponds to a count-rate enhancement $\geq 3\ \sigma$ above the mean of the count-rates measured in the time intervals immediately preceding and following the studied feature. Furthermore to exclude extreme statistical

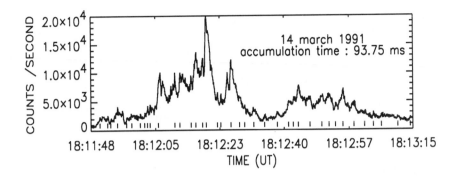

FIG. 1. Count-rate of the 14 March 1991 event in the $\simeq$ 0.10-1.6 MeV range. The dashes indicate the location of the time variations with significant rise times less than 1 second.

fluctuations in time series, an additional constraint is required: the considered feature must obey the above 3 σ criterion for at least two successive values of the accumulation time which is an integer multiple of 31.25 ms. For all the events and for all the values of the accumulation time (multiple of 31.25 ms) comprised between 31.25 ms and 1000 ms, the significance of every increase in the count-rate is thus determined according to the above criterion. For each event, the smallest accumulation time for which there is at least one significant rise of a time structure is considered as the shortest rise time detected in the event (t_{min}). We also consider the total number N_t of the time structures with rise times smaller than one second detected in the 86.6s recorded in this data mode, as well as the accumulation time t_{cm} for which there is a maximum number of significant structures. These last quantities are good indications of the persistency and of the mean characteristics of the fast fluctuations during the first 86.6 s of the event.

RESULTS

Distribution of timescales in a specific event

In this section, the PHEBUS event, which exhibits the largest value of N_t in the whole sample, is studied in details in order to illustrate the method of analysis as well as to derive the characteristics of individual features in a single burst.

Figure 1 shows the time history of the $\simeq$ 0.10-1.6 MeV count-rate during the 14 March 1991 event with an accumulation time of 93.75 ms together with the locations of the time variations with significant rise times less than 1 second which are found in the time profile. The total number N_t of time structures with rise times less than 1 second is found to be 39. The values

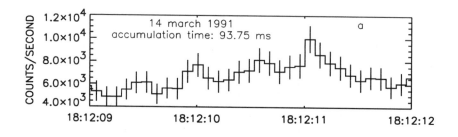

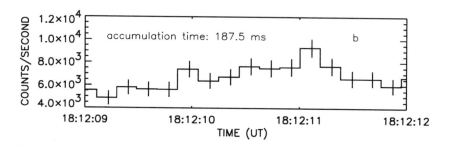

FIG. 2. Count-rate of part of the 14 March 1991 event in the $\simeq$ 0.10-1.6 MeV range.

of the rise times in this event span in fact the whole timescale range between the shortest significant timescale, 93.25 ms, and 968 ms. Figure 2a shows the time history from 18:12:09 UT to 18:12:12 UT with the same accumulation time of 93.75 ms together with the 3 σ error bars defined in the previous section for every point. The time feature at $\simeq$ 18:12:11 UT has clearly a rise time of 93.75 ms thus leading to a value of t_{min} of 93.75 ms. This timescale is however shorter than the real duration of the time feature which could be defined as the time accumulation for which there is both a significant count-rate increase and decrease and which is found here to be 187.50 ms (Figure 2b). It is generally found that the significance level of an identified feature increases with the accumulation time, reaching a maximum when the accumulation time matches its real duration and decreasing for even larger accumulation times. Such a detailed analysis is however difficult to perform for all individual time features since before their significance level starts to decrease, they sometimes merge into neighboring ones. For the cases where a detailed analysis is possible, it is however found that, as in the case of the feature at $\simeq$ 18:12:11 UT, the true duration of the structure is no more than 2 or 3 times 31.25 ms larger than the accumulation time for which its rising becomes significant. Furthermore, this isolated spike exhibits different rise and decay times (see Figure 2a). In that event, among the 39 identified structures, only 5 of them have rise times above 500 ms. Figure 3 shows the

314

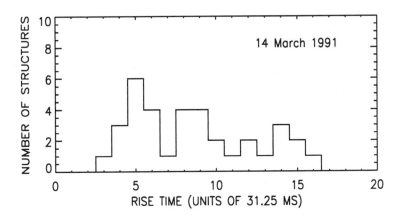

FIG. 3. Distribution of rise times below 500 ms for the 14 March 1991 event.

distribution of the rise times observed below 500 ms. The values found span the whole timescale range between 93.75 ms and 500 ms. The characteristic time t_{cm} defined in the previous section is found to be 437.50 ms.

Distribution of timescales for all the bursts observed by PHEBUS.

Time structures with subsecond rise times are rather systematically detected in the "impulsive" events observed by PHEBUS around 100 keV since 79% of them exhibit values of t_{min} in the range 100 - 900 ms. For 54% of the events t_{min} is less than 500 ms and 26% of them show a more permanent behavior in the event with t_{cm} less than 500 ms. The same analysis was performed on the 29 extended events observed by PHEBUS. It is found that 60% of these events exhibit time features with t_{min} below one second in their rising phase but that only 33% of them exhibit values of t_{min} smaller than 500 ms.

If, regardless of their duration in the $\simeq$ 0.10-1.6 MeV range, all the analyzed events are considered, it is finally found that 73% of them exhibit in their timeprofile shortest rise times t_{min} below 1000 ms. The results of the study performed on all the events are summarized in Figures 4, 5 and 6 which respectively show the distribution of bursts with a value of t_{min} as a function of t_{min}, the distribution of the number of bursts with a given characteristic time t_{cm} as a function of t_{cm} as well as the distribution of the number of bursts exhibiting N_t features as a function of N_t. The distribution of bursts as a function of the shortest rise time t_{min} (Figure 4) can be represented by an exponential function in the 100ms - 900ms range, that is $N(t_{min}) \propto$ exp.(-t_{min}/τ) for t_{min} between 100ms and 900ms where τ=0.47+/-0.05 10^3 ms. On the other hand, Figure 5 shows that the distribution of the number of bursts

315

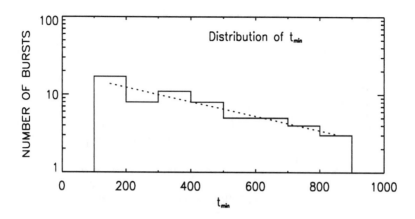

FIG. 4. Distribution of the number of bursts with a value of t_{min} as a function of t_{min}.

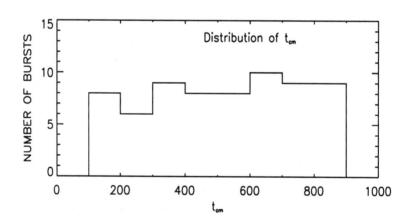

FIG. 5. Distribution of the number of bursts with a value of t_{cm} as a function of t_{cm}.

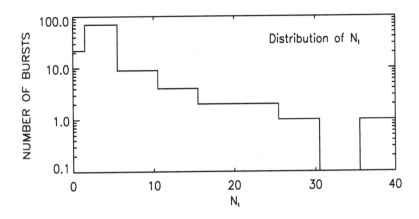

FIG. 6. Distribution of the number of bursts with a value of N_t as a function of N_t.

with a given timescale t_{cm}, characteristic of a more global behavior in the burst, is rather flat in the 100 ms-900 ms time range. Finally, Figure 6 shows that in most events only a limited number of time structures are identified on subsecond timescales.

CONCLUSION

The results of the present analysis can be summarized as follows:

1. 73 % of hard X-ray bursts observed around 100 keV by PHEBUS present time features with rise times in the 100 ms - 1s range.

2. There is a smaller occurrence of rapid time variations (even rise times) in extended events.

3. There is no unique value of rise times of structures in a given burst but rather a distribution of values in a given range. The time profile of an isolated spike is not symmetrical exhibiting different rise and decay times.

4. There is a continuous distribution of the different timescales t_{min}, t_{cm} derived for all the bursts of the sample. The distribution of bursts exhibiting a shortest rise time t_{min} can be represented by an exponential function in the 100-900ms range: $N(t_{min}) \propto \exp(-t_{min}/\tau)$ where $\tau = 0.47 +/- 0.05 \ 10^3$ ms. The distribution of the number of bursts with a given timescale t_{cm} is flat in the 100-900 ms time range. In most events, there is a limited number of features identified on subsecond timescales and the distribution of the number of bursts exhibiting N_t structures is a decreasing function of N_t. This may be a limit due to the sensitivity of the experiment since there is a trend that the events with a large number of identified structures correspond to events with large counting rates.

Some of these results (point 1 and part of point 4) have to be compared with the ones of a recent study performed on the hard X-ray time profiles of solar bursts observed in the burst trigger mode of BATSE in the 25-50 keV and 50-100 keV energy ranges (6). Using a slightly different algorithm to characterize a pulse structure, it is found that 73% of the sample recorded with a burst trigger exhibit significant (5 σ) hard X-ray pulses with duration ranging between 0.3 and 3 seconds at energies in the 25-100 keV range. Furthermore, the e-folding timescale for the distribution of shortest rise times t_{min} found in the present study is of the same order as the one deduced in (6) for the e-folding timescales of the distributions of pulse widths in the 25-50 and 50-100 keV energy ranges. Results of analysis performed on different experiments cannot however be compared in every details since some limits due to the detector sensitivities (depending on the detector sensitive area, the background level as well as on the energy range studied) must be expected in the search of rapid time variations. Only the general trend of the analysis can be compared. Under these conditions, our results are in good agreement with the ones obtained with BATSE (6).

The non-symmetric time profile of an isolated X-ray spike exhibiting different rise and decay times (point 2) is understandable when considering that the decay time and thus the real duration of a spike also reflects energy losses (e.g. collisional) of the accelerated electrons in the interacting medium. The rise time of a spike is thus probably more representative of the non-thermal energy release timescale in the medium. The detection of subsecond rise times is a possible indication that the flare energy release is fragmented, and that an upper limit of the timescales on which the temporal fragmentation may take place is of the order of a few hundred of milliseconds. It also strongly confirms the fact that the electrons produced in these elementary energy release phenomena must interact in dense regions ($\geq 10^{12}$ cm^{-3}). The smaller occurrence of rapid time variations (even rise times) in extended events (point 2) does not necessarily imply that the acceleration timescales are longer than in impulsive events. It may simply indicate that particles interact in a less dense region where their collisional lifetime is longer than the acceleration timescales (see e.g. 14, 15) resulting in a more important smearing of the characteristics of the acceleration.

The existence of subsecond rise times of structures seems to be a general property of the time histories of the bursts observed by PHEBUS around 100 keV. The distribution of bursts as a function of their shortest rise time t_{min} can be represented by an exponential distribution. This is consistent with the energy release being a random process governed by Poisson statistics (6). Furthermore, the fact that there is no unique value of rise times of structures in a given burst as well as no characteristic timescale t_{cm} holding for the many bursts observed - that is no preferred value for the electron acceleration timescale in solar flares- may provide some support to the concept of statistical flare models (see e.g. the review of Vlahos (16) and 17). In these models, where the energy release occurs in many randomly distributed small-scale magnetic structures, it is expected that the typical timescales of electron acceleration should not be characterized by discrete values as in the case of flare models with simple magnetic configurations but rather by continuous distributions of their values, as found in the present study. Finally, our study suggests that the timescales on which the elementary electron acceleration may take place are shorter than 100 ms, this derived upper limit being either due to the sensitivity of the experiment or to the clustering of many individual fragments that is expected from specific models of statistical flare (e.g. 17).

ACKNOWLEDGEMENTS

This research was supported by Centre National d'Etudes Spatiales.

REFERENCES

1. H.F. Van Beek, L.D. de Feiter and C. de Jager, Space Res. **14**, 447 (1974).

2. C. de Jager and G. de Jonge, Solar Phys. **58**, 127 (1978).

3. A.L. Kiplinger, B.R. Dennis, A.G. Emslie, K.G. Frost and L.E. Orwig, Astrophys. J. **299**, 285 (1983).

4. J.C. Brown, J.M. Loran and A.L. MacKinnon, Astron. Astrophys. Letters **147**, L10 (1985).

5. M.E. Machado, A.G. Emslie, G.J. Fischman, C. Meegan, R. Wilson and W. Paciesas, Adv. Space Res. **13**, (9)171 (1993).

6. M.J. Aschwanden, R.A. Schwartz and D.M. Alt, Astrophys. J. **447**, 923 (1995).

7. K. Hurley, M. Niel, R. Talon, I.V. Estulin and V.Ch. Dolidze, Astrophys.J. **265**, 1076 (1983).

8. U.D. Desai, C. Kouveliotou, C. Barat, K. Hurley, M. Niel, R. Talon, G. Vedrenne, I.V. Estulin and V. Ch. Dolidze, Astrophys. J. **319**, 567 (1987).

9. R.J. Murphy, G.H. Share, J.E. Grove, W.N. Johnson and 13 co-authorst, A.I.P. Conf. Proc. **280**, 619 (1993).

10. R. Talon, C. Barat, J.P. Dezalay, N. Vilmer, G. Trottet, R. Sunyaev and O. Terekhov, Adv. Space Res. **13**, (9)171 (1993a).

11. N. Vilmer, G. Trottet, C. Barat, J.P. Dezalay, R. Talon, R. Sunyaev, O. Terekhov and A. Kuznetsov, Space Science Reviews **68**, 233 (1994).

12. C. Barat, F. Cotin, M. Niel, R. Talon, G. Vedrenne, M. Pick, G. Trottet, N. Vilmer, K. Hurley, A. Kuznetsov, R. Sunyaev, and O. Terekhov, AIP Conf. Proc. **170**, 395 (1988).

13. R. Talon, G. Trottet, N. Vilmer, C. Barat, J.P. Dezalay, R. Sunyaev, O. Terekhov, and A. Kuznetsov, Solar Phys. **147**, 137 (1993b).

14. N. Vilmer, Solar Phys. **111**, 207 (1987).

15. N. Vilmer, Adv. Space Res. **13**, (9)221 (1993).

16. L. Vlahos, Space Science Review **68**, 39 (1994).

17. L. Vlahos, M. Georgoulis, R. Kluiving and P. Paschos, Astron. Astrophys.**299**, 897 (1995).

Inferring the Accelerated Electron Spectrum in Solar Flares

R. P. Lin[1] and C. M. Johns-Krull[2]

Space Sciences Laboratory, University of California
Berkeley, CA 94720-7450
[1] *also at Physics Department, University of California, Berkeley*
[2] *now at McDonald Observatory, University of Texas, Austin, TX 78712*

We invert high spectral resolution hard X-ray observations of the 1980 June 27 solar flare to obtain the parent X-ray-producing electron population. Assuming that collisions are the dominant loss mechanism, the accelerated electron population is derived as a function of time through the flare, using a continuity equation. We find that there appear to be two separate components in the accelerated electrons: an impulsive component with a spectrum peaked at 50 keV, and a slowly varying component with a power-law spectrum extending down to 20 keV.

INTRODUCTION

Hard $\gtrsim 20$ keV X-ray observations over the past several decades have shown that the acceleration of electrons to ~ 10-10^2 keV energies commonly occurs in the impulsive phase of solar flares (see reviews 1, 2). The observed hard X-ray emission appears to be bremsstrahlung radiation of fast electrons colliding with the solar atmosphere. If the fast electrons have energies much greater than the thermal energy, kT, of the ambient medium, their energy losses to Coulomb collisions will be $\sim 10^4$-10^5 times their bremsstrahlung losses. As a result, for many flares a large fraction of the total energy released must be contained initially in the fast electrons (3). Thick-target models (4) assume that the primary result of the energy release process is the impulsive acceleration of electrons, probably in the coronal part of a magmetic loop or arcade of loops. These electrons then propagate along the magnetic field lines and deposit their energy predominantly in the lower corona and the chromospheric portions of the loops, heating the ambient gas and leading to evaporation and subsequent radiation in the visible, UV, and soft X-ray bands.

The observed spatial, spectral, and temporal properties of the hard X-ray emission are generally consistent with the quantitative predictions of non-thermal thick-target models. Spatially resolved hard X-ray observations, such as those from the Yohkoh Hard X-ray Telescope (HXT), show a double foot-point structure typically appearing early in the flare, with the two footpoints brightening simultaneously to within a fraction of a second (5). The hard X-ray centroids are located at progressively lower altitudes for increasing photon energies (6). They coincide, spatially and temporally, with H_α and white-light brightenings, implying that non-thermal electrons of tens of keV are interacting with a cool ($T \lesssim 10^5$ K) environment. If the site of the energy release is in the corona, near the tops of flare loops (7), very rapid transport of energy from the energy release site to the interaction region is required, which can only be achieved by streaming fast electrons.

Little else, however, is known about the processes of impulsive energy release and particle acceleration in flares. The shape of the energy spectrum, $F(E,t)$, of the accelerated particles provides a key to the type of acceleration mechanism. In the simplest picture one might expect (i) a power-law spectrum for a stochastic mechanism such as Fermi acceleration; (ii) a spectrum with a peak at an energy corresponding to the total potential drop for a DC electric-field acceleration; or (iii) a Maxwellian electron spectrum for bulk heating. Detailed information on the acceleration and energy release processes can be obtained from examining the temporal variations of $F(E,t)$.

Since the bremsstrahlung cross-section (Figure 1) is accurately known (to about 10% in the hard X-ray range), precise measurements of the solar flare hard X-ray spectrum can, in principle, be inverted to obtain the detailed spectrum of the parent X-ray-producing electrons at the Sun. The scintillation (NaI or CsI) detectors which have been used for all previous spacecraft solar hard X- ray observations are inadequate for this task since their FWHM energy resolution is generally larger than the scale of variation of the sharp features in the flare spectrum. In addition, NaI and CsI scintillators have escape peaks (34 keV for iodine) in the middle of the hard X-ray energy range. The resulting broad and complex response function precludes the direct inference of the incoming photon spectrum. Instead, trial spectra must be assumed, a technique which is known to be imprecise, giving acceptable fits for a relatively wide range of spectral shapes (8). Thus, sharp spectral breaks that are known to exist in the hard X-ray spectrum (see below) are missed.

The high spectral resolution of cooled germanium (Ge) detectors, on the other hand, yields a near delta-function response for continuum measurements above $\sim$10 keV. This makes the direct determination of the X-ray spectrum from the count-rate spectrum possible and eliminates the need to assume an *a priori* spectral shape for the X-rays. The thermal and non-thermal components can be easily and cleanly separated, which is important for determining the low-energy cut-off to the electron spectrum and the total energy in electrons.

The power of such measurements was illustrated by the first (and only) hard X-ray measurement with high-spectral (but no spatial) resolution, which was carried out with a balloon-borne Ge spectrometer for the 1980 June 27 flare. The high spectral resolution immediately led to the discovery of a sharp steepening in the X-ray spectrum below $\sim$30 keV due to thermal bremsstrahlung from a "superhot" $\sim 3.5 \times 10^7$K plasma (9). It also showed that the non-thermal component does not fit a single power law, but rather a double power law with a relatively sharp break at ~50 keV, Such a shape implies a low-energy cutoff or flattening in the X-ray-producing electron spectrum. Here and in (10) we show that with these high spectral resolution hard X-ray observations it is possible to infer the accelerated electron spectrum.

METHOD AND RESULTS

We have developed a numerical inversion technique which derives the parent electron spectrum from the bremsstrahlung X-ray spectrum (11) for optically thin sources. This technique requires no *a priori* assumptions about the form of the electron spectrum. It is able to incorporate any form for the bremsstrahlung photon production cross-section, and it provides accurate assessments of the uncertainties in the electron spectrum based on the uncertainties in the photon measurements. This technique shows explicitly that the parent electron spectrum can be defined with the same energy resolution and over the same energy range as the X-ray observations, subject, of course, to the statistical limitations of the observations. The method is illustrated in Figure 2. The observed X-ray flux in interval 1 must be produced by electrons with energy greater than E_1; it thus constrains the flux of electrons above E_1. Using the cross-section (Figure 1) we can compute the X-ray flux at energies below E_1 emitted by these electrons, and subtract them from the observed X-ray spectrum below E_1. Then the remaining X-ray flux in interval 2 must be produced by electrons with energy between E_2 and E_1; the flux of electrons in that energy interval can then be derived from the cross-section. The method can be continued downward to the lowest energy of the X-ray measurements.

The result of the inversion is the instantaneous X-ray-producing population of electrons, N(E,t) in total number per keV. N(E,t) depends inversely on the ambient ion density n in the emitting region, but is independent of the volume. Figure 3 shows the results of this inversion for the 1980 June 27 flare event.

Once the X-ray producing electron population N(E,t) is obtained, it can be related to the accelerated electron spectrum F(E,t) in units of total number of electrons keV^{-1} s^{-1} by a continuity equation such as

$$\frac{dN(E,t)}{dt} = F(E,t) - \frac{N(E,t)}{\tau_e(E)} - \frac{d}{dE}\left[N(E,t)\frac{dE}{dt}\right] \tag{1}$$

where $N(E,t)/\tau_e(E)$ refers to electron escape from the X-ray-emitting region, and the last term refers to energy change processes. With information or assumptions about the loss processes, the accelerated electron spectrum, F(E,t), can be directly obtained. For example, if Coulomb collisions are assumed to dominate, the appropriate expression for dE/dt is

$$\frac{dE}{dt} = -4.9 \times 10^{-9} n E^{-1/2} (keV s^{-1}) \tag{2}$$

for energy loss in ionized hydrogen (12), where n is in cm^{-3} and E is in keV.

322

Since the Coulomb energy loss is proportional to density n and N(E,t) depends inversely on density, the energy change term in Equation (1) will be independent of density. If escape is neglected ($\tau_e \to \infty$) and Equation (2) is integrated over time with only Coulomb collision loss for dE/dt, we would obtain the thick-target solution for the accelerated electrons. Other energy loss processes such as plasma wave-particle interactions or radiation may be important in certain situations. Because the efficiency for electron escape to the interplanetary medium is generally $\lesssim 0.1\%-1\%$ (1) we will assume escape is negligible.

The observations for the 1980 June 27 flare were divided into 5-s intervals, and then binned into energy channels to provide hard X-ray spectra statistically significant enough for inversion. Each 5-s X-ray spectrum was inverted by the Johns and Lin (11) method to obtain the parent X-ray-producing electron spectra N(E,t). Then the accelerated electron spectra F(E,t) were constructed by numerically solving the continuity Equation (1).

The rapid decay observed in the hard X-ray emission, if attributed to collisional energy loss, imply a minimum density of 10^{10}cm^{-3}. At that density the energy-loss term dominated the time derivative term in Equation (1) at low energies, but the two terms are comparable at higher energies. The computed accelerated electron distribution F(E,t) and its temporal behavior, however, do not depend significantly on the density we choose for the emitting region. Here we neglect the dN/dt term (equivalent to assuming a density $\gtrsim 10^{11}$ cm^{-3}).

Six broad energy channels of F(E,t) are plotted as a function of time in Figures 4d-i. The main spike at ~75 s is clearly seen in the accelerated electrons, with an increase above pre-spike levels of a factor of ~4 at 130–252 keV and ~6 at 78–130 keV. At 44–78 keV and 33–44 keV the increase is of marginal statistical significance and less than a factor of 2. The following spikes at ~100 and ~130 s are barely detectable. An early spike at 35 s, seen only in 44–78 keV accelerated electrons, may be significant. The lowest two energy channels, 16–23 and 23–33 keV, show no evidence at all for impulsive spikes. Thus, the spikes appear to be dominated by higher energy accelerated electrons. The slow rise in the lowest energy channels after ~130 s is due to the onset of the "superhot" thermal component (9) which dominates below 33 keV at late times. Note that the observed hard X-ray emission (Figs. 4a-c) shows impulsive spikes at all energies since the high-energy electrons accelerated in spikes produce photons at all energies equal to or lower than the electron energy.

Besides the spikes, a slowly varying underlying level of accelerated electrons is present in all energy channels throughout the entire flare burst from 35 to ~150 s. Lin and Schwartz (13) found that the X-ray spectrum of this slowly varying component, measured when the spikes are absent, always has the same X-ray power-law index, $\gamma \simeq 3.5$, a much softer spectrum than the spikes. Thus, both the temporal and spectral characteristics of the accelerated electrons suggest that there are two distinct non-thermal components: slowly varying and spikes–although other interpretations are possible. Assuming that the slowly varying component stays approximately constant from ~50 to ~140 s, we can subtract it from the total accelerated electron population to obtain the impulsive electron spectrum. The resulting spectrum (crosses in Fig. 5) averaged over the main spike (~65 to ~85 s) has a peak at ~50 keV and a cutoff below ~40 keV, with only upper limits at lower energies. The spectrum for the slowly varying component (diamonds in Fig. 5) extends in a near power law to 20 keV.

DISCUSSION

The two components may represent two physically distinct flare phenomena: for example the spikes may be due to electrons impulsively accelerated and/or rapidly precipitated into the lower atmosphere, while the slowly varying component could be due to relatively long-lived electrons in the corona (since this component is slowly varying, the ambient density could be much lower). The peaked accelerated electron spectrum of the spike component is suggestive of acceleration by a quasi-static electric fields parallel to the magnetic field (13, 14). Similar acceleration occurs in the Earth's aurora (see (15) for a review), but the potential drops required for this solar flare are about an order of magnitude higher: ~50-100 kV, vs. a few to ~10 kV in aurora.

During the main spike, the rate of energy input in greater than 23 keV electrons is $4.8(\pm2) \times 10^{27}$ and $2.8(\pm1.8) \times 10^{28}$ ergs s^{-1} for the spike and slowly varying components, respectively, assuming a fully ionized ambient medium. Thus, the slowly varying component dominates by about a factor of 6. These energy input rates correspond to 4×10^{34} and $\sim5.3 \times 10^{35}$ electrons s^{-1} respectively above 23 keV.

To completely avoid the superhot thermal component, a conservative estimate can be obtained by using a 33 keV threshold. The total energy in accelerated electrons above 33 keV for this flare is 1.4×10^{30} ergs, assuming an ionized ambient medium, and the total number of accelerated electrons is 1.8×10^{37}, implying an average rate over the ~150 s burst duration of 1.2 $\times 10^{35}$ electrons s^{-1}. Figure 4i shows the rate of energy to be supplied in $\gtrsim 33$ keV accelerated electrons as a function of time by the flare energy release process. Note that this rate is generally slowly varying.

For most of the event, however, the superhot component dominate only below 23 keV. A 23 keV threshold would increase the total energy in accelerated electrons by a factor of ~6. An accurate estimate of the total energy released in all emissions in this flare is not available, but this event is only an importance SB subflare with a GOES M6 soft X-ray burst, although intense microwave and metric radio bursts were detected. Thus the total energy is probably closer to 10^{31} rather than 10^{32} ergs. These energy numbers indicate that the accelerated electrons must carry a very significant fraction of the total energy released in the flare.

Low-resolution measurements of hard X-ray spectra for other flares (16) are consistent with the high-resolution spectra of this event. Thus, these results may well hold more generally. The lack of accelerated electrons below ∼50 keV in the impulsive spike component may be responsible for the soft-hard-soft spectral evolution often reported for hard X-ray bursts (17), since the total electron spectrum would then harden at times of the spikes.

If imaging spectroscopy with high spectral and spatial resolution were available, the evolution of the electrons could be followed in space, time, and energy, e.g., $N(E,\bar{r},t)$ could be obtained. Together with other diagnostic observations to provide density, magnetic field structure, etc., it would then be possible to directly evaluate acceleration, propagation, and loss processes, rather than making assumptions. If the electrons are accelerated by DC electric fields, the spatial distribution of such potential drops could be inferred, and compared to maps of magnetic fields, currents, etc. Since these electrons often contain a significant fraction of the total flare energy, such observations, which are the goal of the High Energy Solar Spectroscopic Imager (HESSI) proposed as a Medium-class Explorer (MIDEX) mission (18), can provide a tremendous leap forward in our understanding of the energy release and particle acceleration processes in solar flares.

ACKNOWLEDGEMENTS

R. Schwartz provided the X-ray data in a form useful for inversion. This research was supported in part by NASA grant NAG5-2351.

REFERENCES

1. R. P. Lin, Space Sci. Rev. **16**, 189 (1974).
2. B. R. Dennis, Solar Phys. **118**, 49 (1988).
3. R. P. Lin and H. S. Hudson, Solar Phys. **17**, 412 (1971); R. P. Lin and H. S. Hudson, Solar Phys. **50**, 153 (1976).
4. J. C. Brown, Solar Phys. **18**, 489 (1971).
5. T. Sakao et al. Kofu Symp., NRO Report, **360**, 275 (1994).
6. Matsushita et al., PASJ, **44**, L89 (1992).
7. S. Masuda et al. Nature **371**, 495 (1994).
8. E. Fenimore et al., AIP Conf. Proc. **77**, 201 (1982).
9. R. P. Lin, R. A. Schwartz, R. M. Pelling, and K. C. Hurley, Astrophys. J. **251**, L109 (1981).
10. R. P. Lin and C. M. Johns, Astrophys. J. **417**, L53 (1993).
11. C. M. Johns and R. P. Lin, Solar Phys. **137**, 121 (1992).
12. B. A. Trubnikov, Rev. Plasma Phys. **1**, 105 (1965).
13. R. P. Lin and R. A. Schwartz, Astrophys. J. **312**, 462 (1987).
14. G. D. Holman and S. G. Benka, Astrophys. J. **400**, L79 (1992); S. Tsuneta, Astrophys. J. **290**, 353 (1985); R. M. Winglee, P. L. Pritchett, and G. A. Dulk,

Astrophys. J. **327**, 968 (1988); R. M. Winglee, A. L. Kiplinger, D. M. Zarro, G. A. Dulk, and J. R. Lemen, Astrophys. J. **375**, 366 (1991).

15. F. S. Mozer, C. A. Cattell, M. K. Hudson, R. L. Lysak, M. Temerin, and R. B. Torbert, Space Sci. Rev. **27**, 155 (1980).

16. G. A. Dulk, A. L. Kiplinger, and R. M. Winglee, Astrophys. J. **389**, 756 (1992); N. Nitta, B. R. Dennis, and A. L. Kiplinger, Astrophys. J. **353**, 313 (1990).

17. S. R. Kane and K. A. Anderson, Astrophys. J. **162**, 1002 (1070).

18. R. P. Lin, B. R. Dennis, A. G. Emslie, R. Ramaty, R. Canfield, and G. Doschek, Proc. COSPAR Symp. on the Fundamental Problems in Solar Activity (1993).

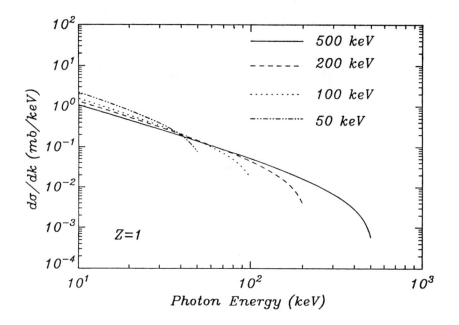

FIG. 1. Differential bremsstrahlung cross sections $d\sigma/dk$ for incident electron energies of 50, 100, 200, and 500 keV (11).

Reprinted by permission of Kluwer Academic Publishers.

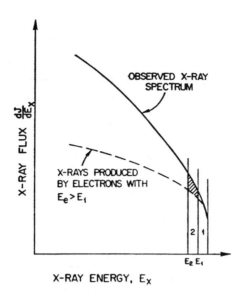

FIG. 2. Schematic illustrating the method of deriving the X-ray-producing electron population, $N(E, t)$. The dashed spectrum, produced by electrons of energies above E_1, can be subtracted from the observed spectrum in interval 2. The remaining X-ray flux in interval 2 (cross-hatched) must be produced by electrons with energies between E_2 and E_1.

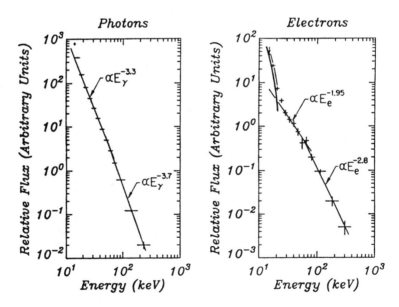

FIG. 3. On the left is the hard X-ray spectrum for the 27 June, 1980 solar flare event observed by HIREX between UT 16:15:36 and 16:1:50.1 (13). On the right is the deconvolved X-ray emitting electron spectrum. The electrons clearly show the thermal superhot component at the lowest energies and the double power law at higher energies (11) .

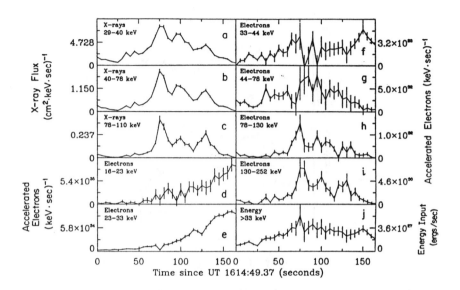

FIG. 4. Panels ($a - c$) show three energy channels of the X-ray flux as a function of time for the 1980 June 27 flare. Panels ($d - i$) show the temporal variations of the accelerated electrons in six energy channels (assuming an ionized ambient medium). Panel j plots the rate of energy release in accelerated >33 keV electrons (10).

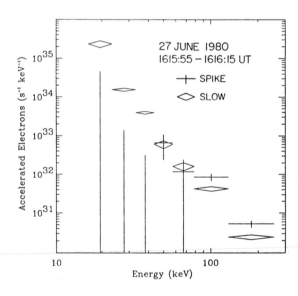

FIG. 5. The spectra of accelerated electrons in the spike (*crosses*) and slowly varying components (*diamonds*) during the main spike between ~65 and ~85 s (10).

328

Solar Hard X-ray Albedo
due to Compton Scattering

T. Bai* and R. Ramaty[†]

*HEPL, Stanford University, Stanford, CA 94305
[†]LHEA, GSFC,NASA, Greenbelt, MD 20771

The photosphere reflects a substantial fraction of hard X-rays impinging on it. Based on Monte Carlo simulations of hard X-ray propagation within the photosphere, we derive the shape and brightness of the albedo patch resulting from Compton reflection. We discuss the information that can be otained from observations of the solar hard X-ray albedo. When the albedo patch is resolved from the primary hard X-ray source, we can learn about the angular distribution of the high-energy electrons that produce the X-rays as well as the height of the primary X-ray source above the photosphere. Taking into account the albedo, however, is important even when the patch cannot be resolved, as the Compton backscatter affects the total flux, the spectrum and the spatial distribution of the observable X-rays.

INTRODUCTION

Because a large fraction of the hard X-rays emitted toward the photosphere is reflected by Compton scattering, it is important to understand the effects of the Compton backscatter to correctly interpret observations of solar hard X-rays. The Compton backscatter is especially important when the energetic electrons which produce the hard X-rays are predominantly beamed down toward the photosphere. In such cases, because of the anisotropy of the bremsstrahlung, the hard X-rays reflected by the Sun may be dominant over the hard X-rays directly emitted toward the observer (1–3). Therefore, without properly considering the backscattered hard X-rays, one could make wrong conclusions regarding the numbers, energy spectra, and angular distributions of the energetic electrons.

In addition to these effects, Compton backscatter also alters the spatial images of the solar hard X-ray emission. When the primary hard X-ray source is located high above the photosphere, the surface brightness of the reflected hard X-rays may be detectable, giving rise to X-ray "albedo patches" (1,3). In view of the fact that deploying a hard X-ray detector with good spatial resolution is regarded as a high-priority objective for future solar missions, it is important to include Compton backscatter in our modeling of solar hard X-ray emission. The reflected hard X-rays can be resolved from the primary hard X-rays when the height of the primary source is moderately larger than its linear size (by a factor of 2 or more) and the primary source is away from the disk center (by about 45 degrees or more). When the reflected source is resolved, we can infer the directivity of the primary X-ray emission, which in turn gives us information about the angular distribution of the energetic electrons emitting the hard X-rays. Detection of the albedo patch could also determine the height of the primary X-ray source.

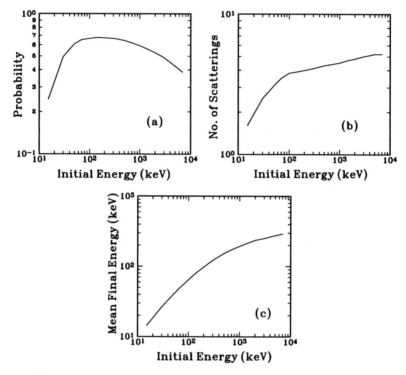

Fig. 1. The escape rate, the mean number of scatterings experienced before escaping, and the average final energy of escaping photons for mono-energetic primary photons.

MODEL CALCULATIONS

We use the Monte Carlo method developed by Bai and Ramaty (3) to calculate the transport of hard X-rays emitted toward the photosphere. We take the primary hard X-rays as unpolarized, but we use the Compton cross section matrix for polarized photons (4). Thus in our simulations the initially unpolarized photons become polarized after the first Compton scattering and their polarization changes after each scattering. We ignore the curvature of the Sun, approximating the photosphere by a planar scattering medium. The curvature of the Sun will influence the total reflectivity of the hard X-rays only when the primary source is very high above the photosphere or when it is very close to the limb. Even when the curvature is important, the effect will mainly be on the faint outskirt of the albedo patch which is generally too faint to be observable with imaging detectors. On the other hand, the brightness of the brightest part the albedo patch, which is directly under the source, does not depend on the Sun's curvature.

We first show results for mono-energetic photons emitted isotropically (Fig. 1). The escape probability (panel a) is the ratio between the number of photons escaping from the photosphere and the number emitted toward the photosphere. We see that the escape probablility is more than 0.6 in the range between 50 and 1000 keV, but it decreases at both low and high energies. At low energies it decreases because of photoelectric absorption. At high energies it decreases because of the strong forward scattering of the photons. The high-energy photons can travel deep into the photosphere, where they can undergo a large number of scatterings leading to the loss of a substantial fraction of their energies before they either are absorbed or escape. The average number of scatterings for the escaping photons (panel b) is less than 10 even at high energies; however, the photons which are finally removed by absorption undergo on average a much larger number of scatterings. Panel (c) shows the mean final energies of the escaping photons. Whereas below about 100 keV the photons emerge without much energy loss, the high energy photons experience large energy losses.

In our subsequent calculations we assume that the primary source is a sphere with radius $r = 0.3H$, where H is the height of the center of the sphere above the photosphere. We let the volume emissivity $F(E, \mu)$ be uniform throughout the sphere at all energies. We further assume that it is mildly anisotropic,

$$F(E, \mu) = A(E) \exp\{-0.25\mu \ln(E/10)\}. \tag{1}$$

Here E is the photon energy in keV, μ is the cosine of the photon emission angle with respect to the vertical, and $A(E)$ is either a single or double power law,

$$A(E) = \begin{array}{ll} C E^{-\delta} & \text{for } E \leq E_0 \\ C E_0 E^{-(\delta+1)} & \text{for } E > E_0, \end{array} \tag{2}$$

where the power-law index δ and the break energy E_0 are taken as free parameters. The primary source is thus isotropic at 10 keV, but becomes increasingly anisotropic at higher energies; for example at 100 keV the ratio between the vertically downward and upward emissions is 3.16.

We first consider the spectrum of the reflected component and its effect on the total emission. Fig. 2 shows spectra, integrated over the solid angle subtended by the source, for the primary, reflected, and combined emissions. We used the angular distribution given by equation (1) and the spectrum given by equation (2) with $\delta = 3$ and $E_0=100$ keV. The range of viewing angles θ (heliocentric angles) used for binning the simulated photons are indicated in the figure. Panel (a) is for a flare near the disk center, panel (d) is for a flare near the limb, and panels (b) and (c) represent intermediate cases. We see that the contribution of the reflected component is most important for flares near disk center. For flares near the limb, the reflected component is suppressed due to scattering and absorption in the photosphere.

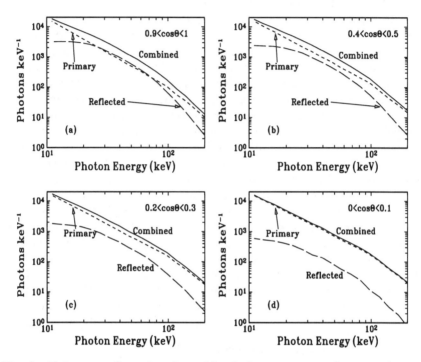

Fig. 2. Primary, reflected and combined photon spectra for an anisotropic primary source (Eq. 1) with a broken power law spectrum (Eq. 2) with $E_0=100$ keV and $\delta=3$; θ is heliocentric angle. The spectra are arbitrarily normalized.

Next we consider the albedo patches produced by the Compton reflection. In Figs. 3, 4 and 5 we plot the surface brightness due to the primary source and its albedo patch projected onto an image plane perpendicular to the line of sight. The three figures correspond to three different heliocentric angles with the photons binned in the indicated ranges of $\cos\theta$. Fig. 6 shows the surface brightness as a fuction of y for $x = 0$, with curves 1, 2 and 3 corresponding to the heliocentric angles of Figs. 3, 4 and 5, respectively.

We see that for large heliocentric angles (Fig. 3 and to some extent Fig. 4 or curves 1 and 2 in Fig. 6) the primary source is well separated from the albedo patch. As the linear separation on the image plane is on the order of the source height H, a detector with 2 arc sec angular resolution [i.e. the planned HESSI instrument (5)] will be capable of resolving the albedo patch from the primary source for sources of height greater than about 1500 km. However, as there is a large contrast (about a factor of 100) between the maximum brightness of the source and that of the albedo patch, the detection of the patch may not be easy.

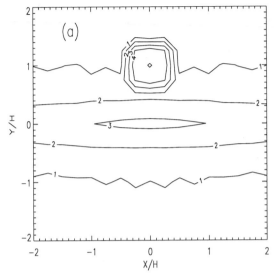

Fig. 3. Equal brightness contours for a primary source centered at $r = 0.3H$ and its albedo patch viewed at heliocentric angles in the range $0.15 < \cos\theta < 0.2$. The source spectrum is a single power law with $\delta = 3.5$ and the brightness plots are for photons in the energy interval from 25 to 75 keV. The contours are labelled logarithmically. Because of the inclination of the image plane the albedo patch is seen squashed along the y-axis.

DISCUSSION AND CONCLUSIONS

We have carried out Monte-Carlo calculation of Compton backscatter of hard X-rays in the solar photosphere. The total reflected hard X-ray emission is strongest relative to the primary emission for flares near disk center. The corresponding albedo patches, however, are centered close to the primary X-ray source, making it difficult to resolve the patch from the primary source. Because of attenuation in the photosphere, the total reflected emission for sources close to the limb is smaller than that for flares near disk center. For these sources, however, the albedo patch can be resolved from the primary source. For an imaging detector with 2 arc sec resolution, albedo patches from sources located at altitudes greater than about 1500 km can be resolved. The brightness contrast between the primary source and the center of the albedo patch is about a factor of 100 (for an assumed mild primary source anisotropy), requiring high sensitivity detectors with large dynamic range to carry out the necessary observations. The detection of the reflected X-rays will offer unique opportunities to learn about the height of the sources and the angular distribution of energetic electrons which produce the hard X-rays.

Even when the albedo patch cannot be resolved, Compton reflection should not be ignored. As it affects the spectral shape, total flux and angular size of the total hard X-ray emission, it must be taken into account in the analysis of solar hard X-ray data.

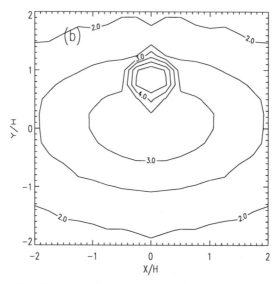

Fig. 4. Same as Fig. 3 except that $0.5 < \cos\theta < 0.55$.

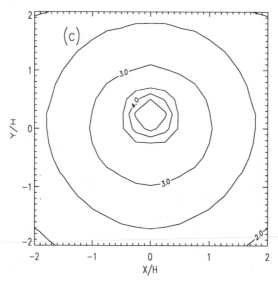

Fig. 5. Same as Fig. 3 except that $0.95 < \cos\theta < 1$.

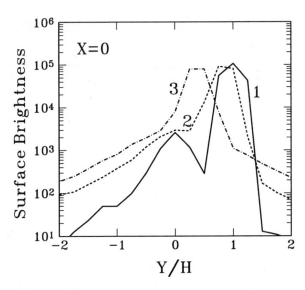

Fig. 6. The surface brightness as a function of y for $x=0$ (x and y are defined in Figs. 3, 4 and 5). The curves labelled 1, 2 and 3 correspond to the heliocentric angles of Figs. 3, 4 and 5, respectively.

REFERENCES

1. J.C. Brown, H. F. van Beek, and A. N. MaClymont, A&A **41**, 395 (1975).
2. S. H. Langer and V. Petrosian, Astrophys. J. **215**, 666 (1977).
3. T. Bai and R. Ramaty, ApJ **219**, 705 (1978).
4. W. H. McMaster, Rev. Mod. Phys. **33**, 8 (1961).
5. B. R. Dennis et al. in High Energy Solar Phenomena—A New Era of Spacecraft Measurements, (AIP: NY), 230 (1994).

What is the Spatial Relationship between Hard X-Ray Footpoints and Vertical Electric Currents?

Jing Li*, Thomas R. Metcalf*, Richard C. Canfield*,
Jean-Pierre Wülser†, and Takeo Kosugi‡

*Institute for Astronomy, University of Hawaii, HI 96822, U.S.A
†Lockheed-Martin Palo Alto Research Laboratories, Palo Alto, CA 94303, U.S.A.
‡National Astronomical Observatory, Mitaka, Tokyo 181 Japan

We examine the spatial relationship between solar flare hard x-ray emission sites observed with the HXT aboard Yohkoh and photospheric electric currents observed at Mees Solar Observatory. Canfield et al (3) concluded that nonthermal electron precipitation tends to occur at the edges of sites of high vertical current, not at their maxima. They did not, however, have very direct observations of the electron precipitation, since they used Hα Stark wing emission. In this work we compare hard x-ray images and vertical current maps in six large M/X-class flares. Our results confirm that electron precipitation sites avoid regions of strong photospheric vertical currents in large flares, and that magnetic mirroring strongly influences the relative HXR brightness of conjugate footpoints.

INTRODUCTION

Knowledge of the spatial relationship between sites of energetic electron precipitation and the distribution of vertical electric currents in the photosphere will help us to understand whether active-region current systems play any direct role in the acceleration of the energetic electrons that precipitate into the chromosphere to produce hard x-rays.

The relationship of Hα flares to sites of high vertical current density was studied in several previous papers. Moreton and Severny (11) studied 30 flares and found that about 80% of the initial brightenings in Hα were coincident with the sites of vertical currents, to within the 6" coalignment accuracy of their data. Hagyard (5) found that bright regions in Hα emission in a flare were located near the principal current system. Their sites of strongest Hα emission were found to be cospatial with the vertical current systems, to within 2" registration accuracy, by Lin and Gaizauskas (8). Finally, a recent study by Romanov and Tsap (12) also showed that the strongest flares were closely associated with vertical currents. Some of the flares were cospatial with current maxima; but most appeared near the edges of currents and oc-

TABLE 1. Summary of 6 flare events.

NOAA Region	UT Date	Location	X-Ray Class[a]	Optical Importance
6919	1991/Nov/15	S13 W19	X1.5	3B
6985	1991/Dec/26	S16 E23	M4.2	1B
7260	1992/Aug/20	N16 W27	M2.9	1B
7518	1993/Jun/07	S09 W30	M5.4	2B
7530	1993/Jul/04	S11 W45	M1.6	1B
7765	1994/Aug/14	S12 W08	M3.9	1N

[a]GOES (Geostationary Operational Environmental Satellite) X-Ray class

casionally between current channels of opposite sign. These studies concluded that the observations supported the current-interruption model (1).

In earlier work we studied the relationship of Hα flare spectroheliograms to the distribution of vertical current density. In the absence of hard x-ray data, Hα spectral images were used to identify the electron precipitation sites at the chromosphere ((3), paper I). Studies of specific events ((7), paper II and (2), paper III) confirmed that sites of intense nonthermal electron precipitation do not overlap sites of strong vertical current at the photosphere. They concluded that the energetic electron precipitation sites tend to occur on the shoulders of vertical current channels, rather than at the vertical current maxima.

In this work, the energetic electron precipitation sites are determined from Yohkoh Hard X-Ray Telescope (HXT; (6)) images of 6 large flares (Table 1). Magnetograms from the Haleakala Stokes Polarimeter at Mees Solar Observatory (MSO) provide vertical current density maps at the photosphere.

METHOD

To study the spatial relationship between flare hard x-ray sites and sites of vertical currents, we chose flares according to three criteria:

(a) The integral photon count of Hard X-Rays should be greater than 150 counts per subcollimator in the H-band (53-93 keV), so that the x-ray flux provides enough information for the synthesis of images. We have used the pixon-based method of Metcalf et al (10) for this purpose.

(b) Vector magnetograms of the flare active regions must have been obtained by the Stokes Polarimeter at MSO at a time differing from that of the flare by an amount that is short compared to the active-region evolutionary time scale. We have used these vector magnetograms to make current density maps following Canfield et al (3) and Metcalf (9).

(c) The active regions should be near the center of the disk, so that current maps are not seriously distorted by projection effects (4).

We identified six active regions (Table 1) observed during the period from October 1991 (when Yohkoh was first available) to December 1994, which

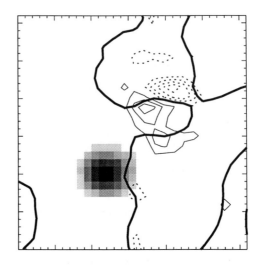

FIG. 1. The underlying (negative) image shows the HXT M1 channel synthesized using the pixon algorithm at 07:48:31 UT on 1993 July 4. Each pixel in the image is 2.47 arc seconds. Solar north is up and east is to the right. The thick solid contour shows the magnetic neutral line derived from the vertical magnetic field. The thin solid contour shows the upward vertical current density and the dotted contour shows the downward vertical current density. The contour levels are at multiples of 1.5σ, where σ is 3.1 mA/m^2 for this event.

satisfy these criteria.

Our analysis is based on three different observations:

(a) The hard x-ray lightcurves in 4 HXT energy channels provide the burst spectrum and its fluctuation with time.

(b) In our analysis, the simultaneous Soft X-Ray (SXR) movie from SXT with the 0.1-mm Aluminum (Al.1) filter helped very much to outline the loop systems around the time of the HXR flare. The simultaneous SXR images with the 119-mm Beryllium (Be119) filter provide a check on the synthesized hard x-ray images in the LO energy channel (14keV - 23 keV).

(c) The magnetograms are used to indicate the magnetic field strength in the particle precipitation sites and to compute vertical current density. The co-aligned HXR images and electric current maps show their morphological relationship at the photosphere.

OBSERVATIONS

One of 6 flares we studied was an M1.6 flare in NOAA active region 7530 on 1993 July 4. Fig. 1 shows the image from the HXT M1 channel at the

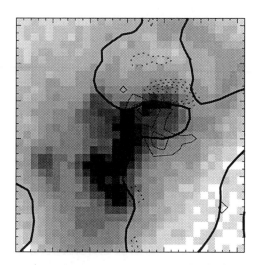

FIG. 2. The underlying (negative) image shows the SXT Al.1 image nearest the HXR image shown in Fig. 1 (07:48:28). The contours are the same as the contours in Fig. 1. Each pixel in the image is 2.47 arc seconds. Solar north is up and east is to the right.

peak of the flare (07:48:28 UT). The HXT image shows a single footpoint. Since this was a weak event, the footpoint is probably unresolved spatially. The magnetic neutral line and the vertical electric current density were derived from a vector magnetogram obtained at Mees Solar Observatory about 9 hours after the flare. Fig. 2 shows an SXT Al.1 image at nearly the same time as the HXT image. The SXR image shows the connection between the HXR footpoint and the current-carrying magnetic field to the north of the HXR footpoint. The Yohkoh data were co-aligned with the Mees data using sunspot images from the Mees White Light Telescope and the known pointing of the Yohkoh instruments. The accuracy of the alignment is about 6 arc seconds. The flare loop does not connect to the region with peak current density but does connect to a point adjacent to the strong current density. The magnetic field at the north footpoint (not seen in HXR) is 800 G while the magnetic field at the south footpoint (seen in HXR) is only 200 G. Since the flare loop connects from a region of strong field to a region of weak field, the loop must be converging as it is traversed from south to north. Hence we expect mirroring of the precipitating electrons at the north footpoint. This explains the asymmetry of the HXR emission at the north and south footpoints. This also explains the asymmetry of the current density. The current density will be strongest at the footpoint with strong magnetic field since the current is more concentrated there.

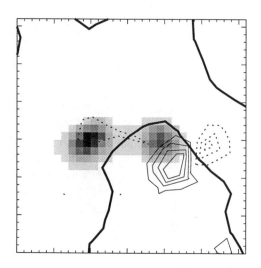

FIG. 3. The underlying (negative) image shows the HXT M2 channel synthesized using the pixon algorithm at 17:35:49 UT on 1994 August 14. Each pixel in the image is 2.47 arc seconds. Solar north is up and east is to the right. The thick solid contour shows the magnetic neutral line derived from the vertical magnetic field. The thin solid contour shows the upward vertical current density and the dotted contour shows the downward vertical current density. The contour levels are at multiples of 1.5σ, where σ is 2.0 mA/m^2 for this event.

We observed an M3.9 flare in NOAA active region 7765 on 1994 August 14. This flare is also one of 6 flares we studied. Fig. 3 shows the image from the HXT M2 channel at the peak of the flare (17:35:49). In this event we see two footpoints in the HXR image. Fig. 4 shows the SXT Al.1 image near the time of the HXR image. The SXR image shows a loop connecting the two HXR footpoints. The SXR source between the HXR footpoints is a loop top source. The magnetic neutral line and the vertical current density were derived from a magnetogram obtained at Mees Solar Observatory about 1 hour before the flare. The accuracy of the co-alignment between the Mees and Yohkoh data was about 4 arc seconds. The magnetic field was somewhat more symmetric between the two footpoints in this event than in the 1993 July 4 event. At the eastern footpoint the magnetic field was 800 G while at the western footpoint the field was 400 G. Hence, in this event there is less mirroring at the eastern footpoint and we see a footpoint in HXR. Since the field is not exactly symmetric at the two footpoints, the western footpoint is stronger in HXR. Again we see that the flare footpoints seen in HXR are adjacent to, but not co-spatial with the strong current density.

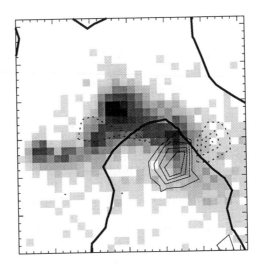

FIG. 4. The underlying (negative) image shows the SXT Al.1 image nearest the HXR image shown in Fig. 3 (17:35:07). The contours are the same as the contours in Fig. 3. Each pixel in the image is 2.47 arc seconds. Solar north is up and east is to the right.

CONCLUSIONS

Hard x-ray emission is the direct signature of precipitation of non-thermal electrons into the solar chromosphere. Our observations lead us to two basic conclusions.

(1) The relative brightness of the nonthermal HXT emission from conjugate magnetic footpoints is very strongly affected by the magnetic field geometry and magnetic mirroring. Electrons approaching the footpoint having the stronger magnetic field tend to be reflected; those approaching the footpoint having the weaker field tend to be stopped there, creating HXR emission. This confirms Sakao's identical finding with 7 HXR flares (13), only one of which (Nov 15, 1991) is included in our study. The footpoint with the weaker field is the one with the lower vertical current density, as one would expect from currents that flow along magnetic fields. All 6 observed events are consistent with this interpretation.

(2) The sites of HXR emission do not coincide with peaks of vertical current density. We therefore conclude that the interruption of the active region current systems observed in our magnetograms does not account for the primary acceleration of energetic electrons in these flares.

ACKNOWLEDGMENTS

This research was supported by NASA through Lockheed contract NAS8-37334 with the Marshall Space Flight Center. Mees Solar Observatory was supported by NASA through grant NAGW-1542.

REFERENCES

1. Baum, P.J., Bratenah, A., and Kamin, G., ApJ, **226**, 286 (1978).
2. de la Beaujardère, J.-F., Canfield, R.C., and Leka, K.D., ApJ, **411**, 378 (1993)
3. Canfield, R.C., de la Beaujardière, J-F., Fan, Y., Leka, K.D., McClymont, A.N., Metcalf, T.R., Mickey, D.L., Wülser, J.-P., and Lites, B.W., ApJ, **411**, 362 (1993).
4. Gary, G.A., and Hagyard, M.J., Sol. Phys. **126**,21 (1990)
5. Hagyard, M.J., Proc. Kunming Workshop on Solar Physics and Interplanetary Traveling Phenomena, **I**, eds. C de Jager and Chen Biao, 179 (1984)
6. Kosugi, T., Makishima, K., Murakami, T., Sakao, T., Donati, Ogawara, Y., Sawa, M., and Shibasaki, K., Sol. Phys., **136**, 17 (1991)
7. Leka, K.D., Canfield, R.C., McClymont, A.N., de la Beaujardière, J.-F., Fan, Y., and Tang, F., ApJ, **411**, 370 (1993)
8. Lin, Y. and Gaizauskas, V., Sol. Phys., **155**, 235 (1994)
9. Metcalf, T. R., Sol. Phys., **155**, 235 (1994)
10. Metcalf, T. R., Hudson, H. S., Kosugi, T., Puetter, R. C. and Pina, R. K., ApJ, submitted (1996)
11. Moreton, G.E., and Severny, A.B., Sol. Phys., **3**, 282 (1968)
12. Romanov, V.A., and Tsap, T.T., Soviet Astron., **34**, 656 (1990)
13. Sakao, T., Ph.D. Thesis, University of Tokyo, (1994)

Solar Coronal Abundances: Some Recent X-Ray Flare Observations

Alphonse C. Sterling

Computational Physics Inc., 2750 Prosperity Ave., Suite 600, Fairfax, VA 22031 USA[1]

I review recent elemental abundance studies from X-ray flare spectra obtained with Bragg crystal spectrometer experiments on board the *SMM*, *P78-1*, and *Yohkoh* spacecraft. Using the line-to-continuum method, data from all three satellites indicate an enhancement of the abundance of low-FIP Ca relative to H. But the average magnitude of the enhancement is somewhat uncertain. Flare-to-flare variations in the enhancement are also seen. Fe flare abundances seem to be close to photospheric values, with differing methods giving somewhat differing values. These findings, in conjunction with results for S, leave open the possibility that H may behave as an intermediate-FIP element or that a more complex characterization may apply. Further studies of the *Yohkoh* data, and studies comparing different analysis methods are needed to clarify these issues.

INTRODUCTION

Over the past decade, there has been an accumulation of evidence indicating that the elemental composition of the upper solar atmosphere differs from that of the solar photosphere. Apparently, the key parameter in determining these composition differences is the *First Ionization Potential* (FIP) of the the elements making up the compositions of the respective regions, with the elements divided into two groups: "low-FIP" elements, where FIP $\lesssim 10$ ev, and "high-FIP" elements, where FIP $\gtrsim 10$ ev. Simply stated, the relative abundance of a given low-FIP element to a given high-FIP element is enhanced in the upper atmosphere compared to the photosphere. Different measurements give various values for the enhancement factor, ranging from 1 (*i.e.*, no enhancement) or 2 up to 10 or more. Recent general reviews of solar abundance studies appear in Meyer (16), (17), (18), and Feldman (7). Saba (25) and Mason (15) discuss some spectroscopic and atomic physics issues regarding UV and X-ray abundance studies. Reames (23) examines the coronal abundance question from solar energetic particle (SEP) observations, and von Steiger (35) reviews solar wind abundances. This paper reviews some of the recent abundance measurements obtained via X-ray spectroscopy

---

[1] Current address: Institute for Space and Astronautical Science, Yoshinodai 3-1-1, Sagamihara, Kanagawa 229, Japan

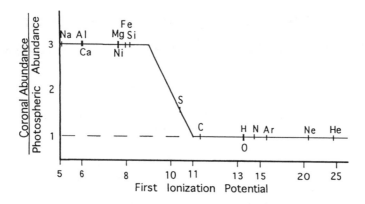

FIG. 1. Figure 1. Ratio of coronal to photospheric abundances as a function of first ionization potential (FIP) of the elements. H is shown "behaving" as a high-FIP element. Various other simplifying assumptions are implicit in this figure, as discussed in the text.

of solar flares, obtained from Bragg crystal spectrometers on board the Solar Maximum Mission (*SMM*), *P78-1*, and *Yohkoh* satellites.

Figure 1 shows a schematic picture of the coronal elemental abundances compared to the photospheric abundances, based on the figure in Feldman's review (7). This picture is oversimplified in a number of ways. For example, it shows the FIP enhancement factor as factor of three for the low-FIP elements over the high-FIP elements. As indicated above, the enhancement factor can vary rather dramatically, and there is now good evidence that the enhancement factor depends on the particular coronal structures being observed (see the general reviews cited above).

Figure 1 plots H assuming it "behaves" as a high-FIP element, meaning that the ratio of some other high-FIP element to H is the same in the corona and photosphere, while the ratio of a low-FIP element to H is enhanced in the corona compared to the photosphere. (Figure 1 is heavily influenced by observations of ratios of low-FIP to non-H high-FIP elements. Information regarding where H should be placed on the vertical axis is relatively scarce.) But, as we shall see, the possibility that H "behaves" as a non-high-FIP element is still open to question. This issue of how H behaves relative to the trace elements is important, since H, of course, makes up the bulk of the solar material by far. Therefore abundances of trace elements relative to H give the absolute abundances of these elements in the solar atmosphere. If H behaves as a high-FIP element, then it means that the absolute abundances of the low-FIP elements are *enhanced* in the upper atmosphere compared to their values in the photosphere, while the absolute abundances of the high-FIP elements are about the same in the upper atmosphere and in the photosphere. If instead, H behaves as a low-FIP element, then it would mean that the absolute abundance of the low-FIP elements in the upper atmosphere are about the same as their values in the photosphere, while the high-FIP elements are *depleted* in the upper atmosphere compared to the photosphere.

Uncertainty in the properties of H stem from the fact that it is difficult to make reliable measurements of heavy elements with respect to H, because these measurements require comparing intensities of spectral lines with a relatively weak and uncertain continuum flux. In contrast, the relative abundances of two heavy elements is generally easier to determine when the line intensities are strong, provided the emissivity functions of the two lines are similar in the relevant temperature regimes (see (25)). So, for example, the abundance ratio of the high-FIP element Ar to that of the low-FIP element Ca in the photosphere is 1.95, (assuming the cosmic abundance value for Ar of 4.47×10^{-6}, which agrees with a value inferred by Feldman et al. (8) and (7) for the photospheric value). This is 3—3.5 times the values found in *P78-1* flare spectra by Doschek et al. (5) and (6) (the factor is even higher according to the findings of Antonucci et al. (1)). This agrees with Figure 1, which shows the coronal abundance ratio of Ca/Ar to be three times larger than the photospheric value, due to a three-fold enhancement in the absolute Ca abundance. If instead H acts as a low- or intermediate-FIP element, then the vertical axis of Figure 1 will slide up accordingly, so that, say, the Ca/H ratio in the corona will be less than three times the photospheric Ca/H abundance ratio value and the coronal Ar/H abundance ratio value will be smaller than the photospheric value. But the Ca/Ar abundance ratio in the corona will still be three times the photospheric value for the ratio. Fludra and Schmelz (11) discuss in more detail some of the consequences of H acting as a non-high-FIP element.

Other simplifications implicit in Figure 1 are that the coronal absolute abundances of all high-FIP elements behave in the same relative way in the corona and in the photosphere. There is evidence, however, that the behavior of Ne may be more complicated; its coronal abundance (relative to other high-FIP elements) appears to be somewhat enhanced over its photospheric value in some (but not all) flares, as discussed by (19), (26), (27), (28), (11), and (24), and a similar situation may hold for Ar (*e.g.*, (11)) See (30) and (29) for suggested explanations of these observations. There is also a simplification with Figure 1 in that the abundance of He is difficult to obtain via spectroscopic means (see (7)), and there is some evidence that it is depleted in the upper atmosphere compared to other high-FIP elements (see (17)).

Here I will examine how Figure 1 should actually look in the case of solar flares. I first present a short discussion of methods for measuring abundances of heavy elements with respect to H in X-ray spectra.

ABSOLUTE ABUNDANCE MEASUREMENT METHODS

Using Bragg X-ray spectra from He-like ions, absolute abundances (*i.e.*, abundances relative to H) are commonly measured by taking the ratio of the intensity of the resonance line ($1s^2\ ^1S_0 - 1s2p\ ^1P_1$ transition, line w in the notation of Gabriel (12)) to continuum flux. A difficulty with this method is that the continuum from Bragg spectrometers can be contaminated with radiation from other sources, such as scattering within the spectrometer, extraneous radiation from the Earth's radiation belts, etc. Only when such spurious radiations are eliminated or properly accounted for can this method be considered. Assuming an optically thin and isothermal plasma, the Ca w-line to continuum intensity ratio is

$$\frac{I_w}{I_{cont}} \propto A_{Ca} \frac{F(T_e)C_w(T_e)}{G_{cont}(T_e)}, \tag{1}$$

where A_{Ca} is the absolute abundance of Ca, $i.e.$, the abundance ratio Ca/H, in the flaring plasma. Also, F is the fraction of the Ca ions in the He-like state compared to the total number of Ca ions $= N(\text{Ca XIX})/N(\text{Ca})$, C_w is the excitation rate coefficient for the w-line, and G_{cont} is the Gaunt factor for the continuum, all as functions of electron temperature, T_e.

In some cases it is preferred not to adopt an a $priori$ assumption that the flaring plasmas are isothermal for the abundance analysis. Fludra et al. (9) developed a method for determining abundances using the distribution of emission measure with temperature (DEM). Their method is based on minimizing the difference between calculated and observed ratios of various spectral line intensities, and between intensities of spectral lines and the continuum.

Fludra et al. (9) found that Ca flare abundance results obtained with the line-to-continuum method and the DEM method give identical results to within a few percent, thus verifying that the isothermal assumption of the line-to-continuum method is acceptable for Ca at typical flare temperatures. This is what is expected, since the Ca w-line-to-continuum flux ratio is a very weak function of temperature for temperatures typically observed in Ca flare spectra. This means that the w-line-to-continuum value is much more strongly determined by the Ca abundance than temperature. For the case of Fe, however, the w-line-to-continuum dependence on electron temperature is not nearly so weak, and some other method, such as DEM, should be considered.

Saba (25) points out an inconsistency in many line-to-continuum measurements to date: Theoretical calculations of the continuum intensity, which generally include free-free, free-bound, and two photon contributions, assume photospheric values for the abundances rather than abundance differences based on FIP or some other criteria. This means that the free-bound contribution to the continuum is likely to be underestimated in the theoretical calculations if the low-FIP elements are enhanced above their photospheric values as, $e.g.$, depicted in Figure 1. This would mean that the flare abundances measured to date would be lower limits. In investigating this concern, estimates by Fludra and Schmelz (11) indicate that the measured flare abundances to date are not consistent with assuming SEP abundance values (where H behaves as a high-FIP element, $e.g.$ (23)) for flare abundances (at 20 MK) in the theoretical calculations. If, on the other hand, the theoretical codes were modified assuming that the low-FIP element abundances in flares are about the same as their photospheric values, the high-FIP elements are depleted in flares (*i.e.*, assuming H behaves as a low-FIP element), and a He abundance from the CHASE experiment (*e.g.*, (13)), then Fludra and Schmelz (11) estimate that the flare abundance measurements to date overestimate the true abundances only by about 15%. These estimates imply that the flare abundance measurements to date are either about correct (in the case where the high-FIPs are depleted), or that they underestimate the amount by which the low-FIP elements are truly enhanced in flares over their photospheric values. So the conclusion of measurements to date that the abundances of heavy elements in flares differs from that of the photosphere is seemingly unchanged by the free-bound contribution issue. That being said, a fully self-consistent treatment of the free-bound continuum contribution and the line-to-continuum (or DEM method) calculated absolute abundances has not yet been carried out.

RESULTS FROM SMM

Abundance results for Ca and Fe from the *SMM* bent crystal Bragg spectrometer have been reported in a number of papers. Sylwester et al. (33) first reported flare-to-flare variations in abundances, and subsequent work has been reported in, *e.g.*, Lemen et al. (14), Fludra et al. (9) and (11), and Sylwester et al. (34). The most complete analysis and summary of the calcium abundances appears in Sylwester et al. (34). They use the line-to-continuum method with Ca XIX spectra and find $<A_{Ca}>= (5.77 \pm 1.41) \times 10^{-6}$, which is about 2.5 $\times A_{Ca(ph)}$, where $A_{Ca(ph)}$ is the photospheric value for the Ca/H abundance ratio, 2.291×10^{-6}. They also find wide variations in the Ca abundance, supporting the earlier findings of Sylwester et al. (33), ranging from about 1 to 4 $\times A_{Ca(ph)}$. In addition, (34) report that flares from the same active region tend to have similar abundance values.

In a preliminary study, Lemen et al. (14) found that the Ca abundance of flares from a single active region tended to increase with age of the active region in some cases. Sylwester et al. (34), while verifying this in two active regions, did *not* find this tendency to hold for flare abundances and active region ages in general. Sylwester et al. (34) also looked for correlations between flare abundances and various characteristics of the flares and/or the active regions from which the flares originate. These characteristics include Hα and *GOES* indices, sunspot area, hard X-ray characteristics, physical location of the flare on the Sun, phase of the solar cycle, and several Ca XIX spectral properties of the flare. They did not find a significant correlation between any of these characteristics and the flaring plasma's Ca abundance. Fludra and Schmelz (11) find Ca abundances of just under 2 $\times A_{Ca(ph)}$ for two *SMM* flares.

Fludra et al. (9), using the DEM method, investigated Fe abundances from *SMM*. They derived the abundances under two different assumptions regarding the ionization balance, *viz.*, assuming the Fe ionization balance of Arnaud and Rothenflug (3), and using a recalculated ionization balance with ionization rates modified so that an un-intuitive temperature dependence of A_{Fe}

347

during the decay phase of flares is removed. Taking the photospheric Fe abundance, $A_{Fe(ph)}$, to be 3.25×10^{-5} (see (7) for a discussion of this photospheric value of Fe) the Fe flare abundances determined by (9) are ~ 1.2—$3 \times A_{Fe(ph)}$ using the ionization balance of Arnaud and Rothenflug (3), and ~ 1.7—$5.3 \times A_{Fe(ph)}$ using the recalculated ionization balance. These results, then, also show flare-to-flare variations in the Fe abundance.

Using a comparison of Ca to Fe resonance line intensities in *SMM* Bragg spectrometer data, Antonucci and Martin (2) suggest that the Ca/Fe abundance ratio varies widely from flare-to-flare. (A similar possibility is noted by Sterling et al. (32), based on *Yohkoh* Ca to Fe resonance line intensities.) Also, using data from a different Bragg spectrometer on *SMM*, Phillips et al. (20) found that the Ar/S abundance ratio in a flare did not differ markedly from the photospheric ratio. This suggests that S, which is near the border between high-FIP and low-FIP elements (*cf.* Fig. 1) behaves more like a high-FIP element for the flare they analyzed. (But the possibility that coronal absolute abundance of Ar is enhanced exists; see Shemi (30)).

RESULTS FROM *P78-1*

Sterling et al. (31) investigated absolute Ca abundances using data from the Bragg spectrometer on the *P78-1* satellite. In the *SMM* Ca XIX spectrometer, electronic discriminators effectively removed much of the radiations which would otherwise contaminate the observed continuum. *P78-1* spectra had no such discriminators, and so it was not possible to make a straight forward application of the line-to-continuum method. Instead (31) applied a correction factor to their line-to-continuum values by normalizing their values to those obtained by *SMM* for a flare that both spacecraft happened to observe simultaneously. They then used the same normalization factor to correct the line-to-continuum values for 25 flares only seen by *P78-1*. As a "double check" on their corrected Ca *w*-line-to-continuum abundance values, (31) also independently calculated for each flare the Ca abundances using the ratio of the intensity of the *P78-1* Ca *w*-line to the intensity of the *GOES* 0.5–4 Å flux at identical times. Although this latter method mixes measurements from two different detectors on two different spacecraft, its advantage is that it does not rely on the *P78-1* Ca XIX contaminated continuum. They found that both methods give about the same values for A_{Ca}: 5×10^{-6} with $1\sigma \approx 1.0 \times 10^{-6}$ from the line-to-continuum method and 5.0×10^{-6} with $1\sigma \approx 1.7 \times 10^{-6}$ from the line-to-*GOES* method, or about twice the photospheric value.

Although not as prominent as in the *SMM* results, *P78-1* also shows flare-to-flare variations of Ca abundance. In the most pronounced case, one flare had $A_{Ca} = 8.0 \times 10^{-6}$, or about 3.5 times the photospheric value. Sterling et al. (31) found the abundance of two "impulsive" (fast rise and fast decay time) flares to be about the same as average flares in their data set.

He-like spectra for Ca, Fe, and S are available from the Bragg crystal spectrometer instrument on board the *Yokoh* spacecraft. Fludra et al. (10) have analyzed these spectra for abundances from seven flares. They use the line-to-continuum method for Ca abundances and the DEM method for Fe abundances. For S abundances, the line-to-continuum method is used when the isothermal approximation is appropriate and the DEM method is used otherwise. Electronic discriminators should have removed any contaminating radiation from the continua used in the analysis. They find $A_S = (8.1\pm1.7)\times10^{-6}$, $A_{Ca} = (3.4\pm0.4)\times10^{-6}$, and $A_{Fe} = (3.7\pm1.0)\times10^{-5}$. These values are about $0.5\times A_{S(ph)}$, $1.5\times A_{Ca(ph)}$, and $1\times A_{Fe(ph)}$, respectively. ($A_{S(ph)}$ is the photospheric S/H abundance ratio, $= 1.622\times10^{-5}$.)

Fludra et al. (10) find that all of their Ca abundance values fall between 2.5 and 4.5×10^{-6} ($\approx 1-2\times A_{Ca(ph)}$). Thus they do not see the relatively high abundances seen by *SMM* or *P78-1*. But the number of events so far sampled from *Yokoh* is small.

Fludra et al. (10) also examined the S abundance from a composite spectrum from active regions integrated over the whole Sun, and found $A_S = 6.1\times10^{-6}$, which is about $0.35\times A_{S(ph)}$.

Employing a totally different method, Phillips et al. (21) examined the abundance of Fe using two Fe lines in the *Yokoh* data. One of these lines, the w-line of Fe XXV, is formed in the flaring plasma, while the other line, the $K\beta$ line, they strongly argue is formed in the photosphere via fluorescence. The ratio of the intensity of these two lines works out to be

$$\frac{I(K\beta)}{I(w)} \approx F(\theta, h) \times \left[\frac{A_{Fe(ph)}}{A_{Fe}}\right], \tag{2}$$

where F is some function of heliocentric angle, θ, and height, h, of the flare above the photosphere. Allowing for errors in determination of h, line intensities, etc., (21) conclude that $1 \gtrsim A_{Fe}/A_{Fe(ph)} < 2$. A very important advantage of this measurement over more standard methods, such as the DEM method, is that it compares the flare and photospheric Fe abundances using spectral lines obtained from the same detector (the Fe XXVI channel, in which the $K\beta$ line appears, and the Fe XXV channel, in which the w-line appears, share a single detector on *Yokoh*).

Despite the carefulness of the Phillips et al. (21) study, their measurement is a difficult one owing to the weakness of the $K\beta$ line in the *Yokoh* Bragg spectrometer. Also, the Fe XXVI channel of the *Yokoh* Bragg instrument, in which the $K\beta$ line appears, has a high background which adds to the difficulty of the measurement. It would be desirable to repeat the same experiment under more favorable circumstances if possible. In a preliminary study, Phillips et al. (22) have done just that, using measurements of the much more intense $K\alpha$ line (which behaves similarly to the $K\beta$ line, and is also most likely of photospheric origin) from Bragg instruments on *SMM* and *P78-1*. Their findings fully corroborate the $K\beta$ results from *Yokoh*, saying that A_{Fe} is within some 30% of $A_{Fe(ph)}$.

DISCUSSION

Line-to-continuum abundance studies of X-ray spectra from *SMM*, *P78-1*, and *Yohkoh* nearly always show that the abundance ratio Ca/H is enhanced in flares compared to the accepted photospheric value for the ratio. Some occasional, interesting flares have photospheric abundances (see, *e.g.*, Feldman (7)), but they are not common.

Less certain from these studies is the magnitude of the enhancement of the flare abundance Ca/H ratio over the photospheric value. *SMM* results, for example, on average are higher than those seen by *Yohkoh*. Moreover, *Yohkoh* has not found flares with enhancements as large as some of those found by *SMM*. In an attempt to explain this Culhane (4) (also see (25)) suggests that the this may be a result of the *Yohkoh* spectrometer being contaminated by active regions outside the flaring region, since that spectrometer views the whole Sun. In contrast, the *SMM* spectrometer had a narrow field of view of 6 arc minutes (FWHM). I am skeptical that this explanation can explain the difference, however, since contributions to the *Yohkoh* Bragg spectrometer Ca XIX channel during non-flaring times are generally extremely small. Some of the difference in results between *Yohkoh* and *SMM* may just be due to low statistics; Sylwester et al. (34) have examined about 150 events, whereas only a small number of *Yohkoh* events have so far been investigated for abundances. Also, only one of the 25 events analyzed by Sterling et al. (31) had an abundance in the neighborhood of $3.5 \times A_{Ca(ph)}$, so it may not be surprising that no high abundance flares have yet been found in the *Yohkoh* data. Further studies will have to be carried out in order to clarify these issue.

Fe/H abundances seem to be close to the photospheric value, based on *Yohkoh* results using both the DEM method (10) and the Kβ (21) methods. Also, the *SMM* Fe abundances are close to photospheric according to the preliminary Kα study of Phillips et al. (22), whereas the DEM method of Fludra et al. (9) gives somewhat higher values. It would be very interesting if the Kα and the DEM methods could be applied to the same set of flares in order to compare the two methods.

In summary, findings in recent years, as discussed here and in Fludra and Schmelz (11), indicate that H may not behave as a high-FIP element in flares. That is, there is evidence against the simple view that the absolute abundance of low-FIP elements are all enhanced in flares compared to their photospheric values, and that the absolute abundances of the high-FIP elements is essentially unchanged in flares compared to the photospheric values. Rather, a situation may hold whereby H is intermediate-FIP, so that low-FIP elements are somewhat enhance and high-FIP elements are somewhat depleted in flares compared to the corresponding photospheric values, as shown in Figure 2. More complex scenarios are also possible, as discussed by Feldman (7) and Fludra and Schmelz (11).

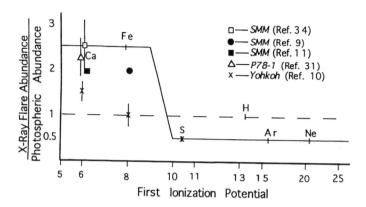

FIG. 2. Figure 2. Ratio of coronal to photospheric abundances as a function of first ionization potential (FIP) of the elements, based on recent X-ray flare observations. H is shown "behaving" as an intermediate-FIP element.

The author acknowledges support from NRL/ONR basic research programs. He also thanks J. Saba and J.-P Meyer for helpful discussions.

REFERENCES

1. Antonucci, E., Marocchi, D., Gabriel, A.H., and Doschek, G.A., A&A, **188**, 159 (1987).
2. Antonucci, E., and Martin, R., ApJ, **451**, 402 (1995).
3. Arnaud, M., and Rothenflug, R., A&AS, **60**, 425 (1985).
4. Culhane, J.L., private communication (1995).
5. Doschek, G.A., Feldman, U., and Seely, J.F., **217**, 317 (1985).
6. Doschek, G.A. and Seely, J.F., ApJ, **348**, 341 (1990).
7. Feldman, U., Physica Scripta, **46**, 202 (1992).
8. Feldman, U. and Widing, K.G., ApJ, **363**, 292 (1990).
9. Fludra, A., Bentley, R.D., Culhane, J.L., Lemen, J.R., and Sylwester, J., Adv. Space Res., **11**, No. 1, 155 (1991).
10. Fludra, A., Culhane, J.L., Bentley, R.D., Doschek, G.A., Hiei, E., Phillips, K.J.H., Sterling, A., and Watanabe, T., Adv. Space Res., **13**, No. 9, 395 (1993).
11. Fludra, A., and Schmelz, J.T., ApJ, **447**, 936 (1995).
12. Gabriel, A.H., M.N.R.A.S., **160**, 99 (1972).
13. Gabriel, A.H., Culhane, J.L., Patchett, B.E., Breeveld, E.R., Lang, J., Parkinson, J.H., Payne, J., Norman, K., Adv. Space Res., **15**, No. 7, 63 (1994).
14. Lemen, J.R., Sylwester, J., and Bentley, R.D., Adv. Space Res., **6**, No. 6, 245 (1986).
15. Mason, H.E., Adv. Space Res., **15**, No. 7, 53 (1995).
16. Meyer, J.-P., Adv. Space Res., **11**, No. 1 269 (1991).
17. Meyer, J.-P., Adv. Space Res., **13**, No. 9, 377 (1993).

18. Meyer, J.-P., in: *Origin and Evolution of the Elements,* N. Prantzos, E. Vangioni-Flam, and M. Cassé, eds., (1993:Cambridge), p. 26 (1993).

19. Murphy, R.J., Ramaty, R., Kozlovsky, B., and Reames, D.V., ApJ, **371**, 793 (1991).

20. Phillips, K.J.H., Harra, L.K., Keenan, F.P., Zarro, D.M., and Wilson, M., ApJ, **419**, 426 (1993).

21. Phillips, K.J.H., Pike, C.D., Lang, J., Watanabe, T., and Takahashi, M., ApJ, **435**, 888 (1994).

22. Phillips, K.J.H., Pike, C.D., Lang, J., Zarro, D.M., Fludra, A., Watanabe, T., and Takahashi, M., Adv. Space Res., **15**, No. 7, 33 (1995).

23. Reames, D.V., Adv. Space Res., **15**, No. 7, 41 (1995).

24. Ramaty, R., Mandzhavidze, N., Kozlovsky, B., and Murphy, R.J., ApJL, **455**, L193 (1995).

25. Saba, J.L.R. 1995, Adv. Space Res., **15**, No. 7, 13 (1995).

26. Schmelz, J.T., ApJ, **408**, 373 (1993).

27. Schmelz, J.T., and Fludra, A., Adv. Space Res., **13**, No. 9, 325 (1993).

28. Schmelz, Adv. Space Res., **15**, No. 7, 77 (1995).

29. Share, G.H., and Murphy, R.J., ApJ, **452**, 933 (1995).

30. Shemi, A., MNRAS, **251**, 221 (1991).

31. Sterling, A.C., Doschek, G.A., and Feldman, U., ApJ, **404**, 394 (1993).

32. Sterling, A.C., Doschek, G.A., and Pike, C.D., ApJ, **435**, 898 (1994).

33. Sylwester, J., Lemen, J.R., and Mewe, R., Nature, **310**, 665 (1984).

34. Sylwester, J., Lemen, J.R., Bentley, R.D., Fludra, A., and Zolcinski, M.-C., ApJ, submitted (1996).

35. von Steiger, R., Wimmer Schweingruber, R.F., Geiss, J., and Gloeckler, G., Adv. Space Res., **15**, No. 7, 3 (1995).

Solar Flare Dynamics as Revealed by *Yohkoh* Observations

G.A. Doschek

E.O. Hulburt Center for Space Research
Naval Research Laboratory
Washington, DC 20375-5352

Solar flare dynamics are reviewed as revealed by the X-ray Bragg Crystal Spectrometer (BCS) package and the Soft X-ray Telescope (SXT) on the *Yohkoh* spacecraft. Doppler motions are detected by BCS as spectral line intensity asymmetries. Motions can be detected by SXT by studying time sequences of images. High speed upflows (400 - 800 km s^{-1}) are correlated to within a few seconds with the onset of hard X-ray bursts. However, for most flares a strong non-Doppler shifted (stationary) spectral line component exists even at flare onset, which is not predicted by standard 1D numerical simulations of chromospheric evaporation into a low density coronal loop. In about 10% of all flares, an intense blueshifted component is present at flare onset. These flares appear to be morphologically complex, but this conclusion is based on poor statistics. The blueshifted plasma shows a longitude dependence, consistent with radial flow. Low speed upflows (< 100 km s^{-1}) appear to occur in some instances before the onset of hard X-ray bursts. In some cases flare loops appear to either twist, expand, or have footpoints that shift in position after the flare rise phase. These motions are slow, e.g., < 50 km s^{-1}. As found from previous missions, during the rise phase of flares, line profiles also reveal nonthermal Doppler broadening that decreases during the rise phase from about 200 km s^{-1} to 60 km s^{-1} or less.

INTRODUCTION

The solar experiments on the *Yohkoh* spacecraft have observed about 600 flares since launch in August, 1991. The X-ray Bragg Crystal Spectrometer (BCS) experiment on *Yohkoh* is about an order of magnitude more sensitive than the Bragg spectrometers flown on the *P78-1* spacecraft, the *Solar Maximum Mission (SMM)*, and the high resolution spectrometers on the *Hinotori* spacecraft. This increased sensitivity has made it possible to investigate flare dynamics as revealed by Doppler broadening and shifts of spectral lines for times much closer to flare onset than was possible with previous instrumentation. In addition, the combination of high spectral resolution Bragg crystal spectrometers and high spatial resolution soft and hard X-ray telescopes has not been available on previous high energy solar physics missions. This has allowed a detailed investigation of the morphology of flares exhibiting large Doppler motions. In this paper I review some of the highlights concerning flare dynamics revealed by the *Yohkoh* observations. There are several summaries of results from previous missions, e.g., Antonucci[1], Doschek[2], Tanaka[3]. I concentrate mainly on bulk motions of plasma, and do not discuss the well-known nonthermal random mass motions inferred from the excess Doppler widths of X-ray lines because the results from *Yohkoh* so far do not add significantly to what is already known about this phenomenon from previous missions.

MULTIMILLION DEGREE UPFLOWS: STATISTICAL RESULTS

Doschek et al.[4] first reported multimillion degree upflows deduced from the profiles of X-ray spectral lines recorded by the Naval Research Laboratory (NRL) spectrometer package (SOLFLEX) on the *P78-1* spacecraft. The upflows, observed as enhanced blue wing emission of spectral lines, were observed only during the rise phase of solar flares. Intense blueshifted spectral components are predicted by theories of chromospheric evaporation, but the *P78-1* (and also *SMM* and Hinotori) spectral lines also exhibited a much more intense non-Doppler shifted component (the stationary component) in the earliest obtained spectra than predicted by numerical simulations of chromospheric evaporation into a low density coronal loop (e.g., Doschek et al.[5]). Supporters of the numerical simulation models argued that the Bragg spectrometers were not sensitive enough to detect the weak spectra in which the blueshifted component would dominate. One of the reasons the BCS experiment for *Yohkoh* was designed to have much more sensitivity than the previously flown instrumentation was to determine if the stationary component would still be present in spectra obtained as close as possible to the onset of the flare. The BCS line profile measurements are for the resonance lines of S XV, Ca XIX, and Fe XXV.

If a single Gaussian distribution is assumed for the blueshifted component, then an average speed for this component can be obtained by deconvolving the entire profile into two Gaussians, a stationary component and a blueshifted component. As found from previous experiments, the intensity of the blueshifted component relative to the stationary component, and the average Doppler speed of the blueshifted component, decrease monotonically in time during the rise phase of the flare. Near flare onset the speed of the blueshifted component, as determined from its wavelength shift from the stationary component, is often about 300 km s^{-1}. It should be noted that the two component Gaussian assumption is invoked for simplicity, and the actual distribution of blueshifted emission might be non-Gaussian. An example of a two component deconvolution of a S XV line profile for a flare with an intense blueshifted component is shown in Figure 1.

A new technique has recently been developed by Newton, Emslie, and Mariska[6] for determining the emission measure of the blueshifted component

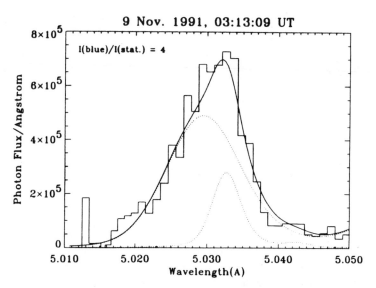

FIG. 1. A rise phase S XV spectrum. The blueshifted component centroid is shifted about 180 km s^{-1} from the stationary component centroid.

as a function of Doppler speed. This technique represents the intensity distribution of the blueshifted component as a continuum of Gaussian distributions, and therefore does not assume a particular distribution such as has been used in previous analyses.

Not long after launch a qualitative survey of about 60 flares (Doschek et al.[7]) revealed that the intense stationary component of Ca XIX was still present at flare onset for most flares, even with the increased instrument sensitivity. Subsequently, Mariska, Doschek, and Bentley[8] extended the Ca XIX survey quantitatively to 219 flares, and reached the same conclusion. A strong stationary component is present at flare onset, for the majority of flares, and in only about 10% of the flares is there a dominant blueshifted component at flare onset. Mariska et al.[8] investigated a number of relationships, e.g., the longitude dependence of the upflows and random mass motions (as revealed by symmetric line broadening beyond the thermal Doppler width of a spectral line). The reader should consult that paper and a following paper by Mariska[9] for details. (The Mariska[9] paper considers S XV and Fe XXV lines as well as the Ca XIX lines). The appearance of an intense stationary component at flare onset for most flares is an important result that must be accounted for in models such as chromospheric evaporation.

The stationary component is in some cases not completely stationary in wavelength. Low speed upflows of about 100 km s^{-1} appear to occur in some instances before the onset of hard X-ray bursts (Cheng, Rilee, and Uchida[10]). Because typical upflows have a continuous range of Doppler speeds ranging from a few km s^{-1} up to about 800 km s^{-1}, it is interesting to separate the higher speed upflows from the lower speeds, and examine the characteristics of the two components separately. The distinction between low and high speeds is somewhat arbitrary. From examination of many spectra, it is reasonable to consider high speed upflows to be 400 km s^{-1} or more. The high speed component is significantly more intense relative to the stationary component in about 10% of all flares. The example in Figure 1 represents such a case.

MULTIMILLION DEGREE UPFLOWING PLASMA: ELECTRON TEMPERATURE

It is possible to obtain an approximate electron temperature for the upflowing plasma by comparing the ratio of the Fe XXV resonance line to the Ca XIX resonance line, which is temperature sensitive (e.g., Sterling, Doschek, and Pike[11]). This can be done for the stationary component and for the blueshifted component. It is not possible to use the standard dielectronic to resonance line ratio technique to obtain temperatures for the upflowing plasma because of severe blending of the dielectronic lines. Results obtained from the Fe XXV/Ca XIX ratio technique for four flares examined by Doschek, Mariska, and Sakao[12] indicate that the average temperature of the upflowing plasma is about the same as the temperature of the stationary component. (In some cases there may be a distribution of temperatures in the upflowing plasma, as found previously by Antonucci, Dodero, and Martin[13] from analysis of *SMM* spectra of an X13 flare.)

MULTIMILLION DEGREE UPFLOWS: CORRELATION WITH HARD X-RAY BURSTS

The onset of the high speed upflows (> 400 km s^{-1}) is temporally correlated to within a few seconds with the onset of hard X-ray emission, as shown by Bentley et al.[14]. An example is shown in Figure 2. The high speed upflows frequently have light curves that roughly resemble the light curves of the hard X-rays, but not in fine details, e.g., the upflows can last longer than the hard X-ray bursts. The hard X-ray data in Figure 2 were obtained from the Hard X-ray Telescope (HXT) on *Yohkoh*. The light curves of the upflows in Figure 2 were obtained by summing the intensity of the spectral profile over wavelengths corresponding to Doppler speeds greater than 400 km s^{-1} (high speed upflows) and over wavelengths corresponding to less than 400 km s^{-1} but greater than the Doppler width of the stationary component in the decay phase.

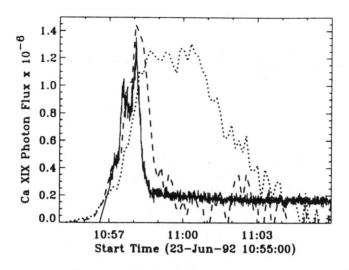

FIG. 2. Light curves of hard X-rays and blueshifted soft X-ray emission. Solid: 23-33 keV flux; dashed: upflows > 400 km s^{-1}; dotted: upflows < 400 km s^{-1}. The light curves are approximately normalized to each other to facilitate comparing them.

MULTIMILLION DEGREE UPFLOWS: UPLOW SITES

SXT and HXT images have been examined by Doschek, Mariska, and Sakao[12] for four events with intense blueshifted spectral components at flare onset. The events are morphologically complex and there appear to be several upflow sources in the flares. The locations of apparent footpoints and upflows are not connected in a simple manner with the locations of the regions that ultimately become the dominant soft X-ray emitters after the rise phase. Thus, while high speed motions clearly are important for events with intense blueshifted spectral components, the standard chromospheric evaporation loop models are not adequate for explaining the observations. One difficulty is that the first SXT images are usually obtained after flare onset. This may explain why no obvious rising fronts of evaporating plasma were seen for the flares discussed by Doschek, Mariska, and Sakao[12]. The SXT images of flares at times when the BCS spectra show dominant blueshifted components do not show obvious moving features, i.e., features with large proper motions. There are frequent brightenings that may represent moving plasma, but the brightenings could also occur in situ. An instrument combining high spatial and spectral resolution is desirable, such as is currently planned for the next Japanese solar physics mission.

LOW VELOCITY MOTIONS

Motions or apparent motions of a few km s^{-1} of entire flare loops have been detected for several flares. Such motions can be detected by co-aligning SXT images. Doschek, Strong, and Tsuneta[15] discussed the apparent expansion of the main loop of an M2.0 limb flare observed on the solar limb. It is very difficult to determine observationally if the apparent motion represents true loop expansion, or instead is the successive activation of higher lying loops. The latter explanation is difficult to accept since the expansion occurs after the rise phase when upflowing plasma (needed to fill the higher lying loops) is no longer present. In the case of a C5.3 disk flare, Doschek, Strong, and Tsuneta[15] detected the apparent motion of one of the main loop footpoints. Feldman et al.[16] observed the apparent loop expansion for two long duration X-class events (LDEs), similar to the expansion of an LDE observed by the NRL spectroheliograph on *Skylab*. Thus, numerical simulation models need to address the role of the magnetic field in a more physical context than has been done up to now. In most models, the field acts primarily as a conduit for plasma flow, but a more physical interaction is implied by the *Yohkoh* observations.

This work was supported by a NASA *Yohkoh* participating science grant and by NRL basic research funds.

REFERENCES

1. E. Antonucci, Solar Phys. **121**, 31 (1989).
2. G.A. Doschek, Astrophys. J. Suppl. **73**, 117 (1990).
3. K. Tanaka, Publ. Astron. Soc. Japan **39**, 1 (1987).
4. G.A. Doschek, U. Feldman, R.W. Kreplin, and L. Cohen, Astrophys. J. **239**, 725 (1980).
5. G.A. Doschek et al., Energetic Phenomena on the Sun (NASA CP-2439), chap. 4 (1986).
6. E.K. Newton, A.G. Emslie, and J.T. Mariska, Astrophys. J. **447**, 915 (1995).
7. G.A. Doschek et al., Publ. Astron. Soc. Japan **44**, L95 (1992).
8. J.T. Mariska, G.A. Doschek, and R.D. Bentley, Astrophys. J. **419** 418 (1993).
9. J.T. Mariska, Astrophys. J. **434**, 756 (1994).
10. C.-C. Cheng, M. Rilee, and Y. Uchida, Proc. of Kofu Symposium, NRO Report No. 360, page 213 (1994).
11. A.C. Sterling, G.A. Doschek, and C.D. Pike, Astrophys. J. **435**, 898 (1994).
12. G.A. Doschek, J.T. Mariska, and T. Sakao, Astrophys. J., in press (1996).
13. E. Antonucci, M.A. Dodero, and R. Martin, Astrophys. J. Suppl. **73**, 137 (1990).
14. R.D. Bentley, G.A. Doschek, G.M. Simnett, M.L. Rilee, J.T. Mariska, J.L. Culhane, T. Kosugi, and T. Watanabe, Astrophys. J. (Letters) **421**, L55 (1994).
15. G.A. Doschek, K.T. Strong, S. Tsuneta, Astrophys. J. **440**, 370 (1995).
16. U. Feldman, J.F. Seely, G.A. Doschek, C.M. Brown, K.J.H. Phillips, and J. Lang, Astrophys. J. **446**, 860 (1995).

Energetics of the DC-Electric Field Model

Dominic M. Zarro* and Richard A. Schwartz[†]

*Applied Research Corporation NASA/GSFC Code 682.3
Greenbelt Maryland 20771
[†]Hughes STX NASA/GSFC Code 682.3
Greenbelt Maryland 20771

We solve the energy balance equation for a coronal loop that is heated simultaneously by field-aligned currents and by runaway electrons accelerated by a DC-electric field. We combine hard and soft X-ray observations from the *Compton Gamma Ray Observatory* and *Yohkoh*, respectively, and compute the temporal variation of the DC-electric field strength during the flare impulsive phase. We find that Joule heating by coronal currents dominates the flare energy balance during the preflare and impulsive phases, while collisional heating by runaway electrons becomes significant only during the hard X-ray impulsive phase.

INTRODUCTION

Quasi-static DC-electric fields parallel to the coronal magnetic field can accelerate thermal electrons until a steady-state current is established (11) (15). Since the collisional drag on the electrons decreases with increasing velocity, the electrons with velocities above a critical velocity will undergo runaway acceleration to super-thermal energies (6). These runaway electrons can produce nonthermal hard X-ray emission via thick-target interactions (3). Electrons with velocities below the threshold for runaway acceleration continue to heat the ambient plasma (via Joule heating) thereby producing thermal soft X-ray emission. The origin of field-aligned current systems is subject to considerable debate. Tsuneta (17) has suggested that such currents may be driven by turbulent downflows from an X-type reconnection point that impact closed loop structures below it.

The DC-electric field model has been applied to the interpretation of observations of several flares (12) (3) (18). These studies have focussed on examining the spectral characteristics of flares heated by field-aligned current systems. This paper explores the energetic consequences of the DC-electric field model by solving the energy balance equation in a loop that is heated simultaneously by field-aligned uniform currents and runaway electrons. In particular, we consider a loop system in which the DC-electric field extends with uniform strength along the loop length, but varies in time during the flare impulsive phase. We use hard X-ray observations to estimate the number flux of electrons that are runaway accelerated by the electric field, and soft X-ray observations to deduce the temperature, density, and geometry of the loop. Using these empirically-derived parameters, we solve for the temporal variation of the electric field strength, and subsequently compute the heating rates due to current dissipation and collisions by runaway electrons.

ENERGY BALANCE EQUATION

For a one-dimensional loop, the energy balance equation is expressed as,

$$\frac{dU}{dt} = Q - R - \frac{dF_c}{dz} - 5nkT\frac{dv}{dz}, \tag{1}$$

where $U = 3nkT$ is the thermal energy per unit volume, Q is the total flare heating rate, $R \simeq 1.2 \times 10^{-19}n^2T^{-1/2}$ is the optically thin cooling rate, $F_c \simeq -10^{-6}T^{5/2}dT/dz$ is the Spitzer conductive heat flux, and the velocity gradient term is the enthalpy flux of convective motions within the loop. We shall assume that the density n and temperature T are uniform along the loop length. This assumption is valid about 20–30 s after heating onset when thermal conduction will have redistributed the heat energy throughout the loop, and hydrodynamic motions will have restored approximate pressure balance within the loop (10).

We simplify the energy balance equation by spatially averaging it with respect to the total loop volume $V = 2A_cL$, where L is the loop half-length and A_c is the constant cross-sectional area. The factor of two accounts for the symmetry of the loop. Integrating along the loop length from the chromospheric footpoint at $z = 0$ to the apex at $z = L$, and across the loop cross-section we obtain,

$$\frac{dU}{dt}V = QV_s - RV - 5nkTv(0)A_c - F_c(0)A_c, \tag{2}$$

where $V_s = 2A_sL \lesssim V$ is the volume in which the flare heating occurs, and $A_s \lesssim A_c$ is the cross-sectional area of the heated volume. Note that the enthalpy and conductive fluxes vanish at the loop apex. These two fluxes are also negligible in the chromosphere where they approximately balance each other and, thus, cancel from the integrated energy equation (10). Note also that the heating volume is less than the total loop volume. In order to maintain stability, a current-heated region is expected to be fragmented into numerous thin channels such that the induction magnetic field of the current-carrying electrons within each channel is below the ambient magnetic field strength (11).

For a loop with parallel DC-electric fields, the flare heating term Q is composed of a current dissipation (i.e., Joule heating) component and a collisional (i.e., nonthermal thick-target heating) component produced by runaway accelerated electrons. The Joule heating rate is given by (11),

$$Q_{curr} = nkT\nu_e(E/E_D)^2 \qquad \text{ergs cm}^{-3}\text{ s}^{-1}, \qquad (3)$$

where $\nu_e \approx 3.2 \times 10^2 nT^{-3/2}$ s^{-1} is the thermal collision frequency (for classical resistivity), E is the electric field strength (assumed uniform along the loop length), and $E_D = 7 \times 10^{-8} nT^{-1}$ volts cm^{-1} is the Dreicer field. The Dreicer field is the field strength at which all the electrons in the plasma undergo runaway acceleration.

The runaway electrons that propagate along the flare loop will heat the plasma by Coulomb collisions, producing thick-target hard X-ray emission as they impact the chromosphere. For a power-law spectrum of electrons, the collisional heating rate (ergs s^{-1} per unit column depth) can be approximated by the thick-target heating function (9),

$$Q_{elec}(N) = \frac{\delta - 2}{6N_c} B\left(\frac{\delta}{2}, \frac{1}{3}\right) \alpha^{-\delta/2}\left[a - b\left(\frac{N}{\alpha N_c}\right)^2\right] F_c, \qquad N < \alpha N_c, \qquad (4)$$

and

$$Q_{elec}(N) = \frac{\delta - 2}{6N_c} B\left(\frac{\delta}{2}, \frac{1}{3}\right) \left(\frac{N}{N_c}\right)^{-\delta/2} F_c, \qquad N > \alpha N_c, \qquad (5)$$

where δ is the electron spectral index, $N = \int ndz$ is the loop column depth (not to be confused with the electron number flux $\dot{N}$), F_c is the energy flux of nonthermal electrons (ergs s^{-1}) with energies above a cutoff E_c, and $N_c = 9.158 \times 10^{16} E_c^2$ is the stopping column depth for electrons of energy E_c. The constants $a, b,$ and α are chosen to ensure that Q_{elec} is continuous at N_c and that the integral of Q_{elec} through the coronal loop into the chromosphere equals F_c. Note that the above formula is a "smoothed" version of the original heating equation given by Brown (4). This smoothed version is more applicable to the case of a distributed coronal acceleration source such as would be produced by DC-electric fields. We compute the total rate of collisional heating by runaway electrons by integrating the smoothed equation along the loop length, thus, $Q_{run} = 2 \int_0^{nL} Q_{elec} dN$.

The electron energy flux is related to the electron runaway number flux $\dot{N}_{run}$ according to,

$$F_c = (\delta - 1)(\delta - 2)^{-1} E_c \dot{N}_{run} \qquad (6)$$

where the runaway flux rate is given by the the formula of Kruskal and Bernstein (13),

$$\dot{N}_{run} \simeq .35 n\nu_e(E_D/E)^{3/8} \exp[-2^{1/2}(E_D/E)^{1/2} - (1/4)(E_D/E)] V_s. \qquad (7)$$

This formula includes electrons that are accelerated out of the thermal distribution as well as electrons that are scattered into the runaway regime by collisions. The low-energy cutoff is related to the critical energy that separates thermal and nonthermal electrons, which is given by $E_{crit} = m_e(E_D/E)v_e^2/2$ where v_e is the electron thermal velocity (11). The critical energy is the threshold above which thermal electrons exceed the frictional force and undergo runaway.

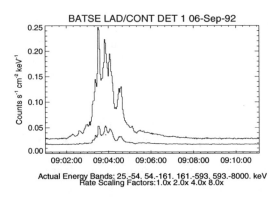

FIG. 1. BATSE-LAD hard X-ray lightcurves for 3 energy bands

ENERGY BALANCE SOLUTION

We numerically solve the loop-integrated energy balance equation for a simple loop flare that was observed jointly by instruments onboard the *Compton Gamma Ray Observatory (CGRO)* and the *Japanese Yohkoh* satellite. The flare commenced at 08:57 UT on 1992 September 6 and was observed in hard X-rays by the *CGRO* Burst and Transient Source Experiment (BATSE) (8) and in soft X-rays by the *Yohkoh* Bragg Crystal Spectrometer (BCS) (5) and Soft X-ray Telescope (SXT) (1). Figure 1 shows the background-subtracted hard X-ray lightcurve in various energy bands observed in one of the sunward-facing BATSE Large Area Detectors (LAD).

We assume that loop heating and electron acceleration occur within regions that have the same characteristic temperature and density as that implied by the observed thermal soft X-ray emission, and set $Q = Q_{curr} V_s + Q_{run}$. Accordingly, the combined hard and soft X-ray observations allow us to constrain many of the terms in the energy balance equation. In particular, we derive the temporal variations of T and emission measure $EM = \int n^2 dV$ from fits of synthetic spectra to the BCS Ca XIX (λ 3.177 Å) resonance line which was observed at 3 sec time resolution. We use high spatial resolution SXT preflare images to infer $L \simeq 3 \times 10^9$ cm and $A_c \simeq 2 \times 10^{17}$ cm^2 from which we compute the loop density $n \simeq (EM/V)^{1/2}$, assuming a filling factor of unity. Because of the full-sun field-of-view of the BCS instrument, the Ca XIX-derived temperature and density are lower limits to the likely higher values in the current heating and acceleration region. Based on the derived densities, the radiative cooling rate is negligible compared to the observed soft X-ray heating rate dU/dt. Further details of the soft X-ray analysis are given in (18).

The BATSE LAD's provide hard X-ray spectra (CONT data) in 16 channels between 10 and > 1000 keV at 2.048 sec temporal resolution. The spectra show a well-defined power-law distibution in the 20-100 keV range. A sample spectrum is plotted in figure 2. From a least-squares deconvolution of the

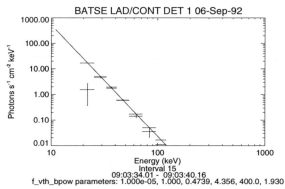

FIG. 2. BATSE-CONT power-law fit for spectrum near the time of the first hard X-ray burst peak

LAD response function, we derive the temporal variation of the amplitude a_1 and spectral index γ of the power-law component. Assuming a thick-target model, these power-law parameters are related to the total number flux of hard X-ray-producing electrons by (14),

$$\dot{N}_{thick} \simeq 3 \times 10^{33} a_1 (\gamma - 1)^2 B(\gamma - 1/2, 1/2) E_c^{-\gamma}, \tag{8}$$

where $B(x, y)$ is the beta function, $\gamma = \delta + 1$, and δ is the spectral index of thick-target electrons. Assuming that the bulk of thick-target electrons are produced by runaways, we set $\dot{N}_{thick} = \dot{N}_{run}$, which allows us to eliminate the acceleration volume V_s from the energy balance equation. The energy balance equation is thus reduced to the two unknowns E and E_c.

The hard X-ray-inferred spectral index for the thick-target electrons is generally harder than the $\sim E^{-1/2}$ energy distribution that is predicted for runaway electrons (3). This discrepancy is due mainly to our inability to accurately deconvolve the thermal and nonthermal electron contributions from the observed power-law spectra and, thus, determine the low-energy cutoff of nonthermal electrons. However, using $E_c = E_{crit}$ and the observed higher δ in equation (8) leads to a gross overestimate for $\dot{N}_{run}$ and a correspondingly large value of V_s that exceeds the observed loop volume. To allow for this effect, we adopt a somewhat higher cutoff that is obtained by equating the loop column depth with the electron stopping depth, $nL = 9.158 \times 10^{16} E_c^{-2}$. This approximation implies that the bulk of observed thick-target hard X-ray emission emanates from the footpoints, being produced only by runaway electrons that have sufficient energy to penetrate to the base of the loop. Using a temporally averaged value of n obtained for the flare as a whole, we deduce a characteristic $E_c \approx 40$ keV. The higher cutoff energy (and, hence, critical energy) also implies the existence of a higher temperature component in the flaring loop. We deduce the temperature of this component below.

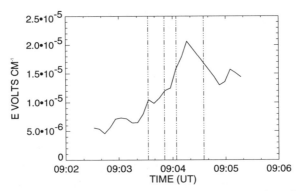

FIG. 3. Variation of the DC-electric field strength derived from loop-integrated energy balance. The vertical dashed lines mark the times of the four hard X-ray burst maxima in the BATSE/LAD lightcurve

RESULTS

Using the adopted E_c, the empirically-derived values of T, n, a_1 and γ, we numerically solve equation (2) for E. Figures 3 and 4 show the resulting variations of E and $\dot{N}_{run}$, respectively. The DC-electric field strength varies from a preflare value of $\lesssim 10^{-5}$ volts cm^{-1} to a maximum of $\simeq 2 \times 10^{-5}$ volts cm^{-1} approximately 50 s after the first hard X-ray burst maximum. The corresponding runaway rate increases monotonically to a peak value of 5×10^{34} electrons s^{-1} at the time of the first hard X-ray burst. The increase in runaway rate is caused by the increase in electric field strength and also the increase in temperature which raises the population of thermal electrons that are available for runaway. The runaway rate decreases after the impulsive phase as the plasma density increases and the associated frictional drag on the electrons is enhanced.

Figure 5 compares the loop-integrated current and runaway electron heating rates with the total heating rate VdU/dt that is required to sustain the observed soft X-ray thermal emission. Current heating dominates the soft X-ray heating during the main flare phase and is significant for 1–2 minutes prior to the first peak of hard X-rays. This strong preflare current heating is believed responsible for the strong Ca XIX stationary component that is often observed before the start of the flare impulsive phase (18). The electron heating rate is negligible during the preflare stage and becomes significant during the impulsive hard X-ray burst phase when $\dot{N}_{run}$ increases. The electron heating is produced by runaway electrons with energies above the 40 keV cutoff energy corresponding to the footpoint stopping depth. Having solved for E, we can deduce the temperature of the high-temperature component by solving $E_{crit} = 40$ keV. Using the peak value of $E = 2 \times 10^{-5}$ volts cm^{-1}, we derive a peak temperature of 4×10^7 K, which is consistent with typi-

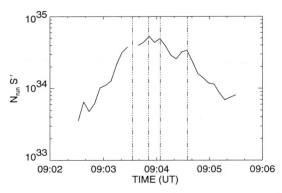

FIG. 4. Variation of runaway rate of electrons above 40 keV

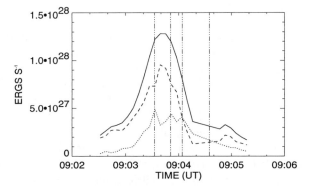

FIG. 5. Comparison between the observed soft X-ray heating rate $V dU/dt$ (solid line) and the loop-integrated heating rates due to current dissipation $Q_{curr} V_s$ (dashed line) and runaway electron collisions Q_{run} (dotted line).

cal "super-hot" temperatures calculated for flares based on spectral fits of low-energy thermal X-ray emission (2).

The runaway-accelerated electrons form a current system that operates in addition to the field-aligned currents. In order for this current system to remain stable, the self-induction magnetic field of the runaway electrons cannot exceed the strength B of the ambient magnetic field. For a cylindrical geometry, this constraint enforces the following upper limit on the volume of the acceleration region (11),

$$ V_j \lesssim \frac{L B^2}{4\pi n^2 e^2} \left(\frac{c}{v_e}\right)^2 \left(\frac{E_D}{E}\right)^2. \tag{9} $$

To satisfy this constraint, the acceleration region must be fragmented into $n_s = V_s/V_j$ multiple filaments. Adopting a typical loop field strength of 100 Gauss, and using $V_s = \dot{N}_{thick}/\dot{N}_{run}$, we deduce the variation of n_s shown

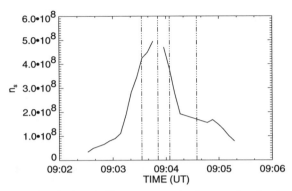

FIG. 6. Number of cylindrical filaments required to maintain the self-induction field of the runaway electron current system below the ambient coronal magnetic field.

in figure 6. The number of filaments increases monotonically from a preflare value of $\sim 4 \times 10^7$ to a maximum of $\sim 5 \times 10^8$ during the impulsive phase. Emslie and Henoux (7) have suggested that such filamented current systems can be closed by cross-field drifting of protons at the chromospheric footpoints.

It is encouraging that our derived values of N_{run} and E are similar to those deduced by Benka and Holman (2) based on spectral fits of hybrid thermal/nonthermal electron energy distributions. By contrast, the degree of current filamentation derived from the energy balance analysis is significantly larger than the $n_s \gtrsim 10^3$ implied by the hybrid model fits. This difference is due mainly to our assumption of a cylindrical filament geometry as opposed to a thin laminar sheet structure which tends to imply factors of 10-100 fewer sheets.

CONCLUSIONS

For a flare that was observed simultaneously in hard and soft X-rays, a simple energy balance analysis has shown that:

- Joule current heating and runaway collisional heating can sustain the observed soft X-ray emission during the preflare and impulsive X-ray phases;

- Assuming classical resistivity, the observed temporal behavior of impulsive nonthermal hard X-ray emission is consistent with a DC-electric field strength that increases by approximately a factor of ten during the hard X-ray rise phase to a maximum value of $E \simeq 2 \times 10^{-5}$ volts cm^{-1};

- The increase in field strength produces up to $\gtrsim 10^{34}$ nonthermal runaway electrons with energies above 40 keV that precipitate to the chromospheric footpoints;

- The assumed 40 keV low-energy cutoff implies the existence of super-hot temperature component.

We conclude that energy balance can be satisfied in a flaring coronal loop system that is heated by field-aligned currents, and that collisional heating by runaway electrons can make a significant contribution to the overall flare energy balance. Our method for solving the energy balance equation provides a novel diagnostic for the strength and temporal variation of the DC-electric field.

ACKNOWLEDGMENTS

This work was funded by *Compton Gamma Ray Observatory* Guest Investigator award NAS5-32791.

REFERENCES

1. Acton, L., *et al.* , *Solar Phys.* **136**, 89 (1991).
2. Benka, S.G., and Holman, G.D., *Ap. J.* **391**, 854 (1992).
3. Benka, S.G., and Holman, G.D., *Ap. J.* **435**, 469 (1994).
4. Brown, J., *Solar Phys.* **18**, 489 (1972).
5. Culhane, L. *et al.* , *Solar Phys.* **136**, 89 (1991).
6. Dreicer, H., *Phys. Rev.* **115**, 238 (1959).
7. Emslie, G., and Henoux, J.-C., *Ap. J.* **446**, 371 (1995).
8. Fishman, G., *et al.* , in *Proc. Gamma Ray Observatory Science Workshop*, ed. W. Johnson (NASA/GSFC) (1989) pp. 2-39.
9. Fisher, G., Canfield, R., and McClymont, A., *Ap. J.* **289**, 414 (1985).
10. Fisher, G., and Hawley, S., *Ap. J.* **357**, 243 (1990).
11. Holman, G., *Ap. J.* **293**, 584 (1985).
12. Holman, G., Kundu, M., and Kane, S., *Ap. J.* **345**, 1050 (1989).
13. Kruskal, M., and Bernstein, I., *Phys. Fluids* **7**, 407 (1964).
14. Lin, R., and Hudson, H., *Solar Phys.* **50**, 153 (1976).
15. Moghaddam-Taaheri, E., and Goertz, C., *Ap. J.* **352**, 361 (1990).
16. Tandberg-Hanssen, E., and Emslie, A., *The Physics of Solar Flares*, Cambridge: Cambridge Univ. Press, 1988.
17. Tsuneta, S., *PASJ*, in press (1995).
18. Zarro, D., Mariska, J., and Dennis, B., *Ap. J.* **440**, 888 (1995).

Hard X-Ray Polarimetry
of Solar Flares with BATSE

M. McConnell*, D. Forrest*, W.T. Vestrand* and M. Finger[†]

*University of New Hampshire, Durham, New Hampshire 03824
[†]Marshall Space Flight Center, Huntsville, Alabama 35812

We describe a technique for measuring the polarization of hard X-rays from solar flares based on the angular distribution of that portion of the flux which is scattered off the top of the Earth's atmosphere. The scattering cross section depends not only on the scatter angle itself, but on the orientation of the scatter angle with respect to the incident polarization vector. Consequently, the distribution of the observed albedo flux will depend on the direction and the polarization properties (i.e., the level of polarization and polarization angle) of the source. Since the albedo component can represent a relatively large fraction (up to 40%) of the direct source flux, there will generally be sufficient signal for making such a measurement. The sensitivity of this approach is therefore dictated by the effective area and the ability of a detector system to 'image' the albedo flux. The 4π coverage of the BATSE detectors on the *Compton Gamma-Ray Observatory* provides an opportunity to measure both the direct and the albedo flux from a given solar flare event. Although the BATSE design (with its large field-of-view for each detector) is not optimized for albedo polarimetry, we have nonetheless investigated the feasibility of this technique using BATSE data.

INTRODUCTION

One of the fundamental properties of electromagnetic radiation is its polarization. Although this property has been exploited extensively for astrophysical studies at longer wavelengths (e.g., radio, optical and, to some extent, x-ray wavelengths), there have been only limited efforts to measure polarization from cosmic sources at hard X-ray energies (E > 30 keV).

Here we investigate how one might use the radiation which is backscattered from the Earth's atmosphere to study source polarization, a technique which we call *albedo polarimetry*. Based on our Monte Carlo simulations, we have estimated the BATSE sensitivity to hard X-ray polarization in solar flares and find significant sensitivity to polarization even for some moderately sized (C- and M-class) events.

The study of the albedo flux will allow us to search for polarization effects within solar flares. The observational evidence for directivity within solar flares requires the presence of polarization in the energetic electron bremsstrahlung emission. Polarization measurements will place important constrains on the energetic electron geometries, thus complementing the spectral measurements.

POLARIZATION MEASUREMENTS OF SOLAR FLARES

The primary goal of solar flare polarimetry is to provide a direct probe of the directivity of the electrons which are accelerated during the flare. Searches for electron anisotropies in solar flares have been motivated by the clues that it can provide about electron acceleration and transport. The usual method for assaying the anisotropy is to search for evidence of directivity in the flare radiation. Until the mid 1980's all the directivity measurements were performed at energies below 100 keV. The problem at these energies is that Compton backscattering of X-rays from the solar atmosphere tends to mask the primary radiation pattern.

The large sample of flares detected at energies greater than 300 keV by SMM GRS allowed, for the first time, a statistical search for directivity at high energies. The greatly reduced Compton reflectivity and increased directionality of the bremsstrahlung cross-section make relatively strong radiation anisotropies a possibility at these energies. The procedure used in these statistical studies was to search for disk center-to-limb variations in flare frequency and spectra. Statistically significant variations from the predictions for isotropic emission are then attributed to radiation anisotropy. Using this technique, analysis of high-energy observations by SMM GRS has provided the first relatively strong evidence for directed emission from flares (1). Data from an independent set of flares detected by instruments on the Venera 13 and 14 spacecraft has also been used to present evidence for radiation anisotropies at hard x-ray energies (2). In this case, the positional spectral variation has also been attributed to bremsstrahlung directivity.

Another promising method for studying the directivity in individual flares is the stereoscopic technique (3), which compares simultaneous observations made on two spacecraft that view the flare from different angles. Combined observations from spectrometers aboard the PVO and ISEE-3 satellites have been used to produce stereoscopic observations of 39 flares that occurred between 1978 and 1980 (4). While the range of measured flux ratios is consistent with the statistical studies (5), the deviations of the ratio from unity show no clear correlation with increasing difference in viewing angles. It has therefore been argued that directivity conclusions based on stereoscopic observations for single flares are not yet possible because of the large error bars (6).

Quantifying the magnitude of the directivity from these statistical measurements is difficult. For example, one needs to know the size-frequency distribution for flares as well as the form of the electron distribution to derive the predicted limb fraction (7). Furthermore, the results only represent an average for the flare sample. Different flaring regions are not likely to have identical geometry. Nor are individual flares likely to have time independent electron distributions. Therefore, to do detailed studies of the particle distributions, one needs a technique that can measure time dependent anisotropies on a flare by flare basis. Polarization measurements are a diagnostic that meet these requirements.

There have been a number of measurements of flare X-ray polarization at low energies (~ 10 keV). The first flare measurements indicated polarization levels in the range of 20-40% (8,9). Although these reports were met with considerable skepticism (10), subsequent observations (11) confirmed the existence and magnitudes of the polarizations ($\sim 10\%$). Later studies measured polarizations of only a few percent at ~ 15 keV for two flares in July 1974 (12). The small but finite polarization is consistent with the predictions for purely thermal emission that contains an admixture of polarized backscattered radiation (13).

The thermal component, with its rather low polarization, tends to dominate the emission from all flares at energies below about 25 keV. In fact, the idea that the emission in this energy band is primarily thermal finds support from more recent polarization measurements (14). These data indicate that the polarization in the 5-25 keV band tends to be less than $\sim 4\%$, consistent with isotropic unpolarized emission that is slightly polarized by backscattering. It has been pointed out that, because of this thermal "contamination", effective polarization measurements can only be performed at higher energies (15).

At higher energies (>100 keV), we expect large linear polarizations based on the anisotropy of the electrons. Linear polarizations of 20% (16) or even larger (13,17) can be expected. Polarization measurements can be expected to provide additional information not only on the degree of isotropy, but also on the exact nature of the electron anisotropy. At present, there is no γ-ray experiment in operation whose major emphasis is the measurement of polarization.

X-RAY ALBEDO POLARIMETRY

A measurement of the linear polarization of a transient event (such as a solar flare) can, in principle, be made by measuring the angular distribution of the albedo flux, i.e., the source flux which is scattered from the Earth's atmosphere prior to reaching the detector. This concept is based on the properties of the Compton scattering of polarized radiation. In particular, this approach relies on the fact that, in the case of linearly polarized radiation, the scattered photon tends to be ejected at right angles to the electric field vector of the incident radiation (18). The atmosphere, as seen from an orbiting satellite, presents a wide range of possible scatter angles for a given source direction. The photon scatter angle will depend on look direction. Hence, the intensity distribution of the albedo flux will exhibit an angular distribution which will depend on the polarization properties of the source radiation.

For the case of an orbiting detector observing some cosmic source, the distribution of the scattered albedo flux across the top of the atmosphere will depend on three parameters: 1) the angular height of the source above the Earth's limb (i.e., the source-earth-detector geometry); 2) the level of linear polarization of the source flux; and 3) the orientation of the bulk polarization vector of the source flux. Therefore, if one knows the direction of the source, then the polarization properties can be determined from the study of the albedo flux.

To illustrate the dependence on the polarization parameters, consider a source of radiation which is observed at the local zenith. In this case, any asymmetries in the observed albedo flux will be due solely to the polarization properties of the incident source flux. These asymmetries will manifest themselves in terms of an azimuthal dependence of the observed albedo flux. For an unpolarized source, the albedo flux will be uniformly distributed in azimuth. For a linearly polarized source, however, the observed albedo flux will exhibit an asymmetry about the azimuthal direction. This asymmetry will take the form of a $\sin^2\theta$ dependence on the azimuthal angle. The magnitude of these variations represents a measure of the polarization fraction, while the plane defined by the azimuthal minima corresponds to the plane of linear polarization in the incident source flux. The largest dependence of albedo distribution on incident polarization occurs for geometries where the source is located well away from the zenith ($30°–40°$). This allows for single scatters at angles near $90°$ where the sensitivity of the Compton cross section to polarization peaks. In these geometries, there will be asymmetries introduced even in the case of an unpolarized source. For example, for events which are close to the Earth's limb, the higher probability of forward scattering will result in a limb-brightening effect in the general direction of the source. A second type of asymmetry will arise from the fact that, for events with a zenith angle in the range of $70°$-$110°$, not all of the visible atmosphere is exposed to direct source flux. A precise measurement of polarization will require that all of these effects be properly acounted for.

We have modeled this scattering process using Monte Carlo simulations, incorporating a version of GEANT which has been modified to handle polarized photons. A library of simulation results has been assembled for photon beams directed at various angles of incidence on a flat slab of (oxygen-nitrogen) atmosphere. These simulations include both polarized and unpolarized photon beams as well as both mono-energetic and power-law source spectra. The effects of multiple scattering are also properly accounted for. The atmosphere of the Earth can then be simulated by combining these data in such a way as to simulate the spherical atmosphere as a superposition of scattering sites, each oriented at various angles with respect to the observer. (This approach is required due to the scaling limitations.) In this way, we have effectively simulated the distribution of the scattered flux for a beam of photons incident on the top of the atmosphere.

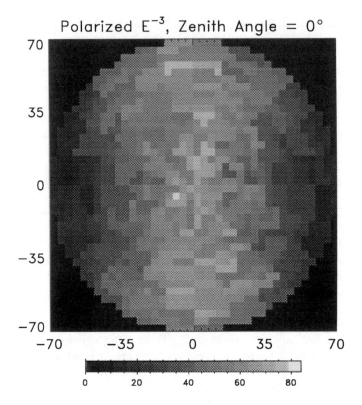

FIG. 1. The albedo distribution (as seen at the spacecraft) corresponding to a 100% polarized E^{-3} spectrum incident from 0° zenith angle. The polarization (E-field) vector lies along the horizontal plane. The scale is a relative measure of the photon flux from each solid angle element.

For an incident E^{-3} power-law spectrum, the energy distribution of the scattered flux (integrated over 2π steradian) peaks in the 50-100 keV range. The vast majority of scattered photons in this energy range originate with photons of initial energy between 50 and 300 keV. For sources near the zenith, the albedo fraction reaches a maximum value of about 40%. Examples of the spatial distribution of the scattered flux, based on our simulations, are shown in Figures 1–3. These figures show the intensity of the scattered flux across the visible disk of the Earth (which subtends an angle of $\sim 70°$ in low Earth orbit) for the case of a 100% polarized source at zenith angles of 0°, 30°, and 60°.

These simulations serve to demonstrate that, in principle, one could study solar flare polarization by mapping out the observed distribution of the flux

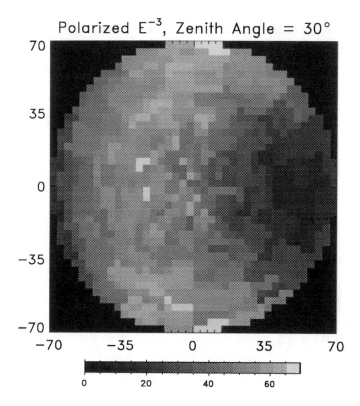

FIG. 2. The albedo distribution (as seen at the spacecraft) corresponding to a 100% polarized E^{-3} spectrum incident from 30° zenith angle. The source lies in the $+x$ direction, with the polarization (E-field) vector lying along the horizontal plane. The scale is a relative measure of the photon flux from each solid angle element.

which is scattered off the atmosphere. A mapping of this distribution at a resolution of a few degrees would certainly provide sufficient detail for such an albedo polarimeter. Unfortunately, we have no way of mapping the albedo flux with such high precision at the present time. However, the BATSE detectors on CGRO (in particular, the Large Area Detectors, or LADs) are capable of providing a crude map of the distribution of the albedo flux. Despite the aliasing effects amongst the detectors, this capability is sufficient for measuring polarization parameters in some of the larger flare events.

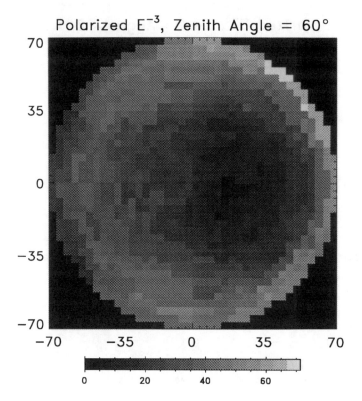

FIG. 3. The albedo distribution (as seen at the spacecraft) corresponding to a 100% polarized E^{-3} spectrum incident from 60° zenith angle. The source lies in the $+x$ direction, with the polarization (E-field) vector lying along the horizontal plane. The scale is a relative measure of the photon flux from each solid angle element.

BATSE SENSITIVITY TO FLARE POLARIZATION

The exact sensitivity of the BATSE to the measurement of flare polarization depends on several factors. These include the following:

1) *Source zenith angle.* The sensitivity will depend on the contrast amongst the different LAD detectors. This depends not only on the contrast within the polarized flux distribution, but also on the contrast (or lack thereof) in the corresponding unpolarized distribution. These effects are such that the sensitivity is near maximum for source zenith angles $\sim 30°$.

2) *Aliasing.* Due to the large FoV and the geometrical spacing of the LAD detectors, there exist certain geometrical arrangements with respect to the polarization vector where the polarization sensitivity will be limited (and perhaps even go to zero). In more general terms, the polarization sensitivity will vary within the polarization parameter space.

3) *Incident source spectrum.* Because the scattered albedo photons tend to result from incident photons > 100 keV, a harder incident spectrum will result in more intense albedo flux and, hence, higher sensitivity. This means, for example, that there will be less sensitivity to a (typically softer) solar flare than to a (typically harder) γ-ray burst of comparable fluence.

4) *Source inensity.* For more intense events, various instrumental effects, most notably pulse pile-up and deadtime, become important factors.

5) *Flux leakage.* We have assumed an LAD angular response which goes to zero at 90°. In other words, we assumed that the leakage of flux through the spacecraft can be neglected. (This is known not to be the case for some of the larger events and may even be the limiting factor in the application of this technique to BATSE data.)

6) *Background level.* The total background level for a given event depends not only on the background *rate*, but also on the duration of the event.

In estimating BATSE sensitivity levels, we consider only the four detectors which face away from the sun and which are exposed only to the albedo flux. We then search for a distribution of events in those detectors which are statistically inconsistent with that expected from an unpolarized event. In practice, this involves a χ^2 comparison between the event distribution derived from albedo simulations for a polarized and an unpolarized solar flare at a given zenith angle.

In order to relate our sensitivity estimates to typical BATSE solar flare events, we chose three events on the basis of zenith angle (< 45°), duration (< 5 minutes) and intensity. Our simulation data was then used to determine the minimum level of polarization which could be observed in each case. The results are shown in Table 1. These results suggest that this technique is capable of reaching reasonable sensitivity levels ($\sim$ 20%) for modest-sized solar flares. For the larger events, sensitivity levels below 10% should be achievable.

BATSE MEASUREMENTS OF FLARE POLARIZATION

At the present time, we have no firm results (either positive detections or upper limits) for any solar flare observed by BATSE. The analysis is presently

375

TABLE 1. Estimated polarization sensitivity levels for three typical BATSE solar flare events. The sensitivity levels reflect the percent polarization required for a 3σ detection.

Trigger No.	Date	Class	Zenith Angle	Δt (secs)	Total Counts	Sensitivity (3σ)
629	5-Aug-91	M1.2	45°	75	252,000	40%
688	16-Aug-91	C9.6	18°	75	694,000	35%
1032	9-Nov-91	M1.4	43°	100	975,000	20%

dominated by several systematic effects which preclude any definitive measurements. These effects include: 1) differences in channel bandwidths amongst the LAD detectors; 2) deadtime effects; 3) leakage of incident flux through the spacecraft; and 4) the need to account for the precise form of the incident spectrum. The ongoing analysis is concentrating on the effort to remove these systematics effects. We are also in the process of setting up our simulation database in a form that can be directly used by software tools developed by the BATSE team. This should greatly facilitate the search for solar flare polarization in the BATSE database.

ACKNOWLEDGEMENTS

This work is supported at UNH by NASA Grant NAG 5-2388 and by NASA contract NAS 5-26645.

REFERENCES

1. Vestrand, W. T., et al., Ap. J. **322**, 1010 (1987).
2. Bogovlov, S.V., et al., Soviet Astr. Letters, **11**, 322 (1985).
3. Catalano, C.P. and Van Allen, J.A., Ap. J., **185**, 335 (1973).
4. Kane, S.R., et al., Ap. J. Lett., **254**, L53 (1982).
5. Vestrand, W. T. and Ghosh, A., 20th ICRC, **3**, 57 (1987)
6. Mctiernan, and Petrosian, V., Ap. J., **359**, 541 (1990).
7. Petrosian, V., Ap. J., **299**, 987 (1985).
8. Tindo, I.P., et al., Sol. Phys., **14**, 204 (1970).
9. Tindo, I.P., et al., Sol. Phys., **27**, 426 (1972).
10. Brown, J.C., McClymont, A.N. and McLean, I.S., Nature, **247**, 448 (1974)
11. Nakada, M.P., Neupert, W.M. and Thomas, R.J., Sol. Phys., **37**, 429 (1974).
12. Tindo, I.P., Shuryghin, A.I. and Steffen, W., Sol. Phys., **46**, 219 (1976).
13. Bai, T. and Ramaty, R., Ap. J., **219**, 705-726 (1978).
14. Tramiel, L.J., Channan, G.A. and Novick, R., Ap. J., **280**, 440 (1984).
15. Chanan, G.A., Emslie, A.G. and Novick, R., Sol. Phys., **118**, 309 (1988).
16. Emslie, A.G. and Vlahos, L., Ap. J., **242**, 359 (1980).
17. Leach, J. and Petrosian, V., Ap. J. Lett., **269**, 715 (1983).
18. Evans, R.D. The Atomic Nucleus, New York: McGraw-Hill Book Co. (1955).

RADIO EMISSION

Submillimeter/IR Solar Bursts from High Energy Electrons

Pierre Kaufmann

CRAAE (Centro de Rádio-Astronomia e Aplicações Espaciais) and NUCATE, State University of Campinas (UNICAMP), Cidade Universitária Zeferino Vaz, B. Geraldo, CEP 13083-592, Campinas, SP, Brazil.

Abstract. There are various observational results which suggest that certain solar bursts have spectral components extending into the submm-w and infrared range. This is essentially unexplored for burst emissions in the continuum. The short time scales of burst structures (tens of milliseconds) correlated to hard X-rays bring serious constraints for interpretation. One possibility is to assume the impulsive phase made of explosions of compact synchrotron sources and self inverse-Compton reducing the electrons energy, producing the correlated spiky hard X-ray component. A new project for a solar submm-w telescope is now being developed in a collaboration between groups in Brazil, Switzerland and Argentina. It is expected to bring important observational clues for the understanding of the initial phase of flares.

THE UNEXPLORED "WINDOW"

The entire submm-IR "window" of solar emissions in the electromagnetic spectrum is essentially unexplored. A number of observations were carried out on the quiet Sun (1-3) and on quiescent regions associated to sunspots (3,4). No flare emissions in the continuum are known. One attempt to measure flares in the far IR was inconclusive (5). There was one qualitative indication of significant emission from a small flare at a submm-wave band (6), which origin was interpreted as being non-thermal (7).

INDICATIONS FROM BURSTS SPECTRAL TRENDS AT µW-MM WAVES

There are several examples of radio emission spectra from solar bursts exhibiting fluxes increasing for shorter mm-waves, suggesting the existence of a component extending into the IR. One early example obtained in 1961 by Hachenberg and Wallis (8) is shown in Figure 1. It indicates a turnover frequency f_m larger than 30 GHz.

This event was initially assumed to have a broadband spectral maximum, interpreted as the result of free-free absorption on gyrosynchrotron radiation (9). Hachenberg and Wallis (8) have suggested that the observed spectra was a composition of several synchrotron emissions with different turnover frequencies, as shown in Figure 1. We also show in Figure 1 synchrotron radiation spectra extending into the IR and optical ranges (10,11), which were primarily intended to explain the highly energetic flare emission in the white light.

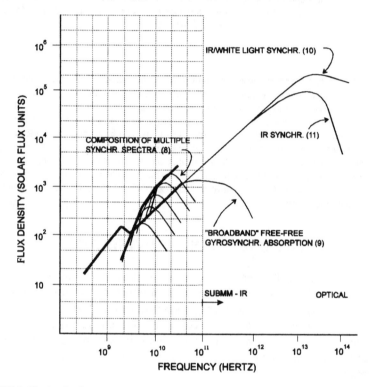

FIGURE 1. First solar burst microwave spectra showing fluxes increasing for frequencies greater than 30 GHz (darker lines) (8). First proposed model spectra are also shown and discussed in the text.

Later it was found that these spectral trends are not unusual at all. In Figure 2 we show many examples of solar burst spectra, several of which extend up to 10^{11} Hz (or 3 mm wavelength).

Fluxes range from few solar fluxes units (1 s.f.u. $= 10^{-22}$ w.m^{-2}.Hz^{-1}), to 10^4-10^5 s.f.u. for the very large events. The great bursts shall obviously have a significant emission contribution in the submm-wave part of the e.m. spectrum, the as exceptionally large event (12) which has a turnover frequency somewhere between 30 and 80 GHz. This would also be the case for the events exhibiting flat spectra (13,14).

Very remarkable are the burst spectra showing larger fluxes for higher frequencies, with minimum or undetectable emissions in the microwaves, as the examples obtained by Shimabukuro (15) and Kaufmann *et al.* (16), with microwaves fluxes below few s.f.u. (i.e. below sensitivities of the solar patrol telescopes). Other spectra indicate one maxima at microwaves, followed by a second rise at mm-waves, suggesting composition of two synchrotron processes.

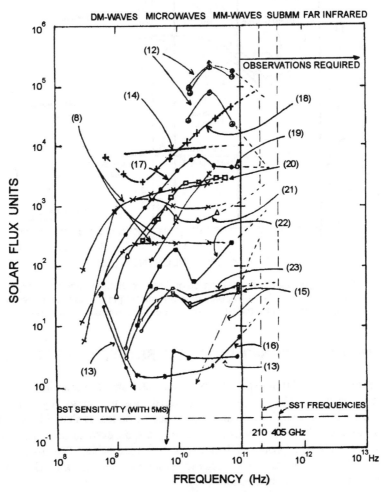

FIGURE 2. Examples of solar bursts observed spectra, indicating the importance of emission expected in the submm-w to far IR range. Dashed vertical lines indicate the operating frequencies of the solar submm-wave telescope project (SST), 210 and 405 GHz, and its sensitivity for 5 ms time constant (horizontal line).

In Figure 3 we show a simplified description of flare continuum emissions from metric-radio to gamma rays wavelengths (24). Various emission models, thermal and non-thermal, may fit into the sub-mm to IR observational gap.

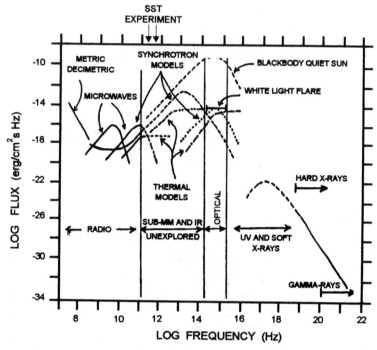

FIGURE 3. Simplified diagram showing solar flare continuum emission for the entire electromagnetic spectrum. Observational data are missing in the submm-IR region. They might bring important clues for the understanding of flare emission processes.

CONSTRAINTS SET BY BURST SHORT TIME-SCALES

The non-thermal bursts spectral extension into the IR range imply in highly energetic electrons (10^6-10^7 eV) accelerated, losing energy by synchrotron mechanism. The time scales for the purely synchrotron losses should be larger than tens of seconds. Since the bursts time scales were believed to be in the range of 10^2-10^3 seconds, the proposed hypothesis was reasonable. The assumption of bremsstrahlung losses at denser regions producing X-rays was also effective enough to explain early observations (25).

However, it has been found that the burst time scales become much shorter as the sensitivity and time resolution of detection devices improve. Based on OSO-V hard X-ray observations, Frost (26) proposed that solar bursts were discontinuous in time, suggesting structures of seconds or less. Similar time structures at hard

X-ray bursts, found by TDRS-1 satellite experiment led to the proposition of the concept of "elementary flare bursts" (27,28). Fast subsecond time structures were found for bursts at metric and decimetric wavelengths (29-31), and microwaves (32). New hard X-ray experiments have suggested the existence of subsecond time scales in solar bursts (33,34). A number of solar events were observed simultaneously at hard X-rays and short microwaves (35,16).

The remarkable event that occurred on 21 May 1984 (16), mentioned before, exhibited radio intensities increasing with frequencies. Very fast time structures at 94 GHz (also observed at 30 GHz) were correlated to hard X-ray emission. The event is shown in Fig. 4. Taking the principal time structure, it has seven well-defined spikes in one second. This structure is correlated in time to X-ray emission in the limit of the resolution, i.e. better than 128 ms.

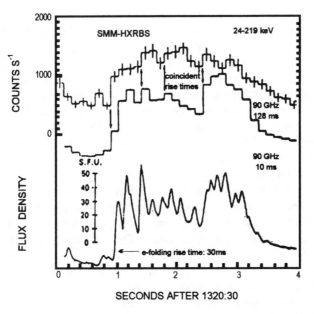

FIGURE 4. The solar burst of 21 May 1984 shows a spectrum with emissions only at mm-waves, and very fast time structures (< 100 ms) correlated to hard X-rays (36).

The fast time scales of bursts structures observed at mm-waves cannot be explained by synchrotron losses. It can be shown that for time scales < 0.1 sec, one should need 10^4 MeV electrons producing emission with a spectral turnover frequency in the X-rays range, which is impossible. On the other hand, the synchrotron emission lifetime could be suppressed by another mechanism, such as bremsstrahlung at denser regions in the solar atmosphere. However, the subsecond time scales for bremsstrahlung losses would require unrealistically high densities

($> 10^{15}$ cm^{-3}). Furthermore, if this was the case no mm-wave radiation could escape from such high density medium, and it would be difficult to explain the close time correlation of the spiky structures at mm-w and hard X-rays.

Finally, the fast subsecond time structures in solar hard X-ray bursts seem to be quite common, according to more recent observations obtained with more sensitive detectors: BATSE on GRO (37) and Phebus on GRANAT (38). These fast spikes superimposed to the burst time profiles may be interpreted in conjunction to their counterpart at the submm-IR wavelengths accordingly to the model described below.

SYNCHROTRON/INVERSE-COMPTON MODEL

In order to reconcile the observational evidences of continuum burst emissions at mm-waves with possible extension into the submm-IR range; the fast subsecond time structures, and the time correlated hard X-ray spikes, we have proposed a self-consistent model which assumes the acceleration of ultrarelativistic electrons (10-10^2 MeV) which energies are reduced by inverse Compton scattering on the synchrotron photons, following the basic approach suggested elsewhere (39).

The idea of inverse Compton action in solar flares to produce X-rays was discussed in the past by several authors (11,40-43). This mechanism, however, was believed to be less efficient compared to collisional losses to produce X-rays, for the long time scales then assumed as typical for bursts (10-10^3 sec), much larger than the subsecond sales found later.

We propose that the fast time structures superimposed to the hard X-rays bursts might be the result of inverse Compton scattering of ultrarelativistic electrons on the photons at the exploding compact synchrotron source which was formed at the first energy conversion phase of the event. The synchrotron emission lifetime is thus suddenly suppressed as the electrons' energy is reduced by the effect. The concept is in fact similar to those used to explain the (relatively) rapid explosive events observed in extragalactic sources (44-46).

In Figure 5 we suggest a qualitative sequence of steps leading to a flare. The "pre-impulsive" explosive source consists in a compact synchrotron source, with 10-10^2 MeV electrons producing an emission spectrum with maximum somewhere in the submm-IR. A substantial fraction of these electrons are de-energized by inverse Compton action on the radiation field made up by the synchrotron photons. The submm-IR time structure is thus shortened and a hard X-ray spike is produced at the same time. The electrons, which energies were reduced to mildly relativistic values, emit in sequence by girosynchrotron mechanism, with maximum in the microwave range, and produce most of the X-rays by bremsstrahlung at the footpoints.

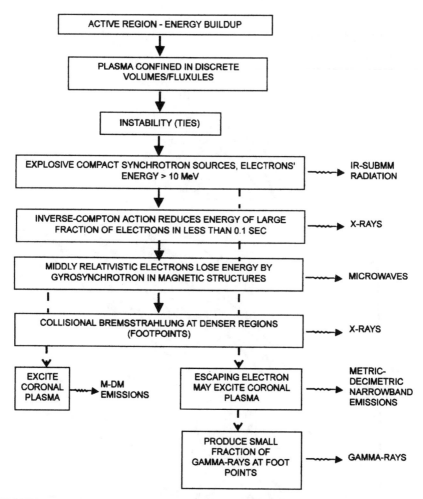

FIGURE 5. A suggested flow diagram showing different and successive phases of a solar flare, with explosive compact synchrotron sources at the beginning, previous to impulsive phase.

In the magnetically complex solar active regions, one would expect to have multiple, pre-impulsive explosive compact synchrotron sources flashing at different times, building up what is usually described as the onset of the impulsive phase of the burst, as described in Figure 6.

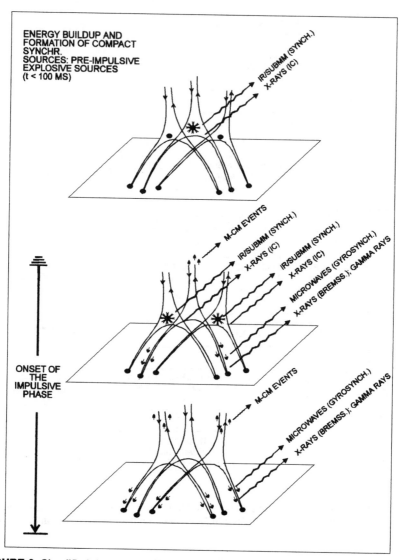

FIGURE 6. Simplified description of an active region, with several explosive synchrotron sources occurring at several sites, originating the solar burst impulsive phase.

This description may fit well to the Masuda *et al.* (47) "loop-top" X-ray event observed by Yohkoh soft- and hard X-ray experiments, which simplified image is shown in Figure 7. Our proposed description is superimposed to it, with the top of the loop hard X-ray source emission suggested to be the result of inverse Compton effect on the pre-impulsive explosive synchrotron sources.

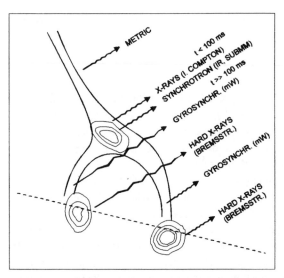

FIGURE 7. At least some of the hard X-ray flare emissions at the top of a loop may be produced by inverse Compton scattering, while the remaining. X-rays are produced by collisions at the footpoints. We overlay our suggested two-phase model to the X-ray images obtained by Yohkoh satellite (47).

A NUMERICAL APPLICATION

The production of compact synchrotron sources at the top of loops, or somewhere higher in the active regions' magnetospheres bring critical conditions for plasma confinement which must be further investigated.

Assuming that the above described sources are formed, the self absorbed compact synchrotron source sizes can be related to the observed flux density S at a frequency $f < f_m$ (the turnover frequency) as follows (48-50).

$$\ell \leq 10^{15.5} \ S^{0.5} \ B^{0.25} \ f^{-1.25} \ d \ cm, \tag{1}$$

where B is the magnetic field, d the distance to the Sun. For the relatively small event shown in Figure 4, $S \sim 10^{-17.2}$ erg/cm^2 s Hz at $f = 10^{11}$ Hz, for B = 500 Gauss we obtain $\ell \leq 10^{6.8}$ cm.

Attributing the observed spikes time scales to the time required by the inverse Compton effect to reduce the electrons energies (39)

$$t_c \approx 10^{51} \ B \ f_m^{-4} \ s, \tag{2}$$

for $t_c \approx 0.1$ s we obtain $f_m \approx 10^{13.4}$ Hz.

The turnover frequency f_m can also be equated to the source size, as derived from Ramaty (51):

$$\ell \approx 10^{15.6} \, f_m^{-1.25} \, [\, S(f = f_m)]^{0.5} \, d \, B^{0.3} , \qquad (3)$$

where $S(f = f_m)$ is the predicted flux at the maximum of the synchrotron spectrum ($\approx 4 \times 10^{-11.2}$ ergs cm^{-2} Hz^{-1} for the example shown in Figure 4). We obtain $\ell \approx 10^{7.2}$ cm, which is consistent to the value found using equation (1).

The total number and density of particles in one source might be evaluated making some assumptions relative to the energy losses with time. The observed synchrotron luminosity is dominated by the flux at the spectral maximum frequency: $S(f = f_m) \times 4\pi d^2$ (ergs^{-1}), where d is the distance of the Sun to the Earth. The peak flux with time for a fast spike is measured at a time t_p such as $t_o < t_p < t_c$, where t_o is the start-time of the observed pulse, and t_c is the time of the end. During this $t_c - t_o$ interval, f_m is expected to shift rapidly, from the IR range into the short microwave range, due to the electrons de-energization by inverse-Compton action. Assuming a linear decay in the electrons' energy (48):

$$E_t \approx 10^{2.7} \, (f_m/B)^{0.5} \qquad (4)$$

from its initial value of $t = t_o$, of about 112 MeV (with $f_m = 10^{13.4}$ Hz, B = 500 gauss), to its value at $t = t_c$, of about 3 MeV (with $f_m = 10^{10.25}$ Hz, B = 500 gauss), at $t = t_p$ the measured flux was due to electrons with energies of about 54 MeV, producing a synchrotron spectrum turnover at about $10^{12.8}$ Hz. We obtain a luminosity of $10^{27.6}$ erg/s which divided by the energy loss of a single electron (52):

$$(dE/dt) = 10^{-14.3} \, B^2 [E(MeV)]^2 \text{ erg/s} \qquad (5)$$

provide an estimate of the total number of particles in one compact source, i.e., $N_t \approx 10^{33}$ particles, which is comparable to the total number of particles in smaller events necessary to explain the observed emissions at microwaves and hard X-rays.

The total energy and the particle density in one source will be time dependent. However, these estimates bring results which are difficult to interpret. For a source size $\ell \approx 10^7$cm, we get a density of about 10^{12} cm^{-3}. For 54 MeV electrons, the total energy at $t = t_p$ is of about $10^{28.9}$ erg, which is 1-2 orders of magnitude larger than energies usually attributed to small or elementary events. At the initial time, $t = t_o$, for electrons' energies ≈ 112 MeV, the spike's energy becomes $10^{29.2}$ ergs. These problems should be more adequately investigated assuming more realistic situations, for example taking into account that the inverse-Compton effect might be effective only in the inner fraction of the compact pre-impulsive synchrotron

THE SST PROJECT

The flare diagnostics in the submm-IR range is therefore highly desirable for the understanding of the primary acceleration processes in flares and burst energetics. To explore this observational gap, it was recently approved the Submm-w Solar Telescope project (SST) with principal resources provided by Brazil's São Paulo State Foundation for Research, FAPESP, complemented by the Swiss solar group from the Bern University's Institute for Applied Physics, and the Argentina space science group from Buenos Aires' Instituto de Astronomia y Fisica del Espacio (IAFE) and from San Juan's Complejo Astronômico El Leoncito (CASLEO) in which site, at Argentinean Andes, the SST will be installed and operated. A preliminary description of the SST was published elsewhere (53).

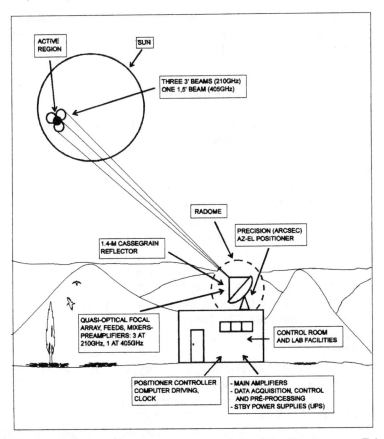

FIGURE 8. A simplified description of the new Solar Submillimeter-wave Telescope (SST project).

The SST Project consists in a 1.5-m cassegrain reflector with surface accuracy better than 30 microns. A cluster of 3 (or 4) front-end horns/radiometers operating at 210 GHz produce partially overlapping beams. The correlated output from the beams allows the determination of the positions of the burst centroid of emission within few arcseconds accuracy, similarly to what has been done at 48 GHz using the Itapetinga 13.7-m antenna (54-56). One 405 GHz will provide a second spectral information in the submm-wave range. The preliminary and simplified description of whole assembly is shown in Figure 8. A complete description of SST will be published in future.

ACKNOWLEDGMENTS

Our participation of this Workshop was partially supported by USRA's Visiting Scientist Program. CRAAE is a joint Center by Agreement between Universities USP, Mackenzie, UNICAMP and the space institute INPE.

REFERENCES

1. Eddy, J.A., Léna, P.J., and MacQueen, R.M., *Solar Phys.* **10**, 330 (1969).
2. Gezari, D.Y., Joyce, R.R., and Simon, M., *Astron. Astrophys.* **26**, 409 (1973).
3. Lindsey, C., Hildebrand, R.H., Keene, J., and Whitecomb, S.E., *Astrophys. J.* **248**, 830 (1981).
4. Degiacomi, K., Kneubühl, F.K., Huguenin, D., and Müller, E.A., *Int. J. Infrared and MM-Waves* **5**, 643 (1984).
5. Hudson, H.S., *Solar Phys.* **45**, 69 (1975).
6. Clark, C.D., and Park, W.M., *Nature* **219**, 922 (1968).
7. Beckman, J.E., *Nature* **220**, 52 (1968).
8. Hachenberg, O., and Wallis, G., *Zeit. für Astrophys.* **52**, 42 (1961).
9. Ramaty, R., and Petrosian, V., *Astrophys. J.* **178**, 241 (1972).
10. Stein, W.A., and Ney, E.P., *J. Geophys. Res.* **68**, 65 (1963).
11. Shklovsky, J., *Nature* **202**, 275 (1964).
12. Ramaty, R., Schwartz, R.A., Enome, S., and Nakajima, H., *Astrophys. J.* **436**, 941 (1994).
13. White, S.M., Kundu, M.R., Bastian, T.S., Gary, D.E., Hurford, G.J., Kucera, T., and Bieging, J.H., *Astrophys. J.* **384**, 656 (1992).
14. Lee, J.W., Gary, D.E., and Zirin, H., *Solar Phys.* **152**, 409 (1994).
15. Shimabukuro, F.I., *Solar Phys.*, **15**, 424 (1970).
16. Kaufmann, P., Correia, E., Costa, J.E.R., Zodi Vaz, A.M., and Dennis, B.R., *Nature* **313**, 380 (1985).
17. Croom, D.L., *Solar Phys.* **15**, 414 (1970).

16. Kaufmann, P., Correia, E., Costa, J.E.R., Zodi Vaz, A.M., and Dennis, B.R., *Nature* **313**, 380 (1985).
17. Croom, D.L., *Solar Phys.* **15**, 414 (1970).
18. Croom, D.L. "Solar Millimeter-Wave Bursts", in *High Energy Phenomena on the Sun* (ed. by R. Ramaty and R.G. Stone), NASA/GSFC Symposium, Sept. 28-30, 1972, *Publ. NASA* SP-342, 1973, 114.
19. Cogdell, J.R., *Solar Phys.* **22**, 147 (1972).
20. Chertok, I.M., Fomichev, V.V., Gorgutsa, R.V., Hildebrandt, J., Krüger, A., Magun, A., and Zaitsev, V.V. "Solar Radio Bursts with a Spectra Flattening at Millimeter Wavelengths", presented at *CESRA Workshop on Coronal Magnetic Energy Release*, Postdam, Germany, 16-20 May, 1994.
21. Zirin, H., and Tanaka, K., *Solar Phys.* **32**, 173 (1973).
22. Akabane, K., Nakajima, H., Ohki, M., Moriyama, F., Miyaji, T., *Solar Phys* **33**, 431 (1973).
23. Shimabukuro, F.I., *Solar Phys.*, **23**, 169 (1972).
24. Kaufmann, P., *Adv. Space Res.* **8**(11), 39 (1988).
25. Acton, L.W., *Nature* **204**, 64. (1964).
26. Frost, K.J., *Astrophys. J.* **158**, L159 (1969).
27. van Beek, H.F., de Feiter, L., and de Jager, C., *Space Res.* XIV, 447 (1974).
28. de Jager, C., and de Jonge, G., *Solar Phys.* **58**, 127 (1978).
29. Dröge, F., *URSI Kleinheubacher Berichte* **19**, 351 (1975).
30. Dröge, F., *Astron. Astrophys.* **57**, 285 (1977).
31. Slottje, C., *Nature* **275**, 520 (1978).
32. Kaufmann, P., Strauss, F.M., Opher, R., and Laporte, C., *Astron. Astrophys.* **87**, 58 (1980).
33. Kiplinger, A.L., Dennis, B.R., Emslie, A.G., Frost, K.J., Orwig, L.E., *Astrophys. J.* **265**, 199 (1983).
34. Hurley, K., Niel, M., Talon, R., Estulin, I.V., and Dolidze, V.C., *Astrophys. J.* **265**, 1076 (1983).
35. Takakura, T., Kaufmann, P., Costa, J.E.R., Degaonkar, S.S., Ohki, K., and Nitta, N., *Nature* **302**, 317 (1983).
36. Costa, J.E.R., and Kaufmann, P., *Solar Phys.* **104**, 253 (1986).
37. Machado, M.E., Ong, K.K., Emslie, A.G., Fishman, G.J., Meegan, C., Wilson, R., and Paciesas, W.S., *Adv. Space Res.*. **13**, 9/175 (1993).
38. Vilmer, N., Trottet, G., Barat, C., Dezalay, J.P., Talon, R., Sunyaev, R., Terekhov, O., and Kuznetsov, A., *Space Sci. Rev.* **68**, 233 (1994).
39. Kaufmann, P., Correia, E., Costa, J.E.R., and Zodi Vaz, A.M., *Astron. Astrophys.* **157**, 11 (1986).
40. Gordon, I.M., *Soviet Astron.* **4**, 873 (1960).
41. Zheleznyakov, V.V., *Soviet Phys. Astron.* **9**, 73 (1965).
42. Korchak, A.A., *Solar Phys.* **18**, 284 (1971).
43. Brown, J.C, *Phil. Trans. R. Soc. Lond.* **A291**, 473 (1976).

44. Kellerman, K.I., and Pauliny-Toth, I.I.K., *Astrophys. J.* **155**, L71 (1969).

45. Jones, T.W., D'Dell, S.L., and Stein, W.A., *Astrophys. J.* **192**, 261 (1974).

46. Dent, W.A., O'Dea, C.P., Balonek, T.J., Hobbs, R.W., and Howard, R.J., *Nature* **306**, 41 (1983).

47. Masuda, S., Kosugi, T., Hara, H., Tsuneta, S., and Ogawara, Y., *Nature* **371**, 495 (1994).

48. Ginzburg, V.L., and Syrovatskii, S.I., *An. Rev. Astron. Astrophys.* **3**, 297 (1965).

49. Pacholczyk, A.G., *Radio Astrophysics*, Freeman, San Francisco, USA, 1970, chap. 3

50. Tucker, W.H., *Radiation Processes in Astrophysics*, MIT Press, USA, 1975, chaps. 3 and 4.

51. Ramaty, R. "Theory of Solar Microwave Bursts", in *High Energy Phenomena in the Sun* (ed. by R. Ramaty and R.G. Stone), NASA/GSFC Symposium, Sept. 28-30, 1972, *Publ. NASA* SP-342, 1973, 188.

52. Wild, J.P., Smerd, S.F., and Weiss, A.A., *Ann. Rev. Astron. Astrophys.* **1**, 291 (1963).

53. Kaufmann, P., Parada, N.J., Magun, A., Rovira, M., Ghielmetti, and H., Levato, H. "The New Solar Submillimeter-wave Telescope Project (SST)", in *Proc. Kofu Symposium on "New Look at the Sun"*, 6-10 Sept. 1993 (ed. by S. Enome and T. Hirayama), Nobeyama Radio Observatory, 1994, 323.

54. Georges, C.B., Schaal, R.E., Costa, J.E.R., Kaufmann, P., and Magun, A. "50 GHz multi-beam receiver for radio astronomy", in *Proc. 2nd Int. Microwave Symp.*, Rio de Janeiro, 1989, p. 447.

55. Herrmann, R., Magun, A., Costa, J.E.R., Correia, E., and Kaufmann, P., *Solar Phys.* **142**, 157 (1992).

56. Costa, J.E.R., Correia, E., Kaufmann, P., Magun, A., and Herrmann, R., *Solar Phys.* **159**, 157 (1995).

Microwave and Hard X-Ray Sources in Two X-Class Limb Flares

Hiroshi Nakajima* and Thomas R. Metcalf[†]

*Nobeyama Radio Observatory, Nobeyama, Minamisaku,
Nagano 384-13, Japan
[†]Institute for Astronomy, University of Hawaii,
2680 Woodlawn Dr., Honolulu, HI 96822, USA

We have analysed two intense, extended microwave and hard X-ray flares which occurred slightly behind the west limb and were accompanied by X-class, long-duration events (LDEs) in soft X-rays. We have found that: (1) Both events have typical soft X-ray properties of LDEs. (2) A major hard X-ray source at 23-33 keV and 33-53 keV is located in the high-temperature region, considerably higher than the corresponding soft X-ray loop, while a 17 GHz microwave source, where electrons at energy $\geq$ several hundreds keV mainly contribute to the 17 GHz emission, nearly coincides with the soft X-ray loop in extent. (3) The electron spectrum, which is derived from the microwave spectrum, is significantly harder than the electron spectrum at energy $\leq$ 100 keV, which is derived from the hard X-ray data. (4) The time profile of total intensity at 17 GHz is delayed by about 25 s with respect to that of the hard X-rays for one of the events.

The above results show evidence that lower-energy ($\leq$ 100 keV) electrons with a softer spectrum are accelerated at the top of the cusp, and that higher-energy ($\gg$ 100 keV) electrons are accelerated at the top of the soft X-ray loop and trapped. Probably, the lower-energy electrons accelerated in the cusp are transferred by a reconnection outflow into the second acceleration region at the top of the soft X-ray loop, and further accelerated to higher energy.

INTRODUCTION

Recently, observations with the Yohkoh Soft X-Ray Telescope (SXT) and Hard X-Ray Telescope (HXT) gave us new information on energy release and particle accleration in solar flares. SXT observations showed that in the late phase of the Long Duration Events (LDEs), a cusp shaped structure appears at the top of a soft X-ray bright loop and that the cusp part, especially its outer boundary portion, has a higher temperature than the other parts of the loop. This is considered to be strong evidence of magnetic reconnection going on at a neutral sheet at the top of the loop (Tsuneta et al. 1993). HXT observations showed that in the impulsive phase of several flares, a loop-top source appears at or above the top of the soft X-ray loop, in addition to the dominant double footpoint sources (Masuda, et al. 1994). This suggests that the loop-top source represents the site where the fast downflow from a reconnection point collides with an underlying closed magnetic loop and forms a shock where electrons are accelerated through transfer of ion kinetic energy to electrons.

In this paper, we present results of microwave and hard X- ray observations of two intense microwave and hard X-ray flares which occurred on the west limb on 1992 June 28 and November 2, using the Nobeyama Radioheliograph images at 17GHz and the Yohkoh HXT and SXT images. Both of these events are so-called extended flares which are characterized by a longer duration ($\geq$ 10 min) and a more gradual temporal variation of microwave and hard X-ray emission as compared with impulsive flares. They are also LDEs from the viewpoint of GOES soft X-ray classification.

DATA ANALYSIS

Both events on June 28, 1992 and November 2, 1992 occurred slightly beyond the west limb at N11, W103 and S23, W95, respectively. The derived longitudes of the flare sites are estimated for the June 28 event from the records of the H- alpha flare location on the visible disk and given for the November 2 event by Feldman et al. (1995) . Both events, respectively, have a duration lasting over 5 hr and over 10 hr, and X1.8 and X9 in the GOES classification. The total fluxes at microwaves and millimeter-waves are quite intense (17 GHz flux density: 3080 sfu for the June 28 event and 30400 sfu for the November 28 event).

The 1992 June 28 Event

The microwave flare can be divided into three phases, i.e., an initial phase during 04:41 UT-0450:30 UT, a main phase with a peak time around 04505:30 UT, and a decay phase as shown in the 17 GHz time profile of Figure 1.

The soft X-ray full-resolution partial frame image is available only in the decay phase and consists of two separate sources, as shown in Figure 2, which later develop into well-defined double arcades. This double arcade system was also observed by H-alpha observations with the Solar Flare Telescope at Mitaka (Courtesy of T. Sakurai). Electron temperature derived using the filter ratio method is slightly less than 10^7 K in the brightest part of the soft X-ray source which is identified as the top of the arcade system, and is more than 1.6×10^7 K above the brightest top of the arcade. The high-temperature region above the arcade top is usually observed in a cusp surrounding the top of a soft X-ray arcade of a LDE and is interpreted as location of hot plasma heated at a slow shock located at the top of the cusp (Tsuneta et al. 1992).

Keeping the soft X-ray structures mentioned above in mind, we make comparisons of microwave, hard X-ray, and soft x-ray images, as shown in Figure

3. Our analyses of microwave and hard X-ray data are mainly concentrated in the initial phase, because no hard X-ray data is available in the main phase. The microwave and hard X-ray images are taken around 0448 UT late in the initial phase, and the soft X-ray image is taken at 0534:48 UT in the decay phase. The hard X-ray image are synthesized by the pixon method (Metcalf et al. 1995). We can clearly see that the dominant source in the hard X-ray M1 (23-33 keV) channel image is located high in the corona, considerably higher than the brightest top of the soft X-ray arcade, and that the microwave source roughly coincides with the soft X-ray arcade in extent. Hereafter, we call this the dominant hard X-ray source as the cusp source. The apparent height of the cusp source is about 40 arcsec or 2.9×10^4 km above the photosphere, while the apparent height of the brightest top of the soft X-ray arcade is about 12 arcsec or 0.9×10^4 km. Therefore, the distance between the high-coronal source and the brightest top of the soft X-ray arcade is about 2.0×10^4 km. If a correction is made for occultation due to the solar limb, the real height of the brightest top of the soft X-ray arcade is estimated to be 2.0×10^4. Although we can also see hard X-ray footpoint sources, they are much weaker than the cusp hard X-ray source.

The location of the microwave source essentially coincides with one or both of the soft X-ray double arcades throughout the initial and main phases. Initially, the microwave source appeared coinciding with the southern part of the soft X-ray arcades. From about 0441 UT, the microwave source expanded to the northern one of the soft X-ray double arcades with increasing brightness, resulting in an elongated, double-source structure similar to the soft X-ray double arcades in the decay phase. The microwave emission in the main phase mainly emanates from the southern part of the soft X-ray double arcades. The peak brightness temperature and polarization degree of the microwave source are, respectively, 8.0×10^5 and about 28 % in the initial phase (0446 UT). At 0449 UT, the microwave spectrum has a turn-over frequency of about 4 GHz and a spectrum index in the optically-thin part of -2.7 which is derived from flux density at 9.4 and 17 GHz. The power-law index of electrons is derived to be -4.4 using the equations given by Dulk (1985). From the electron spectrum index of -4.4 and rather weak magnetic field presumed for the microwave source, we infer that electrons $\geq$ several hundreds keV mainly contribute to the microwave source at 17 GHz.

The hard X-ray spectrum has a power-law index of -6.2 which is estimated from M1 and M2 channels at 0449 UT. The power-law spectrum index of electrons is, thus, derived to be -5.7 by adopting the thin target model. This power-law index is slightly softer that derived from the microwave spectrum.

Figure 4 shows time profiles of hard X-ray and microwave total emission in the initial phase. Both microwave and hard X-ray time profiles are gradual and similar in overall structures, but, surprisingly, the time profile of the 17 GHz emission is delayed not only in onset time but also in overall structure, by about 25 s with respect to those of the M1 and M2 channel hard X-ray emission. Since the dominant hard X-ray emission mainly emanates from the cusp source and the 17 GHz emission emanates from the low-lying closed loops indicated by the soft X-ray double arcades, the observed time delay is due to delay of the time evolution of several hundreds keV electrons in the closed loops with respect to that of several tens keV electrons in the cusp region.

The 1992 November 2 Event

The microwave flare can be divided into three phases, i.e., an initial phase before 0249 UT, a main phase which consists of three peaks around 0250 UT, 0254 UT, and 0257.6 UT, and a decay phase, as shown in Figure 5. In this event, the yohkoh hard and soft X-ray data are available from just after the third peak in the main phase.

Detailed descriptions of the soft X-ray flare are given by Feldman, et al. (1995). Here, we will briefly mention the soft X-ray properties of the November 2 flare. The soft X-ray emission originates from a loop with a projection to the south-west (see left panel in Figure 6). The top of the loop continues to dominate in brightness. The size of the loop increases considerably and the rate of height increase is 9 km/s. The temperature distribution obtained using the filter ratio method shows that the temperature at the loop top is 1.4×10^7 K at 0305 UT, and that the higher temperature is in fainter regions surrounding the bright regions. The inner side of the projection corresponds to a slightly lower temperature region, which is a typical temperature profile of the cusp.

Figure 6 shows overlays of the microwave (0300 UT), hard X-ray (0300 UT), and the soft X-ray (0305:04 UT) images. Here, the hard X-ray image is synthesized by the pixon method (Metcalf et al. 1995). The hard X-ray sources are located at both sides of the projection, well above the brightest top of the soft X-ray loop, while the microwave source is located nearly at the same portion as the soft X-ray loop. This is almost the same situation as seen in the June 28 event. The apparent height of the hard X-ray source is 3.7×10^4 km above the photosphere and 1.7×10^4 km above the center of the brightest top of the soft X-ray loop.

The microwave spectrum at 0300 UT has a turn-over frequency of about 20 GHz and a spectrum index in the optically-thin part of -1.76 which is derived from flux density at 35 GHz and 80 GHz. The power-law index of electrons is estimated to be -3.3.

The hard X-ray spectrum has a power-law index of -7.2 which is estimated from the HXT M1 and M2 channel data at 0300 UT. The power-law index of electrons is, thus, derived to be -6.7 by adopting the thin target model. This power-law index is considerably softer than that derived from the microwave spectrum.

SUMMARY AND CONCLUSION

The main observational facts from study of two long duration events accompanied by extended emission in microwaves and hard X-rays can be summarized as follows.

1. Both events have typical soft X-ray properties of LDEs as development of a well-defined arcade system and a high-temperature region outside of the soft X-ray bright loop, as evidence of magnetic reconnection going on above the soft X-ray loop (arcade).

2. The spectrum of high-energy ($\geq$ several hundreds keV) electrons derived from the microwave spectrum is significantly harder than that of lower-energy (several tens keV) electrons derived from the hard X-ray spectrum.

3. The major hard X-ray source is located in the high-temperature region, considerably higher than the corresponding soft x-ray loop. This suggests that the lower-energy electrons with the softer electron spectrum are accelerated in the upper part of the cusp.

4. The microwave source nearly coincides with the corresponding soft X-ray bright loop in extent. This suggests that the higher-energy electrons with the harder spectrum are accelerated and trapped in the closed loop indicated by the soft X-ray bright loop.

5. From the time delay, about 25 s, of the 17 GHz time profile with the 23-33 keV and 33-53 keV hard X-ray time profiles for the June 28 event, we can derive an apparent velocity, 1000 km/s, of the agent to trigger electrons acceleration at both sites. A possible candidate of the agent is the reconnection flow from a magnetic neutral sheet.

6. The microwave and hard X-ray time profiles are similar to each other, though the microwave and hard X-ray emission emanate from the different sites and have different electron spectra. This must be explained by any proposed model.

Based upon the above observational results, we propose a possible scenario of electron acceleration. Electrons are firstly accelerated to about 100 keV in the upper part of a cusp possibly through shock acceleration at a slow-shock or VxB acceleration around a magnetic neutral sheet. Then, a part of the accelerated lower-energy electrons are transferred to the top of a soft X-ray loop (arcade) by a fast reconnection flow, and are subsequently accelerated to higher energy at a fast shock and trapped in the soft X-ray loop.

REFERENCES

1. U. Feldman, et al., Ap. J.,**446**, 860 (1995).
2. G. A. Dulk, Ann. Rev. Astron. Astrophys. **23**, 169 (1985).
3. S. Masuda, T. Kosugi, S. Tsuneta, and Y. Ogawara, Nature, **371**, 495 (1994).
4. T. R. Metcalf, H. S. Hudson, T. Kosugi, R. C. Puetter, and R. K. Pina, Ap. J., submitted.
5. S. Tsuneta, et al., Publ. Astron. Soc. Japan **44**, L63 (1992).

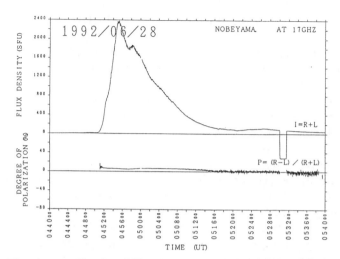

FIG. 1. The time profile of 17 GHz microwave emission for the 1992 June 28 event

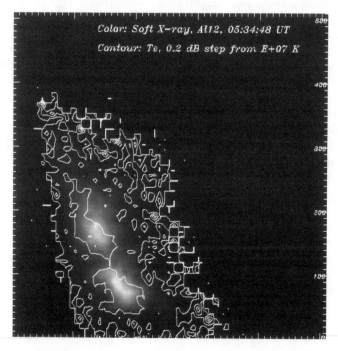

FIG. 2. A Yohkoh SXT Al-filter (12 μm) image of the 1992 June 28 event (0534:48 UT) with electrom temperature (contours) superposed. The temperature is in steps of 0.2 dB from 10^7 K. North is toward the top. The field of view is 2.6x2.6 $arcmin^2$.

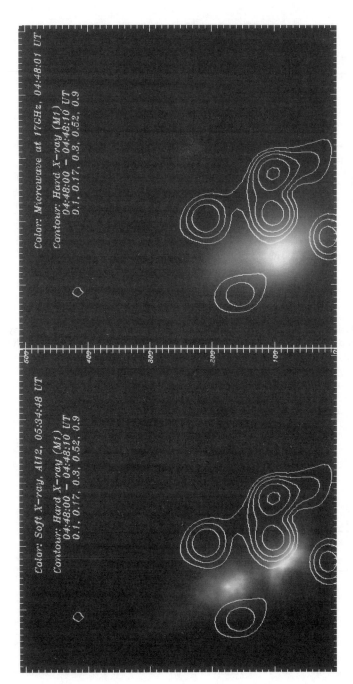

FIG. 3. Overlays of the microwave image (right panel in black and white; 0448:01 UT), the hard X-ray M1 channel (23-33 keV) image (left and right panels in contours; 0448:05 UT), and the soft X-ray image (left panel in black and white; 0534:48 UT) for the 1992 June 28 event. Contour levels in hard X-ray image is 0.1, 0.173, 0.3, 0.52, 0.9 of the peak intensity. North is toward the top. The field of view is 2.6x2.6 *arcmin²*.

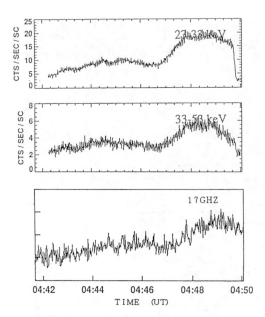

FIG. 4. A comparison of the 17 GHz microwave time profile with the hard X-ray M1 (23-33 keV) and M2 (33-53 keV) channel time profiles in the initial phase of the 1992 June 28 event.

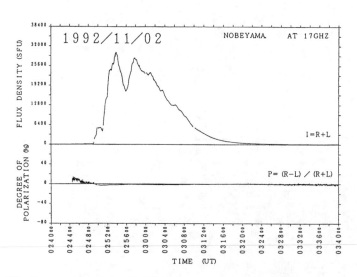

FIG. 5. The time profile of the 17 GHz microwave emission for the 1992 November 2 event.

400

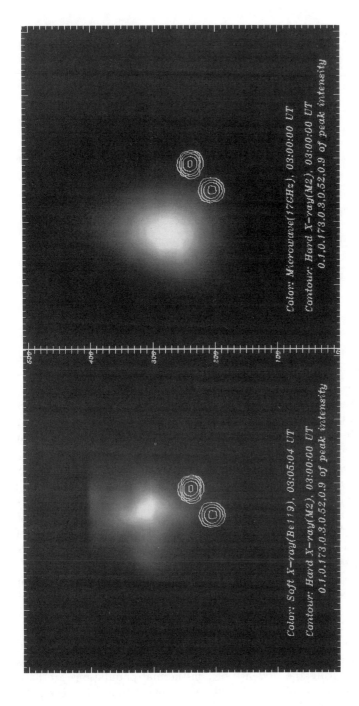

FIG. 6. Overlays of the microwave image (right panel in black and white; 0300:00 UT), the hard X-ray M1 channel (33-53 keV) image (left and right panels in contours; 0300:00 UT), and the soft X-ray image (left panel in black and white; 0305:04 UT) for the 1992 November 2 event. Contour levels in the hard X-ray image is 0.1, 0.173, 0.3, 0.52, 0.9 of the peak intensity. North is toward the top. The field of view is 2.6x2.6 *arcmin²*.

Nonthermal Radio Emission
from Coronal X-ray Structures

Mukul R. Kundu[1], Jean-Pierre Raulin[1] and Nariaki Nitta[2]

[1] *Department of Astronomy, University of Maryland, College Park, MD 20742*
[2] *Lockheed Palo Alto Research Laboratory, Palo Alto, CA 94304*

We have provided evidence that certain coronal X-ray structures such as flaring X-ray bright points and X-ray jets give rise to nonthermal radio emission in the form of metric type III bursts. We have shown an example of a metric type IV/flare continuum being associated with the rupture of a flaring loop-top and the ejection of X-ray emitting material.

INTRODUCTION

Solar X-ray-bright points (XBPs) are compact emitting regions associated with bipolar magnetic fields. Their properties as deduced from Skylab observations have been discussed by (1). At any one time there appear to be dozens of XBPs present on the Sun. Their lifetimes range from a few hours to several days, although only a small number appear to last longer than 2 days. They may be associated with a large fraction of the magnetic flux that emerges to the solar surface. They are known to flare.

Since the launch of the Japanese satellite Yohkoh in 1991, the Soft X-ray Telescope (SXT) has been making excellent images of X-ray-bright points with high time resolution on a regular basis [(2), (3)], and joint studies of XBPs with observations at other wavelengths have been carried out. In contrast to Skylab images, Yohkoh/SXT, has provided more spectacular examples of XBPs and flaring XBPs (3). In particular SXT has revealed that flaring XBPs sometimes have jets associated with them.

From Skylab data it has been known that about 10% of XBPs exhibit a type of sudden, substantial increase in surface brightness which in larger regions would be termed flaring. These flares appear to be impulsive in nature, lasting 2-3 minutes. One of the most important aspects of bright point flares, which has not been studied, is whether or not XBP flares produce nonthermal populations of energetic particles, as do ordinary flares. The two methods best suited for detecting the presence of non-thermal populations are hard X-ray observations and metric-wavelength radio observations. However, the current sensitivity of hard X-ray detectors limits searches for hard X-rays from flaring XBPs. On the other hand, the production of metric radio emission by nonthermal electron beams is very efficient due to the coherent nature of the emission mechanism, and thus such nonthermal electrons might be more easily detectable at radio wavelengths.

In addition to XBP flares, the Yohkoh/SXT has discovered dynamic X-ray jets, or transitory X-ray enhancements with well-collimated motion. These X-ray jets are among the most interesting new discoveries from Yohkoh (4). In many cases, the jets are associated with small flares at or near their apparent footpoints. The motion appears to be a real flow of plasma.

The typical size of a jet, as observed by SXT, is $5 \cdot 10^3$ - $4 \cdot 10^5$ km, the translational velocity is 30-300 km s^{-1}, and the corresponding kinetic energy is about 10^{25} - 10^{28} ergs. Many of the X-ray jets are associated with flares in X-ray bright points (XBPs), emerging flux regions (EFRs), or active regions (ARs).

In this paper, we discuss nonthermal radio emission associated with these two kinds of coronal X-ray structures. We also attempt to delineate the nature of some coronal structures associated with a metric continuum burst.

FLARING XBP'S AND TYPE III BURSTS

Using simultaneousely obtained Nobeyama radioheliograph data along with Yohkoh/SXT data, we found that in addition to thermal soft X-ray emission from the heated plasma XBP flares also produce thermal microwave (17 GHz) radio emission. The microwave emission is gradual and almost as long lasting as the XBP flare; obviously the same thermal plasma produce both emissions (5).

Using the Nançay (France) multifrequency radioheliograph data along with simultaneous SXT data, we found that XBP flares also give rise to nonthermal emission in the form of type III bursts from electron beams. The bursts associated with XBP flares are very weak and not spectacular in appearance, and are observed to be of restricted bandwidth. Figure 1 shows an example of XBP flare-type III association. The X-ray flare emission lasts longer than the radio bursts as in normal flares. Type III's or electron beams appear to be produced preferentially near the onset of impulsive emission (6).

The simple detection of nonthermal type IIIs and thermal microwave emission from flaring XBPs strengthens the association between XBP flares and normal flares, implying that the same physical processes are involved in both. Another important implication of the XBP-type III association is that electrons accelerated to nonthermal energies in XBP flares have access to open field lines. As in normal flares, this raises the question of whether open field lines are intrinsic to the flare process, or whether accelerated electrons can obtain access to open field lines by drifting across closed field lines.

Sometimes long jet-like features have been observed in type III-producing XBP flares. These jets appear to be "static" rather than "dynamic" like Shibata jets. It is possible that such jets suggest the existence of open field lines. Then it is natural to inquire if XBP-associated jets are important for

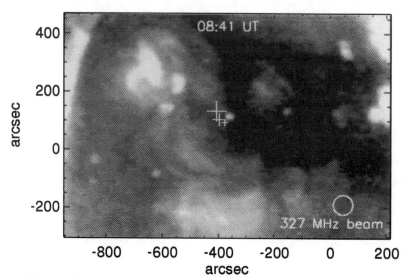

FIG. 1. SXT image showing the XBP and the locations of the radio type III burst observed at 327 MHz (small +). 236 MHz (bigger +) and 164 MHz (biggest +).

type III production.

X-RAY JETS AND TYPE III BURSTS

We have detected type III bursts in association with dynamic coronal X-ray jets observed by Yohkoh/SXT. The type III bursts have been observed with the Nançay multifrequency radioheliograph; they are spatially and temporally coincident with the X-ray jets. The radio locations at different frequencies are aligned along the length of the jet. The type III emission is extremely short-lived relative to the jet process ((7); (8)). Figure 2 shows an example of such association.

The observation of type III bursts in association with X-ray jets implies the acceleration of electrons to serveral tens of keV, along with the heating responsible for the production of soft X-rays. As in XBP flares, this association implies the existence of open field lines in dense coronal structures identified on the sun's disk, along which type III emitting nonthermal electrons propagate.

At the location of the type III bursts we estimate electron densities in the jets which are close to the critical plasma densities ($7 \cdot 10^8$ cm^{-3} at 236 MHz; $6 - 10 \cdot 10^8$ cm^{-3} (SXR)).

The identification of the type III emissions with jets suggests several additional ideas: the location of the low frequency type III bursts on the extension of the soft X-ray jets argues that the coronal jets can be much longer than inferred from soft X-ray measurements; the continuity of jet structure opens the possibility that the electron acceleration may take place considerably lower than the altitude corresponding to the starting frequency of the type III emis-

404

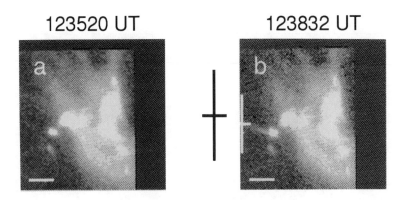

123520 UT   **123832 UT**

FIG. 2. SXT images of AR 7260 before (a), and after (b) the appearance of the jet. The position of the type III radio burst at different frequencies is shown by a white cross (236 MHz) and a black cross (164 MHz). The heavy horizontal bar represents 1 arcmin

sion; we note that the soft X-ray emission and the radio observations provide two independent means of estimating the electron density at the point of observation.

METRIC CONTINUUM BURST AND X-RAY STRUCTURE

We report the first detection of a metric continuum burst (probably of type IV or flare continuum) along with soft X-ray structures as revealed by high spatial resolution imaging observations in both radio and X-rays. The radio event lasted approximately 1.5 hour at 327 - 435 MHz; it was an unpolarized smooth continuum. Figure 3 (a,b) shows a comparison of AR 7773 before and after the occurence of the metric radio continuum. As one can see, by 08:42 UT the active region had a well-observed loop structure $\sim 100 \cdot 10^3$ km in height. The loop seems to undergo a modification of structure near its top just before 09:00 UT, followed by a clear rupture of its top and the ejection of soft X-ray emitting material. Figure 3.b shows that the location of the metric radio sources are spatially coincident with the additional hot plasma observed at the loop-top.

The metric continuum burst has been interpreted as second-harmonic plasma emission due to an anisotropic distribution of electrons (9). Comparison of high spatial resolution imaging observations in X-rays and in radio domains have shown that the continuum radio emission occurs in association with the restructuring of the magnetic field in the underlying active region. During this morphological change, magnetic reconnection may have taken

405

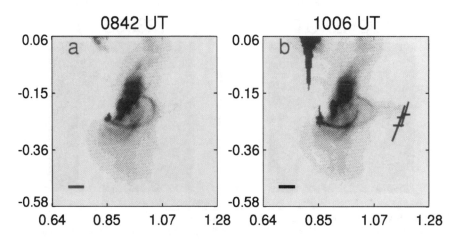

FIG. 3. SXT images of AR 7773 before (a), and after (b) the occurrence of the metric radio continuum, whose locations at 327, 410 and 435 MHz are indicated by crosses. The heavy horizontal bar represents 1 arcmin.

place and accelerated particles may have been produced which are responsible for the radio continuum.

SUMMARY

We have provided evidence that certain coronal X-ray structures such as flaring X-ray bright points and X-ray jets give rise to nonthermal radio emission in the form of metric type III bursts. We point out that these are not the only coronal structures that produce type III's; normal flares occurring in active regions produce type III's as well. We have shown an example of a metric type IV/flare continuum being associated with the rupture of a flaring loop-top and the ejection of X-ray emitting material.

REFERENCES

1. Golub, L., Krieger, A. S., Vaiana. G. S.. Silk. J. K.. Timothy, A. F., *Ap. J.*, **189**. L93, (1974).

2. Nitta, N., Bastian, T. S., Aschwanden, M. J.. Harvey. J. W.. Stro ng, K. T.. *Publ. Astron. Soc. Japan*, **44**, L167, (1992).

3. Strong, K. T.. Harvey, J. W., Hirayama, T., Nitta. N.. Shimizu. T.. Tsuneta. S., *Publ. Astron. Soc. Japan*, **44**. L161. (1992).

4. Shibata, K., Ishido, Y., Acton, L. W., Strong, K. T., Hirayama, T., Uchida, Y., McAllister, A. H., Matsumoto, R., Tsuneta, S., Shimizu, T., Hara, H., Sakurai, T., Ichimoto, K., Nishino, Y., Ogawara, Y., *Publ. Astron. Soc. Japan*, **44**, L173, (1992).

5. Kundu, M. R., Shibasaki, K., Enome, S., Nitta, N., *Ap. J.*, **431**, L155, (1994a).

6. Kundu, M. R., Strong, K. T., Pick, M., Harvey, J. N., Kane, S. R., White, S. M., Hudson, H. S., *Ap. J.*, **427**, L59, (1994b).

7. Kundu, M. R., Raulin, J. P., Nitta, N., Hudson, H. S., Shimojo, M., Shibata, K., Raoult, A., *Ap. J.*, **447**, L135, (1995).

8. Raulin, J. P., Kundu, M. R., Hudson, H. S., Nitta, N., Raoult, A., *Astron. Astrophys.*, in press, (1995).

9. Raulin, J. P., Kundu, M. R., Nitta, N., Raoult, A., in preparation, (1996).

Radio and X-ray Manifestations of a Bright Point Flare

N. Gopalswamy[1], M. R. Kundu[1], Y. Hanaoka[2], S. Enome[2], and
J. R. Lemen[3]

[1] *Department of Astronomy, University of Maryland, College Park, MD 20742*
[2] *Nobeyama radio Observatory, Minamisaku, Nagano 384-13, Japan*
[3] *Lockheed Palo Alto Research Laboratory, Palo Alto, CA 94304*

We have found remarkably different manifestations of a bright point
flare in X-ray and radio (microwave) wavelengths, unlike previous observations. In X-rays, the BP flare was relatively simple while in radio,
the bright point flare had a large scale component and a transient moving component. The large scale structure may be the radio counterpart of large scale structures sometimes seen during X-ray BP flares.
The transient component was also compact and moved away from the
location of the X-ray BP flare with a speed of $\sim 60 km s^{-1}$. The compact source also showed fast time structure which suggests nonthermal
emission mechanism for the transient sources.

INTRODUCTION

A study of bright point (BP) flares is considered useful in understanding
the energy release mechanism in general because of the absence of a multi-
tude of phenomena associated with large flares. The BP studies may also be
useful in understanding the processes of coronal reconnection and heating (1).
Detailed studies of BPs have become possible, thanks to the dedicated solar
instruments such as the ones on board the Yohkoh mission and the Nobeyama
radioheliograph in Japan. The Yohkoh mission has proved to be extremely
useful in studying the small scale energy releases such as BPs (2), transient
brightenings [(3), (4), (5)], and coronal jets (6). Combining X-ray observations with radio data is important in arriving at a better picture of the
BP process such as heating and particle acceleration. Type III radio bursts
at meter-dekameter wavelengths, believed to be due to nonthermal electron
beams, have been found to be associated with at least 10% of the BP flares
observed in X-rays [(7), (8)]. It is not clear if nonthermal radio emission is
produced at high frequencies such as microwaves. Detection of a nonthermal
component at microwaves has important implications to the understanding of
the BP flares since nonthermal emission is indicative of particle acceleration
similar to regular flares. There is no observational evidence of nonthermal
emission from BP flares in microwaves so far. Recently, Kundu et al (2)
have concluded that 17 GHz emission from flaring BPs was thermal. These

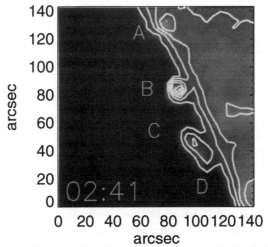

FIG. 1. A string of bright points located near the west limb of the Sun on July 11, 1993. A was just above the limb; B produced an intense flare; C was long-lasting like A; D was seen only occasionally. Contours of X-ray intensity are superposed on gray scale SXT images. The image corresponds to 02:41 UT. North is to the top and east is to the left. X-ray contours show the northwestern limb of the Sun.

authors found that the BP flares seen in radio and X-ray wavelengths had spatial and temporal correspondence. On 1993 July 11, we observed a BP flare in microwaves and X-rays. Unlike previous observations, this BP flare had different manifestations in X-rays and microwaves. A large scale radio structure was associated with the BP flare and the peak emissions in X-rays and in microwaves did not coincide. There were also fast time structures in the radio emission which may be indicative of nonthermal emission. We make a quantitative analysis of this BP flare using the X-ray and radio data.

OBSERVATIONS

X-ray Observations

The BP flare was observed by the Soft X-ray Telescope (SXT – (9)) on board the Yohkoh mission. The SXT was operating in the quiet mode so that only full disk images are available for the present study. The data were obtained with Al.1 and AlMg filters during the flare period so we were able to determine the filter ratios and hence the temperature and emission measure of the BP plasma. The BP flare in question occurred near the northwest limb in a quiet region of the Sun. Fig.1 shows the location of the BP flare activity. There was a group of four BPs (marked A, B, C and D), all of which showed some variability. The BP flare at position B was the most intense and had microwave counterpart. We denote the BPs seen in X-rays as XBP.

409

Radio Observations

The BP flare was observed by the Nobeyama radioheliograph (10) at 17 GHz. The Nobeyama data consist of full disk images in both polarizations obtained every 1 s, with a spatial resolution of 15 arcsecs. Since the BP flare is a weak event, we averaged the data over 10 sec to get a better signal to noise ratio. Fig.2 shows the radio and X-ray fluxes from an area occupied by the XBP flare. Note that the overall duration is the same at both wavelengths. A more detailed comparison is difficult due to the poor cadence of X-ray images. There was a large data gap immediately after the onset of the BP flare until 02:32 UT. Thus we can not determine the exact time of the peak emission in X-rays. We believe that the actual peak is earlier than the one in Fig.2 and is likely to coincide with the radio peak.

When the XBP image was compared with the radio image at 17 GHz, we found that the spatial structure of the BP is very different in the two wavelength domains. Fig.3 shows the superposition of a 10 s snapshot radio map on the X-ray image. We notice that the radio and X-ray emission peaks do not coincide. While there is radio enhancement at the location of the XBP as described in Fig.2, the radio has a much larger maximum located far away from the XBP flare. From Fig.3, we see that the radio BP flare consists of an extended structure starting from the vicinity of the XBP flare and connecting to a distant quiet Sun region. The projected length of the extended structure was $\sim$ 135 arcsec. (From Magnetograms we found that the southern end of the large scale radio structure is a quiet region with a predominant negative polarity). Superposed on this extended extended structure was a compact source with a size of $\sim$ 30 arcsec. The compact source was brighter than the extended structure by 2-3 times and showed a drift away from the location of the XBP flare with a speed of $\sim 60 km s^{-1}$. The radio flux from the compact source and the extended source are much larger than that from the location of the XBP flare.

The drift of the compact source is presented in Fig.4. as the brightness temperature variation of pixels located along the axis of the extended structure. Note that the the radio enhancement starts later as one goes farther way from the XBP. Fig.4 also shows that the enhancement is relatively impulsive for pixels 14 and beyond. At some pixels, there were rapid time variations (up to nearly 70% in some cases) over a period of several seconds (see e.g., the peak of pixel 18). In summary, the radio brightness showed variation over three time scales: (i)the slow variation over the period of the XBP flare which corresponds to the extended radio structure; (ii) the moving compact source which persists at a given location for several minutes and (iii) the fast time structures (tens of seconds) in the brightness temperature of the compact source.

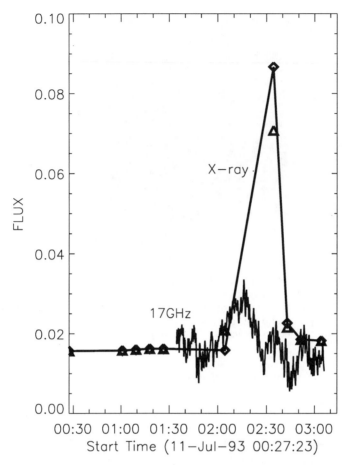

FIG. 2. Flux in soft X-rays and 17 GHz microwaves from an area of 80 pixels occupied by the XBP. The X-ray flux is in units of DNs^{-1} divided by 50000. The radio flux is in units of sfu. We have subtracted a value of .05 sfu from the radio flux to adjust the baseline close to that of the X-ray light curve. The X-ray flux was obtained with the thin aluminum filter. The observed data points are also plotted on the X-ray curve.

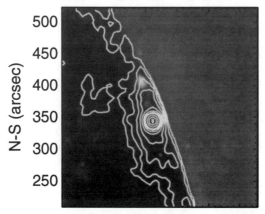

FIG. 3. Superposition of 17 GHz contours on SXT image (gray scale). The contour levels are at 1500, 2000, 2500, 3000, 3500, 4000, 4500, 5000, 6000, 7000, 8000 and 9000 K. The XBP can be seen to the north of the compact radio source. The limb can be seen from one of the lower radio contours. The XBPs C and D are also to the north of the compact radio source. Solar north is to the top and east to the left. Logarithmic display is used to display the X-ray image.

DISCUSSION

We interpret the different manifestations of the BP flare in X-rays and radio as due to the different physical conditions to which the X-ray and radio instruments are sensitive. The SXT is sensitive to hot dense plasmas whereas the radio instrument is sensitive to cool material if the emission is due to free-free process. A detailed analysis by the present authors (11) of the X-ray data has shown that the extended radio structure is the radio counterpart of large scale loop structures sometimes seen in soft X-rays (12). The extended structure seems to be at a temperature less than 1 MK. A large scale structure at such a low temperature is not hydrostatically stable. However, the structure seems to support plasma flow as indicated by the drift of the compact source, and hence can be stable. The physical parameters of the extended structure are such that there is no detectable X-ray emission from it.

The compact source may be explained as thermal free-free emission from plasmoids ejected from the XBP flare through a reconnection process. The temperature of the plasmoids may be similar to that of the large scale structure while the density is relatively higher. The time taken for a plasmoid with a size equal to the size of the compact source to cross a given location is about 6 minutes, similar to the duration of the compact source ($\sim$ 10 min – see Fig.4). However, the faster time structure of the compact source cannot be explained by free-free emission. The main argument against free-free emission for the fast time structures is that the cooling time is of the order

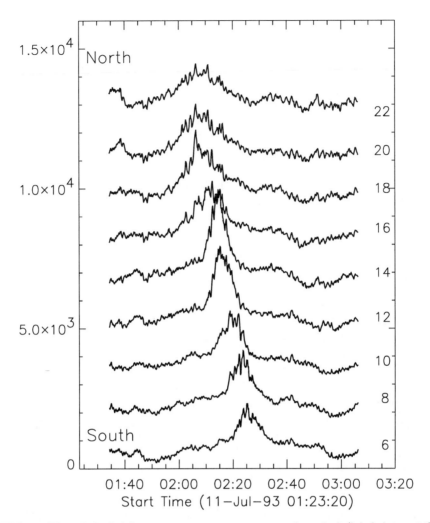

FIG. 4. Plot of the brightness temperature at every other pixel (labeled 6 to 22) located on the axis of the extended radio structure which show the drift of the emission maxima in time. North corresponds to the pixel (22) close to the XBP. Notice several time structures coherent over the size of the compact source. The pixel 22 is closest to the XBP and 6 is farthest. Note that the BPs B, C and D are located to the north of pixel 16 where the brightness curves are different from the rest. We have subtracted n times 1500 K from the actual brightness temperature as one moves away from the top curve.

of several minutes. One way of interpreting the fast structures is by nonthermal gyrosynchrotron emission from mildly relativistic electrons trapped in the plasmoid. Since the BP flare occurred in the quiet region, we expect the magnetic field to be relatively low so the emission has to be in high harmonics of the gyrofrequency. From the standard formulae for gyrosynchrotron emission, we find that nonthermal electrons with a density of about $1. \times 10^6 cm^{-3}$ and a power law distribution with an index of 4 can account for the fast structures in the radio brightness. Since the BP flare occurred in a quiet region, we expect the magnetic field to be around 100 G so that the harmonic number is about 60. We need more observations with spectral information to arrive at firm conclusions. The energetic electrons might have been accelerated in a reconnection process involving the large scale structure and a small loop represented by the XBP.

SUMMARY

We have presented the microwave and X-ray observations of a bright point flare which demonstrate that the flare has different manifestations in the two wavelength domains. The microwave observations reveal structures "invisible" to the X-rays. In particular, a large cool loop structure seen in microwaves did not have any X-ray counterpart. Previously, such loops were observed only in X-rays. This suggests that large scale structures associated with BP flares may have a wide range of physical parameters and they can be detected only with a suitable wavelength range. The large scale structure also supports plasma flow probably originating in the vicinity of the BP flare. Fast time structures in the radio brightness suggests that they may be due to nonthermal particles accelerated during the BP flare.

Acknowledgments We thank N. Nitta for helpful discussions. NG and MRK were supported by NSF grant ATM 16972 and NASA grants NAG-W-1541 and NAG-W-4664 to the University of Maryland, College Park. JRL was supported by NASA contract NAS8–37334 and the Lockheed Independent Research Program. The Nobeyama radioheliograph project was supported by the Ministry of Education and Culture of Japan. The Soft X-ray Telescope on Yohkoh was built at the Lockheed Solar and Astrophysics Laboratory in collaboration with the National Astronomical Observatory of Japan, the University of Tokyo, and the Institute of Space and Astronautical Sciences.

REFERENCES

1. E. Priest, in Proc. of Kofu Symposium, (ed.) S. Enome and T. Hirayama, Nobeyama radio Observatory, Japan, p.93, (1994).
2. M. R. Kundu, K. Shibasaki, S. Enome, N. Nitta, it Ap. J., **431**, L155, (1994a).
3. T. Shimizu, S. Tsuneta, L. W. Acton, J. R. Lemen, and Y. Uchida,it Publ. Astron. Soc. Japan, **44**, L147, (1992).

4. T. Shimizu, S. Tsuneta, L. W. Acton, J. R. Lemen, Y. Ogawara and Y. Uchida, it Ap. J.,**422**, 906, (1994).
5. N. Gopalswamy, T. E. W. Payne, E. J. Schmahl, M. R. Kundu, J. R. Lemen, K. T. Strong, R. C. Canfield, and J. de La Beaujardiere, it Ap. J., **437**, 522, (1994).
6. M. R. Kundu, J.P. Raulin, N. Nitta, H. S. Hudson, M. Shimojo, K. Shibata, and A. Raoult, it Ap. J., **447**, L135, (1995).
7. M. R. Kundu, T. E. Gergely and L. Golub, *Ap. J.*, **236**, L87, (1980).
8. M. R. Kundu, K. T. Strong, M. Pick, J. N. Harvey, S. R. Kane, S. M. White, and H. S. Hudson, *Ap. J.*, **427**, L59, (1994b).
9. S. Tsuneta, L. Acton, M. Bruner, J. R. Lemen, W. Brown, R. Caravalho, R. Catura, S. Freeland, B. Jurcevich and M. Morrison, it Solar Phys., **136**, 37, (1991).
10. H. Nakajima and Radioheliograph Group, *Proc. IEEE*, **82**, 705, (1994).
11. N. Gopalswamy, M. R. Kundu, Y. Hanaoka, S. Enome, and J. R. Lemen, *Ap. J.*, **457**, L1, (1996).
12. K. T. Strong, J. W. Harvey, T. Hirayama, N. Nitta, T. Shimizu, and S. Tsuneta, it Publ. Astron. Soc. Japan, **44**, L161, (1992).

Razin Suppression in
Solar Microwave Bursts

Leila Belkora

University of Colorado, A.P.A.S. Department[1]
Boulder Colorado 80309

We used data from the Owens Valley Solar Array to make a detailed investigation of the gyrosynchrotron spectrum of the flare of July 16 1992. The result of this investigation led us to propose a solution to a long-standing problem in solar microwave bursts, that of the constant peak frequency of the bursts as they evolve in brightness temperature, and the steep slope on the low-frequency side of the spectrum. We propose that the Razin effect is at work, and develop the theory of Razin suppression for solar microwave burst conditions. The Razin effect is the suppression of radiation from an electron in a medium in which the index of refraction is less than unity. We demonstrate that in a medium with density 2×10^{11} cm^{-3} and magnetic field 300 Gauss, conditions not uncommon for solar microwave bursts, the gyrosynchrotron spectrum can be suppressed for frequencies up to at least 10 GHz.

INTRODUCTION

The motivation to study the Razin effect in solar microwave bursts is the difficulty of explaining some well-known features of microwave burst continuum spectra. The flare of July 16 1992 provided a good data set to investigate the problem. The flare was observed at the Owens Valley Solar Array (operated by the California Institute of Technology), a five-element radio telescope. The flare as observed in microwaves had an impulsive rise phase lasting about one minute, an initial decay phase lasting one or two minutes, and a period of low-level emission that persisted for tens of minutes. From the data acquired during this time we were able to determine the brightness temperature spectrum between 6 and 16 GHz. We selected spectra from six time samples, three during the rise phase and three during the decay phase, for further study.

Of particular interest in these data are the steep slope of the spectra on the low-frequency side of the spectral turnover, and the fact that the frequency of maximum emission (or peak frequency ν_{peak}) remains nearly constant (a shift of less than 2 GHz) while the peak brightness temperature varies over nearly two orders of magnitude. Models of solar microwave emission due

---

[1]Current address: P.O. Box 500, Fermilab, Batavia IL 60510

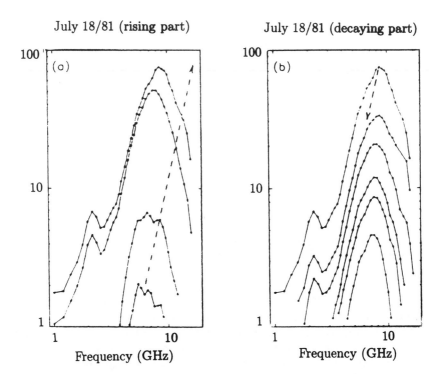

FIG. 1. Figure adapted from Stahli, Gary, and Hurford 1989, showing observed microwave burst spectra and the model predictions of their evolution in time.
Reprinted by permission of Kluwer Academic Publishers.

to the gyrosynchrotron mechanism[2] (1) predict an increase in ν_{peak} as the peak brightness temperature increases and show a shallower slope on the low-frequency side than we observed. Several authors have commented on the fact that observed spectra are often in disagreement with the model predictions on these points, e.g., (2), (3). In the survey of microwave bursts conducted by Stähli, Gary, and Hurford, only 2 events, or 4 percent of the sample, showed a shift in ν_{peak} as large as that expected from theory. Figure 1, adapted from their paper, shows the evolution of one burst during the rise and decay periods, with arrows indicating the extent of the shift in ν_{peak} expected from the Dulk and Marsh (1982) model.

One appealing explanation for the behavior of the spectra is that a low-frequency cutoff is at work. Figure 2 illustrates how a low-frequency cutoff can produce steep slopes on the low-frequency side of the turnover and an apparently constant ν_{peak}. In our analysis of the data we considered whether the Razin effect might provide the required low-frequency cutoff. The Razin effect is the suppression of radiation in a medium with index of refraction less

---

[2]We use the terms gyrosynchrotron or gyroresonance emission for synchrotron radiation at lower energy, $\gamma \sim 2$ or 3, i.e. between cyclotron and synchrotron energies.

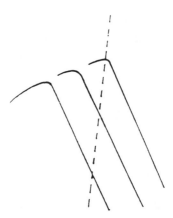

FIG. 2. Sketch to illustrate how a low-frequency cutoff, shown by a dashed line, can produce an apparently constant peak frequency and steep low-frequency slope.

than one. Razin suppression is not a propagation effect; it is a suppression of emission at the source, and as such, as pointed out by Ramaty (4), it leads to a corresponding reduction in the absorption process. This means that medium suppression and gyrosynchrotron re-absorption must be treated consistently in modeling the emission. Another mechanism which could produce a low-frequency cutoff is free-free absorption. We found from analysis of X-ray data for this flare that although free-free absorption is present, it cannot account for the shape and evolution of the microwave spectra (5).

We used a gyrosynchrotron code described in (6) to model the flare emission, taking the Razin effect into account. The code treats the case of a distribution of electrons radiating in a cold, collisionless, magnetoactive plasma. The index of refraction in the plasma depends on the gyrofrequency $\nu_B = \frac{eB}{2\pi mc}$ as well as the plasma frequency $\nu_p = \sqrt{\frac{4\pi n_e e^2}{m}}$. We used the code to evaluate the emission and absorption coefficients, and hence the brightness temperature, resulting from a power-law distribution of electrons with isotropic pitch-angle distribution. The model fits to the data are shown in Figure 3. We found that a low-frequency cutoff due to Razin suppression provides a very satisfactory solution, explaining not only the shape of any individual spectrum, but also the evolution of the spectrum in time. In the following section we show how the Razin effect can account for the low-frequency cutoff.

THE RAZIN EFFECT APPLIED TO SOLAR FLARES

The "classical" derivation of the Razin effect, which is known to English-speaking solar physicists through, for example, the papers of Ginzburg and Syrovatskii (7) and Ramaty (4) assumes that relativistic or nearly-relativistic particles emit synchrotron radiation in a plasma in which the gyrofrequency

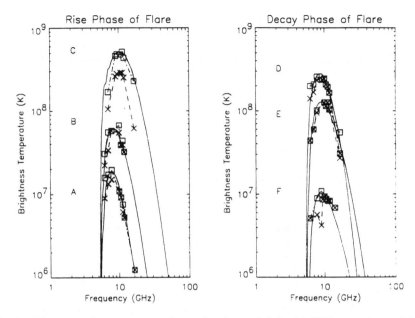

FIG. 3. Microwave burst spectra from the flare of July 16 1992, together with model fits to the data.

ν_B is low compared to the plasma frequency ν_p, so that the index of refraction may be written simply as

$$n^2 = 1 - \omega_p^2/\omega^2. \tag{1}$$

The classical derivation of the Razin effect leads to the following expression for the frequency up to which the radiation may be suppresed:

$$\nu_{affected} \leq \frac{20 n_e}{B} \tag{2}$$

where n_e is the electron density and B the magnetic field in Gauss.

We sought to investigate Razin suppression in the situation which is more relevant to microwave emission from solar flares: non- or mildly-relativistic electrons (Lorentz factor γ near two) and a gyrofrequency which is close to the plasma frequency, so that the full expression for n must be used:

$$n^2 = 1 - \frac{2\nu_p^2 x}{\pm[\nu^4 \nu_B^4 \sin^4\theta + 4\nu^2 \nu_B^2 x^2 \cos^2\theta]^{1/2} - 2\nu^2 x - \nu^2 \nu_B^2 \sin^2\theta} \tag{3}$$

where

$$x = \nu_p^2 - \nu^2. \tag{4}$$

The general procedure, described in more detail in (5), is to consider radiation emitted at low harmonics of the gyrofrequency in the cyclotron (nonrelativistic) limit. Radiation at harmonic m is 2^m electric multipole emission; following Melrose and McPhedran (8) we account for the effect of the dispersive medium by multiplying the power radiated at harmonic m by the factor n^{2m-1}. We then evaluate the Bessel functions in the equation for the radiated power using the Carlini approximations, which are appropriate in this energy range.

In Figure 4 we illustrate the suppression effect with one example in which we used "typical" values for the magnetic field, accelerated and ambient particle densities, and Lorentz factor derived from the best fits to the data. We assume that the particle pitch angle is equal to the viewing angle (angle between the line of sight and the magnetic field); radiation is maximized in this direction. The figure shows the normalized power radiated as a function of harmonic number $m = \nu/\nu_B$, relative to the power radiated in vacuum. The parameters used for this plot were $B = 300$ Gauss, $n_e = 2 \times 10^{11}$ cm^{-3}, $\gamma = 1.5$, and viewing angle and pitch angle both 60^0. The plot shows that the radiation is strongly suppressed for the low harmonics. At harmonic $m = 12$ corresponding to $\nu \sim 10$ GHz, the emission in both the x and o modes (shown by squares and diaonds, respectively) is only about 10 percent that in vacuum. The suppression due to the influence of the medium thus extends far above the plasma frequency, which is 4 GHz or $m \approx 5$ in this case.

EVOLUTION OF FLARE SPECTRA

The action of Razin suppression accounts very well for the behavior of the flare spectra as a function of time. In this section we highlight one result from modeling the data with Ramaty's gyrosynchrotron code, showing that the Razin effect can lead to a constant or nearly-constant ν_{peak}. In contrast to other possible explanations for the constant ν_{peak}, invoking the Razin effect allows us to model the evolution of the spectrum by varying only one parameter, the density of accelerated particles.

To define the ambient medium in the gyrosynchrotron code, the user specifies a uniform magnetic field strength, the cosine of the angle θ between the field and the line of sight, and the so-called Razin parameter $\alpha = 1.5\nu_B/\nu_p$ (9). The α parameter serves as an index for the influence of the medium. The density of the electrons in the ambient medium, which determines the plasma frequency, is set by α, since the field strength and hence the gyrofrequency is independently specified. To define the energy spectrum of the accelerated electrons the user specifies both the exponent δ and the normalization factor A in the electron energy distribution function, and the upper and lower bounds used in the integration over the distribution function, which in the analytic expression is carried out from $\gamma = 1$ to ∞.

In general, a given spectrum may be fit with a magnetic field that is high

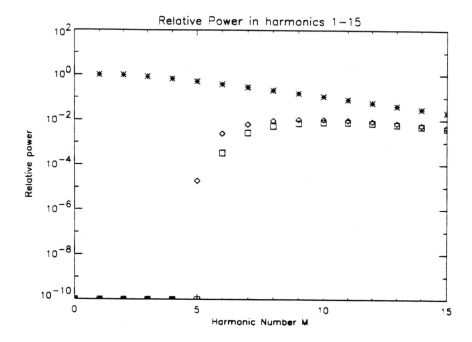

FIG. 4. Power radiated as a function of harmonic number m, where $m = \nu/\nu_B$. The gyrosynchrotron emission in vacuum is shown with asterisk symbols, the x-mode radiation in a dispersive medium with squares, and the o-mode radiation with diamonds.

or low, depending on the energy distribution function and the densities of ambient and energetic electrons. However, the evolution of the spectrum in time is a very strong constraint on the model. Many of the fits that we rejected would require two or more parameters to offset each other in just the right way to keep the peak frequency approximately constant. For example, the variation in ν_{peak} that accompanies a rise and fall in accelerated particle density and peak brightness temperature during the rise and decay phases might be counterbalanced by requiring the magnetic field in the flaring region to vary up and down, also, in such a way as to keep the frequency of maximum brightness temperature constant. Such changes in the flare conditions can be considered unlikely on the grounds of inordinate complexity.

Figure 5 shows the effect of varying only the accelerated particle density, with and without the effect of Razin suppression. The panel on the right shows the spectra with $\alpha = 1$, or a minimal effect of the medium, and the panel on the left shows the spectra strongly affected by the medium, with $\alpha = 0.320$. All other parameters, such as the magnetic field strength or the exponent in the energy distribution function, were kept constant. The figure shows, in the case without Razin suppression, that the peak brightness temperature ranges from about 10^8 to about 2×10^9 K when the accelerated particle density is increased from $10\,\mathrm{cm}^{-3}$ to $10^7\,\mathrm{cm}^{-3}$. This range of peak brightness temperature is too small to accomodate the range in the data. If Razin suppression occurs, with $\alpha = 0.320$ corresponding in this case to an ambient density of $2.6 \times 10^{11}\,\mathrm{cm}^{-3}$, the change in peak brightness temperature is much larger, from 5×10^5 K to 2×10^9 K. The comparison of the two panels shows clearly the effect of Razin suppression: the low-frequency emission is suppressed, so that resonant structures seen on the right are not apparent, and due to the cutoff on the low-frequency side of the spectrum, the frequency of maximum brightness temperature remains constant.

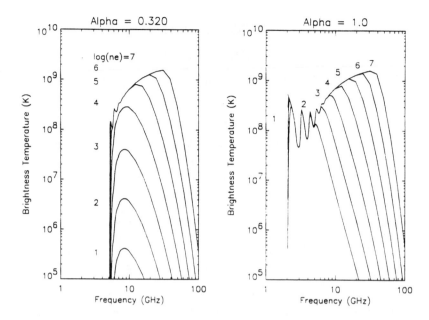

FIG. 5. Effect on the brightness temperature spectrum of varying the density of accelerated particles from 10 to 10^7 cm^{-3}. The left panel shows the effect with Razin suppression, and the right panel shows the effect without Razin suppression.

REFERENCES

1. G.A. Dulk and K.A. Marsh , Ap.J. **259**, 350 (1982).
2. M. Stähli, D.E. Gary, and G.J. Hurford, Solar Phys. **120**, 351 (1989).
3. S.G. Benka and G.D. Holman, Ap.J. **391**, 854 (1992).
4. R. Ramaty, Ap.J. **158**, 753 (1969).
5. L. Belkora. Ph.D. Thesis, University of Colorado (1995).
6. R. Ramaty, R.A. Schwartz, and H. Nakajima, Ap. J. **436**, 941, (1994).
7. V.L. Ginzburg and S.I. Syrovatskii, Ann. Rev. Astron. Astrophys. **3,** 297 (1965).
8. D.B. Melrose and R.C. McPhedran. Electromagnetic Processes in Dispersive Media. (Cambridge: Cambridge University Press) (1991).
9. R. Ramaty and R.E. Lingenfelter, J.G.R. **72**, 879 (1967).

A Summary of Three-Year Observations with the Nobeyama Radioheliograph

Shinzo Enome[1]

Nobeyama Radio Observatory
National Astronomical Observatory
Minamisaku, Nagano 384-13, Japan

The Nobeyama Radioheliograph was completed in two years by 1992 March. It has been put into regular observations in late June 1992 for eight hours a day from 23h UT to 07h UT. It has caught some part of maximum activities of the Sun, but not so large as the flares in 1991 June. Observational results are described for faint structures on the Sun; an introduction to previous reviews is also given. Recent developments of image deconvolution software at Nobeyama are very promising to produce images of high-dynamic range or of high quality. Examples of improved images with Steer algorithm are demonstrated. This image synthsis program will be released at Nobeyama for open use of the observed correlation data. The first light of the Dual-frequency observations is also presented.

INTRODUCTION

Since the start of routine observations in late June, 1992, three years passed. We could fortunately catch the last remnants of the maximum phase of the solar cycle 22. The observed activities, however, were not so high as those of 1991 or the early days in 1992. An introduction of the initial observations has been presented in the Proceedings of IAU Colloquium No. 141 on The Magnetic and Veleocity Fields of Solar Active Regions held at Beijing in 1992 September (2). An introduction of initial results has been decribed in the Proceedings of CESRA Workshop on Coronal Magnetic Energy Releases held at Caputh/Potsdam in 1994 May (3), where emphasis is put on the thermal or gradual phenomena associated with flares based on the published papers mostly in PASJ, 1994, Vol 46, No. 2, Special Features: Initial Results from the Nobeyama Radioheliograph and in the Proceedings of Kofu Symposium, 1994 NRO report No. 360. Non-thermal aspect of radio emission from flares is examined in a summary paper in the Proceedings of the Third China-Japan Seminar on Solar Physics, 1996. In this review radio association with

---

[1]also VSOP Office, National Astronomical Observatory, Mitaka 181, Japan

XBP's, ARTB's, XBP flares and new analyses of radio flare structures in assoication with x-ray flares are described. Emphasis is put on the recent new development of image deconvolution based on the Steer algorithm. Finally, a new result of the dual-frequency observation of a flare at 17- and 34-GHz is briefly introduced.

RADIO ASSOCIATION WITH XBPS, FLARING XBPS, ARTBS AND ABP FLARE

The Yohkoh Soft X-ray Telescope has revealed a variety of activities associated with X-ray bright points (XBP's) with respect to their size, shape, temporal variations, and correlation and evolution to other forms of activities such as jets and transient brightenings [e.g. (13,11,12)]. Several but not many studies have been done on XBP and related issues at the University of Maryland made with data from the Nobeyama Radioheliograph and from the Yohkoh Soft X-ray Telescope. Kundu et al. (9) have reported the first detection of 17-GHz signatures of coronal X-ray-bright points in four BPs out of eight identified XBPs on the SXT image of 1992 July 31. Light curves in radio and X-ray are in some case similar but in other case not similar. The temperature emission measure analyses show that the estimated radio flux densities are in a range from half to one twentieth of the observed values. Thermal mechanism is suggested for the gradual and long-lasting and unpolarized radio emission from the quiet region XBP. Active-region transient brightenings (ARTBs) are known as brighter and XBP and less brighter than sub-flares (12). Radio properties have been examined of active-region transient brightenings from AR 7260 for four events of observations made with the Nobeyama Radioheliograph and Yohkoh/SXT (15). Time profiles are in good agreement in radio waves and X-rays. Predicted radio flux densities are in good agreement with observed values with a factor ranging from 3 to 0.5. Detection of hard X-ray emission assoicateed with one ARTB of GOES class B1 made with CGRO/BATSE, and not for other two cases is discussed whether or not ARTB is a microflare. A large-scale loop structure of 100,000 km has been detected at 17 GHz associated with an X-ray bright point flare (5). The radio emission consists of a large-scale structure as often seen in X-rays [e.g. (13)] and a compact moving source. The XBP flare was located at one end of the radio loop. The differences in the appearance of radio and X-ray images, time profiles, and the computed and observed radio fluxes are explained in terms of a cool plasma loop and a cool plasma flow along this loop. These samples are not enough to establish convicing evidence for small-size activities of the Sun within the range of microflare or less, but they will be sufficient to prove the existence of radio counterparts in these area. It is, however, beyond the limit of sensitivity and spectral information to prove a trace of nonthermal emission from these sources at present.

425

FOOTPOINTS, LOOPS AND LOOP-LOOP INTERACTIONS

Loop configuration has been investigated in terms of asymmetric precip-itation of nonthermal electrons on the basis of observations made with the Yohkoh Hard X-ray Telescope (HXT) and the Nobeyama Radioheliograph in 1993 May citekundu95. The HXT images in one case show two well-separated sources, and the radio images indicate circularly polarized, nonthermal radio emission with opposite polarities from these two sources. This indicates two oppositely directed fields and is consistent with a single-loop model. In the second event there are several sources in the HXT images, which appear to be connected by soft X-ray loops. The strongest hard X-ray source has unpo-larized radio emission, whereas the strongest radio emission lies over strong magnetic fields and is polarized. These features in two cases are summarized as the strongest radio emission is highly polarized and not coincident with the strongest hard X-ray emission. This is consistent with asymmetric loops in which the bulk of the precipitation, and hence the X-ray emission, occurs at the weaker field footpoint. Detailed analyses of AR 7360 appeared in 1992 De-cember have made for four flares with various instruments (6). In this region a small loop emerged near one of the footpoints of a pre-existing large coronal loop. These loops show evidence that that interactions between coronal loops trigger flares, microflares, and plasma flow. All of the four flares observed in this region show that brightenings in the small loop occurred first, and then the large loop flared up. The brightenings in the large loop can not occur by themselves, but must be triggered by the brightenings in the small loop. This fact suggests an existence of interactopms between loops to produce these flares. Many microflares also occurred in the small loop as well as flares. More than half of these microflares are accompanied by plasma ejection phenomena from the small loop into the large loop. The large loops are filled with ejected plasma with a velocity of about 1000 km/s. The author proposes that this plasma ejections are considered as physically same phenomena as the X-ray jets [e.g. (11)].

NEW DUAL-FREQUENCY OBSERVING SYSTEM

Installation of frequency-selective subreflectors and a 34-GHz receiver for each antenna of the Nobeyama Radioheliograph was completed in 1995 Oc-tober after two month discontinuation of observations at 17 GHz. After the start of routine observations the first distinctive event was detected on 1995 November 10 at 17 and 34 GHz. The two sets of observed complex correla-tions were processed by a 16d deconvolution program which employs those correlation data of multiples of 16 time the fundamental baseline lengths (4). A preliminary map of the event is shown in Figure 1, which employs a pro-visional version of the 16d imaging program. The subframes show from the upper left in clockwisely 17 GHz left-handed circular polarizaiton image, 17

GHz right-handed circular polarization image, Yohkoh Soft X-ray image, and 34 GHz intensity image.

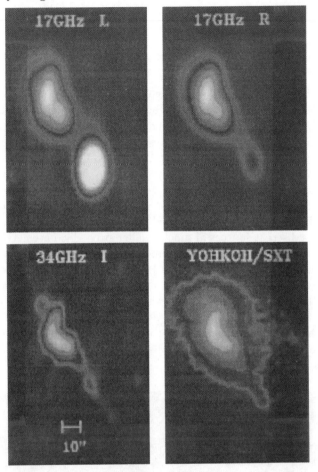

Fig. 1. 1995 November 10 Flare at 17 GHz lcp, rcp, at 34 GHz I, and Yohkoh/SXT.

The radio source consists of two components. The upper or northern component is unpolarized at 17 GHz and the lower or southerm component is almost 100 % polarized. These facts indicate that the northern component emits radio wave by the free-free mechanism and the southern component is due to the gyro-resonance absorption. The northern component clearly shows a loop shape at radio waves and at X-rays with 10 arcsec width and 20 arcsec length. The southern component is permeated by 2000 Gauss magnetic field as indicted by the gyro-resonance theory (10). A weak channeling is traceable in the SXT image and in the 34 GHz image from the northern component to the location of the southern source. It is suggestive that this channel might be the route to feed energy between these two components seen at 17 GHz for their simultaneous light curves within seconds. An extensive analysis of this event is in progress (15).

A NEW DEEP-CLEAN METHOD AND PRELIMINARY RESULTS

The above-mentioned 16d deconvolution procedure employs conventional or classical CLEAN algorithm by Högbom. A new algorithm by Steer et al. is extensively applied to the correlation data from the Nobeyama Radioheliograph (8). The Steer algorithm estimates the shape of the original image by the trim level or a sidelobe-dependent threshold (1) and subtract by convolving the dirty beam with an appropriate loop gain. This algorithm is emphasized to have capability to avoid 'strips' or 'corrugations' or 'dips in CLEANing an extended source. The image of the Sun at 17 GHz is dominated by the quiet Sun with 32 arcminute diameter and with 10,000 degree Kelvin, superposed on which are discreet sources of much smaller size, which are active regions, dark filaments in depression, bright points, etc. It was found that by the proper selection of the trim level and the loop gain we can attain the CLEAN level as low as three times sigma of the instrumental fluctuation, most of which is due to the receiver noise. The current algorithm (6) employs a modification of the conventional CLEAN to fit a Gaussian with a variable width, which has to set the CLEAN level as low as three times the sigma. To demonstrate the effectiveness of the Steer algorithm a number of examples are shown below in Figures 2 to 5.

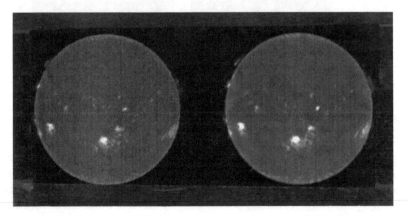

Fig. 2. 1992 September 22, the left panel is 1 hour synthesis by the NRAO aips and the right is 1 sec snap shot by the Steer algorithm.

In Figure 2 on the left is shown a super-synthesis image of the Sun by Tim Bastian, who employed the Clark algorithm of CLEAN in the NRAO aips with one hour integration around the local noon at Nobeyama. On the right is the snapshot of 1 sec integration at the local noon by the Steer algorithm (8). Although the image quality is almost same, there are slight differences between the two images, but it is difficult to deal with these differences in this paper, since the one is one-hour integration and the other is 1-sec integration. The point is that we can reproduce fine structures with the program developed by our group.

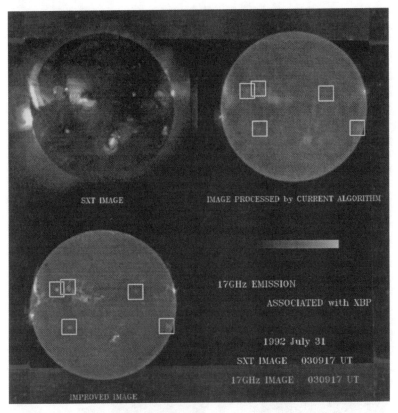

Fig. 3. 1992 July 31 X-ray Bright Points by Yohkoh/SXT, at 17 GHz with the current algorithm and with Steer algorithm

The Figure 3 shows The Yohkoh Soft X-ray (SXT) image on the upper left, 17-GHz image processed with the current algorithm on the upper right, and 17-GHz image processed with the Steer algorithm. There are four boxes in each radio image, in each of which is a radio-counterpart of X-ray bright point as clearly seen in Yohkoh/SXT image. In the improved image we can definitely locate 'radio bright points' without the help of the Yohkoh/SXT image.

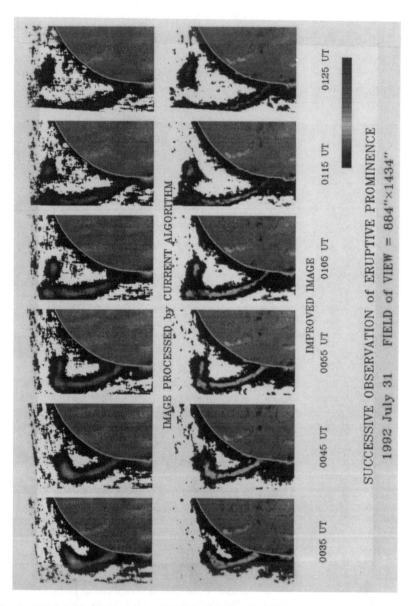

Fig. 4. 1992 July 31 Eruptive Prominence at 17 GHz

Figure 4 shows a time sequence of an eruptive prominence that occurred on 1992 July 31 as it rose high in the corona up to one solar radius. The upper subframes show images processed by the current algorithm and the lower subframes are processed by the Steer algorithm. In the image at 0115 UT in the lower subframe we can recognize unwinding or untwisting structure of the prominence, which is also seen in H-alpha (10).

Finally, Figure 5 shows a sequence of a sudden disappearance of dark filaments (disparition brusque) on 1994 February 20. The upper subframes are made with the current algorithm and the lower subframes are processed with the Steer algorithm. It is well shown that a part of dark filament disappeared and a flare appeared after it, but the other part of the dark filament just disappeared. It should be noticed that the deconvolution algorithm developed by (8) can reproduce dispersion features as seen in these subframes.

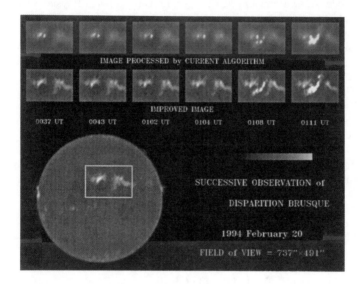

Fig. 5. 1994 February 20 Disparition Bruesque

CONCLUDING REMARKS

The Steer algorithm program will be released soon at Nobeyama Radio Observatory. An extensive application of this program to the observed correlation data for more than three years will probably drastically improve images for phenomena with the brightness of 10,000 degree Kelvin or so. These improved images will be expected to produce new results and enhance the results as described above or in the review papers, which should be revised according to coming papers based on the improved images. It will take some time to complete the 16-d deconvolution program. This program is known to be effective for flares with brightness well above 10,000 degree Kelvin. The solar activity is now rapidly approaching to the minimum between the cycles 22 and 23, therefore, we shall not be able to expect to observe many flares at the dual frequency. This, however, should change with the start of the new cycle.

REFERENCES

1. T. Cornwell and R. Braun, Deconvolution in Synthesis Imaging in Radio Astronomy, Third NRAO Synthesis Imaging Summer School, eds. R. A. Perley, F. R. Schwab, A. H. Bridle, ASP Conference Series **6**, 167 (1989)
2. S. Enome, in Proc. IAU Colloquium No. 141 on the Magnetic and Velocity Fields of Solar Active Regions, eds. Harold Zirin, Gouxiang Ai, and Haimin Wang, ASP Conference Series **46**, 310 (1993)
3. S. Enome, 1995, in Proc. CESRA Workshop on Coronal Magnetic Energy Releases, eds. A. O. Benz and A. Krueger, Lecture Notes in Physics **444**, 35 (1995)
4. K. Fujiki, private communication (1995)
5. N. Gopalswamy et al., Ap. J. **457**, L117 (1996)
6. Y. Hanaoka, Solar Physics, submitted (1996)
7. Y. Hanaoka et al., Proc. Kofu Symposium on New Look at the Sun with Emphasis on Advanced Observations of Coronal Dynamics and Flares, eds. S. Enome and T. Hirayama, NRO Report No. 360, 35 (1994)
8. H. Koshiishi, A Deep-CLEAN Imaging Method Applied to the Nobeyama Radioheliograph and Observations of Polar-Cap Brightenings and Their Association with Coronal Holes, Ph. D. Thesis, The University of Tokyo (1996)
9. M. R. Kundu, et al., Ap. J. **454**, 522 (1995)
10. H. Miyazaki et al., in Proc. Yohkoh Symposium on X-ray Solar Physics from Yohkoh, eds. Y. Uchida, T. Watanabe, K. Shibata, and H. Hudson, Universal Acad. Press, Tokyo, p. 277 (1993)
11. K. Shibasaki et al., Publ. Astron. Soc. Japan **46**, L17 (1994)
12. K. Shibata et al., Publ. Astron. Soc. Japan **44**, L173 (1992)
13. T. Shimizu et al., Pub. Astron. Soc. Japan **44**, L147 (1992)
14. K. T. Strong et al., Pub. Astron. Soc. Japan **44**, L161 (1992)
15. T. Takano et al., private communication (1996)
16. S. M. White et al., Ap. J. **450**, 435 (1995)

THEORY OF PARTICLE ACCELERATION

Selective Enrichment of Energetic Ions in Impulsive Solar Flares

M. Temerin and I. Roth

Space Sciences Laboratory
University of California, Berkeley
Berkeley, CA 94720

We have shown previously that heavy ions as well as ^{3}He can be accelerated to MeV energies by obliquely propagating electromagnetic proton cyclotron waves in impulsive solar flares. Impulsive flares are characterized by an increase of up to four orders of magnitude in the abundance of energetic ^{3}He ions and up to one order of magnitude in energetic heavy ions compared to the abundances in long duration events and in the solar corona. ^{3}He is accelerated when the first harmonic resonance is satisfied, while the heavier ions can be accelerated by the same wave at the second harmonic resonance when the wave frequency is twice the ion gyrofrequency. We postulate that obliquely propagating proton cyclotron waves exist in impulsive solar flares in analogy with the Earth's aurora, where such waves are often the most intense waves present. The heating of ^{3}He and of heavy ions can be substantially more effecient in a nonuniform plasma than in a uniform plasma. The heating efficiencies are nearly equal at higher energies, though at lower energies first harmonic heating is more efficient. The relatively small enrichments of the heavier ions compared to ^{3}He can be qualitatively understood as a result of the damping of the waves by the heavier ions and by the less efficient initial acceleration at lower energies at the second harmonic resonance.

I. INTRODUCTION

Observations of soft X-rays emissions indicate the existence of two classes of flares: impulsive with duration of minutes and gradual with duration of hours (1, 2, 3). In many impulsive solar flares the ^{3}He/^{4}He ratio at energies of a few MeV/nucleon is increased by three to four orders of magnitude over ambient solar coronal densities (4, 5) where a ratio of ^{3}He/ ^{4}He > 1 is sometimes observed, in contrast to a few times 10^{-4} in the solar wind (6) and in the SEP events which characterize coronal abundances (7). The acceleration of ^{3}He requires an efficient mechanism that heats a significant fraction of the coronal pre-flare ^{3}He plasma to MeV energies (8). Flares enriched in ^{3}He are correlated with smaller enrichments in heavier ions (9, 5, 8). However, the amount of heavy ion enrichment is independent of ^{3}He enrichment.

In a recent acceleration model (10, 11) ^{3}He ions are resonantly accelerated to energies of a few MeV/nucleon due to the first harmonic resonance with the electromagnetic proton cyclotron wave, while heavier ions (e.g., Ne, Mg, Si, Fe) are accelerated to similar energies per nucleon due to a second harmonic resonance with the wave. At higher energies ($k_\perp \rho \sim 1$), where ρ denotes the ion gyroradius, the heating rates for both the first and second harmonic interactions are comparable, though they differ at low energy ($k_\perp \rho \ll 1$), where the heating rate at the first harmonic is larger.

In this report we discuss three aspects of our model which we consider of particular interest: (a) the generation of the wave, (b) the actual acceleration mechanism and the importance of the gradients in the magnetic field, and (c) the acceleration of the heavier ions by the same wave.

II. WAVE GENERATION

The generation of electromagnetic proton cyclotron waves presents a problem because some considerations of a standard linear plasma theory may imply that such waves should not reach observable amplitudes. Our main argument in favor of their existence in impulsive solar flares is in analogy with their presence in the so-called 'inverted-V' events in the Earth's aurora (12, 13, 14). Inverted-V events are a frequently observed signature of electron fluxes measured on satellites or rockets passing through the Earth's aurora and are so called because of their inverted-V shape when plotted on energy-time spectrograms. The standard interpretation of these inverted-V events is based on the production of accelerated electrons through quasi-static electric fields parallel to the magnetic field. Typically, the potential drop amounts to a few kV (though it can reach 40 kV in some events) and the latitudinal extension covers a few tens of kilometers. The acceleration occurs in low-plasma density regions at altitudes of several thousand kilometers. From the considerations of a standard plasma theory one would expect that the acceleration of electrons through a parallel electric field would produce a beam of electrons and that such a beam would be unstable to a variety of plasma waves. Indeed, several plasma wave modes are observed on auroral field lines in association with inverted-V events. The most notable waves are (a) auroral kilometric radiation, seen near the electron cyclotron frequency, predominately in the X-mode (15); (b) VLF hiss, observed below the electron plasma frequency on the oblique lower hybrid branch (16); (c) electrostatic ion cyclotron waves, observed above the proton cyclotron frequency (17); and (d) electromagnetic proton cyclotron waves, generated at frequencies below the proton cyclotron frequency and above the highest cyclotron frequency of the next majority ion (18, 13, 19). Both the VLF hiss and the electromagnetic ion cyclotron waves can be generated by electron beams or bump-on-tail instabilities. However, the growth rate of the VLF hiss is much larger than the growth rate of the electromagnetic cyclotron wave. Therefore one might expect that the free energy in the electron distribution would be depleted by the faster growing waves before the electromagnetic ion cyclotron wave could grow to sufficiently large amplitudes.

One possible explanation is that electron acceleration occurs in narrow channels aligned with the magnetic field. Because the group velocity of electromagnetic ion cyclotron waves is almost parallel to the magnetic field, while the group velocities of the other waves can be at large angles to the magnetic field, the other waves may convect out of the instability region before growing to large amplitudes. Lysak and Temerin (20) examined this possibility and found that auroral channels of about 100m are required (in solar flares this would correspond to smaller dimensions in the much stronger magnetic fields and larger plasma densities), smaller than the typical perpendicular wavelength of the waves. A more likely scenario is the generation of the waves by processes related to the nonlinear mechanisms that accelerate the electrons. In the case of the aurora this would mean that the whole parallel potential drop region is unstable to oscillations with frequencies near the ion gyrofrequency. In this case the low wave growth rates are beneficial because they generally imply a low damping rate which allows the waves to propagate long distances along the magnetic field.

III. ACCELERATION MECHANISMS

In our acceleration model ^{3}He and heavy ions are accelerated by obliquely propagating electromagnetic proton cyclotron waves. In a multicomponent ion plasma the single shear Alfvén mode which exists below the ion gyrofrequency splits into several modes corrresponding to each ion gyrofrequency. If we ignore the effect of the very small abundance of ^{3}He on the mode structure, then the electromagnetic proton cyclotron mode is confined between the proton gyrofrequency and the two-ion cutoff frequency, which in coronal plasmas is slightly above the doubly-charged ^{4}He gyrofrequency. The ^{3}He is the only ion with a gyrofrequency in this range. The acceleration process is due to a resonance with this cyclotron wave. The process itself is straightforward: when a wave frequency is close to the ion gyrofrequency, the wave electric field can stay in phase with the ion for a long time and accelerate it to high energies. However, there are actually some important and interesting subtleties involved in the acceleration process which make acceleration in a nonuniform magnetic field, which occurs naturally in many environments, significantly different from accceleration in the often considered idealized case of an uniform magnetic field. This difference is especially striking in the case of a monochromatic wave, which by itself, of course, introduces another idealization. In fact, in a uniform magnetic field a monochromatic wave produces little acceleration even if the gyroresonance condition is satisfied, while a large amount of acceleration is possible even in a slightly nonuniform magnetic field.

Let us consider two cases of an oblique, monochromatic, electromagnetic proton cyclotron wave with a frequency equal or close to the ^{3}He gyrofrequency in a uniform and nonuniform magnetic field. Figure 1 shows the sketch of the geometry in the latter case. In a uniform magnetic field the ^{3}He ion may gain a small amount of energy due to resonance with the wave, while altering slightly its parallel velocity which through the Doppler effect removes the ion from resonance. In order to see why resonance must change the parallel velocity of the ion, it is best to shift into a frame of reference moving with the parallel phase velocity of the wave. In this frame energy is conserved and a change in the ion's perpendicular velocity is compensated by a change in its parallel velocity. In a nonuniform plasma, the change in the parallel velocity due to the resonance is balanced by the magnetic mirror force. This balance creates an interesting situation: the ion becomes trapped at a position and parallel velocity where these forces balance. The energy gain rate is thus determined by the gradient of the magnetic field since the resonance adjusts to balance the mirror force. The mirror force depends on the ion's perpendicular energy; thus the perpendicular energy gain on the average becomes exponential (21)

$$W_\perp = W_o \exp(-v_\phi t/L) \tag{1}$$

where $v_\phi = \omega/k_\parallel$ is the parallel phase velocity and L is the magnetic field gradient.

Figure 2 shows a simulation demonstrating the above points. In this simulation the Lorentz equation is solved numerically in the presence of an obliquely propagating electromagnetic ion cyclotron wave and an inhomogeneous background magnetic field. The background magnetic field is given by $B(x, y, z) = B_o[(1 - z/L)\hat{z} + (x\hat{x} + y\hat{y})/2L]$, with a gradient scale L of 2000 km and magnetic field $B_o = 500$ G, the wave propagates in the $x - z$ plane with an electric field $E_o = 80V/m$ and a perpendicular wavelength $k_\perp$ $= 0.4m^{-1}$, such that the ^{3}He resonance for zero parallel ion velocity occurs at $z=0$. The other electromagnetic components of the wave are given by the dispersion tensor assuming a plasma density of 10^9 cm^{-3} and an ion composition of 90% protons and 10% ^{4}He. Figure 2 shows the energy, parallel position, and parallel velocity of a ^{3}He ion in the simulation. As can be seen the ion finally escapes from the resonance after gaining about 4 MeV/nucleon. The negative parallel velocities are due to relativistic effects. The energy gain is limited since the mirror force eventually dominates the parallel force at the resonance. The mirror force increases linearly with energy, while the parallel component of the force due to resonance has a maximum proportional to $J_1(k_\perp\rho)$, which vanishes when $k_\perp\rho = 3.8$, which is the first zero of the Bessel function. Because of this zero in the Bessel function the maximum energy that a thermal ion can gain has an upper limit in the case of a monochromatic, obliquely propagating electromagnetic ion cyclotron wave, regardless of wave amplitude. However, for parallel propagation the maximum ion energy depends on the wave amplitude. Figure 3 shows a simulation with the same parameters except that the wave's phase velocity is parallel to the magnetic field ($k_\perp = 0$). In this case the ^{3}He ion gained 2.5 GeV. Thus we see that the energy gain in the case of oblique wave propagation is limited mainly by the finite perpendicular wavelength while for parallel propagation the energy gain can be enormous.

While considering a monochromatic wave gives interesting physical insight, it is more realistic to consider broadband waves. We have considered broadband waves before (10) and will not discuss them in detail here except to note some of the main differences as compared to the case of a monochromatic wave. Since a broadband wave consists of different wavenumbers $k_\perp$, the zeroes in the Bessel functions of the different components of the wave do not occur at the same gyroradius ρ and thus there is no absolute limit for the energy gain. For a uniform background magnetic field there can be energy gain since the ions can resonate with different components of the wave (22). In this uniform case the energy gain is limited by the bandwidth of the spectrum while in the nonuniform case it is limited by a balance between the mirror force and the average wave force which can now be quite irregular. The power required to accelerate an ion to a certain average energy is larger for broadband waves and the accelerated ions have a broad distribution in energy.

IV. HEAVY ION ACCELERATION

The observation that impulsive flares are enriched in both ^{3}He and heavy ions events suggests that the same waves may accelerate ^{3}He and heavy ions. However, the lack of correlation in the degree of ^{3}He and of heavy ion acceleration suggests otherwise (9, 8). Resonant stochastic acceleration by low frequency turbulence has been suggested by Miller and Vinas (22) and the cascading of these waves to higher frequency was described by Miller et al. (23). These models imply that heavy ion and ^{3}He enrichments may be due to different physical processes. Here we argue however, that the waves which accelerate ^{3}He also accelerate the heavy ions.

In this regard we wish to make two points. First, due to the similarity of the resonant mechanisms it is very likely that the same waves which accelerate ^{3}He also accelerate energetic heavy ions. In such a case it is necessary to explain why only a small fraction of the heavy ions are accelerated. Whereas the enrichment of ^{3}He can reach four orders of magnitude, the enrichment of the heavy ions is usually less than an order of magnitude, which suggests that at most 0.1% of the heavy ions in the flare plasma are accelerated to MeV energies. Our second point is that this relative lack of acceleration of the heavy ions can be explained by energy considerations together with the less efficient acceleration at low ion energies by the second harmonic process.

Figure 4 demonstrates the first point, i.e, that the same electromagnetic proton cyclotron wave which accelerates ^{3}He can also accelerate the heavier ions. The figure shows a simulation of ^{56}Fe^{+18} ion with the same parameters as those in Figure 2 except that the wave frequency is at twice the gyrofrequency of $^{56}Fe^{+18}$ ion. We observe that the acceleration proceeds to approximately one half of the ^{3}He energy per nucleon. The maximum energy is limited by the zero of the $J_2(k_\perp \rho)$ $(k_\perp \rho \approx 5.1)$ Bessel function rather than the J_1 Bessel function which applies to ^{3}He acceleration. Though the gyroradius is larger, the maximum energy is smaller because the $^{56}Fe^{+18}$ gyroradius is about twice as large for the same energy per nucleon due to the smaller charge-to-mass ratio as compared to ^{3}He. The acceleration processes are similar because in both cases the acceleration rate is controlled by a balance between the resonance and the mirror force. However, at low energies the strength of the resonant interaction is different. Since the resonant interaction of the heavy ion occurs when the wave frequency is twice the ion gyrofrequency, the maximum acceleration rate at low energies where $k_\perp \rho < 1$ is approximately $k_\perp \rho$ times smaller than the maximum acceleration rate at the first harmonic gyrofrequency (21). Thus the resulting heavy ion heating rate at low energies can be much lower than the ^{3}He rate. For broadband waves and for low ion energies the heating rate near the first harmonic is independent of ion energy, while for the second harmonic it is proportional to ion energy.

At higher energies, where $k_\perp \rho \sim 1$, the strength of the interaction at the first and second harmonics are comparable. Thus the more energetic heavy ions in the flare plasma will at least be heated further by the same waves that heat the ^{3}He ions even if some other process is responsible for accelerating the heavier ions from the thermal population. Only for nearly parallel wave propagation would this not be true. In that case $k_\perp \rho$ would be small and the second harmonic acceleration inefficient while the first harmonic acceleration mechanism would still operate. The above discussion summarizes our first point that the same waves that heat ^{3}He at least contribute to the acceleration of the heavy ions.

Our second point, that energy considerations limit the acceleration of heavy ions, can be seen with the aid of Figure 5. This figure shows schematically the distribution of ion gyrofrequencies normalized to proton gyrofrequency in the corona at a temperature of 4 MK. The vertical length of each line is proportional to the nuclear abundance (the relative number of ions weighted by their atomic mass) at that gyrofrequency, based on the thermodynamic equilibrium of charge state distributions for a temperature of 4 MK as given in Arnaud and Rothenflug and Arnaud and Raymond (24) and for elemental (25) and isotopic abundances as given in standard tables. For comparison, the relative nuclear abundance of ^{3}He is indicated at half the ^{3}He gyrofrequency. It is clear from the figure that if the same proportion of the heavy ions as of ^{3}He were accelerated to the same energy per nucleon, then the total wave energy absorbed by the heavy ions would be more than two orders of magnitude larger than that absorbed by ^{3}He. We have already shown that for some acceleration models ^{3}He by itself absorbs a significant fraction of the available wave energy (10), hence the wave would be damped if all the heavy ions were also accelerated to the measured energies. These considerations suggest the following scenerio for the regulation of heavy ion abundances in impulsive flares: Heavy ions are accelerated at the second harmonic resonance. Since the heating rate at low ion energies is proportional to ion energy, the high energy tail of the ion distribution is preferentially accelerated. The acceleration of lower energy ions is inhibited by collisions which result in plasma heating. Waves with large $k_\perp$ are the most effective in the initial heavy ion acceleration, but they are also the most heavily damped. The damping of these waves reduces the number of heavy ions that can be accelerated by moving the minimum energy required for the wave acceleration to overcome collisional loss to higher initial ion energies. If more wave power is available, a larger percentage of the heavy ions can be accelerated. Thus, events with intense waves should have higher relative enrichment in heavy ions while weaker events should result in higher relative enrichment of ^{3}He. The relative abundances of the heavier ions are a more subtle problem. The relative abundances are determined by the distribution of wave power and charge states at the various ion gyrofrequencies.

V. CONCLUSIONS

The acceleration of ^{3}He and of heavy ions is due to a resonant interaction with electromagnetic proton cyclotron wave. The existence of these waves is postulated in analogy to the Earth's aurora where electromagnetic proton cyclotron waves are often observed to be the most intense electromagnetic modes. The enhanced acceleration of the ions is due to the increased residence time in the resonant region, which results from a balance between the wave field and the mirror field. The ratio between the ^{3}He and heavy ion enrichment depends on the wave intensity and on the thermodynamic conditions of the flaring corona.

REFERENCES

1. R. Pallavicini, S. Serio, and G. S. Vaiana, Astrophys. J. **216**, 108 (1977).
2. H. V. Cane, V., R. E. McGuire, and T. T. von Rosenvinge, Astrophys. J. **301**, 448 (1986).
3. D. V. Reames, Astrophys. J. Supp. **73**, 235 (1990); D. V. Reames, Adv. Space Res. **13**, 331 (1993).
4. K. C. Hsieh, and J. A. Simpson, *Astrophys. J. Lett.* **162**, L191 (1970); W. F. Dietrich, Astrophys. J. **180**, 955 (1973); T. L. Garrard, E. C. Stone, and R. F. Vogt, in *High Energy Phenomena on the Sun*, ed. R. Ramaty and R. G. Stone (NASA SP-342), p. 341, (1973).
5. G. M. Mason, J. E. Mazur, and D. C. Hamilton, Astrophys. J. **425**, 843 (1994).
6. M. A. Coplan, K. W. Oglivie, P. Boschsler, and J. Geiss, Solar Phys. **93**, 415 (1984).
7. E. Anders and H. Grevesse, Geochim. Cosmochim. Acta **53**, 197 (1989); J. Geiss and H. Reeves, Astr. Ap. **93**, 198 (1981).

8. D. V. Reames, J. P. Meyer, and T. T. von Rosenvinge, Astron. Astrophys. Suppl. Ser. **90**, 649 (1994).
9. G. M. Mason, D. V. Reames, B. Klecker, D. Hovestadt, and T. T. von Rosenvinge, Astrophys. J. **303**, 849 (1986).
10. M. Temerin and I. Roth, Astrophys. J. **391**, L105 (1992).
11. I. Roth and M. Temerin, Proc. Int'l. Conf. Plasma Physics, Brazil **2**, 393 (1994).
12. D. A. Gurnett, and L. A. Frank, J. Geophys. Res. **77**, 172 (1972).
13. R. L. Lysak, and M. Temerin, Geophys. Res. Lett. **10**, 643 (1983).
14. G. Gustafsson, M. André, L. Matson, and H. Koskinen, J. Geophys. Res. **95**, 5889 (1990).
15. C. S. Wu and L. C. Lee, Astrophys. J. **230**, 621
16. S. R. Mosier and D. A. Gurnett, J. Geophys. Res. **74**, 5675 (1969).
17. P. M. Kintner, M. C. Kelly, F. S. Mozer, Geophys. Res. Lett. **5**, 139 (1978).
18. D. A. Gurnett, and L. A. Frank, J. Geophys. Res. **77**, 3411 (1972).
19. R. E. Erlandson, L. J. Zanetti, M. H. Acuna, A. I. Ericksson, L. Eliasson, M. H. Boehm, and L. G. Blomberg, Geophys. Res. Lett. **21**, 1855 (1994).
20. R. L. Lysak, and M. Temerin, Geophys. Res. Lett. **10**, 643 (1983).
21. I. Roth and M. Temerin, Proc. 24th Cosmic Ray Conf., Rome, **4**, 118 (1995); I. Roth and M. Temerin, submitted to Astrophys. J., (1995).
22. J. A. Miller and A. F. Vinas, Astrophys. J. **412**, 386 (1993).
23. J. A. Miller and D. A. Roberts, Astrophys. J. **452**, 912 (1995).
24. M. Arnaud and R. Rothenflug, Astron. Astrohys. Supp. Ser. **60**, 425 (1985); M. Arnaud and J. Raymond, Astrophys. J. **398**, 394 (1992).
25. D. V. Reames, Adv. Space Res. **15**, 41 (1995).

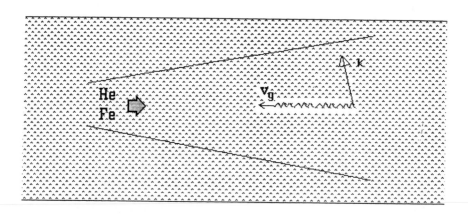

$$B = B_0 \left[(1 - z / L)\hat{z} + (x\,\hat{x} + y\,\hat{y})/2\,L \right]$$

FIG. 1. Schematic description of a segment of the the flaring corona with a small magnetic field gradient.

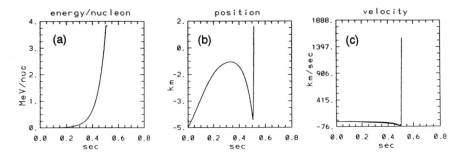

FIG. 2. Simulation of the ^{3}He ion trajectory for L=2000 km , $n = 1. \times 10^9 cm^{-3}$, $B_o = 500G$, $k_\perp = 0.4m^{-1}$, and $E_o = 80V/m$. (a) Energy, (b) parallel position ,(c) parallel velocity.

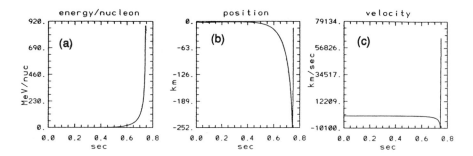

FIG. 3. Simulation of the ^{3}He ion trajectory for parallel propagating waves. L=2000 km, $n = 1. \times 10^9 cm^{-3}$, $B_o = 500G$, $k_\perp = 0.4m^{-1}$, and $E_o = 80V/m$.

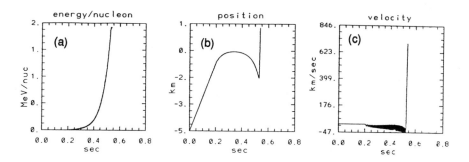

FIG. 4. Simulation of the ^{56}Fe^{+18} ion trajectory for L=2000 km , $n = 1. \times 10^9 cm^{-3}$, $B_o = 500G$, $k_\perp = 0.4m^{-1}$, and $E_o = 80V/m$. (a) Energy, (b) parallel position ,(c) parallel velocity.

443

Coronal abundance-weighted gyrofrequencies

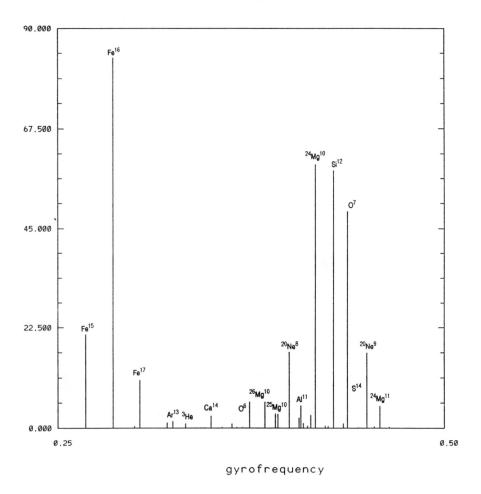

FIG. 5. Distribution of ion abundances vs their gyrofrequencies at a temperature of 4 MK. The main charge states and isotopes are marked.

Acceleration and Heating by Turbulence in Flares

Vahé Petrosian

Center for Space Science and Astrophysics
Stanford University, Stanford, California 94305

It is pointed out that plasma turbulence may play a more major role during solar flares than acknowledged previously. One theoretical and two new observational (from YOHKOH) arguments are presented to support this view.

INTRODUCTION

I would like discuss the role that plasma waves or turbulence can play in solar flares and propose a new paradigm for the impulsive phase of the flares in place of the standard thick target model. The importance of this role in the production of high energy photons has been recognized previously. I would like to present evidence here for a much more widespread and dominant role for plasma waves than has been attributed to them in the past.

It is generally agreed that the source of the flare energy is current carrying magnetic fields. Exactly how this energy is released or how it is dissipated remains controversial. Magnetic reconnection is a possible mechanism for the release and there are several mechanisms for the dissipation of this energy and for the production of the observed radiations. Dissipation can occur via three distinct channels: *1. Plasma Heating; 2. Particle Acceleration; 3. Plasma Turbulence.* Initially it was believed that most of the released energy goes directly into heating. But after the discovery of the non-thermal radiations (microwaves, hard X-rays and gamma-rays) it became clear that some of the energy goes into acceleration of particles. Over the years this has lead to the standard thick target model where most of the energy goes into particle acceleration. These particles then, in addition to producing the non-thermal radiations, heat the plasma via Coulomb collisions. Detailed analysis of the energetics, the spectra and the evidence for the so called Neupert Effect provide support for this model. Here I would like to examine the consequence of the possibility that most of the flare energy is directed through the third channel. The resultant turbulence can both accelerate particle and heat the plasma which could then produce the observed non-thermal and thermal radiations. In what follows I put forward some theoretical arguments and then present two new pieces of evidence in support of this new paradigm.

PARTICLE ACCELERATION

Three different mechanisms have been suggested for particle acceleration in flares [1].

Electric Fields parallel to the magnetic field can accelerate particles. Fields greater than the Dreicer field $\mathcal{E}_D$ lead to runaway particles (unstable particle distributions) and perhaps to turbulence. Sub-Dreicer fields extending over a distance L in a plasma with $kT \simeq 1$ keV can give rise to energy gains of $\Delta E \leq 50\text{keV}(nL/10^{19}\text{cm}^{-2})(\mathcal{E}/\mathcal{E}_D)$, where n is the plasma density. It is, therefore, difficult to accelerate particles to very high energies, especially those needed for the production of gamma-rays (tens of MeV electrons or GeV protons), unless the density becomes very high as in the model proposed by Holman [2]. There are, however, other difficulties with this model [3].

Shocks can also accelerate particles if there exist some scattering agents to force repeated passage of the particles across the shock. This is sometimes referred to as a first order Fermi acceleration mechanism. The rate of energy gain in this process is governed by the pitch angle scattering rate, namely the pitch angle diffusion coefficient $D_{\mu\mu}$. For solar flare conditions the most likely agents for scattering are plasma waves or turbulence.

Stochasic Acceleration by Turbulence, a second order Fermi process, is the third possible mechanism of acceleration. We believe this to be the most likely mechanism in solar flares for the reasons outlined below.

The turbulence needed for scattering can also accelerate particles stochastically with a rate D_{pp}/p^2, where D_{pp} is the momentum diffusion coefficient of particle with momentum p. At relativistic energies this rate is smaller than the pitch angle scattering rate so that acceleration by strong shocks is a possible efficient mechanism. However, for flare condition this added requirement of the presence of strong shocks seems unnecessary because the stochastic acceleration time scale $\tau_a \propto (\Omega_e X)^{-1}$, where Ω_e is the electron gyro-frequency and X is the ratio of energy density of the turbulence to magnetic field energy density, could be a fraction of a second even for $X \simeq 10^{-6}$. Furthermore, as stressed by Jones [4] there is little qualitative difference between the first and second order Fermi acceleration.

A second and more important reason for considering stochastic acceleration stems from our desire to be able to accelerate particles from the background plasma. We [5,6] and others [7] have shown that contrary to the prevalent belief turbulence can accelerate thermal particles within a short time. More importantly, at such low energies the above inequality between the scattering and acceleration rates is reversed [8]. At low energies $D_{pp}/p^2 > D_{\mu\mu}$, so that stochastic acceleration with the rate $\propto D_{pp}/p^2$ becomes more efficient than shock acceleration whose rate is governed by the smaller coefficient $D_{\mu\mu}$. Thus in the presence of turbulence low energy particles are accelerated stochastically whether or not there are shocks present. Shock acceleration could become important once the particles have achieved some high energy, the value of which depends on the plasma and turbulence parameters. However, this is not necessary because the turbulence alone can accelerate particles from thermal energies to very high energies given an appropriate amount and spectrum of turbulence [6].

THE MODEL

We can therefore imagine two different scenarios for particle acceleration in solar flares. In one scenario low energy particles are first accelerated by direct electric fields to 10 or 100 keV and then to higher energies by shocks, thus requiring simultaneous, and perhaps co-spatial, presence of parallel electric fields, shocks and turbulence or some other strong scattering agent. We favor a second much simpler scenario, namely stochastic acceleration by plasma turbulence alone, which as mentioned above can accelerate background plasma particles to high energies. In addition, a variety of particle spectra (power laws with breaks at low and high energies) can be produced in this scenario [9].

We propose, therefore, the third channel described in the Introduction as the main channel for the dissipation of the flare magnetic energy. Presumably this energy is released at the top of a coronal loop and somehow is converted into plasma waves or turbulence. Some of the waves or turbulence produce direct heating and some of their energy goes into acceleration of particles. Because of the efficient pitch angle scattering that will be present, especially once the particles achieve high energies, the accelerated particles will be trapped in the turbulent region and diffuse out gradually. But eventually they escape this region and precipitate down the loop producing radiation and heating as in the thick target model.

This model, in addition to its relative simplicity, provides natural explanation for two new observations by YOHKOH which cannot be explained simply by the standard thick target model.

IMPULSIVE FOOT POINT SOFT X-RAY EMISSION

In the standard thick target model the soft X-ray emission is attributed to thermal radiation from the heated plasma. This emission is expected to occur throughout the loop and increase gradually, reaching its peak at the end of the impulsive phase, obeying the Neupert effect relation [10]. High spatial resolution observations by YOHKOH show, in addition to this component, a component of soft X-ray emission which, like the hard X-ray flux, has an impulsive time profile and originates from the foot points of the loop [11]. This new component has no simple or natural explanation in the standard thick target model. The foot point flux in the event described in the above reference exceeds the flux that is expected from the extrapolation of the hard X-ray flux. This, however, may not be a universal characteristics of this emission [12].

As I have discussed previously [13], it is difficult to explain these observations as either thermal or nonthermal emission resulting from the action of the accelerated electrons. But heating of the foot point plasma during the impulsive phase by an agent with a mean free path much longer than that of the electrons could provide a natural explanation for these observations. Plasma waves or turbulence could play this role. More work is required to determine the amount and spectrum of the waves that are needed for a quantitative explanation of this phenomenon.

LOOP TOP HARD X-RAY EMISSION

Another observation which does not fit the standard thick target model is the discovery by YOHKOH of impulsive hard X-ray emission simultaneously from loop tops and the foot points [14]. This property appears to be fairly common having been seen in six out of ten flares examined [15]. In the thick target model one expects emission from the top of the loops only, more or less throughout the loop, or from the foot points only, depending if $\lambda_{coll}/L = (E/7\text{keV})^2(10^{19}\text{cm}^{-2}/Ln)$ is $\ll$, $\simeq$, or $\gg$ than one, respectively. Here, λ_{coll} is the collision mean free path of the electrons and L is the length of the coronal part of the loop. This is because Coulomb collisions will either prevent the precipitation of the electrons to the foot points giving rise to emission from the top of the loops alone or will be ineffective in stopping the electrons in the corona in which case there will be emission from the foot points alone. However, in the stochastic acceleration model proposed above the waves which accelerate particles can trap them as well, giving rise to a high density of nonthermal electrons at the top of the loop. Therefore, a significant flux of hard X-rays could be produced there before the particles escape the acceleration region. The escaping particles will give rise to the usual radiations at the foot points.

It can be shown that very roughly the ratio of the hard X-ray flux from the top of the loop to that from the foot points is $J_{LT}/J_{FP} = L^2/(\lambda_{coll}\lambda_{scat}) = T_{esc}/\tau_{coll}$, where λ_{scat} is the scattering mean free path of the electrons in the acceleration region, $T_{esc} = L^2/(\lambda_{scat}v)$ is the escape time, and $\tau_{coll} = \lambda_{coll}/v$ is the collision time of the electrons with velocity v. One can obtain a similar relation between the spectra from the foot points and the top of the loop. This we believe is the simplest explanation for these observations. Of course, a more rigorous analysis is required for the determination of the validity of this model. If this turns out to be true then we will be facing the exciting possibility of being able to determine the physical characteristics of the acceleration region from the values and energy dependences of the above mentioned variables.

CONCLUSIONS

The standard thick target model can describe many of the observed characteristics of the impulsive emission of solar flares, but it fails to provide natural explanations for the two new observations described above. In addition, it appears that [16] the so called Neupert effect [10], which in some way is an essential feature of this model, is not universally valid for all flares. Based on these observations and on some theoretical arguments, I have proposed a new paradigm for the acceleration of electrons, and possibly protons, and for the impulsive emission of flares. In this model the energy released is converted to plasma waves or turbulence which accelerate the particles at the top of a loop. The waves also efficiently scatter these particles and consequently trap them in this region, giving rise to hard X-ray emission [14,15]. These electrons eventually escape the acceleration region and precipitate down the loop and radiate and lose their energy at the foot points giving rise to emissions similar to those expected from the standard thick target model. Some of the plasma waves can also diffuse out of the acceleration region and reach the foot points, dissipating there and giving rise to the observed [11] impulsive soft X-ray emission from the foot points.

Of course this division into acceleration and foot point regions is artificial and a more detailed inhomogeneous model is required, with the prospect of determining the characteristics of the acceleration process.

Acknowledgements: I would like to thank T. T. Lee, B. Park and Julia Pryadko for their contributions to this work which is supported by grants NSF ATM 93-1188 and NASA NAGW 1976.

REFERENCES

1. See for example, R. Ramaty, in *Solar Flares* ed. P. A. Sturrock (Boulder: Colorado Assoc. Univ. Press), 177 (1980); M. A. Forman et al., in *Physics of the Sun Vol. 2*, eds. P. A. Sturrock et al., (Boston: Reidel), 249 (1986)

2. G. Holman, these proceedings (1996)

3. G. Emslie, these proceedings (1996)

4. F. C. Jones, ApJ Supplement,**90**, 561 (1994)

5. R. J. Hamilton and V. Petrosian, ApJ, **398**, 350 (1992)

6. R. Dung and V. Petrosian, ApJ, **421**, 550 (1994)

7. J. Miller, these proceedings (1996)

8. J. Pryadko and V. Petrosian, BAAS, **27**, 977 (1995)

9. B. T. Park and V. Petrosian, ApJ, **446**, 699 (1995)

10. W. M. Neupert, ApJ Letters, **153**, L59 (1968)

11. H.S. Hudson et al., ApJ Letters, **422**, L25 (1994)

12. J. M. McTiernan, BAAS, **27**, 988 (1995)

13. V. Petrosian, in the Proc. of Kofu Symposium, eds. S. Enome and T. Hirayama (Japan: Noboyama Radio Observatory Report No. 360), 239 (1994)

14. S. Masuda et al., Nature, **371**, 455 (1994)

15. S. Masuda, Ph. D. Thesis, The University of Tokyo (1994)

16. T. T. Lee, V. Petrosian and J. M. McTiernan, ApJ, **448**, 915 (1995)

Heavy Ion Acceleration by Cascading Alfvén Waves in Impulsive Solar Flares

James A. Miller[*] and Donald V. Reames[†]

[*]*Department of Physics, The University of Alabama in Huntsville
Huntsville AL 35899*
[†]*Code 661, NASA Goddard Space Flight Center
Greenbelt MD 20771*

We propose that the heavy ion abundance enhancements that are observed for impulsive solar flares result from stochastic acceleration by cascading Alfvén wave turbulence. In our model, Alfvén waves are generated at some large scale and nonlinearly cascade to higher wavenumbers and frequencies. As the waves increase in frequency, they will be able to cyclotron resonate with ions of progressively lower energy. For a thermal plasma there will be no damping at low wavenumbers and the waves will freely cascade. However, when the wave frequency becomes close to an ion cyclotron frequency, thermal ions will be accelerated out of the background and to relativistic energies. The first ion species encountered by the waves will be the one with the lowest cyclotron frequency, namely Fe. Due to the low Fe abundance, the waves will not be completely damped and will continue to cascade up to the group of ions with the next higher cyclotron frequency, namely Ne, Mg, and Si. Again, these ions will be accelerated but the waves will not be totally damped. After Ne, Mg, and Si the waves encounter ^{4}He, C, N, and O, which do completely dissipate the waves and halt the cascade. We show that abundance ratios similar to those observed can result from this process.

INTRODUCTION

Solar flares can broadly be classified as either impulsive or gradual. This distinction was initially based upon the time scale of their soft X-ray emission (hence the terminology), and morphologically they are also quite different. Impulsive events are relatively small and confined to the low corona, while gradual events are large and take place in the high corona. Subsequently, many other differences were found (e.g., 1 and 2 and references therein), but perhaps the most significant in terms of particle acceleration processes is the fact that impulsive flares produce dramatic abundance enhancements of ^{3}He, Fe, Ne, Mg, and Si in the energetic particles that are seen directly in space (3) and that interact at the Sun (4), while gradual flares do not.

Ions in gradual flares are likely accelerated directly out of the ambient coronal plasma by a shock that is driven by a coronal mass ejection (CME). This shock sweeps up ambient ions and accelerates them without appreciably changing their abundance ratios at high energies relative to those in the corona. Particles in impulsive flares are likely accelerated in the region of primary energy release, and the abundance enhancements are an important clue to the identity of the ion acceleration mechanism for this class of events. Impulsive flare particle acceleration is in general problematic, given the numbers of both ions and especially electrons that are energized out of the thermal plasma and up to relativistic energies on short time scales, and the enhancements are another severe constraint on any process. However, understanding impulsive flare acceleration is central to high energy flare physics, and it is this class of events that we will consider.

Particle acceleration models fall into three general classes: those that are based upon shocks, those are based upon large-scale DC electric fields, and those that employ stochastic energization. Stochastic acceleration is attractive on several points, and it furthermore offers the opportunity to unify ion and electron acceleration within the context of a single model. Specifically, the picture that is emerging is one in which resonant wave-particle interactions are able to account for the acceleration of a sufficient number of both protons (5) and electrons (6) out of the thermal background and to relativistic energies on the observed time scales in impulsive solar flares. The essence of the proton acceleration model is that Alfvén waves are produced at large scales, cascade to higher frequencies, and eventually cyclotron resonate with and accelerate background thermal protons. The essence of the electron acceleration model is that fast mode waves (the "sister" MHD wave of the Alfvén wave, loosely speaking) are produced at the same large scale, also cascade to higher frequencies, and eventually efficiently Landau resonate with and transit-time accelerate the background thermal electrons.

In this paper we will consider the acceleration of heavy ions, and it is worthwhile to review their basic observational characteristics. We show in Table 1 the impulsive and gradual flare ion abundance ratios. Note that ^{4}He, C, N, and O are not enhanced relative to each other. The Fe/O ratio is about a factor of 8 larger than the coronal value, while the Ne/O, Mg/O, and Si/O ratios increase by a smaller but still significant amount.

There are not many theories that address the acceleration of ^{3}He and heavier ions. Due to its huge enhancement, ^{3}He really needs to be considered in a class by itself. The most attractive theory to date for the large ^{3}He/^{4}He ratio that is seen from impulsive solar flares achieves preferential ^{3}He acceleration by resonance between this ion and electromagnetic hydrogen cyclotron waves that are driven unstable around the ^{3}He cyclotron frequency (7,8). An electron beam is a possible source of free energy for the wave growth. This model was extended (9) to include the preferential acceleration of Ne, Mg, Si, and Fe as well, but failed to produce the maximum observed energies. A secondary acceleration mechanism was thus required. We do not address ^{3}He further,

TABLE 1. Ion abundance ratios in particles of energy $\gtrsim 1\,\mathrm{MeV\,nucleon}^{-1}$ from impulsive and gradual solar flares. Ratios for gradual flares are the same as those in the corona. Data from ref. 2 and references therein.

Ratio	Impulsive Flares	Gradual Flares (Corona)
$^3\mathrm{He}/^4\mathrm{He}$	~ 1 ($\times 2000$ increase)	~ 0.0005
$^4\mathrm{He}/\mathrm{O}$	≈ 46	≈ 55
C/O	≈ 0.436	≈ 0.471
N/O	≈ 0.153	≈ 0.128
Ne/O	≈ 0.416 ($\times 2.8$ increase)	≈ 0.151
Mg/O	≈ 0.413 ($\times 2.0$ increase)	≈ 0.203
Si/O	≈ 0.405 ($\times 2.6$ increase)	≈ 0.155
Fe/O	≈ 1.234 ($\times 8.0$ increase)	≈ 0.155
H/He	~ 10	~ 100

other than to appeal to the electron beam model for its explanation (see also 10).

Qualitatively, the cascading Alfvén wave model will naturally yield the abundance ratios between $^4\mathrm{He}$ and heavier ions. In an actual flare multi-ion plasma, the waves will encounter Fe first, since it has the lowest cyclotron frequency Ω. Iron will be strongly accelerated but is not abundant enough to damp the waves. Consequently, some wave energy cascades to higher frequencies where it encounters Ne, Mg, and Si. The same way, these ions suffer strong acceleration but the wave dissipation is not complete. Some wave energy cascades to reach $^4\mathrm{He}$, C, N, and O. These ions are sufficiently numerous to totally damp the waves and the cascade ceases. Hence, Fe will resonate with the most powerful waves; Ne, Mg, and Si will resonate with waves having less power; and $^4\mathrm{He}$, C, N, and O will resonate with the least powerful waves. Note that if cascading does occur up to $^4\mathrm{He}$, all of the ambient Fe, Ne, Mg, and Si will be energized. To zeroeth order then, Fe acceleration is the most robust, and this ion is expected to be enhanced the most relative to $^4\mathrm{He}$, C, N, and O in the accelerated particles. Enhancement of Ne, Mg, and Si will also occur, but not to the same high degree as Fe. Since $^4\mathrm{He}$, C, N, and O all have the same cyclotron frequency and behave similarly, they should not be enhanced relative to each other.

This general scenario is in accordance with the observations in Table 1, and in the remainder of this paper we quantitatively verify it. We point out that, given the brevity of the paper, we do not perform an exhaustive search of parameter space in order to determine under what conditions the model works. We simply show that it works under one set of realistic conditions. A more detailed study will be performed in the future.

TABLE 2. List of the most important heavy ions in a flare plasma. For ion i, the number density relative to H is f_i, and the most probable charge-to-mass ratio is Q/A. Abundances are coronal and determined from ref. (11). Charge-to-mass ratios are at 3×10^6 K and determined from ref. (12)

Ion	$f_i \times 10^4$	Q/A
Fe	1.32	0.29
Ne	1.29	0.4
Mg	1.73	0.42
Si	1.32	0.43
O	8.54	0.5
N	1.09	0.5
C	4.03	0.5
^{4}He	470	0.5

THE ACCELERATION MODEL

We consider a multi-ion flare plasma consisting of ^{4}He and heavier ions in coronal abundance. We take the temperature to be 3×10^6 K, which was argued for previously (3) on general grounds. A list of these ions, along with their abundances relative to H and their most probable charge-to-mass Q/A ratios are given in Table 2. Since Ne, Mg, and Si have similar values of Q/A, they will encounter the cascading Alfvén waves at about the same time. We thus treat this group (the Ne group) as a single species with relative abundance 4.34×10^{-4}, charge-to-mass ratio 0.42, and mass number 24 (the last two quantities are abundance weighted). For the same reason, ^{4}He, C, N, and O (the He group) are treated as a single species with relative abundance 4.84×10^{-2}, charge-to-mass ratio 0.5, and mass number 4.3. Iron is not grouped. The problem now reduces to finding how Fe, the Ne group, and the He group each respond to cascading Alfvén waves.

The actual environment in which solar flare particle acceleration takes place is likely to be very complicated. We idealize this situation by assuming that acceleration occurs in a single region, which consists of a uniform fully-ionized plasma permeated by a static homogeneous ambient magnetic field $\vec{B}_0 = B_0\hat{z}$. The four components of the plasma are Fe ions, Ne group ions, He group ions, and electrons. The length of the region along the magnetic field is L. Particles escape from the region upon crossing a boundary either at $z = 0$ or $z = L$ (that is, we take the loss cone half angle to be $90°$ at both ends). We assume that a homogeneous spectrum of Alfvén waves exists throughout the acceleration region.

Stochastic Acceleration

The transverse electric field of a plasma wave can strongly affect particle motion through gyroresonant interactions. Such an interaction occurs when the Doppler-shifted wave frequency is an integer multiple of the cyclotron frequency in the particle's guiding center frame, and the sense of rotation of the particle and the field are the same. In this case, depending upon the initial relative phase of the wave and particle, the particle will corotate with either an accelerating or decelerating electric field over a significant portion of its gyromotion, resulting in an appreciable gain or loss of energy, respectively. The condition for resonance to occur is $\omega - k_\parallel v_\parallel - \ell\Omega_i/\gamma = 0$, where γ, $v_\parallel$, and Ω_i are the Lorentz factor, the parallel velocity component, and the gyrofrequency of the ion, and ω and $k_\parallel$ are the frequency and parallel wavenumber of the wave. If the particle and transverse wave electric field rotate in the same sense relative to $\vec{B}_0$ in the plasma rest frame, the harmonic number $\ell > 0$, whereas an opposite sense of rotation requires $\ell < 0$ (for $\omega > 0$).

We assume for simplicity that the Alfvén waves propagate parallel and antiparallel to $\vec{B}_0$. These waves have left-hand circular polarization relative to $\vec{B}_0$ and occupy the frequency range below the hydrogen cyclotron frequency Ω_H. In a multi-ion plasma, there are resonances and cutoffs in the dispersion relation corresponding to each ion species (13). However, we treat Fe and the Ne group as particles that do not affect the dispersion relation. Because of their large abundance, the He group will still produce a resonance at Ω_He and a cutoff at slightly higher frequency. In this case, $\omega = v_\mathrm{A}|k_\parallel|$ when $\omega \ll \Omega_\mathrm{He}$, where v_A is the Alfvén speed. As the frequency increases, the dispersion relation is no longer linear in $|k_\parallel|$, and $|k_\parallel| \to \infty$ when $\omega \to \Omega_\mathrm{He}$. A diagram of the dispersion relation is shown in Figure 1.

We justify our treatment of the Fe and Ne groups by noting that all of these ions will be accelerated, so that their distributions will clearly not remain thermal (either in whole or even for the most part). In this case, the linear Vlasov treatment of the dispersion properties of the plasma is invalid, so that the usual dispersion relation will not hold anyway. The correct treatment of this situation requires a full kinetic plasma simulation. Our simpler approach, however, should be quite accurate, given that the effect of these ions on the waves is relatively small for the turbulence injection rates we need to employ.

For these parallel waves, the gyroresonance condition can only be satisfied with $\ell = +1$, and this is just ordinary cyclotron resonance. Cyclotron resonance will alter both the particle's energy and pitch angle, but pitch angle scattering is much more rapid than acceleration for flare values of v_A (5). The particle distributions can thus be taken to be isotropic. Hence, each particle distribution f can be described by a momentum diffusion equation (eqn. 2.1a of reference 5). The momentum diffusion coefficient $D(p)$ that appears in this equation is obtained from D_{pp} by $D(p) = (1/2) \int_{-1}^{+1} d\mu\, D_{pp}$.

In general, a proton can cyclotron resonate with two waves moving along its direction of motion and one wave moving in the opposite direction (14). However, unless μ is very close to zero, there is only one resonant wave and it is the backward-moving one. As this wave also has the greatest effect on ion motion (by virtue of having the largest amplitude for a power-law spectral

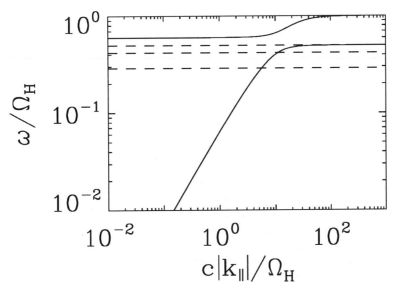

FIG. 1. Dispersion relation of the Alfvén wave branch in a fully ionized plasma consisting of H, ^{4}He, C, N, and O. Note the resonances at Ω_{He} and Ω_H, and the cutoff at $\approx 0.6\Omega_H$. The lowest dashed line marks the cyclotron frequency of Fe, the next highest dashed line is for the Ne group, and the highest dashed line is for the He group. ($v_A = 0.67c$ here.)

density), it is the only one we consider. In light of these simplifications, we also take the Alfvén wave dispersion relation to be $\omega = v_A|k_\parallel|$ for all $\omega < \Omega_{He}$. This is a very good approximation for all particles except those with the lowest energies and $\mu \approx 0$. The linearity of the dispersion relation can also be seen in Figure 1. The wavenumber of the resonant wave for an ion of species i is quickly found to be

$$k_{\parallel r} = \frac{-\varepsilon\Omega_i}{\gamma\left(v_A + \varepsilon v_\parallel\right)} \quad , \tag{1}$$

where $\varepsilon = v_\parallel/|v_\parallel|$. When $v_\parallel = 0$, $\varepsilon = +1$.

The diffusion coefficient D_{pp} is given by equation (B11) of reference (5), using $k_{\parallel r}$ in equation (1). It is then straightforward to calculate $D(p)$. At this point it is convenient to change to dimensionless variables. Momentum $p = \tilde{p}mc$; kinetic energy $E = \tilde{E}mc^2$; speed $v = \tilde{v}c$; frequency $\omega = \tilde{\omega}\Omega_H$; wavenumber $k_\parallel = \tilde{k}_\parallel\Omega_H/c$; length $L = \tilde{L}c/\Omega_H$; and time $t = \tilde{t}\,T_H$, where $T_H = \Omega_H^{-1}$. The spectral density W_T is the total wave energy density per unit wavenumber, and is normalized to the ambient magnetic field energy density $U_B = B_0^2/(8\pi)$, so that $\tilde{W}_T = (\Omega_H/c)W_T/U_B$. That is, $\tilde{W}_T\,dk_\parallel$ is the total wave energy density, in units of U_B, in the normalized wavenumber interval $dk_\parallel$ about $k_\parallel$. The momentum diffusion coefficient for ion species i is then

$$\widetilde{D}(p) = \frac{\pi}{4} \frac{\delta_i^2 \, \widetilde{v}_A^2}{\widetilde{v}} \int_{\widetilde{k}_0}^{\widetilde{k}_m} d\widetilde{k}_{\parallel} \, \frac{1}{\widetilde{k}_{\parallel}} \widetilde{W}_T(\widetilde{k}_{\parallel}) \left[1 - \frac{\widetilde{v}_A^2}{\widetilde{v}^2} \left(1 - \frac{\widetilde{k}_m}{\widetilde{k}_{\parallel}} \right)^2 \right] \quad , \qquad (2)$$

where $\widetilde{k}_0 = \delta_i / [\gamma(\widetilde{v}_A + \widetilde{v})]$ and $\widetilde{k}_m = \delta_i / (\gamma \widetilde{v}_A)$ are the minimum and maximum resonant wavenumbers at momentum p. Here, $\delta_i = Q/A$. Consistent with our assumption above on the linear form of the dispersion relation, we have assumed equipartition between the wave vacuum field energy and the oscillation kinetic energy of the background protons in order to relate the magnetic spectral density W_B to the total spectral density W_T (namely, $W_T = 2W_B$). We also assumed that $W_T(k_{\parallel}) = W_T(-k_{\parallel})$ (see below).

The last element in the momentum diffusion equation is a leaky-box escape term, which takes into account particle escape from the acceleration region. In general, both diffusion and simple transit must be taken into account in order to determine this term (15). However, here we include only transit escape, in which case the characteristic escape time scale is $2L/v$.

Wave Spectral Density

To proceed further, we must specify the form of the spectral density. We assume that Alfvén waves are continuously injected at some wavelength λ and that a Kolmogorov-like nonlinear cascade transfers the wave energy to smaller scales. It is dissipated there by accelerating ions. The precise mechanism by which waves are generated in flares is unknown. MHD waves can easily be formed by a large-scale restructuring of the flare magnetic field, which presumably occurs at least in the initial energy release. In this case, λ is expected to be on the order of the flare size, say 10^7–10^9 cm. Values of λ much smaller than this would require a microinstability. The corresponding wave injection wavenumbers are $\pm k$, where $k = 2\pi/\lambda$.

Injection, cascading, and damping are described by the diffusion equation in wavenumber space

$$\frac{\partial W_T}{\partial t} = \frac{\partial}{\partial k_{\parallel}} \left(D_{\parallel\parallel} \frac{\partial W_T}{\partial k_{\parallel}} \right) - \gamma(k_{\parallel}) W_T + S, \qquad (3)$$

where $D_{\parallel\parallel}$ is a coefficient that depends upon W_T and can be determined for Kolmogorov cascading (5), $\gamma(k_{\parallel})$ is the damping rate due to accelerating ions, and S is the volumetric wave energy injection rate per unit $k_{\parallel}$. Since the spectral density is symmetric about $k_{\parallel} = 0$, we only consider $k_{\parallel} > 0$. Here $S = Q\delta(k_{\parallel} - k)$, where $Q = \widetilde{Q} U_B \Omega_H$ is the volumetric wave energy density injection rate Q at either k or $-k$. The total rate of wave energy injection in both parallel and antiparallel waves is therefore $2Q$.

The damping rate due to ion species i can be found from conservation of energy (see argument in reference 5) to be

$$\gamma_i(k_\parallel) = \frac{mc^2}{U_B} \frac{\pi}{4} \delta_i^2 v_A^2 \frac{1}{k_\parallel} \int_{E_0}^{E_m} dE \ N(E) \frac{1}{p} \left[1 - v_A^2 \left(1 - \frac{k_m}{k_\parallel} \right) \right] \quad , \quad (4)$$

where m is the ion mass and $N(E)$ is the number density of ions per unit kinetic energy E. The quantities E_0 and E_m are the minimum and maximum energies an ion can have in order to resonate with a wave of wavenumber $k_\parallel$, and are given by solutions to $k_\parallel = \delta_i / [\gamma(v_A + v)]$ and $k_\parallel = \delta_i / (\gamma v_A)$, respectively.

The total wave damping rate is due to accelerating Fe, the Ne group, and the He group. Hence $\gamma(k_\parallel) = \gamma_{Fe} + \gamma_{Ne} + \gamma_{He}$. In summary, ion acceleration and wave cascading are described by three momentum diffusion equations (one for each ion or ion group) and one wave diffusion equation, in which the damping depends upon the three ion energy spectra. The system is therefore highly coupled and nonlinear. We transform the momentum diffusion equations to Fokker-Planck equations in $\log(E)$ space (to solve for $N(E)$ directly), transform the wave diffusion equation to $\log(k_\parallel)$ space, work in terms of the dimensionless variables described above, and solve all of the equations using the Crank-Nicholson method with iteration (this is further described in reference 5).

RESULTS

We take $B_0 = 500\,\text{G}$, $v_A = 0.036c$, and assume Fe, Ne group, and He group abundances as given above. The length of the acceleration region is 6.3×10^7 cm. Waves are injected at $\lambda = 10^7$ cm, although as demonstrated in reference (5) the ion energy spectra and acceleration time scales are independent of λ. The injection rate is given by $\tilde{Q} = 4 \times 10^{-9}$ for time $t \leq 6 \times 10^6 T_H$ and by $\tilde{Q} = 0$ thereafter. This corresponds to a total injection of $480\,\text{ergs cm}^{-3}$ of wave energy over 1.25 s.

The general behavior of the waves and ions is very similar to the wave/proton system of reference (5), except that Fe is encountered first and that complete dissipation is on the He group. We cannot show the extensive amount of data that results from the simulation in the space here, but note that there are essentially three phases to acceleration, characterized by the behavior of the spectral density:

(1) Initial cascading, when spectral energy is transferred from low to high $k_\parallel$ and a power law is established throughout the inertial range. The end of this phase is marked by wave penetration into the Fe dissipation range and the initial acceleration of a large number of Fe ions to suprathermal energies. The top panel in Figure 2 shows the evolution of the Fe distribution when the

457

waves have achieved a high enough frequency to just begin accelerating the Fe ions.

(2) Quasisteady cascading, when the spectral density is approximately static at frequencies below Ω_{Fe}. During this phase, the normalization $\widetilde{W}_0$ of the spectral density $\widetilde{W}_T = \widetilde{W}_0 |\tilde{k}_\parallel|^{-5/3}$ can be determined analytically and is given by $\widetilde{W}_0 = [3\sqrt{2}\widetilde{Q}/(5\tilde{v}_A)]^{2/3}$. As far as the particles are concerned, Fe ions are accelerated to the highest energies and the waves have cascaded up to the He group and begun to accelerate some of these ions out of the tail. The bottom panel in Figure 2 shows the He group distribution when acceleration just begins. Waves in all three dissipation ranges are replenished almost as quickly as they are dissipated by acceleration.

(3) Decay, when there is no further wave energy injection, and the spectral density decreases at all wavenumbers; acceleration effectively ceases shortly after the beginning of this phase.

Note that during all phases, all of the ambient Fe and Ne group ions are accelerated, while only those He group ions in the tail are. At $t = 8 \times 10^6 T_H$, acceleration has effectively stopped, and above $1\,\text{MeV nucleon}^{-1}$ there are 3.5×10^5 Fe ions, 9.9×10^5 Ne group ions, and 3.0×10^6 He group ions that have escaped per cm^3. Let these densities be represented by n_{Fe}, n_{Ne}, and n_{He}. Then the energetic Fe/O ratio is given by $(n_{\text{Fe}}/n_{\text{He}})(4.84 \times 10^{-2}/f_O)$, where 4.84×10^{-2} is the He group abundance relative to H and f_O is the oxygen abundance relative to H. The last factor corrects for the fact that ^4He, C, N, and O were treated together. Hence, Fe/O equals 6.6. The Ne/O ratio is given by $(n_{\text{Ne}}/n_{\text{He}})(f_{\text{Ne}}/4.34 \times 10^{-4})(4.84 \times 10^{-2}/f_O)$, where 4.34×10^{-4} is the abundance of the Ne group relative to H and $f_{\text{Ne}} = 1.29 \times 10^{-4}$ is the abundance of just Ne. Hence, Ne/O is about 5.5, and Si/O and Mg/O will be similar.

CONCLUSIONS

We have considered Alfvén wave cascading in a plasma consisting of Fe, Ne group ions, and He group ions. From a comprehensive quasilinear code that takes into account wave cascading, wave damping by the accelerated ions, ion acceleration, and ion escape from the acceleration region, we have begun to explore quantitatively the detailed behavior of the system. We showed above that for a particular choice of wave injection rates and acceleration region lengths, the Fe/O ratio in the energetic particles is about 6.6, while the Ne/O, Mg/O, and Si/O ratios are about 5.5. All of these ratios are above the observed values in Table 2, but can be lowered by slightly lowering Q. Also, the model does predict that Fe/O is greater than Ne/O. We conclude on the basis of these preliminary results that cascading along with cyclotron resonant acceleration is a viable mechanism for yielding the observed heavy ion abundance enhancements in impulsive solar flares.

There is no dichotomy between preacceleration and main acceleration: the

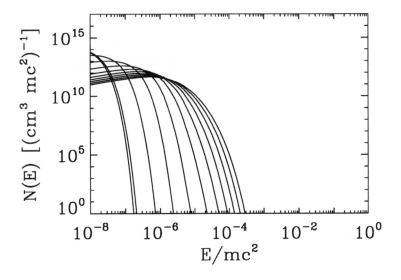

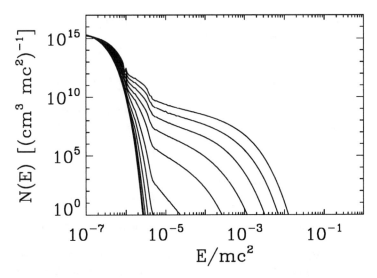

FIG. 2. Top panel: The Fe spectrum at 11 equally spaced times between 2.7×10^5 and $3.7 \times 10^5 T_H$. The leftmost curve is the initial thermal distribution. Bottom panel: the He group distribution at 11 equally spaced times between 2×10^6 and $4 \times 10^6 T_H$. The leftmost curve is the initial thermal distribution. He group acceleration occurs later than Fe acceleration since some of the Ne group ion must escape to allow cascading to the He group.

waves accelerate ions directly out of the background plasma and up to tens of MeV nucleon^{-1} on very short time scales. The model also yields energy spectra that can be compared with observations. We did not show escaping particle spectra at late times, but they look similar to the proton spectra obtained in reference (5) for Kolmogorov cascading. As such, they are too hard at low energies and too soft at high energies to produce a good fit with the observed power laws. There are other many factors that may influence the shape of an ion spectrum from its origin in the acceleration region to the point where it is observed, and the model should not be ruled out at this time on the basis of this poor agreement.

ACKNOWLEDGEMENTS

This work was supported by the NASA Cosmic and Heliospheric Physics Program through grant NAGW–4378.

REFERENCES

1. Reames, D. V., *Adv. Space Res.*, **13**, No. 9, 331 (1993).
2. Miller, J. A., *EOS*, **76**, No. 41, 401 (1995).
3. Reames, D. V., Meyer, J.-P., and von Rosenvinge, T. T., *Astrophys. J.*, **90**, 649 (1994).
4. Murphy, R. J., Ramaty, R., Kozlovsky, B.-Z., and Reames, D. V., *Astrophys. J.*, **371**, 793 (1991).
5. Miller, J. A., and Roberts, D. A., *Astrophys. J.*, **452**, 912 (1995).
6. Miller, J. A., LaRosa, T. N., and Moore, R. L., *Astrophys. J.*, in press (10 April) (1996).
7. Temerin, M., and Roth, I., *Astrophys. J.*, **391**, L105 (1992).
8. Miller, J. A., and Viñas, A. F., *Astrophys. J.*, **412**, 386 (1993).
9. Miller, J. A., Viñas, A. F., and Reames, D. V., "Heavy Ion Acceleration and Abundance Enhancements in Impulsive Solar Flares", in *Proc. 23rd ICRC* (Calgary), 1993, p. 17.
10. Litvinenko, Y. E., "On the Formation of the Helium-3 Spectrum in Impulsive Solar Flares", this volume.
11. Reames, D. V., "Energetic Particle Observations and the Abundances of Elements in the Solar Corona", in *Proc. 1st SOHO Workshop* (Annapolis), 1992, ESA SP–348.
12. Shull, J. M., and Van Steenberg, M., *Astrophys. J. Suppl.*, **48**, 95 (1982).
13. Smith, R. L., and Brice, N., *J. Geophys. Res.*, **69**, 5029 (1964).
14. Steinacker, J., and Miller, J. A., "Proton Gyroresonance with Parallel Waves in a Low-Beta Solar Flare Plasma", in *Particle Acceleration in Cosmic Plasmas*, 1992, New York: AIP, pp. 235–8.
15. Lenters, G., and Miller, J. A., "Charged Particle Diffusive Transport", this volume.

Accelerated Particle Composition in Impulsive Events : Clues to the Conditions of Acceleration

Jean-Paul Meyer

Service d'Astrophysique, CEA,DSM,DAPNIA,
Centre d'Etudes de Saclay, 91191 Gif-sur-Yvette, France

The heavy element composition anomalies systematically found in the impulsive, ^{3}He–rich, solar energetic particles events are reviewed. These anomalies imply, in a quasi-model-independent way, that the particles are accelerated predominantly out of gases with temperatures in the $\sim$ 2.5 to $\sim$ 5 MK range, *i.e.* typical active region, not flaring loop, temperatures. Existing models for selective ion acceleration by plasma waves are briefly overviewed. A specific model to account for the heavy element enhancements is presented, in terms of the damping of electromagnetic He cyclotron waves by interaction with ^{4}He ions, which sets constraints to the source gas temperature and to the acceptable rate of wave cascading. The high charge states observed among the energetic particles are discussed.

INTRODUCTION

This paper is primarily devoted to the heavy element composition anomalies, which are observed as a rule among the $\sim$ MeV energetic particles accelerated during the extremely common *impulsive*, ^{3}He–rich, solar energetic particle (SEP) events. Contrary to *gradual* SEP events, in which essentially solar wind material is being accelerated far away from Sun in the interplanetary medium, *impulsive* events are believed to accelerate material very close to the solar surface, in the immediate environment of a flaring loop, see *e.g.* (23).

I will *(i)* summarize the composition observations, *(ii)* describe the inferences on the source gas temperature that can be derived from these observations in a quasi-model-independent way, *(iii)* very briefly overview the proposed models for selective ion acceleration by plasma waves, *(iv)* present and discuss a specific model for selective heavy ion acceleration in terms of wave damping by ^{4}He ions, and *(v)* summarize all these points.

OBSERVED IMPULSIVE EVENT COMPOSITIONS

A thorough statistical analysis of the energetic ion composition in a large number of impulsive SEP events has been performed by (23). Particle-averaged compositions have also been provided by (22), which also covers rarer elements. Additional data, especially on isotopic anomalies among the accelerated ions, have been provided by (11). Largely based on (23), these observational results can be summarized as follows :

(i) The accelerated ions have been extracted from a gas with typical "coronal" composition, as evidenced by Fig. 2 and 3. This coronal composition, also observed in the slow solar wind and in *gradual*-event SEP's, differs from that of the photosphere : heavy elements with first ionization potential (FIP) $\lesssim 10$ eV are enhanced by a factor of ~ 4.5 relative to those with higher FIP, see *e.g.* (15), (16). In addition, He is depleted by a factor of ~ 2 relative to heavier high-FIP elements. This coronal will be our reference composition here.

(ii) Huge enhancements of ^{3}He/^{4}He by factors of up to ~ 20000 (actually observable only when the enhancement exceeds a factor of ~ 200 !) are commonly observed in impulsive event SEP's (see (23) and ref. therein).

(iii) Among the heavy elements, much more modest enhancements are being observed, which are *not* correlated with the ^{3}He enhancements (13), (23). The accelerated particles show on the average no enhancement of ^{4}He, C, N and O relative to each other (Fig. 1 and 2), mean enhancements of Ne, Mg and Si by roughly equal factors of ~ 2.5 to ~ 3.5 relative to CNO (Fig. 3 and 4), and a mean enhancement of Fe by factors of ~ 7.5 and of ~ 2.8 relative to CNO and to NeMgSi, respectively (Fig. 4). Studies by (13) and (11), however, suggest that O is sometimes moderately enhanced relative to C. But the more comprehensive investigation of (23) (and (22)'s average O/C ratio) shows that this is definitely not the more common situation.

(iv) The variability of these enhancements is surprisingly limited : factors of ~ 2.6, ~ 2 and ~ 1.5 around the mean value for Fe/CNO, NeMgSi/CNO and Fe/NeMgSi, respectively, and factors of $\lesssim 1.5$ for Ne/Mg, Ne/Si and Mg/Si (all at the 95 % probability level ; Fig. 3 and 4). Only the ^{4}He/C ratio shows a large spread (Fig. 2).

(v) There seems to be extra enhancements, by factors on the order of ~ 2, of the rarer species Na, Al, ^{22}Ne, and ^{25,26}Mg, as compared to the neighboring dominant ones, ^{20}Ne, ^{24}Mg, and ^{28}Si (22), (11).

(vi) We have only a single recent observation of the charge states Q of impulsive event SEP's (10). It suggests that Si is essentially fully stripped Si^{+14}, but, in view of the extremely poor charge resolution, a predominance of He-like Si^{+12} cannot be strictly ruled out. The data on Fe suggest a broad distribution of charges between $Q = 15$ and 25, peaking at $Q = 22$, and with an average charge $\overline{Q} = 20.5$. As we shall see, these high observed charge states, pose a problem. They will be discussed below.

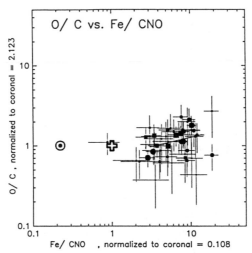

FIG. 1. The O/C ratio in impulsive event SEP's, plotted vs. the Fe/CNO ratio, which best describes the heavier element enhancements relative to coronal composition. Each solid dot represents one individual solar event. The size of the dot is inversely proportional to the statistical error on the abundance ratios. The "Sun" symbol represents the photospheric abundance ratios, and the open cross the coronal ones (to which all ratios are normalized). The average energetic particle O/C ratio is consistent with coronal (= photospheric), and there is no statistically significant spread of the O/C ratio around this value, nor a significant correlation with Fe/CNO. Adapted from (23).

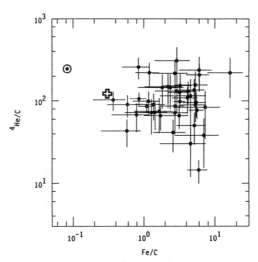

FIG. 2. Similar to Fig. 1, but for the ^{4}He/C ratio ; (in abscissas, Fe/C, which is about equivalent to Fe/CNO ; not normalized to coronal ratios). The average impulsive event SEP ^{4}He/C ratio is consistent with the coronal, *not* the photospheric, ratio. Contrary to O/C (Fig. 1), there is large spread of the ^{4}He/C ratio around this value, but no correlation with Fe/C. This spread could reflect a spread of the He abundance in the source gas. From (23).

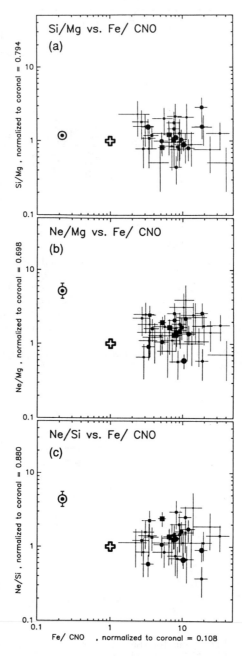

FIG. 3. Same as Fig. 1, but for the Si/Mg, Ne/Mg, Ne/Si ratios. The three ratios are close to the coronal, *not* the photospheric, ratios (with a small excess of Ne, by a factor of ∼ 1.4). The non-statistical spread is quite limited (a factor of ∼ 1.4 around the average, at the 95 % confidence level). There is no correlation with Fe/CNO. From (23).

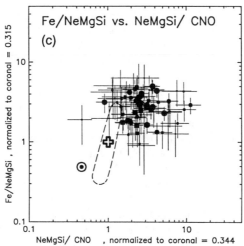

FIG. 4. Similar to Fig. 1, but for the global Fe/NeMgSi ratio vs. the global NeMgSi/CNO ratio. (The corresponding ratios observed in *gradual* SEP events have also been indicated by the dashed enclosure.) One can see the typical NeMgSi enhancement factor of ~ 2.6 relative to CNO, and the additional enhancement Fe factor of ~ 2.8 relative to NeMgSi, yielding the factor of ~ 7.5 average enhancement of Fe relative to CNO (or ^{4}He). The 95 % non-statistical spread on the NeMgSi/CNO ratio is about a factor of ~ 2, and that on Fe/NeMgSi about a factor of ~ 1.5. The Fe/NeMgSi and NeMgSi/CNO ratios are not correlated. (The apparent continuous trend from photospheric to coronal to impulsive event SEP ratios is spurious, due to a mixture of high-FIP Ne and low-FIP Mg and Si in NeMgSi). Adapted from (23).

HEAVY ELEMENT COMPOSITION : BASIC, QUASI-MODEL-INDEPENDENT INTERPRETATION

As we have seen, ^{3}He and heavy elements are enhanced by vastly different orders of magnitude in impulsive events, and their enhancements are not correlated. This suggests that their enhancements may be due to different causes, plausibly to the action of different types of plasma waves. This allows separate studies for the ^{3}He and the heavy element enhancements. In the following, I will concentrate on the latter only.

Temperature of the source gas

The sharp discontinuities in the energetic particle heavy element enhancements vs. Z would be very difficult to understand in terms of selection effects during the later phases of the acceleration. I will therefore assume that the accelerated particle composition is primarily controlled by their extraction out of the thermal pool. I will then take a single hypothesis : *the acceleration efficiency depends on the ion charge-to-mass ratio Q/A in the source gas only*. This will be very generally the case if the ions are accelerated by plasma waves. This implies that ions with the same Q/A ratio behave alike, *i.e.* that *selective acceleration is possible only between elements whose predominant ions have different values of Q/A* .

Fig. 5 shows the averaged Q/A ratio of key elements, vs. temperature T, for the dominant isotope ($A = 2Z$, except for Fe). It is based on the ionization *equilibrium* calculations of (2) and (1). See comments in the figure caption. Fig. 5 will be our basic tool to set limits to the acceptable temperatures of the source gas of the energetic ions.

The upper limit to the acceptable T's is derived from the enhancements of Ne, Mg, and Si relative to ^{4}He,C,N,O, and from their roughly equal magnitudes (Fig. 3 and 4). Above $\sim$ 2 MK, C,N,O are fully stripped, with $Q/A = 0.50$. At *very* high temperatures, say above $\sim$ 10 or 20 MK, Ne,Mg,Si are all fully stripped, as well, so that no enhancement of these elements relative to C,N,O could be understood, if the ions were extracted out of gases at such high temperatures. Actually, already above $\sim$ 5 MK, Ne becomes predominantly fully stripped while Mg and Si do not until significantly higher temperatures are reached. So, if most of the acceleration took place out of gases at $\gtrsim$ 5 MK, Ne would be significantly less enhanced than Mg and Si. The opposite is observed (if anything, Ne is slightly *more* enhanced than Mg and Si, Fig. 3). So, only source gas temperatures $T \lesssim$ 5 MK are acceptable.

The lower limit to the acceptable T's is derived from the *lack* of an enhancements of O relative to C or ^{4}He (Fig. 1 and 2). C and ^{4}He are essentially fully stripped, with $Q/A = 0.50$, at all temperatures beyond $\sim$ 1.5 MK. Now, O starts to be significantly non-fully stripped for $T \lesssim$ 3.5 MK. But we cannot entirely exclude that the H-like ion O^{+7} might possibly *not* be preferentially accelerated, because its Q/A value of 0.44 is not far from 0.50. But He-like O^{+6}, if present, is certainly preferentially accelerated, since its $Q/A = 0.38$ lies further from 0.50 than those of He-like Ne^{+8}, Mg^{+10}, and Si^{+12} (0.40 to 0.43), which *must* get preferentially accelerated to account for the excesses of Ne, Mg and Si (in the acceptable temperature range, below $\sim$ 5 MK, these He-like states are largely dominant !). The fraction of He-like O^{+6} becomes quite significant below $\sim$ 2.5 MK. So, if most of the ions were accelerated out of gases at $T <$ 2.5 MK, O would be enhanced relative to C or ^{4}He, again in contradiction with observation.

On this basis, (23) have concluded that the impulsive event energetic particles are accelerated out of a gas with temperature somewhere in the $\sim$ 2.5 to $\sim$ 5 MK range.

In this range of temperatures, Fe is predominantly in its Ne-like Fe^{+16} state, with values of $Q/A \sim 0.28$ which are much lower than those for element up to Si (Fig. 5). It is therefore not surprising to have Fe enhancements which differ markedly from those for Ne,Mg,Si, and to have this difference *not* correlated with the magnitude of the NeMgSi enhancement (Fig. 4).

These $\sim$ 2.5 to $\sim$ 5 MK temperatures obtained for the source gas of the escaping accelerated particles are totally at variance with flaring loop temperatures (commonly $\gtrsim$ 10 MK). They are, by contrast, quite consistent with the typical $\sim$ 2.5 to $\sim$ 3.5 MK temperatures of the active region (AR) gas which surrounds the hot flaring gas itself (see ref. in (26)). The strict condition that little source gas lies below $\sim 2.5 \times 10^6$ K might possibly be understood in terms of the shape of the radiative loss function (3). Note that the broad γ-ray line study by (21) suggests that the composition of the trapped accelerated particles is, by and large, similar to that of the escaping ones. This suggests that both escaping and trapped particles are accelerated, either in the AR gas *surrounding* the flaring loop itself, or within the flaring loop *before* it got heated.

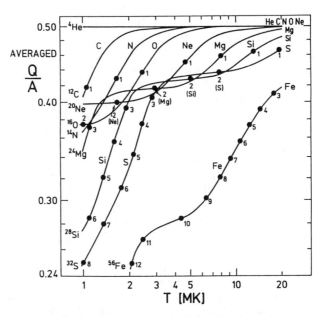

FIG. 5. The "averaged" charge-to-mass ratio Q/A of various elements, versus temperature T, for the dominant isotope ($A = 2Z$, except for Fe). It is based on the ionization equilibrium calculations of (2) and (1). The average is taken over the various ionization states present at a given temperature. Each solid circle denotes an actual, discrete ionization state, labeled by its number of attached electrons (*e.g.* "1" for H–like O^{+7}, Ne^{+9}, ..., "2" for He–like O^{+6}, Ne^{+8}, ...) ; its ordinate is the ion's exact Q/A ratio, and its abscissa the *approximate* temperature at which its abundance peaks. The conspicuous plateaus of the averaged Q/A vs. T associated with 2– and 10–electron ions reflect the particular stability of these He– and Ne–like electronic structures, which remain dominant over wide ranges of temperatures. Adopting (2)'s instead of (1)'s calculation for the ionization of Fe would not change this picture significantly. From (23).

The observed charge states problem

There is a clear cut conflict between these derived Fe charge states $Q \sim 16$ in the source gas and the high Fe charge states ($\overline{Q} = 20.5$) observed by (10) in the energetic particles, which largely correspond to gas temperatures > 10 MK

467

totally inconsistent with the Ne,Mg,Si enhancements. Note that (10)'s data for Si, indicating that it is essentially fully stripped ($Q/A = 0.50$), seem to conflict even more directly with the enhancement of Si relative to C or ^{4}He. There are two possible ways out of this contradiction.

One is that (10)'s sole recent measurement be incorrect. As discussed in (23), its charge resolution is very poor, at least for Si. Further, similar charge state measurements on *gradual* events hint that its charge scale for Fe might possibly be systematically shifted [1].

Note also that we have one earlier Fe charge state measurement in impulsive events, by (14), which yielded, in addition to lower charge states, a predominance of charges $Q = 16$ to 18. So, (10)'s measurement certainly needs duplication.

Otherwise, if (10)'s measurement turns out to be correct, the ions must have been further stripped *after* they have been kicked off from the thermal distribution, after the first stages of acceleration. Various ways by which this late stripping might take place have been discussed by (23). Stripping due to interaction with the traversed gas (8) does not seem to be able to do the job : at the low ~ 1 Mev/n energies observed by (10), the equilibrium charge states are lower than the observed ones. Stripping due to energetic particle trapping in the progressively heating flare environment does work either : the trapping times are shorter than the common heating times. On the other hand, ^{3}He-rich events are strongly associated with fast electron beams and Type III bursts ; selective ^{3}He and possibly heavy element acceleration by waves generated by these beams is actually being considered (27) (19), (20).

---

[1]Similar charge state observations have been performed by the same instrument on *gradual* event SEP's, with much higher statistical accuracy. They might help evaluate (10)'s results on *impulsive* events. On *gradual* events, this instrument has yielded a mean Fe charge of $\overline{Q} = 14.9 \pm 0.9$ around ~ 1 MeV/n (9). For this same gradual event Fe mean charge, an early measurement by (5) had yielded $\overline{Q} \sim 11.6$ around ~ 1 MeV/n, while three recent experiments, based on the geomagnetic cutoff technique, have found values of $\overline{Q} = 11.0 \pm 0.2$ (12), also around ~ 1 MeV/n, and $\overline{Q} = 14.2 \pm 1.4$ (28) and $\overline{Q} = 15.2 \pm 0.7$ (7) at much higher energies > 15 MeV/n. The signification of the differences between these various values is not clear. But they might indicate a true increase of the Fe charge state with energy, which would imply a stripping during a longer propagation of the higher energy ions ; it has been noted by (7) that, at high energies, the traversal of very small amounts of material would be sufficient to significantly increase the charge states ($\ll 0.6$ mgcm^{-2} at 50 MeV/n). As noted by (12), low initial gradual event energetic particle charge states $\overline{Q} \sim 11$, as observed by (5) and (12), would also be entirely consistent with the mean charge state $\overline{Q} \sim 11$ observed in the slow Solar Wind, the source material of the particles (6), (29). If (5) and (12)'s value $\overline{Q} \sim 11$ turned out to be correct around ~ 1 MeV/n, it could suggest a systematic error by ~ 4 charge units affecting (9)'s charge scale for Fe for gradual events, that should affect the results of this experiment for impulsive events (10) as well. The *impulsive* event average charge $\overline{Q} = 20.5 \pm 1.2$, corresponding to $T \sim 10$ to 12 MK (1) (2), would then have to be be shifted down to $\overline{Q} \sim 16.5 \pm 1.2$, suggesting a predominance of Ne-like Fe^{+16}, and corresponding to T anywhere in the range between ~ 2.3 and 7.5 MK (1) (2), consistent with the NeMgSi enhancements.

As discussed by (19), the fast beam electrons could perform the late ionization, over time scales comparable with the acceleration time scales.

MODELS FOR SELECTIVE ION ACCELERATION
BY PLASMA WAVES – GENERAL

Resonant interaction of thermal ions with plasma waves is generally considered the most plausible process for selectively accelerating some ions relative to others, out of the thermal gas. Waves fulfilling the gyroresonance condition with some ion i, with gyrofrequency $\Omega_i \propto Q_i/A_i$, may, indeed, be more abundant than waves fulfilling it for some other ion. Numerous scenarios have been developed along this line (see ref. in (19), (26), (30)). The existence of relative enhancements between ions with fairly close values of Q_i/A_i requires the presence of *rather sharp structures* in the wave spectrum. These may be understood in terms of two types of processes :

(i) Wave generation around specific frequencies : the huge enhancements of ^{3}He almost certainly require the injection of large amounts of wave energy specifically near the gyrofrequency of the rare ^{3}He^{+2} ion. Several types of plasma instabilities have been proposed to produce appropriate plasma modes, largely based on the fact that ^{3}He^{+2}, with $Q/A = 0.67$, is the *only* ion with a value of Q/A intermediate between those of ^{4}He^{+2} and of protons : electrostatic ion cyclotron waves by (4), electrostatic two-ion hybrid waves by (24), electromagnetic ion cyclotron waves by (27).

Among these mechanisms, wave generation by energetic electron beam instability, first proposed by (27) and developed by (19) and (20), seems particularly attractive. It accounts for the observed association between ^{3}He enrichments, high electron fluxes and type III bursts. Contrary to most earlier scenarios, it is consistent with the observed, unenhanced, ^{4}He/p ratios. In addition, the observed high Fe and Si charge states (10), if confirmed, may result from ionization by the very beam electrons responsible for the waves, *after* the ions have been kicked off from the thermal pool. Along this line, (19) and (20) have explored a complete model with a nonrelativistic electron beam drifting along the magnetic field, which produces quasi-perpendicular proton electromagnetic ion cyclotron waves (EMIC), capable of accelerating ^{3}He, quasi-perpendicular shear Alfvén waves, capable of accelerating heavy ions, as well as R-X waves and electrostatic waves. In (20), these authors have attempted to account for the moderate enhancements of heavy elements, as well as for the huge ones of ^{3}He. But the waves considered may not be able to accelerate the ions all the way to the observed energies (~ 10 MeV/n), so that a two-stage process seems required (17). Also there exists, as yet, no analysis considering higher orders of resonance, which become crucial for the relevant quasi-perpendicular waves. Note also that the appropriate waves can be generated only in underdense plasmas, with $\omega_p < \Omega_e$, while the relevant, actual AR plasmas seem to commonly lie on either side in the neighborhood of the underdense-overdense limit $\omega_p = \Omega_e$ (see (26) and ref. therein).

On the other hand, (30) has recently provided a new development of (4)'s approach in terms of electrostatic ion cyclotron waves, removing the earlier requirement of a high ^{4}He/p ratio in this type of models. Since the required electrostatic waves are also generated by electron beams, this model, like (19)'s, accounts for the association between ^{3}He enrichments, high electron fluxes and type III bursts, as well as for the late stripping of the accelerated ions (for which the author, however, considers another alternative : the late heating of the flaring plasma, in the case of non-typical, *extremely* impulsive flares). Direct acceleration to the observed MeV energies is not possible, so that a two-step acceleration process is required.

(ii) Wave damping around specific frequencies :   Sharp structures in the wave spectrum may also result, not from the wave generation mechanism, but from the selective damping of some frequencies in an originally smooth wave spectrum. Smooth injected turbulence spectra near the flare site can be inferred from wave spectra measured in the interplanetary space during impulsive flares. They extend over a broad range of wave numbers, probably due to wave cascading from turbulence injected by large scale fluctuations on the solar surface, and cover all propagation angles θ with the magnetic field direction (see (18), (26) and ref. therein). Such smooth injected wave spectra, in which the energy is not concentrated over a limited range of frequencies, probably cannot account for the huge ^{3}He enhancements. But the selective damping of waves by interaction with specific ions of the gas could play a major role in producing the much more moderate heavy element composition biases. As we shall see, wave energy transfer to the very abundant ^{4}He^{+2} ions could be essential in this context.

As we shall see, these mechanisms can be effective only if the waves in the plasma are *not* predominantly quasi-perpendicular, which is indeed the case for waves associated with the general turbulence of the medium. No condition is required on the ^{4}He/p ratio. The general turbulence should be able to accelerate the ions directly to the observed energies, in one step (18) (25). While the relevant, precise conditions for wave damping, of course, smoothly depend on n_e and B, such selective acceleration processes should basically work equally well for overdense and underdense plasmas, *i.e.* for both $\omega_p < \Omega_e$ and $> \Omega_e$. Such models, however, do not by themselves offer an explanation for the high Fe and Si charge states observed by (10) and yet to be confirmed (see discussion above).

In brief, entirely different types of scenarios have been recently proposed to explain the heavy element enhancements. Which applies may largely depend on which type of waves is actually dominant in the flare environment : electrostatic ion cyclotron waves (30), electron beam generated shear Alfvén waves (27) (19) (20), or waves generated by cascading of the general turbulence (18), (26). Of course, in actual reality various types of waves may play a role. But it should be remembered that electron beam generated shear Alfvén waves can shape the heavy element composition only through their *injected* spectral shape, *not* due to damping, since they are quasi-perpendicular.

A WAVE DAMPING MODEL FOR
SELECTIVE HEAVY ELEMENT ACCELERATION

I will now describe in more detail the scenario recently proposed by (26) to account for the impulsive event heavy element composition in terms of damping of the electromagnetic He cyclotron waves associated with the general turbulence, by interaction with the ions in the gas, and particularly with the abundant $^4\text{He}^{+2}$ ions. It is based on a general analysis of wave-particle interaction in the warm AR plasma surrounding an impulsive flare.

In this first approach, (26) have simply investigated the effect of damping on an initially smooth wave spectrum, assuming an instantaneous wave injection for simplicity. So, this study does not include steady wave cascading. It, however, allows to pinpoint the essential effects of wave damping on the composition, that should still play a basic role when wave cascading is being considered. The link between (26)'s study and an approach including cascading will be discussed below.

This study also considers only the extraction of the ions out of the thermal pool, not their higher energy acceleration. As discussed above, only this first stage should be capable of producing the observed sharp discontinuities in the energetic particle heavy element enhancements vs. Z. Higher energy acceleration can produce only smooth Q/A-dependent biases.

Qualitative discussion

The Doppler condition for gyroresonant interaction between a plasma wave and an ion of species i reads, for the first harmonic and for non-relativistic ion speed,

$$\omega = \Omega_i + v_z \cos\theta \cdot k , \tag{1}$$

where $\Omega_i \propto Q_i/A_i$ is the ion's gyrofrequency, v_z the ion's velocity component along the field direction, ω and k the wave frequency and wave number, and θ the angle of the direction of wave propagation relative to the field direction. For given wave propagation angle θ and ion velocity component v_z, this equation describes a straight line with a slope $v_z \cos\theta$ in the (k,ω) plane, as the locus for resonant energy transfer from the waves to an ion i with velocity component v_z.

We will be interested in the width of the range of wave frequencies ω, around Ω_i, over which the waves will be effectively damped by transferring their energy to the ions of a particular species i. Eq. 1 indicates that, for a given wave number k, this damping width depends on :
– The range of ion velocities v_z, over which *the density of ions i with velocity component v_z in the gas* is large enough to significantly damp the waves in specific conditions. The density of ions i at each particular v_z, hence the width of this range of v_z's, depends *(i)* on the total density of the ion species i in the gas, or, for a given hydrogen density n_H, on the ion abundance, and *(ii)* on

the shape of the ion Maxwell distribution for v_z ; the smaller the ion mass A_i and the higher the temperature T, the broader this v_z distribution. Let us denote by $v_{z,lim,i}$ the limit of the range of speeds $|v_z|$ for which the density of ions i is sufficient to produce a significant damping, say, by an e-folding factor, under some specific conditions.

– The wave propagation angle θ ; the damping width is maximal for $\cos\theta = 1$, *i.e.* for parallel waves ; for quasi-perpendicular waves, $\cos\theta \approx 0$, so that the frequency range of wave damping is very narrow ; the subsequent discussion will therefore not apply for quasi-perpendicular waves, such as the shear Alfvén waves considered by (19) and (20).

Now, k is not fixed. The width of the range of damped wave frequencies increases linearly with k. Therefore, each sufficiently abundant ion present in the gas produces a "damping region" in the (k,ω) plane, which is roughly delimited by the two lines

$$\omega = \Omega_i \pm v_{z,lim,i} \cos\theta \cdot k , \qquad (2)$$

or rather a damping "Valley", within which the dearth of waves becomes more and more severe as one approaches $\omega = \Omega_i$ for a given k, as a reflection of the shape of the Maxwell distribution for the v_z's.

To sum up, the larger the ion abundance, the smaller its mass, the higher the temperature, and the smaller the wave propagation angle θ, the broader this "damping region" in the (k,ω) plane, *i.e.* the range of frequencies ω around Ω_i over which waves get damped for a given wave number k.

Now, we are interested in the damping of the waves which can accelerate heavy ions in $\gtrsim 2.5$ MK gases (cf. above), which all have values of $0.25 \lesssim Q_i/A_i \leq 0.50$ or, in terms of gyrofrequencies, $0.25 \lesssim \Omega_i/\Omega_p \leq 0.50$, where Ω_p is the proton gyrofrequency. In this range of frequencies, the wave damping is largely dominated by the energy transfer to $^4\text{He}^{+2}$ ions, with $(Q/A)_{^4\text{He}^{+2}} = \Omega_{^4\text{He}^{+2}}/\Omega_p = 0.50$, in view of the comparatively huge ^4He abundance, and of its low mass, which both tend to produce a very broad He damping region or "Helium-Valley". This wave damping by $^4\text{He}^{+2}$ provides a physical significance for the concept of ions with $Q_i/A_i = 0.50$ being comparatively less efficiently accelerated than those of lower Q_i/A_i , and for the existence of a finite range of values of Q_i/A_i around 0.50, within which this reduced acceleration efficiency remains in effect (for a propagation angle $\theta \not\approx 90°$).

To get an evaluation of the actual frequency width of the damping by $^4\text{He}^{+2}$, the above considerations about the damping regions in the (k,ω) plane have to be associated with a determination of the dispersion relation $\omega(k)$ for the relevant left-polarized component of $^4\text{He}^{+2}$ cyclotron waves in the plasma under consideration. Essentially, the waves remain undamped for all frequencies below the frequency ω_{lim} at which the dispersion relation $\omega(k)$ intersects the line $\omega = \Omega_{^4\text{He}^{+2}} - v_{z,lim,^4\text{He}^{+2}} \cos\theta \cdot k$.

472

Now, this was all qualitative. The interplay between wave damping and the plasma dispersion relation $\omega(k)$ actually depends on the plasma density n_H, on the plasma temperature T, on the field B, on the wave propagation angle relative to the field θ, and on the time available for damping Δt (whose significance in terms of wave cascading will be discussed below).

In their study, (26) have quantitatively investigated the wave damping for typical AR plasma densities and fields of $n_H = 10^{10}$ cm^{-3} and $B = 100$ G, realistic coronal abundances of all major heavy elements, temperatures T between 2 and 20 MK, and for parallel and oblique waves with propagation angles between $\theta = 0°$ and $70°$. The dispersion relation $\omega(k)$ has been calculated in such plasmas (with the impact of all heavy ions for $\theta = 0°$, in an e-p-He plasma for $\theta \neq 0°$). The wave damping by electrons, protons, ^{4}He^{+2}, and by heavier ions for $\theta = 0°$, has been quantitatively evaluated, with particular emphasis on ^{4}He^{+2}. The e-folding damping time scale τ_d has been evaluated along the dispersion relation for various conditions, allowing an evaluation of the broadening of the He–Valley with temperature T and time Δt after wave injection (assumed instantaneous). For $\theta = 45°$ and a damping duration Δt of 10^{-4}, for instance, the He–Valley extends down to $\omega_{lim}/\Omega_p \approx 0.450$ and 0.435, for $T = 3$ and 6 MK, respectively : of course, the He–Valley broadens with temperature. For Δt of 10^{-2} s, these figures become $\omega_{lim}/\Omega_p \approx 0.420$ and 0.395. It can be noted that the He–Valley broadens only by a factor of ~ 1.6 as the damping duration is multiplied by a factor of 100. This reflects the very steep slope of the Maxwell tail for high v_z's : a large change in the particle differential density is obtained for a small change in v_z.

Such limits of the He–Valley have to be confronted with the values of $\Omega_i/\Omega_p = Q_i/A_i$ for the various heavy ions present in the plasma at various temperatures, to determine the ranges of temperatures T and damping durations Δt that allow to meet the observed energetic particle elemental composition (for a wide distribution of propagation angles, say, mainly between $\theta = 0°$ and $45°$). The key constraints from observations are those pointed out by (23) and discussed above : the lack of enhancement of O relative to C or ^{4}He (Fig. 1 and 2), and the comparable enhancements of Ne, Mg and Si relative to CNO (Fig. 3).

For very short damping durations $\Delta t < 0.1 \times 10^{-3}$ s, the He–Valley tends to be very narrow at any given temperature. Then, we can't find a temperature for which we have a Ne enhancement without an O enhancement. The lack of an O enhancement, indeed, requires that the Valley be wide enough to have predominant O^{+7}, with its $Q/A = 0.44$, lie within the Valley ; this requires $T > 3.8$ MK. But the Ne enhancement requires that the fraction of Ne^{+9} be small, since, with $Q/A = 0.45$, this ion also lies within the Valley ; Ne^{+8}, with $Q/A = 0.40$, must be dominant, and this requires $T < 3.8$ MK. In brief, at the low temperatures at which Ne is enhanced, most of the O lies outside the He–Valley.

For very long damping durations $\Delta t > 5 \times 10^{-3}$ s, the He–Valley tends to be very broad at any given temperature. Then, we can't find a temperature for which we have a Si enhancement without an O enhancement. The lack of an O enhancement, indeed, requires that the fraction of O^{+6}, which always lies far outside the Valley with its low $Q/A = 0.38$, be small ; this requires $T > 2.6$ MK. But the Si enhancement requires that the Valley be narrow enough to have predominant Si^{+12}, with its larger $Q/A = 0.43$, lie outside the Valley ; this requires $T < 2.6$ MK. In brief, at the high temperatures at which O is not enhanced, most of the Si lies within the He–Valley.

So, there exists only a limited range of typical durations of damping, between $\Delta t \sim 0.1$ and $\sim 5 \times 10^{-3}$ s, for which the width of the He–Valley is such that the observed energetic particle composition can be met (for $\theta = 0°$ to 45°). For such time scales, there exists an acceptable temperature range within which O^{+7} lies within the Valley, the O^{+6} fraction is small, while Si^{+12} lies outside the Valley and the Ne^{+8} fraction is large, as required. This range depends on the exact value of Δt, but it extends at most from ~ 2.6 to ~ 3.8 MK, as imposed by the constraints on the O^{+6} and Ne^{+8} fractions at the lower and at the upper bound of this range, respectively. In addition, any significant *additional* acceleration, *e.g.* of Fe, taking place in hotter gases in the ~ 3.8 to ~ 6.5 MK range, which would not contribute any selectively accelerated C,N,O and Ne and Si ions, would be accompanied by an additional acceleration of Mg ; it would therefore conflict with observation. When the uncertainties on the O^{+6} and Ne^{+8} equilibrium ionization calculations (2) are folded in, the above range can be extended, at the most, to ~ 2.4 to ~ 4.5 MK. This range of temperatures, obtained in the framework of a specific model for the forming of the composition biases as a result of damping of $^4He^{+2}$ cyclotron waves, is very similar to that obtained in (23)'s basic analysis, which had first indicated that the ions are accelerated out of AR gases.

All these results have been obtained with nominal values of $n_H = 10^{10}$ cm^{-3} and $B = 100$ G. But a number of recent AR investigations suggest that $n_H = 0.1$ to 0.4×10^{10} cm^{-3} and $B = 50$ to 400 G may be more typical values for AR plamas (see (26) and ref. therein). While the acceptable gas temperature range is not sensitive to these plasma conditions, the acceptable damping time scales Δt are, and may be some ~ 10 times larger than estimated above, *i.e.* more typically in the range $\Delta t \sim 1$ to $\sim 50 \times 10^{-3}$ s.

Discussion – Injected wave cascading and energy

In (26)'s study, an instantaneous injection of the wave energy has been assumed, for simplicity. This is probably not realistic, first of all due to wave cascading, and possibly also due to continued injection of large scale, low-k turbulence. Most probably, the He–Valley will be, to some degree, continuously replenished with waves.

If cascading is very intense, the He–Valley essentially cannot form, or at least will remain very narrow, limited to the peak of the Maxwell distribution for v_z. On the other extreme, without any cascading, the He–Valley may broaden more or less indefinitely, as the waves with frequencies further and further from $\Omega_{^4He^{+2}}$ progressively get damped by the rarer and rarer $^4He^{+2}$ ions along the Maxwell tail.

So, wave cascading is probably what controls the available time scale for wave damping Δt. Therefore, the above conclusions on the acceptable Δt's actually set *constraints to the acceptable rates of wave cascading.*

To be more specific about the significance of the time interval Δt, we have to establish a link between two extreme descriptions : *(i)* (26)'s approximate description in terms of an instantaneous wave injection, in which Δt is defined, and *(ii)* a more realistic description in terms of a steady wave injection. In both situations, the injected wave energy varies smoothly with ω, *i.e.* is roughly frequency independent in the neighborhood of $\Omega_{^4He^{+2}}$. In the framework of a steady wave injection *(ii)*, wave energy is simultaneously being steadily dissipated, at a rate which peaks at $\omega = \Omega_{^4He^{+2}}$ and decreases away from this frequency, following the shape of the Maxwell distribution for the differential ion density vs. v_z. Here we will consider the frequencies $\omega < \Omega_{^4He^{+2}}$. If the rate of wave injection is not too large, the injection and dissipation rates cancel out at some frequency ω_{cut}, where damping effectively ends. Back to an approximate "equivalent" description in terms of an instantaneous injection *(i)*, the broadening of the He–Valley will effectively stop at the time (Δt) when the He–Valley has broadened down to this frequency ω_{cut}. So, *(26)'s approximate "damping duration" Δt is the time interval after which replenishment of the He–Valley by cascading prevents its further broadening.* Obviously, the higher the steady wave injection rate, the narrower the maximal width of the He–Valley, and the shorter the equivalent damping duration Δt.

The He–Valley may, by contrast, be effectively broadened *(i)* in the case of very strong turbulence, for which the resonance condition is relaxed, *(ii)* if suprathermal tails are present in the gas, and *(iii)* due to wave propagation in an inhomogeneous field, since $\Omega_{^4He^{+2}} \propto B$ (26).

On the other hand, the He–Valley, of course, can exist only if the wave damping by $^4He^{+2}$ ions is, indeed, severe. This requires that the wave energy injected near the $^4He^{+2}$ gyrofrequency, instantaneously or over a time scale $\sim \Delta t$, be small as compared to that required to accelerate *all* the $^4He^{+2}$ ions in the volume under consideration. This condition is certainly fulfilled, apart from energy considerations, because a much more stringent condition is imposed : since NeMgSi, with neighboring gyrofrequencies, are depleted *relative to Fe* among the accelerated particles, we can be sure that by far *not all* NeMgSi's in the gas are being accelerated. So, the waves are able to accelerate only a small fraction of these elements, which are ~ 1000 times less abundant then ^4He. A fortiori, they won't be able to accelerate all $^4He^{+2}$'s.

Considering the other extreme case, that of comparatively small injected wave energies, one should consider the possibility of a significant damping of the waves, not only by ^4He's, but also by the rare heavy ions themselves ; (if wave cascading is important, damping by Fe at low-k, low-ω will then act as a first wave filter). In addition to the effect of the He–Valley, each individual ion enhancement in the energetic particles will then be larger if its own source gas abundance is smaller, provided that its gyrofrequency $\Omega_i \propto Q_i/A_i$ is sufficiently separated from those of neighboring ions. This constitutes an hypothesis to be explored, to possibly account for the observed extra enhancements of the ~ 10 to ~ 20–fold rarer Na, Al, ^{22}Ne, and ^{25,26}Mg, as compared to the neighboring dominant species ^{20}Ne, ^{24}Mg, and ^{28}Si, which have comparable abundances in the coronal gas (22), (11).

SUMMARY

In the energetic particles escaping impulsive SEP events, ^{3}He is commonly enhanced by huge factors of up to ~ 20000, while C,N,O are not enhanced at all, Ne,Mg,Si are enhanced by factors of ~ 3, and Fe by factors of ~ 7.5 relative to ^{4}He. The trapped energetic particles seem to have a similar heavy element composition. The ^{3}He and heavy element enhancements are not correlated. The rarer species Na, Al, ^{22}Ne, and ^{25,26}Mg are enhanced by factors of ~ 2 relative to the neighboring dominant ones. One observation has yielded surprisingly high charge states $Q \sim 20.5$ for the energetic Fe ions ; it requires confirmation.

With the sole assumption that the energetic particle composition is controlled by the ion charge-to-mass ratio Q/A in the source gas at equilibrium, this observed composition implies that the heavy elements are accelerated predominantly out of gases with temperatures in the ~ 2.5 to ~ 5 MK range, *i.e.* with typical active region, not flaring loop, temperatures. This implies that both escaping and trapped particles are accelerated, either in the active region gas *surrounding* the flaring loop itself, or within the flaring loop *before* it got heated. These temperatures are totally inconsistent with the currently observed Fe charge states ; either the observation is invalid, or the ions get further stripped *after* they have been extracted from the thermal pool.

Models for selective ion acceleration by plasma waves have been briefly overviewed. They are mainly of two types. *(i)* Models in terms of electron beam generated plasma waves ; the waves are *generated* over specific, appropriate, frequency ranges ; such models may account simultaneously for the ^{3}He and the heavy ion enhancements, as well as for the currently observed high ion charge states (*late* stripping by the beam electrons) ; but they probably cannot accelerate the ions to $\sim$ MeV energies in one single step, so that another process is requires to boost the ions to these energies. *(ii)* Models in terms of the general ambient turbulence, and of its cascading, yielding a smooth wave frequency spectrum ; the waves are *damped* over specific, appropriate, frequency ranges ; such models deal with the heavy ion acceleration only (the huge, uncorrelated ^{3}He enhancements must then be produced by another process), and they do not account for the currently observed high ion charge states ; but they should be able to fully accelerate the ions in a single step.

I have then investigated in more detail the formation of the heavy element enhancements in this latter type of models, due to the damping of electromagnetic He cyclotron waves by interaction with the very abundant ^{4}He^{+2}

ions in the gas. In the assumed smooth, injected wave frequency spectrum, an *"He–Valley" of dearth of waves* is thus formed, whose width depends on the plasma temperature (Doppler condition for the resonant wave-particle interaction, considering the ion thermal velocities), and on the characteristic time scale for the He–Valley replenishment due to wave cascading. The observed elemental composition, together with the ionization balance in the gas, imposes that some heavy ions remain unenhanced relative to ^{4}He, hence have their gyrofrequency lie *within* the He–Valley, and that others be enhanced, and hence have their gyrofrequency lie *outside* the He–Valley. These requirements allow to impose constraints to the source gas temperature, which are very similar to those first obtained on more general grounds ($T \sim 2.4$ to 4.5 MK), and on the rate of wave cascading into the He–Valley, which must suppress further broadening of the He–Valley after time scales somewhere in the range between $\Delta t \sim 0.1$ and $\sim 5 \times 10^{-3}$ s (which could be shifted by a factor of ~ 10 upward, depending on the density and field in the active region gas). These processes take place indifferently in under- and overdense plasmas, provided that the available waves are not predominantly quasi-perpendicular. The observed specific excess of of the rarer species in the Ne to Al range might suggest that wave damping by the dominant species in this range is significant.

This paper is essentially based on work performed in collaboration with Don Reames, Adriane Steinacker, Jürgen Steinacker, and Tycho von Rosenvinge, which has been initiated while I was visiting at the Laboratory for High Energy Astrophysics at NASA/Goddard Space Flight Center, thanks to a National Research Council–GSFC Senior Research Associateship. I also wish to thank Monique Arnaud, Nancy Brickhouse, Jeff Brosius, Jacques Dubau, Joe Mazur, Jim Miller, Julia Saba, and Ken Widing for discussions regarding various aspects of this work.

REFERENCES

1. Arnaud, M., & Raymond, J., ApJ **398**, 394 (1992).
2. Arnaud, M., & Rothenflug, R., A&A Suppl. **60**, 425 (1985).
3. Cook, J.W., Cheng, C.C., Jacobs, V.L., & Antiochos, S.K., ApJ, **338**, 1176 (1989).
4. Fisk, L.A., ApJ **224**, 1048 (1978).
5. Gloeckler, G., Sciambi, R.K., Fan, C.Y., & Hovestadt, D., ApJ **209**, L93 (1976).
6. Ipavich, F.M., Galvin, A.B., Geiss, J., Ogilvie, K.W., & Gliem, F., in : *Solar Wind Seven*, E. Marsch & R. Schwenn eds., (Pergamon Press), p. 369 (1992).
7. Leske, R.A., Cummings, J.R., Mewaldt, R.A., Stone, E.C., & von Rosenvinge, T.T., ApJ **452**, L149 (1995).
8. Luhn, A., & Hovestadt, D., ApJ **317**, 852 (1987).
9. Luhn, A., Hovestadt, D., Klecker, B., Scholer, M., Gloeckler, G., Ipavich, F.M., Galvin, A.B., Fan, C.Y., & Fisk, L.A., *19th Intern. Cosmic Ray Conf.*, La Jolla, **4**, 241 (1985).
10. Luhn, A., Klecker, B., Hovestadt, D., & Möbius, E., ApJ **317**, 951 (1987).
11. Mason, G.M., Mazur, J.E., & Hamilton, D.C., ApJ **425**, 843 (1994).

12. Mason, G.M., Mazur, J.E., Looper, M.D., & Mewaldt, R.A., ApJ **452**, 901 (1995).

13. Mason, G.M., Reames, D.V., Klecker, B., Hovestadt, D., & von Rosenvinge, T.T., ApJ **303**, 849 (1986).

14. Ma Sung, L.S., Gloeckler, G., Fan, C.Y., & Hovestadt, D., ApJ **245**, L45 (1981).

15. Meyer, J.P., in : *Origin and Evolution of the Elements*, N. Prantzos, E. Vangioni-Flam & M. Cassé eds., (Cambridge Univ. Press), p. 26 (1993) ; or Adv. Space Res. **13**, (9), 377 (1993).

16. Meyer, J.P., in : *Cosmic Abundances*, S.S. Holt & G. Sonneborn eds., ASP Conf. Series (Astr. Soc. of the Pacific), in press (1996).

17. Miller, J.A., private communication (1995).

18. Miller, J.A., & Roberts, D.A., ApJ **452**, 912 (1995).

19. Miller, J.A., & Viñas, A.F., ApJ **412**, 386 (1993).

20. Miller, J.A., Viñas, A.F., & Reames, D.V., 23rd *Intern. Cosmic Ray Conf.*, Calgary, **3**, 13 (1993) ; *ibid.*, **3**, 17 (1993).

21. Murphy, R.J., Ramaty, R., Kozlovsky, B., & Reames, D.V., ApJ **371**, 793 (1991).

22. Reames, D.V., Adv. Space Res. **15**, (7), 41 (1995).

23. Reames, D.V., Meyer, J.P., & von Rosenvinge, T.T., ApJ Suppl. **90**, 649 (1994).

24. Riyopoulos, S., ApJ **381**, 578 (1991).

25. Smith, D.F., & Miller, J.A., ApJ **446**, 390 (1995).

26. Steinacker, J., Meyer, J.P., Steinacker, A., & Reames, D.V., submitted to ApJ (1995).

27. Temerin, M., & Roth, I., ApJ **391**, L105 (1992).

28. Tylka, A.J., Boberg, P.R., Adams, J.H., Beahm, L.P., Dietrich, W.F., & Kleis, T., ApJ **444**, L109 (1995).

29. von Steiger, R., & Geiss, J., in : *Cosmic Winds and the Heliosphere*, J.R. Jokipii, C.P. Sonnett, & M.S. Giampapa eds., (U. of Arizona Press), p. 1 (1995).

30. Zhang, T.X., ApJ **449**, 916 (1995).

Particle Acceleration by DC Electric Fields in the Impulsive Phase of Solar Flares

Gordon D. Holman

NASA/Goddard Space Flight Center
Laboratory for Astronomy and
Solar Physics, Code 682
Greenbelt, MD 20771

DC electric fields provide an attractive means of accelerating electrons and ions out of the thermal plasma in the impulsive phase of flares. They also provide simultaneous direct heating of the flare plasma. The physical mechanisms for producing runaway electrons and ions are reviewed, and the conditions required for obtaining observed particle fluxes, energies, and acceleration time scales are presented. Possibilities for distinguishing acceleration models and mechanisms, and the difficulties inherent in this, are discussed. Some suggestions for future observational and theoretical work are presented.

INTRODUCTION

The energy released in flares is understood to derive from non-potential magnetic fields and, therefore, electric currents in the solar atmosphere. This energy release may occur in many different geometries, including those associated with large-scale magnetic reconnection. The plasma heating and particle acceleration leading to the observed emissions from flares must result directly from these currents or from secondary processes driven by them. It is therefore important to understand the conditions under which these currents can produce the required heating and acceleration.

A complete theory for particle acceleration in flares must demonstrate how electrons and ions are energized out of the thermal plasma. It must provide particle fluxes, energies, and acceleration time scales that are consistent with observations. A complete model must also provide particle energy and abundance distributions that are consistent with observations, and explain the close association between heating and particle acceleration observed in flares.

Electric currents in a plasma are accompanied by straight-forward physical mechanisms that heat and accelerate the plasma particles: Joule heating and the acceleration of runaway particles by the electric fields that drive the currents. These closely coupled processes provide an attractive means for generating the energetic particles produced in the impulsive phase of flares. The currents can also, either directly or indirectly, induce the growth of plasma wave turbulence. Depending upon the conditions in the plasma and the wave properties, this turbulence can also accelerate particles. The direct electric field acceleration of runaway particles and the stochastic acceleration of particles by resonant wave modes provide the most compelling mechanisms for impulsive-phase particle acceleration (see (15) for a review). It is therefore important to obtain a thorough understanding of both processes.

The directness and relative simplicity of the electric field acceleration of runaway particles contribute to the attractiveness of the process. Solar flare data place substantial constraints on the conditions required for the mechanism to be significant, however. The physics of electron and ion runaway, the successes and weaknesses of models based upon the mechanism, and these important constraints are discussed in the following sections.

THE RUNAWAY PROCESS

In the presence of an electric field, particles in a plasma can be accelerated out of the thermal distribution and form a nonthermal tail of energetic particles. This occurs because the collisional drag force on a particle falls off with increasing velocity while the electric force is velocity independent. Therefore, for particles with high enough velocities the electric force exceeds the drag force and the particles are able to "run away" from the thermal distribution (11).

Equating the electric and collisional drag forces allows the critical speed above which runaway will occur to be calculated for a given electric field strength. Particles with speeds below this value remain thermalized and drift with the thermal current. Runaway can occur as long as there is a component of the electric field along the magnetic field, or in a neutral sheet, where the magnetic field changes direction across the sheet. The results are quite different for electrons and ions, however.

Electrons

The collisional drag force on an electron in a (fully ionized) hydrogen plasma is plotted in Figure 1 as a function of the speed of the electron normalized to the thermal speed of the plasma electrons ($\sqrt{2kT/m_e}$). The dashed curve is the drag force due to protons, the dotted curve is the drag force due to electrons, and the solid curve is the total drag force. The drag force is normalized to eE_D, where E_D, the Dreicer field, is the electric field strength for which the critical speed (above which runaway occurs) is equal to the electron thermal speed.

Since the drag force is equal to the electric force eE at the critical speed, Figure 1 is also a plot of the threshold for electron runaway. The horizontal arrow in Figure 1 shows, for example, that when $E = 0.1E_D$ the critical speed $v_c \sim 4v_{the}$ and electrons above this speed are accelerated out of the thermal plasma. New particles are constantly scattered into the runaway

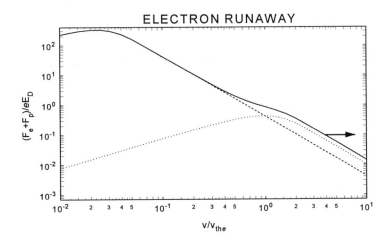

FIG. 1. The collisional drag force on a test electron in ionized hydrogen.

regime ($v > v_c$) by collisions. The highest energy gained by the electrons is determined by the electric field strength and the distance over which it applies.

The drift speed of the thermal current and v_c both approach v_{the} as E approaches E_D. Therefore, in principle the entire distribution will run away when $E = E_D$. In actuality, however, wave growth can be induced that increases the effective drag force (and resistivity) within the current channel before E reaches E_D. If this enhanced drag force behaves formally in the same manner as classical Coulomb collisions, the Dreicer field is replaced by an "anomalous" Dreicer field and the electric field strength can be much higher without the entire distribution running away.

Ions

The drag force on an ion is substantially different from that on an electron. Ion runaway would be negligible for the drag force in Figure 1 when E is on the order of or less than E_D. Since the ion thermal speed is $\sqrt{m_i/m_e}$ times smaller than the electron thermal speed, there would be a negligible fraction of the thermal ions in the runaway regime. As shown in Figure 2, however, the drag force on an ion has a minimum between the ion thermal speed and the electron thermal speed. The result is that a significant tail of suprathermal ions can form, but the maximum ion speed is somewhat less than the electron thermal speed.

Figure 2 illustrates the critical ion speeds and electric field thresholds relevant to ion runaway. The threshold condition rather than the actual drag force (normalized drag force — see Ref. (8)) is plotted. The lower peak in the drag force is due to ions (primarily hydrogen and helium in the solar

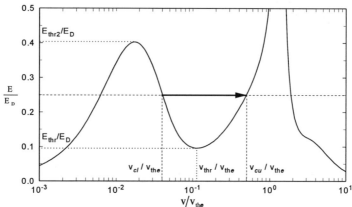

FIG. 2. The normalized collisional drag force on $^{12}C^{+6}$, showing the location of the critical ion speeds and electric field thresholds.

atmosphere) while the upper peak is due to electrons. As for electron runaway, there is a critical ion speed v_{cl} above which runaway occurs. Unlike for electrons, however, there is also a maximum ion speed v_{cu}. The electron drag force effectively prevents the ions from exceeding the electron drift speed. This is illustrated by the dashed lines and arrow in Figure 2 for $^{12}C^{+6}$ with $E = 0.25E_D$.

Also unique to ions is the existence of a threshold electric field strength E_{thr} below which runaway does not occur. This is seen in Figure 2 to be $\sim 0.1E_D$ for $^{12}C^{+6}$. As can be seen in Figure 3, E_{thr} is lower for higher Z ions. Another threshold, E_{thr2}, is shown in Figure 2. When the electric field strength exceeds E_{thr2}, the entire population of that ion in the current channel can be accelerated up to v_{cu}.

It is interesting that, because of the dominance of the electron drag, all of the (positive) ions except protons are accelerated against the electric field, i.e., in the same direction as the electrons. Protons are accelerated in the direction of the electric field as expected. Protons have the highest acceleration threshold, $\sim 0.5E_D$ (see Figure 3). It is also interesting that protons would not be accelerated in a pure hydrogen plasma. The presence of other ions is crucial.

FLARES

Particle Flux

A critically important quantity deduced from flare observations is the flux of accelerated particles. This flux places tight constraints on the acceleration

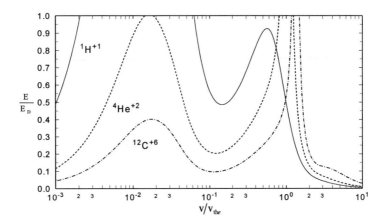

FIG. 3. The normalized collisional drag forces on a proton, fully ionized ^{4}He, and ^{12}C.

model. For the electric field acceleration of runaways, the constraints are both quantitative and qualitative.

The particle flux is best known for electrons from hard x-ray observations. It is less well known for protons and other ions from γ-ray observations. The biggest uncertainty is in the flux of accelerated particles at low energies. Since the number of particles increases with decreasing particle energy over the range of energies accessible with existing observations, the total flux is quite sensitive to the form of the distribution at low energies. X-ray spectra have typically provided information about nonthermal electrons with energies down to ~ 20 keV. Observations of γ-ray lines have provided information about ions with energies above ~ 30 MeV. These indicate electron fluxes of at least 10^{34} electrons s^{-1} (3) and proton fluxes of at least 10^{32} protons s^{-1} (15).

The electron flux of 10^{34} electrons s^{-1} is several orders of magnitude higher than can be carried in a single current channel (7). To avoid an unacceptably high induction magnetic field, typically 10^4 or more pairs of oppositely directed current channels are required. It is possible to accelerate the particles in a single current channel if they rapidly escape the channel so that the net current never becomes unacceptably large. However, this requires an electric field strength that is much higher than can be obtained with classical Coulomb collisions (see below).

The flux of runaway particles, $\dot{N}$ (particles s^{-1}), is related to the rate at which thermal particles are scattered into the runaway regime, γ_{run} (s^{-1}), through

$$\dot{N} = \gamma_{run} n s V_J, \tag{1}$$

where n is the (thermal) particle density, s is the number of current channels, and V_J is the volume of each channel. At least one dimension of V_J (one dimension for a current sheet geometry, two for a cylindrical geometry) is constrained by Ampere's Law to be quite small, typically ~10–100 cm. The runaway rate is sensitive to the value of the critical speed v_c (v_{cl} for ions) relative to the thermal speed v_{th} of the runaway species. It is roughly given by

$$\gamma_{run} \approx \nu_{coll} \exp\left[-\frac{1}{2}\left(\frac{v_c}{v_{th}}\right)^2\right], \tag{2}$$

where ν_{coll} is the collision frequency for the runaway particle at v_c.

The ratio of v_c to v_{th} is proportional to $\sqrt{E_D/E}$. Taking the plasma electrons and protons to have $T = 4 \times 10^7$ K, $n = 3 \times 10^{11}$ cm^{-3}, $s = 10^4$, and $V_J = 10^{20}$ cm^3, an electron flux of 10^{35} s^{-1} is obtained when $E \approx 0.1E_D$ and a proton flux of 10^{32} s^{-1} is obtained when $E \approx 0.6E_D$. All of these values are plausible and consistent with observations.

Maximum Particle Energies

Electrons. For classical resistivity (Coulomb collisions) and sub-Dreicer electric fields, the velocity to which electrons can be accelerated is limited by the value of the Dreicer field and the distance over which the electrons can be accelerated. The Dreicer field depends upon the density and temperature of the plasma: $E_D \approx 6 \times 10^{-8} n/T$ V cm^{-1}. For n as high as 10^{12} cm^{-3} and T as low as 2×10^6 K, $E_D = 0.02$ V cm^{-1}. Therefore, for electrons making a single pass through a current channel 10^9–10^{10} cm long, energies up to 10–100 MeV can reasonably be attained.

These electron energies are more than adequate to produce flare x-ray and microwave bursts. Nevertheless, they fall short of the highest energies produced in large flares. In the presence of anomalous resistivity, however, which is likely to be generated in the current channels as E approaches E_D, much higher energies are possible. In the presence of well developed ion acoustic turbulence, for example, electron energies exceeding 10 GeV are feasible.

Electrons may also be scattered out of the current channels, allowing them to interact with many channels. Once an electron is scattered out of the channel in which it originates, it may either gain or lose energy in subsequent channels. Some of the particles could gain substantially more energy by interacting with many current channels.

The electrons may be accelerated in a single, 10^4–10^5 km long current sheet if they escape the sheet with energies exceeding 100 keV in a distance less than 1 km (9). A small fraction of the electrons could remain in the sheet and be accelerated to energies exceeding 10 GeV. This requires an electric field strength of at least 1 V cm^{-1}, well above the classical Dreicer field. Such an electric field could in principle be sustained with well developed anomalous resistivity (such as ion acoustic turbulence), giving sub-Dreicer behavior relative to the anomalous Dreicer field. Alternatively, the basic form of Ohm's law may be violated and the plasma would be in the difficult-to-model super-Dreicer regime.

Ions. Since the electron drag force limits the ion speed to around the electron drift speed, the highest ion energies are ~ 1 MeV nucleon^{-1}, corresponding to a speed of $\sim 10^9$ cm s^{-1}. This is well above thermal, but not high enough to provide the ion energies measured in space or inferred for a γ-ray flare. However, it is high enough to be a source of seed particles to another mechanism such as stochastic acceleration by Alfven waves. Alfven waves require particles with velocities on the order of the Alfven speed or higher and, therefore, do not resonate with thermal ions. Electric field acceleration can provide these suprathermal ions.

Can direct electric field acceleration alone provide the more energetic ions above ~ 1 MeV nucleon^{-1}, without requiring an additional mechanism? For classical, sub-Dreicer conditions the answer appears to be no. In the presence of anomalous resistivity or general super-Dreicer conditions, however, it is clear that it is possible to have the potential drop required to accelerate both electrons and ions to GeV energies. The problem for the ions is getting them beyond the electron drag. (In a reconnecting current sheet the maximum energy of the particles also depends upon the structure of the magnetic and electric fields within the sheet — see Ref. (13).)

The effective drag on the ions may be quite different in the presence of a super-Dreicer field and the accompanying plasma turbulence that is driven under these circumstances. Under these conditions it may be possible for the ions to overcome the electron barrier. Alternatively, the entire electron distribution in the current channels may be accelerated. In this case the ions can gain higher energies along with the accelerating electrons. A third possibility is that some of the ions can gain enough total momentum to overcome the drag through multiple scatterings in and out of the current channels.

Acceleration Time

Electrons. The time required for a DC electric field E (V cm^{-1}) to accelerate an electron from energy W_1 to W_2 (keV) is

$$t_{acc} = \frac{1.71 \times 10^{-5}}{E} \left[\sqrt{\left(\frac{W_2}{511} + 1\right)^2 - 1} - \sqrt{\left(\frac{W_1}{511} + 1\right)^2 - 1} \right] \text{ s.} \quad (3)$$

The distance traveled by the electron in this time is $1000(W_2 - W_1)/E$ cm. The time required to accelerate an electron from 25 keV to 100 keV and from 25 keV to 1 MeV as a function of electric field strength is shown in Figure 4. The top (dotted) curve is truncated below 10^{-4} V cm^{-1}, where the acceleration distance becomes greater than 10^{10} cm.

The analysis of flare x-ray data indicates electron acceleration times as short as 10 ms (1). Figure 4 shows that this can be accomplished with sub-Dreicer electric fields.

If the acceleration occurs along the full length of the flaring region, the low-energy electrons will arrive and emit x-rays before the higher energy electrons.

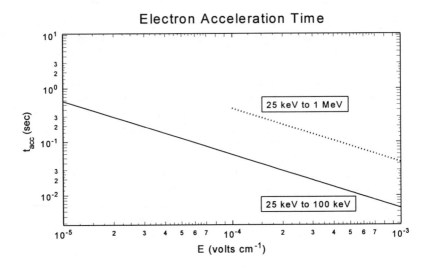

FIG. 4. Electron acceleration time plotted as a function of electric field strength.

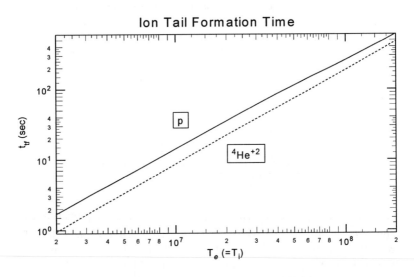

FIG. 5. Characteristic time scale for suprathermal proton and alpha particle tails to form, plotted as a function of plasma temperature (see text).

This is because the lower energy electrons originate closer to the thick-target interaction region. This is contrary to what has been found to be most common for x-ray spikes, however. The time delays for most of the spikes indicates that the highest energy electrons arrive first, consistent with free streaming (1). This suggests that for these spikes the electron acceleration occurs high in the corona, on time scales that are less than the free-streaming time differences ($\sim$ tens of milliseconds).

Ions. The time scale for the suprathermal ion tail to form is not well defined, since the collisional drag on the ions cannot be neglected and the ions approach their maximum velocity asymptotically (8). The time required for a proton and an alpha particle to be accelerated from $1.01 v_{cl}$ to $0.99 v_{cu}$ is plotted in Figure 5 as a function of the plasma temperature. This is for a thermal proton density of 3×10^{11} cm^{-3} and an electric field strength that maintains the runaway rate at 10^{-3} s^{-1} (see (8)).

Observations of γ-ray bursts indicate that protons are sometimes accelerated on time scales as short as a few seconds (5). As Figure 5 indicates, this is possible if the temperature is not much higher than typical coronal temperatures. Acceleration times of a second or less would be difficult to achieve with sub-Dreicer electric fields. If the plasma temperature in the current channels is on the order of 4×10^7 K or greater, however, v_{cl} is high enough that protons can be picked up directly from the thermal distribution by Alfven waves, for example. The acceleration time would then be determined by this wave-particle process rather than by the electric field acceleration.

X-ray and Microwave Spectra and Ion Abundances

The combination of runaway electron acceleration and direct Joule heating provides an attractive alternative to the standard nonthermal thick-target model for flare hard x-ray spectra (3) (10). It provides a physical interpretation of the x-ray spectra, including the low energy cut-off in the nonthermal electron distribution. In addition to the "super hot" plasma seen after the rise phase of some flares, it predicts that plasma with temperatures as high as 10^8 K is present early in the flare as well. The model relaxes the electron number and energy flux requirements over the pure nonthermal model. It provides information about physical parameters in the flaring region that can be related to other data. It also provides a physical explanation for the difference between hot thermal and impulsive flares (20).

The model is consistent with flare microwave spectra as well. It provides explanations for several interesting features of these spectra (2).

Direct electric field acceleration can enhance ion abundances in the accelerated particles (8). It is not apparent that it can produce all of the specific enhancements typical of energetic flare particles, however. Since resonant plasma waves are expected to be driven unstable as the electric field strength approaches the Dreicer field, it is likely that the ion abundances will be affected through interaction with these waves. The interplanetary ions from impulsive flares appear to arise from plasma at coronal temperatures rather than the hotter flare temperatures (19). If they are accelerated in the same location and at the same time as the impulsive phase electrons, then the impulsive phase particle acceleration must occur outside the hot loops seen in x-rays.

A limitation to deriving physical information from flare x-ray spectra is demonstrated in Figure 6, from Ref. (12). The top frame shows two fits to a flare x-ray spectrum from the Compton Gamma-Ray Observatory. Each fit consists of an isothermal component and a model runaway electron distribution. The two runaway electron distributions are plotted in the bottom frame. The two distributions are identical at high energies and give about the same total flux of accelerated electrons, but otherwise are quite different. They give significantly different values for $\mathcal{E}_{cr}$ (the electron energy corresponding to v_c), the total potential drop in the current channels, and other derived physical parameters. They produce virtually identical x-ray spectra, however. The upshot is that with x-ray spectra alone some physical quantities cannot be uniquely determined, but can only be limited to a rather broad range of parameter space.

CONCLUSIONS

The relatively simple sub-Dreicer electric field acceleration of runaway particles can provide the required flux of electrons and ions on the measured time scales. It provides a clear mechanism for energizing both electrons and ions out of the thermal plasma, and explains the observed relationship between particle acceleration and heating without requiring that most of the flare energy be released into nonthermal particles. Models based upon the mechanism allow us to derive physical information about the flaring region.

Particle acceleration by sub-Dreicer fields requires the presence of many oppositely directed, thin current channels. Such a configuration is consistent with ideas about coronal heating (17) and with "avalanche" models for solar flares (14). That nature can produce such a configuration is indicated by *in situ* observations of many oppositely directly currents in the earth's auroral zone (6). These currents may be driven on the sun in the convection zone where mass motions dominate the magnetic field pressure, or possibly in the corona by cross-field currents or flows (21) (22). Currents that are initially unbalanced are likely to close in the chromosphere (4).

Electrons and ions cannot be accelerated by sub-Dreicer electric fields to energies much greater than ~ 1 MeV with a single traversal of a current channel. Higher energies require either many interactions with the current channels or electric field strengths that exceed the classical Dreicer field. The behavior of a plasma in such high electric fields has not been established.

It is possible for the required flux of electrons to be accelerated in a single current sheet if the electric field strength is $\sim$ a few volts cm^{-1}, much greater than the classical Dreicer field, and they rapidly escape the sheet (on a scale length less than 1 km). Producing rapid x-ray spikes in a single current sheet may be a problem, however. Most of the sheet must respond on the 10–100

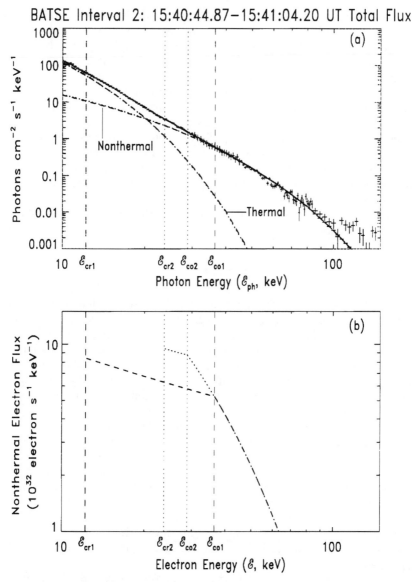

FIG. 6. Indistinguishable x-ray spectra (*top*) produced by two different runaway electron distributions (*bottom*).

ms time scale to produce the spike electron flux, but the long induction time associated with such a single, large scale current sheet may prevent this from happening. The induction time for a system of multiple, oppositely directed currents, on the other hand, puts a significant lower limit on the number of current pairs but does not exclude these rapid fluctuations (4).

Although direct (sub-Dreicer) electric field acceleration enhances the abundance of ions relative to protons, it is not apparent how the specific observed relative ion abundances could be obtained. Since plasma turbulence is a natural consequence of currents driven by electric fields comparable to the Dreicer field, it is likely that the specific abundance enhancements result predominantly from the interaction of the ions with resonant plasma modes that are generated in the process. It is yet to be determined, however, if this can be accomplished without raising the plasma temperature in the ion acceleration region above the 2–4 MK coronal temperatures that are apparently required by the observations.

Next Steps

The best prospect for distinguishing particle acceleration models, much less distinct physical mechanisms, is with a combination of simultaneous data providing quantitative diagnostics of both accelerated particles and plasma heating. The most direct information about accelerated electrons and the hottest plasma in flares is provided by x-ray observations above ~ 1 keV. Radio observations provide a crucial complement to this. The most direct information about impulsive phase ion acceleration is provided by observations of γ-ray lines. Observations of interplanetary ions from impulsive flares provide a crucial complement to the γ-ray line observations.

X-ray and γ-ray observations with high spectral resolution are a necessary component for constraining acceleration models. It is important that spectral resolution be high so that steep spectra and spectral changes be distinguishable and that the photon spectra derived from the count rate observations not be significantly model dependent. The modeling of these high-resolution spectra allows valuable information about the acceleration region to be derived. Spectra alone, however, have limited utility for distinguishing models. This will be substantially facilitated by the addition of photon-energy-dependent imaging and high time resolution.

High-resolution x-ray spectra that continue to energies well above 100 keV will provide an important test of acceleration models. This diagnostic of high-energy electrons and the transition to relativistic energies will provide a more stringent test of the models. A flattening of the spectrum at these energies (beyond relativistic effects) would indicate the presence of a second acceleration mechanism.

X-ray and γ-ray timing measurements also provide a stringent test of acceleration models. Accelerating a substantial flux of particles becomes considerably more challenging on time scales less than ~ 10 ms for electrons, and less than ~ 1 s for ions.

An important piece of information for acceleration models is the total flux of accelerated ions, especially protons. Since γ-ray and interplanetary particle observations have only provided information about particles with energies above ~ 10 MeV, the total energy that an acceleration model must put into ion acceleration is not known. A recent study of flare γ-ray data has indicated that the energy in accelerated protons above ~ 1 MeV may be substantially larger than the values obtained for protons with energies above 10 MeV (18). At present the most promising possibility for obtaining quantitative information about the flux of accelerated protons with energies below 1 MeV is to search for an enhancement of the red wing of hydrogen Lyman-α during the early rise of flares (16).

Of immediate interest on the theoretical side is a better determination of the runaway electron distribution function in a finite-length current channel. Existing numerical work has only been done for an infinitely long channel. This work is important both for direct application to x-ray data and for studies of the stability of the runaway distribution.

Of greatest importance on the theoretical side is the existence of competing models that are physical, quantitative, and directly address solar flare data. Significant progress has been made in this respect since the days of the *Solar Maximum Mission*. Hopefully, the next decade will provide a substantial increase in the quality and quantity of both high-resolution flare data and theoretical models. Certainly the capability is at hand for accomplishing both.

Acknowledgements. I thank Brian Dennis for his comments on this paper, and the *Compton* Gamma Ray Observatory Guest Investigator Program for partial support of this work.

REFERENCES

1. M. J. Aschwanden, this proceedings (1995); M. J. Aschwanden, R. A. Schwartz, and D. M. Alt, Ap. J. **447**, 923 (1995).
2. S. G. Benka, and G. D. Holman, Ap. J. **391**, 854 (1992).
3. S. G. Benka, and G. D. Holman, Ap. J. **435**, 469 (1994).
4. A. G. Emslie and J.-C. Henoux, Ap. J. **446**, 371 (1995).
5. D. J. Forrest and E. L. Chupp, Nature **305**, 291 (1983).
6. R. A. Hoffman, M. Sugiura, and N. C. Maynard, Adv. Space Res. **5**, No. 4, 109 (1985).
7. G. D. Holman, Ap. J. **293**, 584 (1985).
8. G. D. Holman, Ap. J. **452**, 451 (1995).
9. G. D. Holman, M. R. Kundu, and S. R. Kane, Ap. J. **345**, 1050 (1989).
10. G. D. Holman, and S. G. Benka, Ap. J. **400**, L79 (1992).
11. H. Knoepfel and D. A. Spong, Nuclear Fusion **19**, 785 (1979).
12. T. A. Kucera, P. J. Love, B. R. Dennis, G. D. Holman, R. A. Schwartz, and D. M. Zarro, Ap. J., in press (1995).

13. Yu. E. Litvinenko and B. V. Somov, Solar Phys. **158**, 317 (1995); B. V. Somov, this proceedings (1995).

14. E. T. Lu and R. J. Hamilton, Ap. J. **380**, L89 (1991).

15. J. A. Miller, G. A. Emslie, G. D. Holman, P. J. Cargill, B. R. Dennis, T. N. LaRosa, R. M. Winglee, S. G. Benka, and S. Tsuneta, J. Geophysical Res., submitted (1995).

16. F. Q. Orrall and J. B. Zirker, Ap. J. **208**, 618 (1976).

17. E. N. Parker, Ap. J. **330**, 474 (1988).

18. R. Ramaty, N. Mandzhavidze, B. Kozlovsky, and R. J. Murphy, Ap. J. **455**, L193 (1995); R. Ramaty, this proceedings (1995).

19. D. V. Reames, J. P. Meyer, and T. T. von Rosenvinge, Ap. J. Supplement **90**, 649 (1994).

20. S. Tsuneta, Ap. J. **290**, 353 (1985).

21. S. Tsuneta, PASJ, in press (1995).

22. R. M. Winglee, G. A. Dulk, P. L. Bornmann, and J. C. Brown, Ap. J. **375**, 382 (1991).

Reconnection and Acceleration to High Energies in Flares

B.V. Somov

Astronomical Institute, Moscow State University,
Universitetskii Prospekt 13, Moscow 119899, Russia

Acceleration of protons and heavy ions in a reconnecting current sheet (RCS), which forms as a consequence of coronal transient, CME, or filament eruption in the corona, is advocated as a possible mechanism of relativistic particle generation during the late phase of γ-ray flares. The important feature of the mechanism suggested is that neither the maximum energy nor the acceleration rate depend upon the particle mass.

INTRODUCTION

Particle acceleration to high energies in flares is the subject of a great deal of study but the mechanisms of acceleration still baffle the full theoretical understanding (1). The existence and importance of acceleration by shock waves in large flares is beyond doubts. However there are flares in which shock acceleration seems to be unsuitable, for example, for interpretation of the delayed component of γ-ray emission (2). A shock is already too high in the corona by the time the delayed component appears.

The purpose of this paper is to illustrate that the acceleration of protons and heavy ions during the late phase of large γ-ray flares to GeV energies can occur in a RCS, formed behind a rising coronal transient (e.g. 3), CME, or an eruptive prominence. In principle, the electric field, generated in such structures by fast change of magnetic field, can be as high as 10 V cm^{-1}; this field is the fastest and easiest means of acceleration to relativistic energies (4). In practice, however, various effects act mainly to decrease the acceleration efficiency. Some of them are reviewed in next Section.

MAGNETICALLY 'NON-NEUTRAL' RCS

Syrovatskii (5) considered the MHD process of a current sheet formation in the vicinity of a 2D neutral line of magnetic field under action of a DC electric field $\mathbf{E}_0$. He has found a 'dynamic' dissipation mechanism which converts magnetic energy of a 'neutral' current sheet into kinetic energy of fast particles. According to (5) the mean energy $\mathcal{E}$ of accelerated particles in solar flares can be as high as 10^9 eV.

The problem of fast particle motion in a magnetic field $\mathbf{B}_0$, which changes direction inside the RCS, and in the electric field $\mathbf{E}_0$ related to reconnection was considered several times.

First, Speiser (6) found particle trajectories near the 'neutral' plane where the magnetic field equals zero. Formally, a particle can spend an infinite time near such a plane a can take an infinite energy from the electric field. In the real conditions of the solar atmosphere such a situation does not exist. Usually the magnetic field in the 'reconnecting' plane (i.e., the RCS) has non-zero transverse component $\mathbf{B}_\perp$ related with reconnection and non-zero longitudinal component $\mathbf{B}_\parallel$. The last one is related to the fact that reconnection takes place at the so-called separators (e.g. 7). So real current sheets are *magnetically* 'non-neutral'.

Speiser (6) showed that even a very small transverse field $B_\perp = \xi_\perp B_0$ ($\xi_\perp \ll 1$) changes the particle motion in a such a way that after a finite time the particle leaves the RCS having a finite energy. This energy is not sufficient in the context of solar flare (8).

Second, it was shown in (8) that the longitudinal field $B_\parallel = \xi_\parallel B_0$ increases considerably the efficiency of electron acceleration in the RCS. 100 keV electrons can be accelerated in this way. However, the longitudinal magnetic field cannot influence the motion of relativistic protons. Moreover, a non-zero transverse field $B_\perp$ radically restricts the energy of protons: the energy gain $\delta\mathcal{E}$ cannot exceed 20 MeV if a typical value of $\xi_\perp$ is assumed according to the high-temperature turbulent current sheet model (7).

Therefore, the relativistic energies cannot be reached after a single 'interaction' of a proton with the RCS. That is why it was proposed in (9) that the protons interact with the RCS more than once, each time gaining a relatively small amount of energy $\delta\mathcal{E}$. The cumulative effect is the required relativistic acceleration.

ELECTRICALLY 'NON-NEUTRAL' RCS

The factor that makes a positively charged particle return into the RCS is the transverse electric field $\mathbf{E}_\perp$ which always exists as a consequence of electric charge separation in the vicinity of the RCS (10). In this sense, the RCS is also *electrically* 'non-neutral'.

On the one hand, the charge separation gives rise to the potential ϕ, which is not large: $\phi \approx kT$. On the other hand, the accelerated protons leave the RCS almost along its plane. This property is a characteristic feature of the Speiser's mechanism (6). So even a modest transverse electric field $\mathbf{E}_\perp$ may considerably influence the motion of protons because they move almost perpendicular to this field.

The largest energy $\mathcal{E}_{\max}$ attainable is determined by the condition that the potential ϕ is just enough to prevent a proton from leaving the RCS. In other words, the transverse electric field $\mathbf{E}_\perp$ can cancel the perpendicular momentum. This gives the maximum energy (see 9)

$$\mathcal{E}_{\max} = \frac{\phi}{\xi_\perp^2} \left[1 + \left(1 - \xi_\perp^2 + \frac{\xi_\perp^4 (mc^2)^2}{\phi^2} \right)^{1/2} \right] \approx \frac{2\phi}{\xi_\perp^2} \,. \tag{1}$$

Formula (1) shows that protons can actually be accelerated to GeV energies in the high-temperature RCS (7): for instance $\mathcal{E}_{\max} \approx 2.4$ GeV provided the current sheet temperature $T = 10^8$ K. Even larges energies can be reached in the RCS with a smaller transverse magnetic field (11) provided the RCS model and its applicabilty region allow one to take such a small value of $\xi_\perp$; see, however, Chapter 3 in (7).

ACCELERATION TIME

According to (9) the particle's energy grows with time as

$$\mathcal{E} \approx \frac{2}{\pi} c e E_0 \left(\frac{E_0}{B_\perp} \right) t \,. \tag{2}$$

Hence the account of the actual magnetic field structure in the RCS considerably diminishes the energy gain rate as compared with the ideal case $B_\perp = 0$ or, which is the same, with the simple acceleration by DC electric field without magnetic field (e.g. 12). In the last case

$$\mathcal{E} \approx c e E_0 \, t \,.$$

The diminishing factor equals

$$\frac{2}{\pi} \frac{E_0}{B_\perp} \approx \frac{2}{\pi} \frac{v}{c} \frac{B_0}{B_\perp} \approx \frac{2}{\pi} \frac{v}{c} \frac{1}{\xi_\perp} \le 10^{-1}$$

for the acceleration conditions considered in (9); here v is a typical velocity of plasma inflow into the RCS, i.e. the reconnection rate.

It also follows from (2) that the acceleration time

$$t_{\text{ac}} \approx 0.03 \left(\frac{\mathcal{E}}{1\,\text{GeV}} \right), \; \text{s} \,. \tag{3}$$

This result clearly demonstrates the possibility of very fast acceleration of protons by dint of direct electric field $\mathbf{E}_0$ in the non-neutral RCS.

DISCUSSION

The acceleration mechanism under consideration invokes the direct electric field $\mathbf{E}_0$ in the RCS. This is quite ordinary approach in studies of the impulsive phase of flares (5,13,14). Litvinenko and Somov (9) suggested that the extended acceleration of protons to relativistic energies during the late phase of large γ-ray flares occurs by the same mechanism in RCSs formed below erupting prominences. The time of RCS formation in this scenario corresponds to the observed delay of the second phase of acceleration after the first, impulsive one (2).

Some additional arguments in favour of particle acceleration in RCSs during the second phase have to be, at least, mentioned above

- Already early radio observations of flares (15,16) were indicative of electron acceleration at the cusps of helmit magnetic structures in the corona, similar to observed by the SXT at the Yohkoh. These are exactly the places where RCSs are expected to form.

- The observed γ-ray emission consists of separate peaks with a characteristic duration of 0.04–0.1 s (17). So, the shockwave mechanism (e.g. 18) seems to be too slow (for more detail, see 9).

- The acceleration by Langmuir turbulence inside the RCS in the helmet structure (19) is also too slow.

The direct electric field inside the RCS provides not only the maximum energy but also the necessary acceleration rate. High velocities (up to the coronal Alfvén speed) of plasma motion in coronal transients, filament eruptions and other CMEs imply a large direct electric field in the RCS (4). This is the reason why the acceleration mechanism considered is so efficient.

There is another important feature of this mechanism. Formulae (1) and (2) show that neither the maximum energy nor the acceleration rate depend upon the particle mass. Hence, in principle, the mechanism can be responsible for the preferential acceleration of heavy ions during flares. This property, as well as questions on the ionic charge states and accelerated ion spectrum, should be carefully considered somewhere else.

CONCLUSION

Though MHD shocks are usually thought to be resposible for the relativistic generation of protons and heavy ions in γ-ray flares (e.g. 20), another mechanism – the direct electric field acceleration in non-neutral RCS (for a review, see (21)) – is necessary at least in flares with fast variability of γ-ray emission.

ACKNOWLEDGEMENTS

The author wishes to acknowledge Dr. R. Ramaty for offering him the opportunity to attend the Workshop. This work was supported by the Russian Fond of Fundamental Researches.

REFERENCES

1. R. Ramaty, Summary of this Workshop.

2. V. V. Akimov, P. Ambroz, A. V. Belov, A. Berlicki, I. M. Chertok, M. Karlicky, V. G. Kurt, N. G. Leikov, Yu. E. Litvinenko, A. Maggun, A. Minko-Wasiluk, B. Rompolt, B. V. Somov, Solar Phys., in press (1996).

3. S. I. Syrovatskii, Solar Phys. **76** 3 (1982).

4. B. V. Somov, Bull. Acad. Sci. USSR, Phys. Ser. bf45, No. 4, 114 (1981).

5. S. I. Syrovatskii, Soviet Astronomy – AJ **10**, 270 (1966).

6. T. W. Speiser, J. Geophys. Res. **70**, 4219 (1965).

7. B. V. Somov, Physical Processes in Solar Flares, Dordrecht: Kluwer Academic Publishers, 1992.

8. Yu. E. Litvinenko and B. V. Somov, Solar Phys. **146**, 127 (1993).

9. Yu. E. Litvinenko and B. V. Somov, Solar Phys. **158**, 317 (1995).

10. E. G. Harris, Nuovo Cimento **23**, 115 (1962).

11. P. C. H. Martens, Astrophys. J. **330**, L131 (1988).

12. S. G. Benka and G. D. Holman, Astrophys. J. **435**, 469 (1995).

13. C. de Jager, C., Adv. Space Res **10**, No. 9, 101 (1990).

14. J. I. Sakai, Solar Phys. **140**, 99 (1992).

15. I. D. Palmer and S. F. Smerd, Solar Phys. **26**, 460 (1972).

16. R. T. Stewart and N. R. Labrum, Solar Phys. **27**, 192 (1972).

17. A. M. Gal'per, V. M., Zemskov, B. I. Luchkov, Yu. V. Ozerov, V. Yu. Tugaenko, and A. M. Khodarovich, Pis'ma v ZhETP **59**, 145 (1994).

18. S. A. Colgate, Solar Phys. **118**, 1 (1988).

19. H. Q. Zhang and E. L. Chupp, Astrophys. Space Sci. **153**, 95 (1989).

20. T. Bai and P. A. Sturrock, Ann. Rev. Astron. Astrophys. **27**, 421 (1989).

21. B. V. Somov, Fundamentals of Cosmic Electrodynamics, Dordrecht: Kluwer Academic Publishers, 1994.

On the Formation of the Helium-3 Spectrum in Impulsive Solar Flares

Yuri E. Litvinenko

Institute for the Study of Earth, Oceans, and Space,
University of New Hampshire, Durham, NH 03824-3525

The resonant interaction with oblique electromagnetic ion-cyclotron waves is the most promising mechanism for selective acceleration of ^{3}He ions in some impulsive solar flares. At the same time, the properties of the observed particle spectrum remain unexplained, in particular the spectral steepening or break at energies < 10 MeV that leads to an increase in the effective power-law spectral index at higher energies. An analytical solution of the Fokker-Planck equation, describing the resonant wave-particle interaction, shows that the steepening is unlikely to stem from the action of the acceleration mechanism alone. It is argued that the Coulomb energy losses at energies greater than a few MeV may be large enough to provide the observed spectral break. Its position is determined by the balance between energy gain by acceleration and energy loss. Therefore, the position of the break may serve as a diagnostic tool for the study of the acceleration mechanism.

INTRODUCTION

One of the most curious phenomena occuring during impulsive solar flares is the enrichment of the accelerated particle flux by the ^{3}He isotope with respect to ^{4}He (see (1) for a recent review of observational data). Under typical conditions of the solar atmosphere, the density of ^{3}He ions is only $\approx 5 \cdot 10^{-4}$ of that of ^{4}He. Since a typical concentration of ^{4}He in flaring regions is 10^9 cm^{-3} ($\approx 10\%$ of the hydrogen concentration), it means that the number of ^{3}He ions in the flare volume $V \approx 10^{26}$ cm^3 is $5 \cdot 10^{31}$. A significant portion of these is accelerated during a ^{3}He-rich flare. Enhancements in the ^{3}He/^{4}He ratio by a factor of $10^3 - 10^4$ are observed for accelerated particles. This is because about $10^{30} - 10^{31}$ ^{3}He ions are accelerated during a typical event up to energies of a few MeV, so that their total energy can be $> 10^{25}$ erg. The acceleration is known to occur in a time less than 10 s, implying an energy gain rate of a few hundred keV per second. Measurements of ^{3}He energy spectra showed them to be power-laws with spectral indices of about 2 and steepening at an energy < 10 MeV (2).

This note attempts to explain some peculiarities of ^{3}He spectra in impulsive flares. The new physical effect introduced into consideration is the influence of Coulomb losses on the particles which counteracts the acceleration at high energies and thus gives rise to characteristic breaks in the ^{3}He spectrum.

MECHANISM OF HELIUM-3 ACCELERATION

The acceleration model that appears to be the most promising is that of ^{3}He acceleration by oblique electromagnetic hydrogen ion-cyclotron waves, which are excited in a narrow frequency range by unstable electron beams or currents at the Landau resonance (3,4). For an ion with mass m_i and charge q_i to interact strongly with a wave, the linear resonance condition must hold:

$$\omega - k_\parallel v_\parallel = n\Omega_i, \tag{1}$$

where ω is the wave frequency, $k_\parallel$ is the component of the wave vector $\mathbf{k}$ parallel to the external magnetic field $\mathbf{B}$, $v_\parallel$ is the parallel particle velocity, $\Omega_i = q_i B/(m_i c) \equiv Z\Omega_H/A$ is its gyrofrequency, Z and A are its charge and mass numbers, and n is an integer. The resonant interaction is most effective for $n = 1$. The electromagnetic ion-cyclotron waves are excited most efficiently in the frequency range $0.6\,\Omega_H < \omega < 0.8\,\Omega_H$ (4,5). The ^{3}He isotope, with its ratio $Z/A = 2/3$, is the only ion under coronal conditions whose gyrofrequency lies in this range. Therefore, only this ion can be selectively accelerated by resonating with the waves (assuming that the term $k_\parallel v_\parallel$ can be ignored for the oblique waves).

Hereafter, I consider only the effect of the ion-cyclotron wave turbulence on the particle distribution, setting aside the question of how the waves are generated. This question and the properties of the wave spectrum in the flaring plasma were thoroughly studied in (4,5).

The particle average energy gain rate is estimated by using the quasilinear plasma theory (6) or by directly integrating the equation of particle motion. The nonrelativistic result in the limit of large parallel phase velocities, $\omega/k_\parallel \gg v_\parallel$, is as follows:

$$\left\langle \frac{dw}{dt} \right\rangle = 32\pi^2 \frac{q^2}{m} \int dk \sum_n W(k, n\dot{\Omega}) \frac{n^2 J_n^2(k\rho)}{(k\rho)^2}. \tag{2}$$

Here w is the particle kinetic energy, ρ is the particle gyroradius, J_n are the Bessel functions, and W is the turbulence energy density per unit frequency and wave vector interval. To simplify the notation, the quantities without a subscript, like w, q, m, Ω and ρ, are assumed hereafter to pertain to ^{3}He.

Note that for typical values of k, $k\rho = 1$ corresponds to an energy on the order of 0.1 MeV. Hence, when studying acceleration to energies on the order of a few MeV, it is reasonable to use the asymptotic expansion of J_n valid for large arguments. The integration is then performed to give for the principal $n = 1$ resonance (3)

$$\left\langle \frac{dw}{dt} \right\rangle \approx 8\pi\sqrt{2m}\, q^2 \frac{W_{\text{turb}}}{\Delta\omega} \frac{\Omega^3}{\langle k^3 \rangle} w^{-3/2} \equiv D_0 w^{-3/2}. \tag{3}$$

Here W_{turb} is the total density of turbulent energy, $\Delta\omega$ is the frequency interval occupied by the waves. Equation (3) shows that the acceleration can occur fast enough to provide the selective energization of ^{3}He ions in impulsive solar flares. One gets a simple estimate by defining the acceleration time $t_{ac} = w/d_t w$ and adopting the following typical values: $B = 100$ G, $\Delta\omega = 0.2\,\Omega_H$, $\langle k \rangle = 5 \cdot 10^{-3}$ cm^{-1} (3,4). Then the estimate

$$W_{turb} \approx 4.17 \cdot 10^{-6} \frac{(w/1\text{MeV})^{5/2}}{(t_{ac}/1s)} \left(\frac{100G}{B} \right)^2 \frac{\text{erg}}{\text{cm}^3}, \tag{4}$$

shows that in order to impulsively ($t_{ac} \approx 1$ s) accelerate the ions to energies of a few MeV, a reasonably low turbulence level on the order of 10^{-5} erg/cm^3 is required. It is important that the corresponding rms wave electric field is about 5 V/cm, a value confirmed by direct measurements of electric fields in the solar atmosphere (7). In what follows, the value of $W_{turb} = 1.13 \cdot 10^{-5}$ erg/cm^3, corresponding to $E_{rms} = 5$ V/cm, is used.

The acceleration model should explain not only the energy gain rate, but also the shape of the particle spectrum. Using Equation (3) and taking into account diffusion in energy space, one can obtain the Fokker-Planck equation for the distribution function $f = f(w,t)$ in the form (cf. (8) for the case $k\rho \ll 1$):

$$\frac{\partial f}{\partial \tau} = \frac{\partial}{\partial w} \left(w^{-1/2} \frac{\partial f}{\partial w} \right), \tag{5}$$

where $\tau = \int D_0 dt$ is not necessarily proportional to t.

The Green's function of the last equation is found by standard methods:

$$G(w, w_0, \tau) = \frac{2(ww_0)^{3/4}}{5\tau} \exp\left(-\frac{4}{25} \frac{w^{5/2} + w_0^{5/2}}{\tau} \right)$$

$$\times I_{-3/5} \left(\frac{8(ww_0)^{5/4}}{25\tau} \right). \tag{6}$$

This formula can be considerably simplified in the event of acceleration of initially low-energy particles: $w_0 \to 0$. In this limit, the distribution function is

$$f(w, \tau) = \int_0^\infty f_0(w_0) G(w, w_0, \tau)\, dw_0$$

$$\approx \frac{(5/2)^{1/5}}{\Gamma(2/5)} \frac{n}{\tau^{2/5}} \exp\left(-\frac{4}{25} \frac{w^{5/2}}{\tau} \right), \tag{7}$$

irrespective of the initial distribution $f_0(w_0)$. Here n is the total density of accelerated particles. As discussed in Introduction, observations imply that $n < 5 \cdot 10^5$ cm^{-3}.

Solution (7) to the Fokker-Planck equation is in good agreement with the numerical results (9). It predicts the spectrum to be flat below a cut-off energy $\approx \tau^{2/5}$ and to rapidly fall off above it. The cut-off energy is time-dependent and can have any value. This is in contradiction to observations.

First, a typical ^{3}He spectrum is softer at low energies than it follows from solution (7). Second, the spectrum is not as steep at high energies as formula (7) predicts. Reames *et al.* (2) approximated an observed spectrum as $\sim w^{-2}$ below the cut-off energy and as $w^{-3.5}$ above it. Möbius *et al.* (10) used an approximation involving the Bessel function, with the high-energy asymptotic form $\sim w^{-3/8} \exp(-\text{const } w^{3/4})$. Recent data from the SAMPEX spacecraft (11) also indicate that the ^{3}He spectrum falls off in a manner similar to that reported in (10). Finally, the observed cut-off energy itself is not arbitrary; its value is close to 2 MeV/nucleon (1,11). The acceleration mechanism alone cannot explain these features of the spectrum. A possible additional physical mechanism is presented below.

FORMATION OF THE SPECTRUM

Effect of Coulomb Energy Losses

Equation (3) shows that the ^{3}He energy gain rate decreases with increasing energy. The decrease is so fast that Coulomb losses can become important for sufficiently large energies. It is this effect that can be responsible for the shape of the ^{3}He spectrum.

Butler and Buckingham (12) calculated the Coulomb loss rate for a particle of speed v, moving in a plasma with electron temperature T_e and concentration n_e:

$$P(v) = \frac{8\sqrt{\pi}\, e^2 q^2}{\sqrt{2k_B T_e m_e}}\, n_e \ln\Lambda\, F\left(\frac{v}{\sqrt{2k_B T_e/m_e}}\right).$$ (8)

Here $\ln\Lambda \approx 20$ is the Coulomb logarithm and the function F is defined as

$$F(x) = \frac{1}{x}\int_0^x \exp(-x^2)\, dx - \left(1 + \frac{m_e}{m}\right)\exp(-x^2).$$ (9)

The maximum of F is reached at $x \approx 1.5$. For ^{3}He, this corresponds to the power loss $P_{max} \approx 0.36$ MeV/s at an energy of about 1 MeV for typical coronal conditions ($n_e = 10^{10}$ cm^{-3} and $T_e = 10^6$ K). For larger energies, F is a decreasing function. However, it falls off less rapidly than the energy gain rate (3) and eventually overcomes the latter.

Because the energies of at least a few MeV are of primary interest to us here, let us use the asymptotic expansion valid for $x \gg 1$:

$$F(x \to \infty) \approx \frac{\sqrt{\pi}}{2x}.$$ (10)

Equating the energy gain rate (3) to the loss rate (8), with account taken of (10), gives a formula for the characteristic ^{3}He kinetic energy:

$$w_0 = 4\frac{W_{turb}}{\Delta\omega}\frac{\Omega^3}{\langle k^3\rangle}\frac{m_e}{n_e e^2 \ln\Lambda}.$$ (11)

Substitution of the above-given numerical values gives $w_0 = 5.3$ MeV, a value very close to the observational result of 2 MeV/nucleon.

The effect of Coulomb losses should also be important in the formation of the particle spectrum. To see this, let us rewrite the Fokker-Planck equation (5) with due regard to the losses:

$$\frac{\partial f}{\partial t} = \frac{\partial}{\partial w}\left(D_0 w^{-1/2}\frac{\partial f}{\partial w} + P(w)f\right). \tag{12}$$

Again, this equation is considerably simplified in the limit of large energies, reducing in the time-independent case to the following:

$$\frac{\partial}{\partial w}\left(w_0 w^{-1/2}\frac{\partial f}{\partial w} + w^{-1/2}f\right) = 0, \tag{13}$$

with the obvious solution

$$f(w) = \text{const } \exp(-w/w_0). \tag{14}$$

This solution closely agrees with the Bessel function fit (10) used to describe the ^{3}He spectrum. Formula (14) seems to be more adequte for interpretation of the observational data than Equation (7), derived by ignoring the energy losses.

Possible Generalizations

The form of distribution (14) essentially depends upon the energy gain rate (3). The latter has been calculated in the quasilinear regime. One obvious way to generalize solution (14) is to consider the energy gain rate of the form $dw/dt \sim w^{-\alpha}$ with an arbitrary α. It is known, in particular, that ion acceleration may be more efficient in strong turbulence than the quasilinear approach predicts: α decreases from $3/2$ to $1/4$ as the turbulence level increases (13).

Generalizing the calculations above, one can easily modify Equation (12) and solve it to get the particle spectrum

$$f(w) \sim \exp\left[-\text{const}\,\frac{w^{\alpha-1/2}}{(\alpha - 1/2)}\right]. \tag{15}$$

Note that this solution does not tend to zero at large energies if $\alpha < 1/2$. This means that the Coulomb losses cannot balance the energy gain due to strong turbulence. In this case, the break in the spectrum should be absent and the observed particle spectra should resemble power-laws. It is of interest, though, that the observed ion spectra (of both ^{3}He and heavier ions) can be fitted by formula (15) with $\alpha = 1.25$ (10) and $\alpha = 1.075$ (14).

Returning to Equation (12), an analytical description of the particle spectrum for low energies can be obtained by using an analytic approximation for F, valid for both small and large arguments:

$$F(x) \approx \frac{2x^2}{3 + 4x^3/\sqrt{\pi}}. \tag{16}$$

Then the stationary solution to Equation (12) is

$$f(w) = \text{const } \exp\left[-\frac{w}{w_0} + \frac{2}{\sqrt{3}}\frac{w_1}{w_0}\arctan\left(\frac{2\sqrt{w/w_1}-1}{\sqrt{3}}\right)\right]$$
$$\times \left(\frac{1 - \sqrt{w/w_1} + w/w_1}{(1 + \sqrt{w/w_1})^2}\right)^{w_1/(3w_0)}, \tag{17}$$

where $w_1 = (3\sqrt{\pi}/4)^{2/3}mk_B T_e/m_e$.

A still more accurate solution can be obtained by numerically integrating Equation (12). I intend to use it, together with the analytical approximations, for comparison with the spectra that will be provided by the instruments on board SOHO spacecraft. Other effects, such as particle escape from the acceleration region, will be also incorporated into the solution.

CONCLUDING REMARKS

The upshot of the above calculations is that the Coulomb losses can counteract the acceleration due to the ion-cyclotron waves at energies of a few MeV, thus giving rise to the observed exponential spectrum of the ^{3}He ions in impulsive solar flares.

The largest uncertainty of this treatment results from the ambiguity in the choice of parameters, first of all the turbulence level W_{turb}. To eliminate the ambiguity, the parameters should be determined from a self-consistent solution to the problem of generation and dissipation of the turbulence (k, $\Delta\omega$, W_{turb}) and from a model for the structure and dynamics of the flare energy source (n_e and T_e). As an example of such an approach, consider the energy balance of the turbulence:

$$\frac{dW_{\text{turb}}}{dt} = 2\gamma W_{\text{turb}} - n\left\langle\frac{dw}{dt}\right\rangle. \tag{18}$$

One can use this relation in the steady-state case, together with Equations (3) and (11), to eliminate the particle energy and get a relation between the number density of accelerated ions and the turbulence level:

$$n = \frac{\gamma}{4\pi\sqrt{2m}\,q^2}\left(\frac{\Omega^3}{\Delta\omega\,\langle k^3\rangle}\right)^{1/2}\left(\frac{4m_e}{n_e e^2 \ln\Lambda}\right)^{3/2}W_{\text{turb}}^{3/2}. \tag{19}$$

Here $\gamma \approx 10^{-3}\Omega_H$ is the growth rate of the instability (4). It follows from this equation that the turbulence level must be $1.6 \cdot 10^{-5}$ erg/cm^3 in order to obtain the typically observed ^{3}He density of 10^5 cm^{-3}. This value of W_{turb} is almost equal to the one used above.

Therefore, by considering conservation equations, the parameters used in the model can be unambiguously determined and hence the ^{3}He spectra in impulsive flares can be construed to reflect the balance between the acceleration by the ion-cyclotron turbulence and energy loss due to Coulomb collisions. Comparison of the model predictions with observations should place additional constraints on the parameters used. In particular, the absence of the energy break in ion spectra in some flares is an indication of a more efficient energy gain by the ions. As Equation (3) shows, this can be due to a smaller value of the wave vector. Strong turbulence effects cannot be excluded either.

ACKNOWLEDGMENTS

I am grateful to Prof. M. A. Lee for numerous useful discussions. This work was supported by the NSF grant ATM-9215279.

REFERENCES

1. D. V. Reames, J. P. Meyer, and T. T. von Rosenvinge, Astrophys. J. Supplement **90**, 649 (1994).
2. D. V. Reames, I. G. Richardson, and K.-P. Wenzel, Astrophys. J. **387**, 715 (1992).
3. M. Temerin and I. Roth, Astrophys. J. **391**, L105 (1992).
4. J. A. Miller and A. F. Viñas, Astrophys. J. **412**, 386 (1993).
5. D. W. Forslund, J. M. Kindel, and M. A. Stroscio, J. Plasma Phys. **21**, 127 (1979).
6. R. C. Davidson, *Methods in Nonlinear Plasma Theory*, Academic Press: New York (1972).
7. P. Foukal and S. Hinata, Solar Phys. **132**, 307 (1991).
8. P. A. Sturrock, Phys. Rev. **141**, 186 (1966).
9. J. A. Miller, A. F. Viñas, and D. V. Reames, *Proc. 23rd International Cosmic Ray Conf.*, **3**, 13 (1993).
10. E. Möbius, M. Scholer, D. Hovestadt, B. Klecker, and G. Gloeckler, Astrophys. J. **259**, 397 (1982).
11. G. M. Mason, J. E. Mazur, and D. C. Hamilton, Astrophys. J. **425**, 843 (1994).
12. S. T. Butler and M. J. Buckingham, Phys. Rev. **126**, 1 (1962).
13. R. L. Lysak, M. K. Hudson, and M. Temerin, J. Geophys. Res. **85A**, 678 (1980).
14. J. E. Mazur, G. M. Mason, and B. Klecker, Astrophys. J. **448**, L53 (1995).

Charged Particle Diffusive Transport

Geoffrey Lenters and James A. Miller

Department of Physics, The University of Alabama in Huntsville
Huntsville AL 35899

We consider the transport of protons in a spatially-bounded magnetized plasma containing a homogeneous distribution of Alfvén wave
turbulence. The protons interact strongly with the waves via cyclotron
resonance, and are in turn pitch-angle scattered and spatially diffused
throughout the plasma. Upon traversing one of the boundaries, a
proton escapes from the region. We solve numerically the diffusion
equation describing this process for the number of trapped particles as
a function of time. We find that a leaky-box approximation is in poor
agreement with the exact result when the turbulence injection rate is
small, which is likely the prevailing condition after particle acceleration
and during subsequent transport. On the other hand, this approximation nevertheless yields acceptable results when used in acceleration
models.

INTRODUCTION

The nature of charged particle motion in a magnetized turbulent plasma
is of fundamental importance in determining how a given acceleration mechanism will behave under solar flare conditions, as well as how the resulting
energetic particles are transported throughout the flare and deposit their energy. Regardless of the precise mechanism(s) responsible for acceleration, an
impulsive solar flare loop is likely to be a chaotic environment, with one or
more wave modes present that can affect the movement of particles throughout the loop and hence their production of observable radiation. For example,
many studies of proton transport/gamma-ray production in solar flares have
employed a loop model in which the coronal region is filled with a spectrum of
Alfvén waves (see refs. 1 and 2 for a concise review). These waves pitch-angle
scatter the protons into the loss cone, after which they promptly interact in
the denser chromosphere or photosphere and produce (i) nuclear de-excitation
lines between $\approx$ 4 and 8 MeV, (ii) neutrons that may thermalize and yield the
2.223 MeV line, and (iii) pions that decay and contribute to the high-energy
gamma-ray continuum.

The level of turbulence determines the scattering rate, which in turn determines the time scale for interaction and concomitant radiation production.
As the turbulence level increases from zero, scattering populates the loss cone
more aggressively and the time scale for radiation production decreases. When
the turbulence reaches a level where the particles are scattered through an an-

gle equal to the loss cone on about a transit time across the loop, the loss cone is kept fully populated and the interaction time scale reaches a minimum (3,4). This state is sometimes referred to as "saturation". Beyond this point, further increases in the wave level inhibit motion through the corona, the transport becomes diffusive, and the interaction time scale increases. Pitch-angle scattering may thus be responsible for the short radiation decay times observed in some flares, and flare-to-flare variations in the level of wave turbulence may also account for the range of observed decay time scales for gamma-ray emission. It has even been suggested that time-varying turbulence may account for the complicated emission properties of some flares (e.g., 5).

References (1)–(5), as well as some more mentioned therein, describe solar flare transport models that employed a Monte Carlo technique, in which discrete particles were followed from an initial energy until they either interacted or fell below a particular threshold energy. In these models, scattering and its effect on particle motion has been treated exactly, in the sense that diffusion coefficients for a given turbulence level were used to change the particle's pitch angle at several points along its motion through the coronal portion of the loop. Transport can also be described by an appropriate diffusion equation for the phase space particle density, and this has been explored in models which neglect scattering by waves but include that due to Coulomb collisions (6,7).

While wave-particle interactions may play a critical role in transport, they nearly completely determine the characteristics of the energetic particles in stochastic acceleration models. In this case, waves are intimately involved with the particles throughout the latter's history. In studies of stochastic particle acceleration (e.g., 8, 9, 10, 11,12), the emphasis has been on understanding the behavior of the particles in energy space and not on the details of their transport. Consequently, transport has been dealt with only very approximately, and a standard practice has been to treat particle escape from the acceleration region by the addition of a "leaky-box" loss term to the Fokker-Planck or momentum diffusion equations. This approach has allowed effort to be focused on the general properties of the mechanism, and the picture that is emerging is one in which resonant wave-particle interactions are able to account for the acceleration of a sufficient number of both ions (11) and electrons (13) out of the thermal background and to relativistic energies on the observed time scales in impulsive solar flares.

It is now appropriate to examine in more detail the assumptions and simplifications which underlie these stochastic acceleration models. In this paper, we will consider only one, but it is the one common to all of the above models; namely, the treatment of escape by a leaky-box loss term in the Fokker-Planck equation. We will compare the particle escape rate obtained by this method with that obtained by an exact numerical solution of the diffusion equation. Since the details of the scattering depend upon the wave mode (e.g., Alfvén, whistler, lower hybrid) and the type of particle (ion or electron), both must be specified for a comparison. For the sake of brevity we consider only one case: protons interacting with Alfvén waves. While our conclusions hold

506

only for this situation, they at least illustrate the potential issues for other wave/particle combinations.

PROTON TRANSPORT

The actual environment in which solar flare particle acceleration takes place is likely to be very complicated. We idealize this situation by assuming that acceleration occurs in a single region, which consists of a uniform fully-ionized H plasma permeated by a static homogeneous ambient magnetic field $\vec{B}_0 = B_0\hat{z}$. The length of the region along the magnetic field is L. The results are independent of the precise geometry of the region, but one could imagine it to be the coronal portion of a flare loop. Particles escape from the region upon crossing a boundary either at $z = 0$ or $z = L$ (that is, we take the loss cone half angle to be 90° at both ends). These boundaries could mark the transition from the ionized corona to the denser partially-ionized chromosphere, where interactions between energetic and ambient particles occur. A spectrum of plasma waves is superposed upon the background magnetic field. Such an idealized acceleration region is consistent with earlier acceleration models.

We are interested here in transport effects within, and escape from, the acceleration region. It is therefore sufficient to consider pitch-angle scattering and transit along $\vec{B}_0$, and employ the diffusion equation that describes the evolution of the particle distribution function $f(z, \mu, t)$. Here, $f(z, \mu, t)$ is the number density of particles per unit z and per unit pitch-angle cosine μ at time t. This equation is given by (e.g., 14)

$$\frac{\partial f}{\partial t} + v_\parallel \frac{\partial f}{\partial z} = \frac{\partial}{\partial \mu} \left(D_{\mu\mu} \frac{\partial f}{\partial \mu} \right) \quad , \tag{1}$$

where v is the particle speed, $v_\parallel = \mu v$ is the particle velocity component along $\vec{B}_0$, and $D_{\mu\mu} = \langle \Delta\mu\Delta\mu \rangle/(2\Delta t)$ is the pitch-angle diffusion coefficient. We now turn our attention to this coefficient.

Pitch-Angle Scattering

The transverse electric field of a plasma wave can strongly affect particle motion through gyroresonant interactions. Such an interaction occurs when the Doppler-shifted wave frequency is an integer multiple of the cyclotron frequency in the particle's guiding center frame, and the sense of rotation of the particle and the field are the same. In this case, depending upon the initial relative phase of the wave and particle, the particle will corotate with either an accelerating or decelerating electric field over a significant portion of its gyromotion, resulting in an appreciable gain or loss of energy, respectively. The condition for this interaction to occur is $\omega - k_\parallel v_\parallel - \ell\Omega/\gamma = 0$, where γ

and Ω are the Lorentz factor and gyrofrequency of the particle, and ω and $k_\parallel$ are the frequency and parallel wavenumber of the wave. If the particle and transverse wave electric field rotate in the same sense relative to $\vec{B}_0$, the harmonic number $\ell > 0$, whereas an opposite sense of rotation requires $\ell < 0$ (for $\omega > 0$).

Alfvén waves have been extensively employed in both transport and acceleration models. We consider for simplicity Alfvén waves propagating parallel and antiparallel to $\vec{B}_0$. These waves have left-hand circular polarization relative to $\vec{B}_0$ and occupy the frequency range below the hydrogen cyclotron frequency Ω_H. If $\omega \ll \Omega_H$, the dispersion relation is given by $\omega = v_A|k_\parallel|$, where v_A is the Alfvén speed. As $\omega \to \Omega_H$, the dispersion relation is more complicated and $|k_\parallel| \to \infty$ near Ω_H; waves in this region are sometimes called electromagnetic ion cyclotron waves, but they lie on the same branch as the Alfvén waves. For these parallel waves, the gyroresonance condition for any ω can only be satisfied with $\ell = +1$, and this is just ordinary cyclotron resonance. Cyclotron resonance will alter both the particle's energy and pitch angle. However, for solar flare values of v_A, the wave magnetic field is more than ten times larger than the electric field, the wave will be predominantly magnetic in nature, and the dominant effect of resonance will be pitch-angle scattering. Significant acceleration can still occur, but requires a time scale much longer than that needed for the appreciable redistribution of particles in pitch angle (11).

In general, a proton can cyclotron resonate with two waves moving along its direction of motion and one wave moving in the opposite direction (15). However, unless μ is very close to zero, there is only one resonant wave and it is the backward-moving one. As this wave also has the greatest effect on proton motion (by virtue of having the largest amplitude for a power-law spectral density), it is the only one we consider. In light of these simplifications, we also take the Alfvén wave dispersion relation to be $\omega = v_A|k_\parallel|$ for all $\omega < \Omega_H$. This is a very good approximation for all particles except those with the lowest energies and $\mu \approx 0$. The wavenumber of the resonant wave is quickly found to be

$$k_{\parallel r} = \frac{-\varepsilon \Omega_H}{\gamma \left(v_A + \varepsilon v_\parallel \right)} \quad , \tag{2}$$

where $\varepsilon = v_\parallel/|v_\parallel|$ indicates the direction of motion of the proton. Notice that this expression nicely avoids the problem of having $|k_{\parallel r}| \to \infty$ when $|\mu| \to 0$, which occurs when the first (i.e., ω) term in the gyroresonance condition is neglected.

The pitch-angle diffusion coefficient is given by equation (B12) of reference (11), using $k_{\parallel r}$ in equation (2). At this point it is convenient to change to dimensionless variables. Momentum $p = \widetilde{p}mc$; kinetic energy $E = \widetilde{E}mc^2$; speed $v = \widetilde{v}c$; frequency $\omega = \widetilde{\omega}\Omega_H$; wavenumber $k_\parallel = \widetilde{k}_\parallel\Omega_H/c$; length

$L = \widetilde{L}c/\Omega_{\mathrm{H}}$; and time $t = \widetilde{t}\, T_{\mathrm{H}}$, where $T_{\mathrm{H}} = \Omega_{\mathrm{H}}^{-1}$. The spectral density W_{T} is the total wave energy density per unit wavenumber, and is normalized to the ambient magnetic field energy density $U_{\mathrm{B}} = B_0^2/(8\pi)$, so that $\widetilde{W}_{\mathrm{T}} = (\Omega_{\mathrm{H}}/c)W_{\mathrm{T}}/U_{\mathrm{B}}$. That is, $\widetilde{W}_{\mathrm{T}}\, d\widetilde{k}_{\parallel}$ is the total wave energy density, in units of U_{B}, in the normalized wavenumber interval $d\widetilde{k}_{\parallel}$ about $\widetilde{k}_{\parallel}$. We then have that

$$\widetilde{D}_{\mu\mu} = \frac{\pi}{4}\frac{(1-\mu^2)}{\widetilde{p}^{\,2}}\widetilde{W}_{\mathrm{T}}(\widetilde{k}_{\parallel \mathrm{r}})\,(\widetilde{v}+|\mu|v_{\mathrm{A}})^2\,\frac{1}{v_{\mathrm{A}}+|\widetilde{v}_{\parallel}|} \quad , \tag{3}$$

where $D_{\mu\mu} = \widetilde{D}_{\mu\mu}\Omega_{\mathrm{H}}$. Consistent with our assumption above on the linear form of the dispersion relation, we have assumed equipartition between the wave vacuum field energy and the oscillation kinetic energy of the background protons in order to relate the magnetic spectral density W_{B} of reference (11) to the total spectral density W_{T} (namely, $W_{\mathrm{T}} = 2W_{\mathrm{B}}$).

Wave Spectral Density

To proceed further, we must specify the form of the spectral density. We assume that Alfvén waves are continuously injected at some wavelength λ_{i} and that a Kolmogorov-like nonlinear cascade transfers the wave energy to smaller scales. It is dissipated there by accelerating protons (11). The precise mechanism by which waves are generated in flares is unknown. MHD waves can easily be formed by a large-scale restructuring of the flare magnetic field, which presumably occurs at least in the initial energy release. In this case, λ_{i} is expected to be on the order of the flare size, say 10^8–10^9 cm. Values of λ_{i} much smaller than this would require a microinstability. The corresponding wave injection wavenumbers are $\pm k_{\mathrm{i}}$, where $k_{\mathrm{i}} = 2\pi/\lambda_{\mathrm{i}}$. The spectral density in the inertial range at equilibrium is $\widetilde{W}_{\mathrm{T}} = \widetilde{W}_0|\widetilde{k}_{\parallel}|^{-5/3}$, where

$$\widetilde{W}_0 = \left(\frac{3\sqrt{2}}{5\widetilde{v}_{\mathrm{A}}}\widetilde{Q}\right)^{2/3} \tag{4}$$

and the volumetric wave energy density injection rate Q at either k_{i} or $-k_{\mathrm{i}}$ is $\widetilde{Q}U_{\mathrm{B}}\Omega_{\mathrm{H}}$. The total rate of wave energy injection in both parallel and antiparallel waves is therefore $2Q$. The power law spectral density is applicable only in the inertial range, and may not apply in the wavenumber range sampled by most of the particles. We neglect this effect and employ the power law for all $k_{\parallel}$, which is consistent with earlier treatments.

The spectral density normalization and hence the scattering rate thus depends only upon the rate of wave energy injection. The normalization factor $\widetilde{W}_0$ can also be found from the total energy density in either forward or backward propagating waves $\widetilde{U}_{\mathrm{T}} \approx (3/2)\widetilde{W}_0\widetilde{k}_{\mathrm{i}}^{-2/3}$ and the injection wavenumber.

509

This approach has typically been used before, but can lead to unphysical conditions in the model region. For example, if the largest scale in the turbulence is assumed to be the maximum gyroradius of a 10 GeV proton (see 1), then $\widetilde{k}_i = 0.57$. If $U_T \sim 1 \, \mathrm{ergs\,cm^{-3}}$ ("strong scattering" in ref. 1) and $B_0 \approx 100$ G, then $2Q \approx 1300 \, \mathrm{ergs\,cm^{-3}\,s^{-1}}$, which is unacceptable. Since scattering and acceleration rates depend upon only W_0, and this can be expressed in terms of only Q, it is best to work with Q rather than the total energy density. Of course, the objection concerning unphysically large values of Q is mute if wave generation occurs at every scale in the spectrum and not just at some maximum value accompanied by nonlinear interactions. This possibility, however, is not attractive given the large dynamic range of the turbulence.

RESULTS

Equation (1) can be solved approximately using an eigenfunction expansion (14). However, a direct numerical solution is both straightforward and extremely fast. We use the method of operator splitting, and employ explicit upwind differencing in the z dimension and fully-implicit finite differencing in the μ dimension (16). The upwind differencing scheme is the only one which was stable for the advective term and also conserved particles. (The addition of numerical viscosity helped the stability of other techniques but did not conserve particles over the long time scales that we are interested in.) A constraint on the timestep Δt is imposed by the Courant condition in the explicit scheme, which requires that Δt be smaller than the grid crossing time of the fastest signal in the problem. In this case, the Courant condition reduces to $\Delta t \leq \Delta z / v_\parallel < \Delta z / v$, where Δz is the z gridpoint spacing. (The implicit method is unconditionally stable.)

We write equation (1) in terms of the dimensionless quantities in the last two sections and use a 150x150 grid for its numerical solution. Finer grids do not change any of the following results. We take $L = 10^9$ cm, $B_0 = 500$ G, $v_A = 0.036c$, and $f(z, \mu, 0)$ to be constant. The total number density of protons in the region at time t is $N(t) = \int d\mu \, dz \, f(z, \mu, t)$, and we normalize $f(z, \mu, 0)$ such that $N(0) = 1 \, \mathrm{cm^{-3}}$. In Figures 1(a) and 1(c) we show $N(t)$ as a function of dimensionless time for protons of 5 and 30 MeV, respectively, with dimensionless injection rate $\widetilde{Q} = 2 \times 10^{-9}$ for both energies. (Note that $t\Omega_H = 10^7$ corresponds to 2.09 s.)

Previous analytical estimates of escape have approximated the region by a "leaky-box", with the evolution of $N(t)$ given by $dN(t)/dt = -N(t)/T_{lb}$, where T_{lb} is the leaky-box escape time scale. For very low turbulence energy densities, particles escape by simple transit through the region and equation (1) can be averaged (11) to yield $T_{lb} = T_t$, where $T_t = 2L/v$ is the transit escape time. This approximation also requires that the particles be isotropic in the region. For very high levels of turbulence, the particles can change direction many times before escaping and transport is diffusive. In this limit,

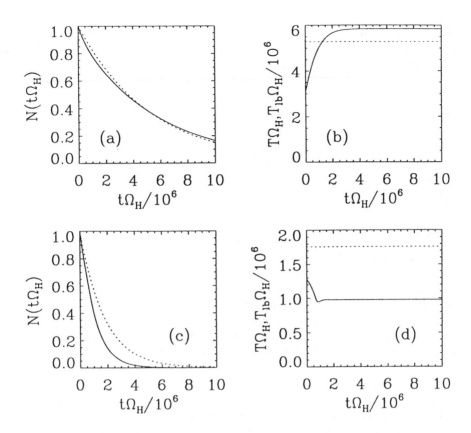

FIG. 1. (a) Total number density of trapped 5 MeV protons. Solid line: numerical result. Dotted line: leaky-box approximation. (b) Instantaneous e-folding decay time for 5 MeV protons. Solid line: numerical result. Dotted line: leaky-box approximation. (c) Same as (a) but for 30 MeV protons. (d) Same as (b) but for 30 MeV protons.

$T_{\mathrm{lb}} = T_{\mathrm{d}}$, where the diffusive escape time $T_{\mathrm{d}} = L^2/(8\kappa_\parallel)$. The parallel diffusion coefficient can be related to $D_{\mu\mu}$ by $\kappa_\parallel = (v^2/2)\int_0^1 d\mu \, (1-\mu^2)^2/D_{\mu\mu}$. For intermediate levels of turbulence, T_{lb} could either be taken to be the larger of T_{t} and T_{d}, or $T_{\mathrm{t}} + T_{\mathrm{d}}$ (1,11). We use the later expression. For a Kolmogorov spectral density, $\kappa_\parallel$ cannot be determined analytically and we thus employ numerical integration.

The dotted lines in Figures 1(a) and 1(c) show the evolution of $N(t)$ in the leaky-box approximation. For 5 MeV protons, $T_{\mathrm{d}} \approx 2.4 \times 10^6 T_{\mathrm{H}}$ and $T_{\mathrm{t}} \approx 3 \times 10^6 T_{\mathrm{H}}$, and their sum yields a characteristic escape time that is in excellent agreement with the exact numerical result. Note that the alternative expression for T_{lb} mentioned above (namely, T_{lb} being equal to the larger of either T_{d} or T_{t}) would be a relatively poor approximation, and would lead to particles remaining trapped much longer than they actually are. The leaky-box approximation does not fare as well at 30 MeV, and overestimates the escape time by about 75%. This can be seen by a direct comparison of the two times: The solid line in Figure 1(b) and 1(d) is the instantaneous e-folding decay time $T = -(d\ln N/dt)^{-1}$ of the numerical solution of equation (1), while the dotted lines are the approximation $T_{\mathrm{lb}} = T_{\mathrm{t}} + T_{\mathrm{d}}$. The exact e-folding times initially depend on time but eventually achieve a steady-state value. This value is about a factor of 0.6 smaller than T_{lb} at 30 MeV, but is only about a factor of 1.1 larger than T_{lb} at 5 MeV. At 30 MeV, however, $T_{\mathrm{t}} \approx 1.3 \times 10^6 T_{\mathrm{H}}$ and $T_{\mathrm{d}} \approx 5 \times 10^5 T_{\mathrm{H}}$, so that taking T_{lb} to be the larger of the two would yield better agreement in this case.

The general validity of the leaky-box approximation can be determined by plotting T_{lb} and the steady-state value of the actual escape time T as a function of the turbulence injection rate Q. We show in Figure 2(a) these results for 5 MeV protons, along with the transit time at this energy. For $\widetilde{Q} \lesssim 2 \times 10^{-12}$, T increases rapidly with decreasing $\widetilde{Q}$ and does not agree well with $T_{\mathrm{lb}} \approx T_{\mathrm{t}}$. In this regime, pitch-angle scattering is negligible and particles escape the region mostly by simple transit. However, scattering is too small to keep the protons isotropic over an escape time, and the above expression for the transit escape time cannot be expected to be valid (recall that isotropy was assumed in its derivation). The long steady-state escape time scale that is observed results from the fact that protons with large $|\mu|$ escape quickly while those with relatively small $|\mu|$ are not appreciably scattered into smaller angles and consequently remain in the region for a long time. Essentially, the transit time escape expression $2L/v$ averages over μ, and seriously underestimates the typical transit time once high-$|\mu|$ particles have escaped.

As Q increases, the escape time decreases, and reaches a minimum when $\widetilde{Q} \approx 5 \times 10^{-11}$. While approaching the minimum, T is equal to $T_{\mathrm{lb}} = T_{\mathrm{t}}$ at $\widetilde{Q} \approx 2 \times 10^{-12}$. This corresponds to "saturated scattering" (1,11). Here, the particles are kept isotropic but scattering is not large enough to substantially inhibit their transit through the region, so that $2L/v$ is an accurate expression for the escape time. It had previously been thought that saturated scattering

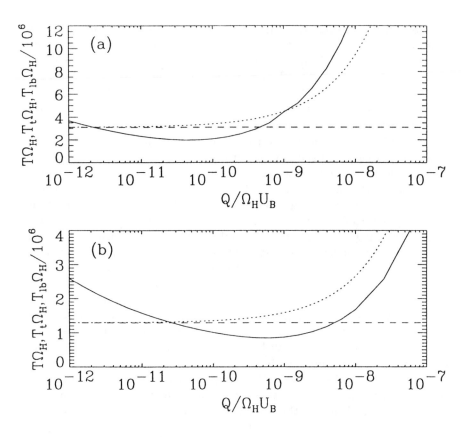

FIG. 2. The e-folding decay time of the trapped-proton number density as a function of the turbulence injection rate. Solid line: steady-state value of the exact numerical solution. Dotted line: leaky-box escape time. Dashed line: transit time. (a) 5 MeV protons. (b) 30 MeV protons.

resulted when T_t was equal to the time scale to be scattered through the loss cone, which in this case is $90°$. Approximating this pitch-angle scattering time scale T_μ by $(\Delta\mu)^2/D_{\mu\mu}$, where $D_{\mu\mu}$ is evaluated at $\mu = 0.5$ (the average of $D_{\mu\mu}$ is also close to this value) and $\Delta\mu = 1$, we find that $T_\mu = 5 \times 10^7 T_H$ in the saturated case. This time scale is ≈ 16 times longer than the transit time scale. Hence, a much lower level of pitch angle scattering is required for T_t to be an accurate approximation of the escape time.

For $\widetilde{Q} = 5 \times 10^{-11}$, when the escape time is a minimum, $T_\mu \approx 5.8 \times 10^6 T_H$. This value is closer to the transit time T_t, but T is not given by T_t. For $\widetilde{Q} \gtrsim 5 \times 10^{-11}$, scattering also keeps the protons isotropic, but does so on a time scale short compared to T_t. Hence, a proton will probably change its direction before encountering a boundary and diffusive effects become important. Higher turbulence injection rates yield higher scattering rates and more efficient trapping, and T increases steadily with Q. The diffusive escape term in T_{lb} also becomes larger and dominates at high Q, producing a T_{lb} that also steadily increases. The difference between T and T_{lb} ranges from $\approx 50\%$ at the minimum to $\approx 16\%$ at $\widetilde{Q} = 8 \times 10^{-9}$. There is near perfect agreement when $\widetilde{Q} \approx 10^{-9}$ (cf. Figs. 1(a)–(b)), but this appears to be fortuitous. There may be a question concerning why T_t and T_d cannot be added "in parallel" instead of "in series" to produce a T_{lb}. The reason is clear in the strong scattering regime: addition in parallel would yield a T_{lb} less than T_t, which would be in horrible disagreement with the exact result.

The same quantities for 30 MeV protons are shown in Figure 2(b). In this case, the increase of the actual escape time T over $T_{lb} \approx T_t$ at low Q is more evident, and due to the same effect as in the 5 MeV case. Saturated scattering, when $T = T_{lb} \approx T_t$ occurs when $\widetilde{Q} \approx 2.5 \times 10^{-11}$. Here, the pitch angle scattering time is about $7.1 \times 10^6 T_H$, which again is much longer than the transit time. The minimum of T is at $\widetilde{Q} \approx 8 \times 10^{-10}$, where $T_\mu \approx 7 \times 10^5 T_H$. This scattering time scale is relatively close to T_t, but, as above, T is not given by T_t. For higher turbulence injection rates, diffusive confinement becomes more important and particles typically change their directions before reaching a boundary. Hence, both T and T_d (along with T_{lb}) subsequently increase. For $\widetilde{Q}$ greater than that at saturation, the discrepancy between T and T_{lb} ranges from $\approx 100\%$ at the minimum of T to $\approx 40\%$ at $\widetilde{Q} \approx 2 \times 10^{-8}$ (which corresponds to $2000 \text{ ergs cm}^{-3} \text{s}^{-1}$ of waves).

CONCLUSIONS

After an initial time interval, we have shown that $N(t)$ decays exponentially with an e-folding time of T. The leaky-box approximation then reduces to setting this decay time equal to $T_{lb} = T_t + T_d$. For low values of Q, the leaky-box expression is generally not an accurate approximation for use in modeling particle transport and gamma-ray production. This results from T_{lb} being about equal to $T_t = 2L/v$, and the latter expression not taking into account

514

the fact that particles cannot be kept isotropic over an escape time scale. For example, with $\widetilde{Q} \approx 10^{-12}$ (corresponding to $2Q \approx 0.1\,\mathrm{ergs\,cm^{-3}\,s^{-1}}$), which may be encountered after energy release and during transport, the difference between T and T_{lb} is $\approx 80\%$ at 30 MeV, and becomes much worse at higher energies. (It clearly becomes much worse for lower values of Q as well.) Transport in this regime must either be treated with a Monte Carlo simulation or by an exact solution of the transport diffusion equation. Near saturation, however, the leaky-box approximation could be adequate.

For higher values of $\widetilde{Q}$ (say $\sim 10^{-9}$), the acceptability of the leaky-box time scale for particle escape from a magnetized plasma depends upon the context in which it is employed. In this regime, we showed that the difference between T and T_{lb} is typically a few tens of percent for the proton energies involved in gamma-ray line production. In stochastic acceleration models, this few tens of percent error in the leaky-box expression is within the uncertainties of other parameters in the model (e.g., the precise nature of the spectral density). Hence, the leaky-box approximation can be used safely in studies of particle acceleration in conjunction with either a Fokker-Planck (8,10) or momentum diffusion equation (11).

In the above discussion and in the derivations of T_{t} and T_{d}, we assumed initially homogeneous particle distributions inside the region. We point out that the leaky-box approximation can still remain accurate even when this is not the case. Should particles be confined initially to a smaller region within the overall volume (corresponding, for example, to a localized area of acceleration), the leaky-box approach can remain a good approximation of escape. For example, releasing 5 MeV protons from $z = L/2$ with $\widetilde{Q} = 2 \times 10^{-9}$, we find that the steady-state escape time scale $T = 5.8 \times 10^{6} T_{\mathrm{H}}$, which is the same result and excellent agreement found for the homogeneous case (cf. Figs. 1(a)–(b)).

We are currently parameterizing T in terms of E and Q, to produce a highly accurate expression that can be used in any transport or acceleration model. Also, having established the relative ease of numerically solving equation (1), we will consider more realistic flare scenarios. Equation (1) can be generalized to include a spatially dependent magnetic field (and thus magnetic mirroring), energy losses (which adds another dimension), a spatially dependent background density, and Coulomb collisions. Transport in a loop can then be treated with a diffusion equation rather than a Monte Carlo simulation, and a time-dependent, high-resolution, low-noise particle spectrum can be calculated in about a minute or two as opposed to hours, which is the time scale encountered using Monte Carlo codes. Acceleration from wave particle interactions (or some other process) can also be included, for a comprehensive and self-consistent acceleration/transport model of solar flares.

ACKNOWLEDGEMENTS

This work was supported by the NSF Solar Terrestrial Research Program through grant ATM–9415379 and by the NASA Cosmic and Heliospheric Physics Program through grant NAGW–4378.

REFERENCES

1. Ramaty, R., and Mandzhavidze, N., "Theoretical Models for High Energy Solar Flare Emissions", in *High Energy Solar Phenomena–A New Era of Spacecraft Measurements*, 1994, New York: AIP, pp. 26–44.
2. Ramaty, R., Miller, J. A., Hua, X.-M., and Lingenfelter, R. E., *Astrophys. J. Suppl.*, **73**, 199–207 (1990).
3. Hua, X.-M., Ramaty, R., and Lingenfelter, R. E., *Astrophys. J.*, **341**, 516–532 (1989).
4. Miller, J. A., and Ramaty, R., *Astrophys. J.*, **344**, 973–990 (1989).
5. Guglenko, V. G., et al., *Astrophys. J. Suppl.*, **73**, 209–211 (1990).
6. Leach, J., and Petrosian, V., *Astrophys. J.*, **251**, 781–791 (1981).
7. Lu, E., and Petrosian, V., *Astrophys. J.*, **327**, 405–416 (1988).
8. Miller, J. A., Guessoum, N., and Ramaty, R., *Astrophys. J.*, **361**, 701–708 (1990).
9. Steinacker, J., and Miller, J. A., *Astrophys. J.*, **393**, 764–781 (1992).
10. Hamilton, R. J., and Petrosian, V., *Astrophys. J.*, **398**, 350–358 (1992).
11. Miller, J. A., and Roberts, D. A., *Astrophys. J.*, **452**, 912–932 (1995).
12. Park, B. T., and Petrosian, V., *Astrophys. J.*, **446**, 699–716 (1995).
13. Miller, J. A., LaRosa, T. N., and Moore, R. L., *Astrophys. J.*, in press (10 April) (1996).
14. Earl, J. A., *Astrophys. J.*, **425**, 331–342 (1994).
15. Steinacker, J., and Miller, J. A., "Proton Gyroresonance with Parallel Waves in a Low-Beta Solar Flare Plasma", in *Particle Acceleration in Cosmic Plasmas*, 1992, New York: AIP, pp. 235–8.
16. Press, W. H., Flannery, B. P., Teukolsky, S. A., and Vetterling, W. T., *Numerical Recipes*, Cambridge: Cambridge Univ. Press, 1986, ch. 17.

RAPPORTEUR PAPERS

X-ray Aspects of the High Energy Solar Physics Workshop: Rapporteur Paper I

Brian R. Dennis

Laboratory for Astronomy and Solar Physics,
NASA Goddard Space Flight Center, Greenbelt, MD 20771
Internet: Brian.R.Dennis.1@gsfc.nasa.gov

This paper is a summary of the presentations on X-ray observations and electron acceleration mechanisms at the workshop. The emphasis is on dispelling myths, starting with the original **Solar Flare Myth** and including others, both serious and not so serious.

INTRODUCTION

In this summary of the workshop, I will concentrate on the hard X-ray observations and discussions of electron acceleration, leaving most of the γ-rays and ion acceleration to Reuven Ramaty. As you read this summary, please keep in mind the statement by Leveret Davis as quoted by Peter Sturrock in his review of the IAU Colloquium at Iguazú Falls (49) -

"Anyone who will agree to give a summary talk does not have the intelligence to do the job."

One theme that seemed to pervade this workshop and indeed has dominated the discussions at many other recent solar meetings has been the question of myths and how to dispel them. Given the presence at the workshop of the Astronomer Royal for Scotland, John Brown, and his magical after-dinner speech at the banquet, the major myth that must be dispelled here is the following:

The position of Astronomer Royal for Scotland is important.

Lest the reader suspect sour grapes or some intra-British rivalry, let me hasten to add that I greatly admire the efforts by the current Astronomer Royal for Scotland to fully exploit the potential of his newfound connection to royalty, and I hope that his endeavors will lead to greater interest in and support for astronomy in general and solar physics in particular.

THE SOLAR FLARE MYTH

Our concern with myths began, of course, with the paper by Jack Gosling (12) and the introduction of the term "Solar Flare Myth" into our lexicon. This myth can be briefly stated as follows:

Solar flares cause the major transient disturbances in the near-Earth space environment.

In other words, the myth is that solar flares are responsible for the high energy particles and magnetic disturbances that have such diverse effects as endangering astronauts and satellites in space and causing power outages and communications problems on Earth. It is now generally accepted that, in fact, coronal mass ejections (CMEs) are responsible for most of these phenomena. As Don Reames (42) points out, "Except for the 'sudden ionospheric disturbances' caused directly by photons, flares are not 'geo-effective.'" Also, contrary to earlier beliefs, flares do not cause CMEs. Significant controversy still surrounds this subject, however, as evidenced by the "Great Debate" in a recent issue of Eos (17), (32), (42). The concern seems to have more to do with the perceived "importance" of flares vs. CMEs and the relative amounts of money that should be spent on studying them rather than in trying to understand the physics involved in each case. While it may be incorrect to say that one "causes" the other, Hudson (17) states that, "Flares, CMEs, and geomagnetic storms all start out in the dynamics of the solar magnetic field, which is the fundamental object of study." It would seem that both flares and CMEs are the result of rearrangements of the coronal magnetic field and that they are often "associated" with one another in some way. Consequently, much can be learned from studying them both and exploring their relationship to one another.

GAMMA-RAYS FROM A CME?

Presentations by Ed Cliver and Tom Vestrand at the workshop touched on the connection between flares and CMEs. They discussed observations of γ-rays that possibly originated from ions accelerated in a CME shock and directed back towards the Sun (7) (57). Despite the fact that the flare on 1989 September 29 occurred some $10°$ over the limb, strong fluxes of γ-rays were observed, including the 2.223-MeV neutron-capture deuterium line. This line is not generally seen from limb flares since it is produced deep in the photosphere where the neutrons must thermalize before they can be captured by protons to produce deuterium nuclei in an excited state. These nuclei decay to produce the 2.223-MeV line but the γ-rays cannot escape in the direction of an observer tangential to the solar surface because of the large absorption.

The observation of the line in this over-the-limb flare can only mean that accelerated ions must have interacted to produce the neutrons on the visible disk. But how did they get there from the flare site? One possibility is that they were accelerated in the shock of the associated CME as it expanded outwards. If this were the case, then the γ-ray spectrum would provide direct information on the shock-accelerated particles. Boris Somov argued that shock accelerated ions do not produce the delayed component of γ-ray emission seen in some events since the shock is already too high in the corona by the time this component appears (48). Natalie Mandzhavidze and Reuven Ramaty (29) (40) have also expressed doubt about the role of shocks in accelerating the particles that produce the observed γ-rays in other events, citing timing and composition inconsistencies. An alternate explanation is that there was a large magnetic loop connecting the over-the-limb flare with a site on the visible disk (57). In that case, the particles may not be related to the CME shock at all. One way to separate these two explanations in the future will be with imaging information. For shock acceleration, the 2.223-MeV line should come from a large diffuse area whereas with the magnetic loop hypothesis, the γ-ray source should be much more compact.

The discussion following Vestrand's presentation served to resolve an error in interpreting the γ-ray spectral data made by Ramaty. The raw count-rate spectrum shows a surprisingly strong 2.223-MeV line (57). Ramaty compared the flux in this line with the measured intensity in the nuclear line region from 4 to 6 MeV and believed that the neutron capture line was, in fact, enhanced in relative intensity compared to the average for a disk flare. What Ramaty failed to take into account, despite the detailed explanation in the original paper, was that the early impulsive phase of the flare was not observed because the Solar Maximum Mission (SMM) was in the South Atlantic Anomaly at the time and the Gamma Ray Spectrometer was not collecting data. Consequently, a correction must be applied to account for the delay in the production of the 2.223-MeV line. Once this is taken into account, the corrected intensity relative to the flux in other lines is normal for a disk flare. Since we know that to err is human, the error in interpretation by Ramaty provides us with rare ammunition to dispel the myth that

Reuven Ramaty is omnipotent, omnipresent, and omniscient.

Certainly Reuven's energy, enthusiasm, insight, and involvement throughout this workshop provided no further help to dispel this myth. Perhaps, in fact, it is no myth at all.

THE FLARE SIZE DISTRIBUTION

Another myth addressed at the workshop was that

There is no end to the flare size distribution.

In particular, we a talking about a cutoff for the largest flares. The flare size distribution can be represented by a power law over many orders of magnitude (8), although there does seem to be a paucity of the largest flares. The evidence for any cutoff is statistically very weak, however, in both the SMM and ISEE/ICE data sets. Sharad Kane reported new observations of a giant flare that suggest the size distribution might extend to even more powerful flares than had previously been suspected, perhaps so large, in fact, that a single active region could not have provided all of the energy (19). Kane estimates that as much as 10^{34} ergs may have been released to accelerate electrons to >25 keV during this flare whereas the total available magnetic energy in the active region was $< 4 \times 10^{32}$ ergs. Clearly, it is important to measure such large flares with instruments that do not suffer from the saturation effects that have plagued all previous observations. Kane's instrument was somewhat better off than most in that it was on Ulysses and so a factor of ~ 3 further away from the Sun than an instrument in Earth-orbit. Nevertheless, it still suffered significant saturation effects and Kane had to make controversial (but plausible) assumptions to obtain the large energy estimate.

Terry Kucera has presented other evidence relating to a possible end or high energy cutoff in the flare size distribution (21). She has looked at the size distribution of the over 12,000 X-ray flares observed with the Hard X-ray Burst Spectrometer (HXRBS) on SMM and plotted their peak counting rates as a function of the size of the sunspots in the active region from which they originated. She finds evidence for a cutoff in the size distribution of flares from active regions that have sunspots with areas of <500 millionths of the visible hemisphere. For these selected flares, the number of flares detected with a peak counting rate of over 10,000 counts s^{-1} was more than 3σ below the number expected from an extrapolation of the power-law distribution of smaller flares. Since this cutoff occurs at such a low counting rate, it is unlikely that pulse pileup or any other form of detector saturation could have caused it. Taking this result at face value, it would seem that an active region does have a maximum energy that it can release during a flare as would be predicted by the avalanche model (28). Furthermore, Kucera et al. (21) show that if we assume that the available energy in an active region is proportional to the sunspot area, then the cutoff for the largest regions is at the level of the largest flares seen during the almost 10-years of HXRBS observations. Since the number of such large flares is so small, any cut off in the size distribution has low statistical significance. The flare observed by Kane would be within an order of magnitude of the expected cutoff level for the size of the active region from which it probably originated. (It was over the limb as seen from the Earth).

TABLE 1. Properties of Impulsive and Gradual Events

	Impulsive	Gradual
Particles:	Electron-rich	Proton-rich
^{3}He/^{4}He	~1 (×2000 increase)	~0.0005
Fe/O	~1.234 (×8 increase)	~0.155
H/He	~10	~100
Q_{Fe} (mean charge state)	~+20	~+14
Duration	Hours	Days
Longitudinal Cone	< 30°	~ 180°
Radio Type	III, V (II)	II, IV
X-ray Type	Impulsive	Gradual
Coronagraph	-	CME (96%)
Solar Wind	-	IP Shock
Event Rate (maximum number per year)		
Particle Events	~1000	~10
CMEs	-	~500
Hard X-ray Flares	~4000	~10

IMPULSIVE AND GRADUAL EVENTS

Of fundamental importance in understanding transient solar events is to distinguish between impulsive and gradual events. We are still guilty of blurring the distinction between them, perhaps because the observational differences are not always clear for any given event but also because many events have characteristics of both types at different times during their evolution. Table 1 adapted from (32) lists the major characteristics of these two distinct classes of events.

We heard little at the workshop specifically about gradual or Type-C X-ray flares (50) (51) (55), even though they are more likely to be associated with CMEs (6). Indeed, the following myth has developed from the lack of published results to date:

Yohkoh has not seen any gradual flares.

It is, of course, true that Yohkoh has not seen many large gradual flares since it was not launched until August 1991, too late to observe the sequence of five very large flares in June 1991. Nevertheless, several large gradual flares have been observed with Yohkoh but apparently the analysis is complicated by the high intensities and the large extent of the sources in hard X-rays. Masato Yoshimori showed observations of three γ-ray line flares in 1991 (Oct. 27, Nov. 15, and Dec. 3) and reported rather controversial evidence for redshifted Be/Li lines produced in $\alpha - \alpha$ interactions. It was not clear if these events had any of the characteristics of Type-C flares since they apparently showed double footpoint sources in hard X-rays. Interestingly, the hard X-ray images appeared the same at the times of four different γ-ray peaks even though the last two peaks showed significant line emission but the first two did not.

Hiroshi Nakajima showed observations of two X-class events by both Yohkoh and Nobeyama that appeared to be classic large gradual flares (37). Because both flares were over the west limb, he was able to show that the hard X-ray and 17-GHz microwave sources were from different locations in the corona. They also indicate different electron spectra. He suggests that this implies two stages of acceleration, with the initial acceleration to ~100 keV taking place above the soft X-ray loop and the subsequent acceleration at the top of the loop. Trapping in the loop is invoked to explain the ~25-s delay in the microwave emission of one of the flares.

IMPULSIVE EVENTS

With the dispelling of "The Solar Flare Myth," there is a danger that the following new myth will take its place:

We don't understand impulsive flares but who cares.

Well, Reames cares. He has gone on record, stating that "The physics of particle acceleration in impulsive flares is no less interesting because the events are not geo-effective" (42). Indeed, understanding particle acceleration is one of the most important problems in plasma physics and astrophysics. It is particularly challenging in impulsive flares because of the rapidity with which such a large number of particles are accelerated so efficiently. One of the great successes of flare observations over the last two cycles has been the enormous strides that have been taken in the observations of the impulsive phase in many different energy or wavelength regions. The continued study of impulsive flares, with our new abilities to resolve the high energy processes on relevant physical scales in space, time, and energy, will hopefully dispel this insidious myth before it gets off the ground.

The hard X-ray imaging observations with Yohkoh's HXT, in particular, have been very illuminating in confirming the predominant footpoint nature of the hard X-ray sources in most flares. Takeo Kosugi indicated that 20 flares in a sample of 28 flares had double or multiple sources on opposite sides of the magnetic neutral line (20) (43). Nariaki Nitta reported a similar fraction from an independent study (38). The time difference in the emission from the two footpoints was less than a few tenths of a second, i.e., so short that accelerated ions could not possibly explain such synchronism unless the acceleration site is always exactly equidistant (within <1000 km) from the footpoints. Furthermore, the brighter of the two sources and the one with the harder spectrum is always the one from the footpoint with the weaker magnetic field, when that could be measured (25) (43). These results make a very strong case that the 15 - 100 keV footpoint emission observed with HXT is produced by precipitating electrons - more electrons precipitate at the footpoint with the weaker field while more are mirrored in the corona at the other footpoint. This is also consistent with the result that the most intense microwave source is the one over the footpoint with the stronger magnetic field (23), (59).

Several avenues suggest that there are two different components to the nonthermal electrons that produce the hard X-rays. Bob Lin reported on finding an impulsive component with a spectrum peaked at 50 keV and a more slowly varying component with a power-law spectrum extending down to 20 keV (27). This result was obtained by inverting the high spectral resolution hard X-ray observations of the 1980 June 27 flare (26) using a continuity equation. Markus Aschwanden also found two distinct components based on the temporal fluctuations of the hard X-ray emission observed with BATSE (2) (3). One component was characterized by rapid fluctuations on timescales of <1.5 s, and by energy-dependent delays with the higher energy emissions leading the lower energy emissions (see below). The other component had more gradual fluctuations on timescales of several seconds to minutes with the opposite energy-dependent delays, i.e., the the higher energy emissions lagged the lower energies. The spiky emission may be due to electrons impulsively accelerated by a quasi-static electric fields parallel to the magnetic field as happens in the Earth's aurora (27). The slower component is suggestive of a different process, possibly trapping of the electrons in the corona with the resultant collisional energy losses, and may be related to the delays often seen in the microwave emission relative to the X-rays (2). Nicole Vilmer showed that a large percentage (73%) of all flares detected with the BGO detectors of the PHEBUS instrument on GRANAT showed the spiky fluctuations with rise times between 0.1 and 1 s at around 100 keV (58). She suggested that this is an indication that the flare energy release is usually highly fragmented and that the electrons must interact in regions with a density $> 10^{12}$ cm^{-3}.

Observations of thermal emissions can also show what processes are involved in impulsive flares. George Doschek summarized the Yohkoh BCS results on the Doppler-shifted Ca XIX line emission from over 200 flares (9). Observations of a blue-shifted component coincident with the hard X-ray emission in most flares indicate high-speed upflows (400 - 800 km s^{-1}) consistent with chromospheric evaporation following electron precipitation, although lower-speed upflows are sometimes seen prior to the start of the hard X-ray emission. However, the usually stronger, Doppler-broadened (corresponding to velocities up to 200 km s^{-1}) but unshifted line emission, present even at flare onset, remains enigmatic.

Hugh Hudson presented new Yohkoh evidence for the superhot component (26) with the rather surprising result that in at least two flares its source is not cospatial with the rest of the hard X-ray emission (18). He reported that some 75 events have been identified in the Yohkoh data base with superhot characteristics, which he defined as having a significant fraction of the plasma at a temperature exceeding 30×10^6 K as determined from the Fe XXV and XXVI emissions and the hard X-ray continuum. The emission measure can be up to 10% of that in soft X-rays. It is not clear if the components with different temperatures occupy different spatial locations or if the source has a non-Maxwellian electron distribution function. These results currently serve only to deepen the mystery of the superhot component (or property of the flare, as Hudson prefers to call it) but the inclusion of the imaging information from SXT and HXT in the analysis of these events promises further insight.

"Gopal" Gopalswamy and Mukul Kundu both showed observations of another kind of impulsive event that originates in X-ray bright points (11) (24). They showed correlated radio and X-ray observations of such bursts and argued that nonthermal electron distributions are produced in these short-duration events as in the normal active-region flares.

THE ACCELERATION SITE

The location of the electron acceleration site may have been determined by observations of some flares that showed a coronal hard X-ray source in addition to the footpoint sources (30). This coronal source appears to be located above the top of the loop seen in soft X-rays in certain limb flares where this can be resolved without foreshortening. Remarkably, this above-the-loop-top source has a similar time profile to the footpoint sources, and the spectrum is very hard, although slightly softer than the footpoint spectrum, in six of ten limb flares observed (20). The nature of this coronal source has excited theoreticians and model builders since it provides new information on the energy release site. Its interpretation and its relation to the hot compact source within the soft X-ray loop seen in many flares remain uncertain, however.

Aschwanden's evidence of energy-dependent time delays in the hard X-ray emission discussed above supports the HXT observations of a site of electron acceleration high in the corona (2) (3). He interprets the delay of the softer X-rays compared to the harder photons in the impulsive spikes as resulting from the longer travel time from the acceleration site of the lower velocity electrons that produce the softer X-rays compared to the faster electrons that produce the higher energy photons. From the magnitude of the delay (a few tens of ms measured using cross-correlation techniques with observations having only 64-ms time resolution!), he is able to calculate the distance that the electrons must have traveled from their acceleration site to the footpoints, assuming that they were all accelerated simultaneously. He finds that the site of electron acceleration is at an altitude some 60% higher in the corona than the top of an assumed semicircular loop joining the HXT footpoints (3). Note however that this only applies to the spiky emission with timescales shorter than 1.5 s.

While these remarkable observational results do not tell us the answers to the fundamental questions concerning the energy release and particle acceleration processes in impulsive flares, they do point to the X-points or current sheets above the soft X-ray loops as being the most likely sites where the processes take place. At least five people (Cliver, Enome, Forbes, Kosugi, and Somov) discussed models with the energy release resulting from magnetic reconnection at an X-point above a loop system (see also (6), (45), (46), (53)). Kosugi showed tantalizing observational evidence for the second, higher loop in Shibata's model and there are several SXT observations showing beautiful cusps on soft X-ray loops (56). We seem to have a consensus that this is the correct structure to describe most, if not all, flares. Boris Somov said that

"After Yohkoh, magnetic reconnection is an observational fact," but Gordon Holman would only admit to "strong observational evidence." Perhaps not coincidentally, the same magnetic structure has been used to explain the early stages of CMEs, including their often helical nature (1) (12). This suggests that CMEs and flares, both impulsive and gradual, have the same basic origins in a neutral sheet reconnection-type geometry (45).

A related myth discussed at the meeting is that

Flares occur at sites of strong photospheric currents.

Dick Canfield used SXT and HXT images and vector magnetograms from the Mees Solar Observatory for six large flares to show that the observed bright hard X-ray footpoints do not coincide with sites of high vertical current densities (25). He concluded that the model in which the flare energy is released by the interruption of the predominant pre-existing currents is not consistent with the observations. Presumably the measured currents are long-lived phenomena and may have little to do with the flare itself, even though there is much other evidence indicating that flares tend to occur in regions with large current systems (5).

THE ACCELERATION PROCESS

As for the acceleration process itself, the best thing to do is read the recent review prepared by a group of experts in the field (33). Controversy still rages intensely on this subject, as evidenced by the sometimes heated debate that ensued between two of the chief protagonists at the meeting, the usually mild-mannered Gordon Holman and Jim Miller. The little known myth that

Jim Miller has been to the site of particle acceleration on the Sun

was quickly dispelled by a personal denial from Jim. Holman's effort to dispel the myth that

Direct electric field acceleration was disproved as a viable flare mechanism 20 years ago

also made some headway as he, Dominic Zarro (61), and Terry Kucera (22), further expounded on the continued successes of the thermal/nonthermal model of simultaneous Joule heating and runaway electron acceleration by parallel electric fields (13) (15) (52). However, the evidence seems to be stacked against the idea that the field could be extended along the full length of the loop between the observed bright footpoints. A more likely scenario is that the field extends along the top of an arcade of loops as described by Holman et al. (16).

Gordon Emslie also spoke about this model, worrying about the huge numbers of electrons that must be accelerated and the huge number of small oppositely-directed field channels that must exist to obey Ampere's and Faraday's laws. He presented a new idea to achieve current closure at the footpoints that may remove one impediment to the model (10). He suggested that some of the precipitating electrons combine with protons in the partially-ionized chromosphere at a density of $\sim 10^{12}$ cm^3 to produce neutral hydrogen atoms that can drift across the field lines and re-ionize in a neighboring oppositely-directed field channel some 1-10 m away. Because of their large gyroradius (compared to electrons) and their collision mean free path, the newly freed protons can again cross the field lines back to the original channel to complete the circuit. In this way, the electron drift current is closed at the loop footpoints. What could produce the highly fragmented electric fields that drive the currents is not known, however, but several possibilities exist (54) (60).

Holman also discussed the acceleration of ions by a parallel electric field (14). Surprisingly, except for protons, the ions are accelerated "backwards" since they are dragged along by the drifting electrons until they reach the thermal electron drift velocity. Because of the large electron-ion drag forces, ions (including protons) cannot run away to the highest energies unless the applied electric field greatly exceeds the Dreicer field. Thus, this mechanism, at least under classical conditions, can only serve to pull the ions out of the thermal distribution to energies up to $\sim$1 MeV per nucleon, but this is sufficient to provide seed particles for other mechanisms that can accelerate them to higher energies. Somov has considered much stonger super-Dreicer fields in the complex geometry of the magnetic and electric fields in a reconnecting current sheet and has shown how ions could be accelerated to GeV energies on the time scale of the late phase of γ-ray flares (48). However, in his case, it is not clear how the particles could be accelerated out of the thermal distribution. Neither Somov nor Holman have explained how the measured ion abundances can be obtained using their proposed acceleration mechanisms.

Vahé Petrosian and Miller both discussed stochastic acceleration. Petrosian suggested that most of the flare energy could go initially to generate plasma turbulence, rather than to heating or particle acceleration (39). He argued that stochastic acceleration by turbulence alone can both heat the plasma and accelerate particles from the thermal distribution to very high energies in a short enough time, although shock acceleration could become important at some high energy.

Miller discussed his stochastic acceleration model that involves the simultaneous acceleration of both ions and electrons to the highest energies with no need for the pre-acceleration of seed particles out of the thermal distribution (34) (35) (36). He uses the concept of nonlinear cascading in which a continuous broad-band spectrum of plasma waves, such as shear Alfven waves and fast mode waves, are produced from the long-wavelength MHD turbulence generated, for example, by large-scale restructuring of the magnetic field during the primary flare energy release. Once this broad-band spectrum of waves is set up, a particle can be accelerated for a while by a wave with the appropriate phase velocity or resonant condition and then be picked up by another wave that is in resonance with the now faster particle. There is a net gain in the total energy of the particles over time resulting in stochastic acceleration. The types of waves and how they are generated are open to question, but the model can provide detailed heavy-ion abundance predictions for comparison with observations. The predicted energy spectra differ from the measured spectra but, as the authors point out, "There are many other factors that may influence the shape of an ion spectrum from its origin in the acceleration region to the point where it is observed..."

Miller got unexpected support from Gerry Share and Ramaty in dispelling the long-held belief that

Ion acceleration is energetically unimportant.

Share published comprehensive results on 19 gamma-ray flares observed with the Gamma Ray Spectrometer on SMM (44) and showed that the Ne line at 1.634 MeV was more intense relative to the other lines than would be expected using standard abundances and the usual Bessel-function form for the proton spectrum. However, the cross-section for the production of this Ne line is significant down to about 2 MeV and allows the proton spectrum to be estimated down to lower energies than had previously been possible. Ramaty et al. (41) have shown that the most likely explanation for the surprising intensity of the line is that significantly more protons must be accelerated to between 1 and 10 MeV than are indicated by the Bessel-function spectrum. In fact, good agreement is achieved by assuming that the power-law spectrum determined from the ratio of other γ-ray lines extends down to $\sim$1 MeV. This increase in the flux of lower energy protons results in a corresponding increase in the estimate of the total energy in accelerated protons. The preliminary indications are that the energy in ions may be comparable to the energy in electrons. If this conclusion is borne out by further analysis, then it has important consequences for flare models and compounds the difficulty in understanding such an apparently efficient acceleration mechanism for both electrons and ions.

Even Miller and Holman could agree that no viable acceleration model can any longer ignore the acceleration of ions. George Simnett should get some satisfaction from this new result since he has long been advocating the energetic importance of protons (47), albeit primarily those with energies below $\sim$1 MeV. Stay tuned for another "Great Debate" in an upcoming issue of Eos on the question of electron versus proton energy transport in flares, with Emslie squaring off against Simnett, and Peter Cargill acting as moderator.

FUTURE OBSERVATIONS

This summary would not be complete without mention of exciting new observations that can be expected in the near future from advanced instrumentation both in space and on the ground. Several people described new capabilities that will come on line in time for the next solar maximum at the turn of the century. Shinzo Enome described the, now operational, upgrade to the Nobeyama interferometer to operate at 34 GHz. Mark McConnell described the possibility of measuring hard X-ray polarization by looking back at the Earth and observing the distribution of the albedo photon flux (31). With data from the Large Area Detectors of the BATSE instrument on the *Compton Gamma Ray Observatory*, it should be possible to detect a polarization level below 10% at energies above 100 keV for the larger flares. Pierre Kaufmann discussed "fully-funded" plans to build a submillimeter telescope at El Leoncito in the Argentinean Andes to provide multiple-beam dynamic imaging of solar bursts, with three partially-overlapping 3-arcminute beams at 210 GHz and a single 1.5-arcminute beam at 405 GHz. Stephen White indicated that the Berkeley-Illinois-Maryland Array (BIMA) should be up to nine dishes by the end of 1995 with the possibility of obtaining polarization measurements sometime in the future.

And of course, Bob Lin mentioned the High Energy Solar Spectroscopic Imager (HESSI) that we all hope will be selected as the first MIDEX for launch in the year 2000. It will provide the first high resolution imaging spectroscopy in hard X-rays with 2 arcsecond angular resolution from 2 to >200 keV, ~1 keV energy resolution, and sub-second time resolution, plus high resolution γ-ray imaging and spectroscopy up to 20 MeV. One example of the new results that could be obtained with HESSI was given by Taeil Bai, who discussed the results of his Monte Carlo simulations of Compton scattered X-ray emission (4). He showed that imaging the reflected X-rays with the 2-arcsecond angular resolution and 100:1 dynamic range capability of HESSI will provide unique information on the height of the sources above about 1500 km and on the angular distribution of the electrons that produce the hard X-rays.

CONCLUSION

I will close this summary paper with a quote from Saku Tsuneta (as polished by Hugh Hudson) (54) since it provides a focus for our efforts during the next solar maximum and is a clear rallying cry for HESSI.

> **"The understanding of electron acceleration in solar flares is now within our reach owing to the high quality hard and soft X-ray images provided by Yohkoh as well as the high quality spectral data of the 1981 June 27 event with germanium detectors... Based on these data, the direction of future hard X-ray instrumentation in solar physics is clear.**

> **The spatially-resolved observations with high energy resolution in the 10-100 keV range are crucial to solve the particle acceleration problem."**

I am grateful to the many people at the workshop and elsewhere who have tirelessly and patiently explained to me their work and the work of others in words and at a speed that I can understand. Although I take full responsibility for the contents of this summary paper, I particularly appreciate the helpful comments and suggestions made by Markus Aschwanden, Dick Canfield, Ed Cliver, Gordon Emslie, Gordon Holman, Terry Kucera, Reuven Ramaty, and Don Reames.

REFERENCES

1. Anzer, U. and Pneuman, G.W., Solar Phys., **79**, 129 (1982).
2. Aschwanden, M.J., this volume.
3. Aschwanden, M.J., Wills, M.J., Kosugi, T., Hudson, H.S., and Schwartz, R.A., Ap.J., submitted (1996).
4. Bai, T. and Ramaty, R., this volume.
5. Canfield, R.C., de La Beaujardiere, J.-F., Fan, Y., Leka, K.D., McClymont, A.N., Metcalf, T.R., Mickey, D.L., Wuelser, J.-P., and Lites, B.W., Ap.J., **411**, 362 (1993).
6. Cliver, E.W., Dennis, B.R., Kiplinger, A.L., Kane, S.R., Neidig, D.F., Sheeley, N.R., Jr., and Koomen, M.J., Ap.J., **305**, 920 (1986).
7. Cliver, E.W., Kahler, S.W., and Vestrand, W.T., Proc. 23rd Int. Cosmic Ray Conf., Calgary, **3**, 91 (1993).
8. Crosby, N.B., Aschwanden, M.J., and Dennis, B.R., Sol. Phys., **143**, 275 (1993).
9. Doschek, G.A., this volume.
10. Emslie, A.G. and Hénoux, J.-C., Ap.J., **446**, 371 (1995).
11. Gopalswamy, N., Kundu, M.R., Hanaoka, Y., Enome, S., and Lemen, J.R., this volume.
12. Gosling, J.T., JGR, **98**, 937 (1993).
13. Holman, G.D., Ap.J., **293**, 584 (1985).
14. Holman, G.D., Ap.J., **452**, 451 (1995).
15. Holman, G.D. and Benka, S.G., Ap.J. Letters, **400**, L79 (1992).
16. Holman, G.D., Kundu, M.R., and Kane, S.R., Ap.J., **345**, 1050 (1989).
17. Hudson, H.S., Eos, **76**, 405 (1995).
18. Hudson, H.S. and Nitta, N., this volume.
19. Kane, S.R., Hurley, K., McTiernan, J.M., Sommer, M., Boer, M., and Niel, M., Ap.J. Letters, **446**, L47 (1996).
20. Kosugi, T., this volume.
21. Kucera, T.A., Dennis, B. R., Schwartz, R.A., and Shaw, D., Ap.J., submitted (1996).
22. Kucera, T.A., Love, P.J., Dennis, B. R., Holman, G.D., Schwartz, R.A., and Zarro, D.M., Ap.J., submitted (1996).
23. Kundu, M.R., Nitta, N., White, S.M., Shibasaki, K., Enome, S., Sakao, T., Kosugi, T., and Sakurai, T., Ap.J., **454**, 522 (1995).
24. Kundu, M.R., Raulin, J.-P., and Nitta, N., this volume.
25. Li, J., Metcalf, T.R., Canfield, R.C., Wülser, J.-P., and Kosugi, T., this volume.
26. Lin, R. P., Schwartz, R.A., Pelling, R.M., and Hurley, K., Ap.J. Letters, **251**, L109, (1981).
27. Lin, R.P. and Johns-Krull, C.M., this volume.
28. Lu, E.T., Hamilton, R.J., McTiernan, J.M., and Bromund, K.R., Ap.J., **412**, 841 (1993).

29. Mandzhavidze, N. and Ramaty, R., Nuclear Physics, **33**, 141 (1993).

30. Masuda, S., Kosugi, T., Hara, H., Tsuneta, S., and Ogawara, Y., Nature, **371**, 495, (1994).

31. McConnell, M., Forrest, D., Vestrand, W.T., and Finger, M., this volume.

32. Miller, J.A., Eos, **76**, 401 (1995).

33. Miller, J.A., Emslie, A.G., Holman, G.D., Cargill, P.J., Dennis, B.R., LaRosa, T.N., Winglee, R.M., Benka, S.G., and Tsuneta, S., JGR, submitted (1996).

34. Miller, J.A., LaRosa, T.N., and Moore, R.L., Ap.J., submitted (1996).

35. Miller, J.A. and Reames, D.V., this volume.

36. Miller, J.A. and Viñas, A.F., Ap.J., **412**, 386 (1993).

37. Nakajima, H. and Metcalf, T.R., this volume.

38. Nitta, N., this volume.

39. Petrosian, V., this volume.

40. Ramaty, R. and Mandzhavidze, N., Kozlovsky, B., and Skibo, J.G., Adv. Space Res., **13**, 275 (1993).

41. Ramaty, R., Mandzhavidze, N., Kozlovsky, B., and Murphy, R.J., Ap.J. Letters, **455**, L193 (1995).

42. Reames, D.V., Eos, **76**, 405 (1995).

43. Sakao, T., PhD thesis, U. Tokyo, (1994).

44. Share, G. H. and Murphy, R.J., Ap.J., **452**, 933 (1995).

45. Shibata, K., Proc. 2nd SOLTIP Symp., STEP GBRSC News, (T. Watanabe, ed.) **5**, 85 (1995).

46. Shibata, K., Masuda, S., Shimojo, Hara, H., Yokoyama, T., Tsuneta, S., Kosugi, T., and Ogawara, Y., Ap.J., **451**, L83 (1995).

47. Simnett, G.M., Space Sci. Rev., **73**, 387 (1995).

48. Somov, B.V., this volume.

49. Sturrock, P.A., Lecture Notes in Physics, **399** (1991).

50. Tanaka, K., Proc. IAU Coll. #71, Activity in Red Dwarf Stars, eds. P.B. Bryne and M. Rodono, p. 307 (1983).

51. Tsuneta, S., Proc. Japan-France Seminar on Active Phenomena in the Outer Atmosphere of the Sun and Stars, eds. J.C. Pecker and Y. Uchida (CNRS and L'Observatoire de Paris), pp. 243-260 (1983).

52. Tsuneta, S., Ap.J., **290**, 353 (1985).

53. Tsuneta, S., Ap.J., **456**, 840 (1996).

54. Tsuneta, S., Publ. Astron. Soc. Japon, **47**, 691 (1995).

55. Tsuneta, S., Takakura, T., Nitta, N., Ohki, K., Tanaka, K., Makashima, K., Murakami, T., Oda, M., Ogawara, Y., and Kondo, I., Ap.J., **280**, 887 (1984).

56. Tsuneta, S., Hara, H., Shimizu, T., Acton, L.W., Strong, K.T., Hudson, H.S., And Ogawara, Y., Publ. Astron. Soc. Japan, **44**, L63 (1992).

57. Vestrand, W.T. and Forrest, D.J., Ap.J. Letters, **409**, L69 (1993).

58. Vilmer, N., Trottet, G., Verhagen, H., Barat, C., Talon, R., Dezalay, J.P., Sunyaev, R., Terekhov, O., and Kuznetsov, A., this volume.

59. Wang, H., Gary, D.E., Zirin, H., Schwartz, R.A., Sakao, T., Kosugi, T., and Shibata, K., Ap.J., **453**, 505 (1995).

60. Winglee, R.M., Dulk, G.A., Bornman, P.L., and Brown, J.C., Ap.J., **375**, 382 (1991).

61. Zarro, D.M. and Schwartz, R.A., this volume.

Implications of Solar Flare Charged Particle, Gamma Ray and Neutron Observations: Rapporteur Paper II for the High Energy Solar Physics Workshop

Reuven Ramaty[1] and Natalie Mandzhavidze[1,2]

[1] *Laboratory for High Energy Astrophysics, Goddard Space Flight Center, Greenbelt MD 20771*
[2] *Universities Space Research Association*

Solar flares provide a unique site for the study of energy release and particle acceleration in astrophysics. This paper summarizes the charged particle and gamma ray aspects of the Workshop on High Energy Solar Physics held at the Goddard Space Flight Center in August 1995. The X-ray aspects of the Workshop are summarized in another paper. Here are our highlights: (i) both the charged particle, and now also gamma ray observations have shown that stochastic acceleration due to gyroresonant interactions with plasma waves is very important in solar flares; (ii) CME driven shocks accelerate the particles observed in space from gradual flares; (iii) accelerated MeV/nucl ions can contain a large fraction of the energy released in flares; (iv) GeV ions are either trapped at the Sun or accelerated for hours; (v) nuclear gamma ray spectroscopy has become a tool for abundance determinations in the solar atmosphere; (vi) behind-the-limb flares can provide imaging information for solar gamma ray emission.

I. INTRODUCTION

High energy solar physics is the study of the processes of energy release and particle acceleration in the magnetized and highly dynamic solar atmosphere. The High Energy Solar Physics Workshop was devoted to the study of these problems. Particle acceleration is a widespread phenomenon in astrophysics, but only in the case of the Sun can both the accelerated particles and the high energy radiations that they produce be simultaneously observed. Consequently, a variety of papers on both particle and photon observations were included in the workshop, along with theoretical papers on acceleration theories and the analysis and interpretation of the data. In this article we discuss the implications of solar flare charged particle, gamma ray and neutron observations. The X-ray aspects are summarized in Brian Dennis' rapporteur paper and the radio aspects are discussed in both papers. We have divided our article into three broad topics: acceleration, transport and abundances. We include in each section the relevant data, the pertinent theories, and our own, admittedly somewhat biased, interpretations of the results. The opening talk at the Workshop was presented by Ed Chupp (1), who provided a unique view of the history of the field, its development and its current status.

II. ACCELERATION

A. Impulsive vs. Gradual Particle Events

There is a significant body of interplanetary particle observations which strongly suggests the existence of at least two classes of solar energetic particle (SEP) events (2–5). Impulsive events, for which the associated soft X-ray emission is of relatively short duration, have large electron-to-proton ratios (e/p), large $^3He/^4He$, large heavy ion (particularly Fe) to C or O ratios, and charged states corresponding to temperatures which significantly exceed the $(1-2) \times 10^6$K coronal temperature. Gradual events, for which the soft X-rays last longer, exhibit smaller e/p, and have heavy ion abundances, $^3He/^4He$, and charged states which are consistent with coronal abundances and temperatures. In addition, there is a clear association of the gradual events with coronal mass ejections (CME). This classification, and additional data supporting it, were reviewed at the Workshop by Kahler (7), Meyer (8), and Reames (9). Cliver (6) has expanded this classification to four classes by including information from the gamma ray observations. The data presented by Leske (10) and Tylka (11) extended the information on charge states to high energies (up to 600 MeV/nucl). Dröge (12) discussed SEP electron spectra which have also served to distinguish between impulsive and gradual events. Dröge's results, based on updated instrumental response functions, confirm the previous findings (13).

B. Shock vs. Stochastic Acceleration

It is reasonable to associate the above two classes of SEP events with different acceleration mechanisms. Because of the strong association with CMEs, as well as because of a variety of other reasons, particles in gradual events are probably accelerated out of coronal gas by the CME driven shock [e.g. (7,9)]. The new high energy charge state data imply that this acceleration is not limited to just the low energy particles but extends to high energies as well.

On the other hand, particle acceleration in impulsive events is probably due to gyroresonant interactions with plasma waves. Relativistic electron acceleration by such interactions, particularly with whistler waves, can be quite efficient (14). Temerin and Roth (15) and Miller and Viñas (16) have shown that both the ^3He and heavy ion enhancements could be due to acceleration by turbulence. Miller and Viñas have further shown that the turbulence could be generated by nonrelativistic electron beams. At the Workshop, Miller

(17) discussed the possibility of preferential acceleration of heavy ions due to the cascading of Alfvén waves, generated at long wavelengths, to shorter wavelengths where they can resonate with the thermal particles. Thus, no preacceleration is needed in this model. An alternative model for the preferential acceleration of heavy ions, but not of ^{3}He, was presented by Meyer (8). Temerin (18) discussed the possibility of selective acceleration of ^{3}He and heavy ions (at the first and second harmonics, respectively) by electromagnetic ion cyclotron waves. Litvinenko (19) showed that Coulomb losses could produce a break in the ^{3}He energy spectrum. The implied acceleration rate, however, is much lower than that inferred from gamma ray observations; moreover, the predicted break in the ^{3}He spectrum has not been observed.

While, the above processes seem to provide reasonable explanations for the observed abundance anomalies, there still are no model calculations that can account for the observed energy spectra. There are also no simple explanations of the high charge states in impulsive events, even though it is possible that the ions, originally accelerated from a $\sim 3 \times 10^6$K plasma, are further collisionally ionized by the same electron beam which produces the turbulence (16). The important role played by plasma turbulence in the flare energy release and particle acceleration was also discussed by Petrosian (20). He attributed two recently discovered X-ray phenomena, impulsive foot point soft X-ray emission and loop top hard X-ray emission, to effects associated with plasma turbulence.

C. The Interacting Particle: Stochastic Acceleration

Independent information on particle acceleration comes from observations of gamma rays and neutrons which probe the particles which interact at the Sun. Analysis of the gamma ray data strongly suggests that these interacting particles have a common origin with that of the particles observed in impulsive events (21,22). The e/p ratio derived from gamma ray line to continuum data is even higher than the e/p observed in SEPs from impulsive flares (22). Furthermore, already in the analysis of the 1991 April 27 flare, there was an indication for ^{3}He/^{4}He, Mg/O and Fe/O enhancements in the accelerated particles (23). More recently we analyzed data from 9 SMM flares (24) and found that enhanced ^{3}He and heavy ion abundances are needed to produce sufficient neutrons to account for the 2.22 MeV line observations (25,26). Furthermore, the analysis of the behind-the-limb flare of 1991 June 1, from which gamma ray lines were observed with GRANAT/Phebus (27), required a strong heavy ion enhancement (28). Interestingly, this enhancement increased with time as the flare progressed. Both the April 27 and June 1 flares, as well as several other flares for which e/p, ^{3}He/^{4}He and heavy ion enhancements were found, are gradual flares. Thus, it appears that, independent of whether the flare is impulsive or gradual, the interacting particles are accelerated by the same mechanism as that which accelerates the particles in impulsive flares. This issue was also discussed at Workshop by Cliver (6).

D. Energy Content in Accelerated Ions

Another very important result from the gamma ray analysis concerns the extension of the ion spectrum to low energies. Based on SMM data (24), we showed that to account for the strong 1.634 MeV Ne line, the ion spectrum has to be an unbroken power law down to about 1 MeV/nucl (25,26). This implies an energy content in ions around 10^{32} ergs, which is comparable to the energy content in deka-keV electrons. While the gamma rays don't provide information on ions below 1 MeV/nucl, the spectrum must bend somewhere below this energy because otherwise the energy contained in the ions would exceed the total available flare energy. This spectral information, and the apparent equipartition of energy between electrons and ions, should have important consequences on acceleration theories which were not yet explored. Both Holman's DC electric field acceleration model (29) and Miller's gyroresonance model (17) could produce flat ion spectra up to about an MeV with steeply falling spectra at higher energies. With subsequent acceleration by other processes, such scenarios could be consistent with the new findings based on the gamma ray data.

III. TRANSPORT

A. Long Duration Gamma Ray Events: Trapping vs. Continuous Acceleration

The increased sensitivity of several gamma ray detectors, and the occurrence of large flares while these instruments were observing the Sun, have shown that flares can produce gamma rays for very long periods of time. The most striking example of such long duration emission is the 1991 June 11 flare from which 50–2000 MeV gamma rays were observed with the EGRET instrument on CGRO for 8 hours (30,31) and 2.22 MeV line emission was seen for over 5 hours with CGRO/COMPTEL (32). Another example is the 1991 June 15 flare from which 30–3000 MeV gamma rays were seen with GAMMA-1 for about 2 hours (33) and 2.22 MeV line emission was seen for 5 hours (32). The report of the detection of long duration 2.22 MeV line emission from flares with COMPTEL was one of the highlights of the workshop (32,34). Nuclear line emission was observed from the 1991 June 4 flare for about 2 hours (35).

Two limiting possibilities exist as to the nature of the particles which produces these long duration emissions. The particles could either be accelerated in an impulsive phase and subsequently trapped at the Sun (36,37) or be accelerated continuously over the duration of the emission. Continuous acceleration up to high energies, based on the Kopp–Pneuman model (38), was discussed previously (39) and at the Workshop by Cliver (6). In addition, Somov (40) showed that it is possible to accelerate particles to GeV energies in reconnecting current sheets under special conditions. An argument against trapping, based on the similarity of the combined pion decay–nuclear line time profile and microwave time profiles, was presented for the 1991 June 15 flare (39,41).

There is of course also an intermediate possibility, i.e. that the particles are accelerated episodically and subsequently trapped between the acceleration episodes. Evidence for acceleration episodes producing interacting particles with different spectra was presented by Rieger (42). The detailed analysis of the 1991 June 11 flare presented at the Workshop (31) showed that there were at least three distinct emission phases in this flare, with the third phase being the most extended, lasting for about 7 hours. Much of the data for this flare came from CGRO/EGRET observations presented at the Workshop by Schneid. The possibility that the interacting ions remained trapped during this extended phase cannot be ruled out (31). As we have shown (31), this would require a transport model in which the ions are trapped in a low density coronal region by a strongly convergent magnetic field, so that their energy spectrum stays constant over long time periods, and precipitate into denser subcoronal regions where they interact before they can mirror and be reflected back into the trapping region. This scenario (31) allows the interacting particles to maintain a constant energy spectrum and is different from the previous trapping model that we developed (43,36). The constant energy spectrum is implied by the 2.22 MeV line to pion decay flux ratio which remains constant for long periods of time (hours). To avoid the escape of the GeV ions from the loops due to drifts it is required that the loops be either large enough or twisted (44).

B. Thin vs. Thick Target Interactions

The paradigm that solar flare gamma ray emission is due to thick target interactions of the accelerated particles has generally been accepted by most workers in the field. However, two examples of thin target emission, and an argument (12) that at least part of the interactions are thin target, were presented at the Workshop. The first example is the 1991 June 1 flare for which gamma ray spectroscopic evidence was found supporting thin target interactions (28). As this flare was located behind the limb, the thick target interaction region was occulted so that the observed gamma ray emission was produced in the corona. As mentioned in the previous section, the gamma ray data from this flare required a strong heavy ion enhancement. The contribution of the heavy ions relative to that of protons and α particles is greater in a thin target than in a thick target because of the larger Coulomb energy losses of the heavy ions. The hard X-ray observations of the 1991 June 1 flare with detectors on Ulysses showed that this flare was one of the largest on record (45). It is because of its very high intensity that the coronal contribution of the June 1 gamma ray emission could be observed. The second example is the 1991 March 26 flare (46). The observed 30–300 MeV gamma ray emission with a power law spectrum strongly suggest a primary electron bremsstrahlung origin for this emission. Because at these high energies the bremsstrahlung is strongly collimated along the direction of the electrons, and because the flare was located near the center of the disk, the radiating electrons had to move upward in the solar atmosphere to produce the observed emission. Akimov (46) suggested that the simplest explanation involves the acceleration of electrons in a region of density $\sim 10^{12}$ cm^{-3} and the production of bremsstrahlung as these electrons move upward in the solar atmosphere. The electrons which produce bremsstrahlung gamma rays at tens of MeV also produce synchrotron radiation in the millimeter range. Millimeter wave observations were discussed by Kaufmann (47). Refinements in the theory of gyrosynchrotron radiation in solar flares were presented by Belkora (48).

C. Spatial Extent and Location of the Gamma Ray Emission

For flares close to the limb of the Sun, the 2.22 MeV line, being formed below the photosphere, is attenuated relative to the prompt nuclear deexcitation lines which are formed higher in the atmosphere. The predicted (49) limb darkening of the 2.22 MeV line has been confirmed by observations, e.g. (50). Even though strong attenuation is expected for flares at or behind the limb, 2.22 MeV line emission was observed (51) from the 1989 September 29 flare located at a heliocentric angle of about 100°. Because of the expected attenuation, this line must have been produced by charged particles interacting on the visible hemisphere of the Sun. It was suggested (52) that these particles were accelerated by a coronal shock over a large volume producing an extended gamma ray emitting region visible from the Earth even if the optical flare was behind the limb. Alternatively, the accelerated particles could have been transported in large scale magnetic loops that connect the impulsive acceleration site to the visible hemisphere of the Sun. In either scenario, the resulting gamma ray emitting region must extend over tens of degrees of heliocentric angle. However, because of the well observed limb darkening of the 2.22 MeV line, which requires attenuation, such extended gamma ray emitting regions are probably more the exception rather than the rule. For most flares, therefore, the gamma ray emitting regions should have small sizes. Future imaging gamma ray observations, e.g. with HESSI (53), could test this hypothesis. The September 29 flare appears to be an exception also from another point of view. As discussed in the previous section, for most flares the gamma ray production is due to particles from impulsive flare acceleration. The September 29 flare may provide an example of gamma ray production following acceleration by a CME driven shock.

The gamma ray spectroscopic observations have also provided information on the location of the interaction site. The analysis (25,26) of 19 SMM flares (24) has shown that for the ambient medium the abundance ratios of the low FIP (first ionization potential) to high FIP elements are essentially coronal and are enhanced relative to photospheric values. This implies that the interaction region is located above the photosphere. However, because the Ne abundance was found to be higher than its coronal value, the gamma rays should be produced in a subcoronal region. This issue was further explored by Share (55) who presented new data showing a positive correlation between

$3\gamma/2\gamma$ and the low FIP to high FIP gamma ray fluence ratios (here $3\gamma/2\gamma$ is the positronium annihilation continuum to the 511 keV line flux ratio). The simplest explanation of this correlation is that deeper in the atmosphere the positronium is broken up by collisions (56) while at the same time the low FIP to high FIP element abundance ratio is closer to photospheric. The temperature of the annihilation region can also affect $3\gamma/2\gamma$. However, this dependence appears to be inconsistent with the positive correlation mentioned above because $3\gamma/2\gamma$ is expected to decrease with increasing temperature. The temperature could be independently inferred from the width of the 511 keV line. The SMM data, however, is inconclusive (55). Future HESSI Ge detector observations, which will have much better energy resolution than the SMM spectrometer, may allow the determination of the temperature and density of the positron annihilation site. The combination of this information with abundances determined from nuclear line spectroscopy will lead to a better understanding of the nature of the FIP effect in the solar atmosphere.

Information on the gamma ray production site was presented by Yoshimori (54) using hard X-ray imaging from YOHKOH. Based on the similarity in the onset times of one of the hard X-ray foot point sources and the gamma ray emission, he argued that the gamma rays were probably produced at that foot point.

D. Interplanetary Transport

In addition to the transport of the interacting particles at the Sun, the question of particle transport in interplanetary space was also considered at the Workshop. Dröge (12) showed that electrons from the same event observed by widely separated spacecraft (in both radial distance and longitude) have similar energy spectra suggesting that interplanetary and/or coronal transport are energy independent. Comparisons between SEP energy spectra and the energy spectra of the interacting particles were carried out for the 1990 May 24 and 1991 June 11 flares. For the former, Leon Kocharov (57) carried out a multiwavelength study including Hα, radio and gamma ray emissions, as well as SEP data and performed calculations for coronal and interplanetary transport. He concluded that the SEP could have had a common origin with the interacting particles in one of the phases of this flare. For the June 11 flare, we compared the best fitting shock acceleration spectrum obtained by Smart (58) from GLE observations of this flare with the interacting particle spectrum obtained by fitting the pion decay gamma ray spectrum observed with EGRET (31). We found that a common origin for the two particle populations cannot be ruled out.

Interplanetary transport calculations for neutron decay electrons were carried out by Ruffolo (59) for the 1980 June 21 flare suggesting that such electrons may have been observed from this flare. Neutron decay electrons complement the other methods of solar neutron detection: neutron decay protons, direct neutron detections on the ground with neutron monitors and in space, and observations of the 2.22 MeV line. Such combined observations allow the study of neutron production spectra and directivity over a broad energy range. Neutron monitor observations were reported by Muraki (60). The derived neutron fluxes from such observations still suffer from uncertainties in the neutron transport calculations in the atmosphere and the response functions for these detectors [see (61) for details].

Cane (62) and Shea (63) discussed unusual SEP events. Cane discussed the relationship between the longitudinal extent of interplanetary shocks determined from in situ observations and requirements based on SEP intensity-time profiles. The suggested interpretation of the observations is based on the idea that the longitudinal extent of shocks evolve (become smaller) as they propagate outwards from the Sun. Shea discussed several GLE events with unusually short duration ($\sim$10 min) and anisotropic spikes. The nature of this phenomenon is not understood. A discussion on GLEs was also presented by del Peral (64).

IV. ABUNDANCES

In additional to the information discussed above on the nature of the interacting particles and their interaction site, abundance studies based on gamma ray spectroscopy provide information on solar abundances (24–26) which is complementary to that obtained from atomic spectroscopy. We have already seen that gamma ray spectroscopy has provided confirmation for the FIP bias originally discovered using SEP observations. In the absence of gamma ray imaging observations, the location of the gamma ray production site is not known, although several indirect arguments suggest that the site is located fairly deep in the atmosphere. If this is indeed the case, then one has to conclude that the FIP bias sets in somewhere between the photosphere and corona. The one surprising result from gamma ray spectroscopy concerns the abundance of Ne. The analysis yielded Ne/O greater than the commonly accepted photospheric and coronal value of 0.15 (65). Similar to the SMM results discussed above, the CGRO/OSSE data presented by Murphy (35) for the very intense 1991 June 4 flare also shows a well defined, strong ^{20}Ne line at 1.634 MeV. The long duration and high intensity of the gamma ray emission from this flare allows the study of the time variation of the low FIP to high FIP line ratios. At the 17% probability level, the data shows that this ratio increases with time. A possible implication is that the interaction region moves upwards as the flare progresses. There is very clear evidence for the FIP effect in X-ray and EUV data [e.g. (66)], as well as in SEP data [e.g. (5)]. At the Workshop Sterling (67) reviewed the previous X-ray spectroscopic abundance studies and presented new YOHKOH data on abundances. For reasons that are not yet fully understood, the FIP bias is not as prominent in the YOHKOH data as in the previous studies.

V. SUMMARY

Our conclusions are the following:

(i) Stochastic acceleration due to gyroresonant interaction with plasma waves is being established as the most promising mechanism for particle acceleration in impulsive solar flares. This conclusion follows from the abundance anomalies (^{3}He and heavy ions) that are clearly present in the charged particle data and also in the interacting particles which produce the gamma rays. The particles observed from gradual flares are most likely accelerated by CME driven shocks. With one possible exception identified so far, these particles do not produce significant gamma ray emission.

(ii) Ion acceleration is energetically very important in solar flares. This conclusion follows from the analysis of the gamma ray data, in particular gamma ray line emission (from ^{20}Ne) resulting from nuclear reactions with low energy thresholds (near 1 MeV/nucl). The comparison of the energy content in MeV/nucl ions and deka-keV electrons suggests energy equipartition between accelerated ions and electrons. The implied advantages, or disadvantages, of stochastic acceleration vs. DC electric field acceleration have not yet been explored.

(iii) Solar flares can produce gamma ray emission, including gamma rays from pion decay, for very long periods of time (many hours). These extended emission periods include distinct episodes of particle acceleration. Between these episodes the particles could either remain trapped in closed coronal magnetic structures or be continuously accelerated. An important observational diagnostic is the particle energy spectrum which extends to GeV energies and appears to remain constant in time. This requires either a well tuned acceleration process, which can generate particles at a decreasing rate but constant energy spectrum, or trapping in a low density region with precipitation and gamma ray production outside this region.

(iv) Gamma ray spectroscopy has become another tool for solar atmospheric abundance determinations. The gamma ray analysis has confirmed the previously discovered FIP effect in the solar atmosphere. Further spectroscopic studies, including observations of the 511 keV positron annihilation line and its accompanying positronium continuum, will provide unique information on accelerated particle transport and the structure of the solar atmosphere.

(v) Gamma ray emission was observed from two large behind-the-limb flares. These observations indicate the existence of thin target gamma ray sources, as well as spatially extended sources. In one of these there is strong evidence for impulsive flare acceleration, while in the other shock acceleration could play the dominant role. Imaging observations are needed to further study the implications of such events.

We wish to acknowledge D. V. Reames for useful comments on this paper.

REFERENCES

1. E. L. Chupp, this volume.
2. H. V. Cane, R. E. McGuire, and T. T. von Rosenvinge, ApJ **301**, 448 (1986).
3. T. Bai, ApJ **308**, 912 (1986).
4. D. V. Reames, ApJ (Suppl.) **73**, 235 (1990).
5. D. V. Reames, J-P. Meyer, and T. T. von Rosenvinge, ApJ (Suppl.) **90**, 649 (1994).
6. E. W. Cliver, this volume.
7. S. W. Kahler, this volume.
8. J.-P. Meyer, this volume.
9. D. V. Reames, this volume.
10. R. A. Leske et al., this volume.
11. A. J. Tylka, this volume.
12. W. Dröge, this volume.
13. D. Moses, W. Dröge, P. Meyer, and P. Evenson, ApJ **346**, 523 (1989).
14. J. A. Miller and J. Steinacker, ApJ **399**, 284 (1992).
15. M. Temerin and I. Roth, ApJ **391**, L105 (1992).
16. J. A. Miller and A. F. Viñas, ApJ **412**, 386 (1993).
17. J. A. Miller and D. V. Reames, this volume.
18. M. Temerin and I. Roth, this volume.
19. Y. E. Litvinenko, this volume.
20. V. Petrosian, this volume.
21. N. Mandzhavidze and R. Ramaty, Nucl. Phys. B **33**, 141 (1993).
22. R. Ramaty, N. Mandzhavidze, B. Kozlovsky, and J. G. Skibo, Adv. Space Res. **13**, (9)275 (1993).
23. R. J. Murphy, R. Ramaty, B. Kozlovsky, and D. V. Reames, ApJ **371**, 793 (1991).
24. G. H. Share and R. J. Murphy, ApJ **452**, 933 (1995).
25. R. Ramaty, N. Mandzhavidze, B. Kozlovsky, and R. J. Murphy, ApJ, **455**, L193 (1995).
26. R. Ramaty, N. Mandzhavidze, and B. Kozlovsky, this volume.
27. C. Barat et al., ApJ, **425**, L109 (1994).
28. G. Trottet et al., this volume.
29. G. D. Holman, this volume.
30. G. Kanbach et al., A&A (Suppl.) **97**, 349 (1993).
31. N. Mandzhavidze, R. Ramaty, D. L. Bertsch, and E. J. Schneid, this volume.
32. G. Rank et al., this volume.
33. V. V. Akimov et al., 22nd Internat. Cosmic Ray Conf. Papers **3**, 73 (1991).
34. J. M. Ryan and M. M. McConnell, this volume.
35. R. J. Murphy et al., this volume.
36. N. Mandzhavidze and R. Ramaty, ApJ **396**, L111 (1992).
37. N. Mandzhavidze, R. Ramaty, V. V. Akimov, and N. G. Leikov, 23rd Internat. Cosmic Ray Conf. Papers **3**, 119 (1993).
38. R. A. Kopp and G. W. Pneuman, Solar Phys. **50**, 85 (1976).
39. V. V. Akimov et al., 1993, 23rd Internat. Cosmic Ray Conf. Papers **3**, 111 (1993).
40. B. V. Somov, this volume.

41. G. E. Kocharov et al., 23rd Internat. Cosmic Ray Conf. Papers **3**, 107 (1993).

42. E. Rieger, this volume.

43. N. Mandzhavidze and R. Ramaty, ApJ **389**, 739 (1992).

44. Y.-T. Lau and R. Ramaty, Solar Phys. **160**, 343 (1995).

45. S. R. Kane, ApJ **446**, L47 (1995).

46. V. Kurt, V. V. Akimov, and N. G. Leikov, this volume.

47. P. Kaufmann, this volume.

48. L. Belkora, this volume.

49. H. T. Wang and R. Ramaty, Solar Phys. **36**, 129 (1974).

50. E. L. Chupp, Physica Scripta, **T18**, 15 (1987).

51. W. T. Vestrand and D. J. Forrest, ApJ **409**, L69 (1993).

52. E. W. Cliver, S. W. Kahler, and W. T. Vestrand, 23rd Internat. Cosmic Ray Conf. Papers , **3**, 91 (1993).

53. B. R. Dennis, this volume.

54. M. Yoshimori et al., this volume.

55. G. H. Share, R. J. Murphy, and G. J. Skibo, this volume.

56. C. J. Crannell, G. Joyce, R. Ramaty, and C. Werntz, ApJ **210**, 582 (1976).

57. L. Kocharov et al., this volume.

58. D. F. Smart and M. A. Shea, this volume.

59. D. Ruffolo, W. Dröge, and B. Klecker, this volume.

60. Y. Muraki and S. Shibata, this volume.

61. N. Mandzhavidze, in Cosmic Ray Conference (Calgary), eds. D. A. Leahy, R. B. Hicks, and D. Venkatesan, (World Scientific), 157 (1994).

62. H. V. Cane, this volume.

63. M. A. Shea and D. F. Smart, this volume.

64. L. I. Miroshnichenko et al., this volume.

65. J-P. Meyer, in Origin and Evolution of the Elements, eds. N. Prantzos et al. (Cambridge), 26 (1992).

66. K. G. Widing and U. Feldman, ApJ **442**, 446 (1995).

67. A. C. Sterling, this volume.

Author Index

A

Adams, J. H., Jr., 96
Akimov, V. V., 237
Anttila, A., 246
Aschwanden, M. J., 300

B

Bai, T., 329
Barat, C., 153, 311
Beahm, L. P., 96
Belkora, L., 416
Bennett, K., 219
Bertsch, D. L., 225
Bloemen, H., 219
Boberg, P. R., 96

C

Cane, H. V., 124
Canfield, R. C., 336
Chupp, E. L., 3
Cliver, E. W., 45
Cummings, J. R., 86

D

Debrunner, H., 219
del Peral, L., 140
Dennis, B. R., 519
Dezalay, J. P., 153, 311
Dietrich, W. F., 96
Doschek, G. A., 353
Dröge, W., 78, 116

E

Enome, S., 408, 424

F

Finger, M., 368
Forbes, T. G., 275
Forrest, D., 368

G

Gallegos-Cruz, A., 140
Gopalswamy, N., 408
Grove, J. E., 184

H

Hanaoka, Y., 408
Holman, G. D., 479
Hudson, H. S., 285

J

Johns-Krull, C. M., 320
Johnson, W. N., 184
Jung, G. V., 184

K

Kahler, S. W., 61
Kaufmann, P., 379
Kinzer, R. L., 184
Klecker, B., 116
Kleis, T., 96
Kocharov, L., 246
Kosugi, T., 267, 336
Kovaltsov, G., 246
Kozlovsky, B., 172

Kundu, M. R., 402, 408
Kurfess, J. D., 184
Kurt, V., 237
Kuznetsov, A., 153, 311

L

Leikov, N. G., 237
Lemen, J. R., 408
Lenters, G., 505
Leske, R. A., 86
Li, J., 336
Lin, R. P., 320
Litvinenko, Y. E., 498
Lockwood, J., 219

M

Mandzhavidze, N., 153, 172, 225, 533
Matsuda, T., 210
McConnell, M. M., 200, 219, 368
Metcalf, T. R., 336, 393
Mewaldt, R. A., 86
Meyer, J.-P., 461
Miller, J. A., 450, 505
Miroshnichenko, L. I., 140
Morimoto, K., 210
Muraki, Y., 256
Murphy, R. J., 162, 184

N

Nakajima, H., 393
Nitta, N., 285, 294, 402

P

Pérez-Peraza, J., 140
Petrosian, V., 445

R

Ramaty, R., 153, 172, 225, 329, 533
Rank, G., 219
Raulin, J.-P., 402

Reames, D. V., 35, 450
Rieger, E., 194
Rodríguez-Frías, M. D., 140
Roth, I., 435
Ruffolo, D., 116
Ryan, J. M., 200, 219

S

Saita, N., 210
Schneid, E. J., 225
Schönfelder, V., 219
Schwartz, R. A., 359
Share, G. H., 162, 184
Shea, M. A., 106, 131
Shibata, S., 256
Skibo, J. G., 162
Smart, D. F., 106, 131
Somov, B. V., 493
Sterling, A. C., 343
Stone, E. C., 86
Strickman, M. S., 184
Suga, K., 210
Suleiman, R., 219
Sunyaev, R., 153, 311

T

Talon, R., 153, 311
Temerin, M., 435
Terekhov, O., 153, 311
Torsti, J., 246
Trottet, G., 153, 311
Tylka, A. J., 96

U

Usoskin, I., 246

V

Vainio, R., 246
Vashenyuk, E. V., 140

Verhagen, H., 311
Vestrand, W. T., 368
Vilmer, N., 153, 311
von Rosenvinge, T. T., 86

W

Wülser, J.-P., 336

Y

Yoshimori, M., 210

Z

Zarro, D. M., 359
Zirin, H., 246

AIP Conference Proceedings

	Title	L.C. Number	ISBN
No. 340	Strangeness in Hadronic Matter (Tucson, AZ 1995)	95-77477	1-56396-489-9
No. 341	Volatiles in the Earth and Solar System (Pasadena, CA 1994)	95-77911	1-56396-409-0
No. 342	CAM -94 Physics Meeting (Cacun, Mexico 1994)	95-77851	1-56396-491-0
No. 343	High Energy Spin Physics Eleventh International Symposium (Bloomington, IN 1994)	95-78431	1-56396-374-4
No. 344	Nonlinear Dynamics in Particle Accelerators: Theory and Experiments (Arcidosso, Italy 1994)	95-78135	1-56396-446-5
No. 345	International Conference on Plasma Physics ICPP 1994 (Foz do Iguaçu, Brazil 1994)	95-78438	1-56396-496-1
No. 346	International Conference on Accelerator-Driven Transmutation Technologies and Applications (Las Vegas, NV 1994)	95-78691	1-56396-505-4
No. 347	Atomic Collisions: A Symposium in Honor of Christopher Bottcher (1945-1993) (Oak Ridge, TN 1994)	95-78689	1-56396-322-1
No. 348	Unveiling the Cosmic Infrared Background (College Park, MD, 1995)	95-83477	1-56396-508-9
No. 349	Workshop on the Tau/Charm Factory (Argonne, IL, 1995)	95-81467	1-56396-523-2
No. 350	International Symposium on Vector Boson Self-Interactions (Los Angeles, CA 1995)	95-79865	1-56396-520-8
No. 351	The Physics of Beams Andrew Sessler Symposium (Los Angeles, CA 1993)	95-80479	1-56396-376-0
No. 352	Physics Potential and Development of $\mu^+ \mu^-$ Colliders: Second Workshop (Sausalito, CA 1994)	95-81413	1-56396-506-2
No. 353	13th NREL Photovoltaic Program Review (Lakewood, CO 1995)	95-80662	1-56396-510-0
No. 354	Organic Coatings (Paris, France, 1995)	96-83019	1-56396-535-6
No. 355	Eleventh Topical Conference on Radio Frequency Power in Plasmas (Palm Springs, CA 1995)	95-80867	1-56396-536-4

	Title	L.C. Number	ISBN
No. 356	The Future of Accelerator Physics (Austin, TX 1994)	96-83292	1-56396-541-0
No. 357	10th Topical Workshop on Proton-Antiproton Collider Physics (Batavia, IL 1995)	95-83078	1-56396-543-7
No. 358	The Second NREL Conference on Thermophotovoltaic Generation of Electricity	95-83335	1-56396-509-7
No. 360	The Physics of Electronic and Atomic Collisions XIX International Conference (Whistler, Canada, 1995)	95-83671	1-56396-440-6
No. 361	Space Technology and Applications International Forum (Albuquerque, NM 1996)	95-83440	1-56396-568-2
No. 362	Two-Cneter Effects in Ion-Atom Collisions (Lincoln, NE 1994)	96-83379	1-56396-342-6
No. 363	Phenomena in Ionized Gases XXII ICPIG (Hoboken, NJ, 1995)	96-83294	1-56396-550-X
No. 364	Fast Elementary Processes in Chemical and Biological Systems (Villeneuve d'Ascq, France, 1995)	96-83624	1-56396-564-X
No. 365	Latin-American School of Physics XXX ELAF Group Theory and Its Applications (México City, México, 1995)	96-83489	1-56396-567-4
No. 366	High Velocity Neutron Stars and Gamma-Ray Bursts (La Jolla, CA 1995)	96-84067	1-56396-593-3
No. 367	Micro Bunches Workshop (Upton, NY, 1995)	96-83482	1-56396-555-0
No. 368	Acoustic Particle Velocity Sensors: Design, Performance and Applications (Mystic, CT, 1995)	96-83548	1-56396-549-6
No. 371	Sixth Quantum 1/f Noise and Other Low Frequency Fluctuations in Electronic Devices Symposium (St. Louis, MO, 1994)	96-84200	1-56396-410-4
No. 372	Beam Dynamics and Technology Issues for + - Colliders 9th Advanced ICFA Beam Dynamics Workshop (Montauk, NY, 1995)	96-84189	1-56396-554-2
No. 374	High Energy Solar Physics (Greenbelt, MD 1995)	96-84513	1-56396-542-9